First printed edition, October 2004
Published by Brainstorm Fantasia™, Inc.,
P.O. Box 4280, Pittsburgh, PA 15203, U.S.A

Printed at Copies at Carson,
the location of Copies at Carson, Inc.,
1315 East Carson St. Pittsburgh, PA 15203

ISBN#: 0-9761810-0-2

Introduction

A. The purpose of this book is to provide material for a bridge course between an advanced precalculus level and an algebraic structures course (usually taken in the student's third or fourth year). Typically such a course should be taken immediately after or soon after a student has taken a first semester calculus course.

How much material from this book should be used for a course of this nature, is entirely up to the instructor and it will depend heavily upon the precise design and scope of the intended course. The actual structure of this book lends for flexibility in designing a course out of this book. Normally, a course that uses this book as its textbook should be taken in the student's second (or sophomore) year, although there is no reason why a first or third year student should not take this course. A first year student with an exceptionally strong background would be well-suited for this course. Also, keep in mind that different colleges and universities have different types of requirements for mathematics and computer science majors; and they have a wide variety of required and elective courses for math and computer science majors.

It should be pointed out here, that the main audience of this book are mathematics and computer science majors. Having said that however, any student (both in the formal and in the broader sense of the word) with a deeper and more involved interest in mathematics, may find this book as a gateway for further studies and the experiencing of the world of mathematics.

B. Salient points about this book.

1. The last Section (just before the Exercise Set) of every Chapter (the book contains seven Chapters) is entitled, "Practice Problems". These are exercises solved in great detail, to the point that sometimes the reader might think that the author is pedantic. One has to remember that this is done with the expressed purpose of teaching the student techniques of solving problems and successfully carrying out proofs without omitting any details, unless they are very obvious; the idea here is that for the majority of the student readers of this book, this will be their first contact and experience with mathematical proofs. In essence the Practice Problems provide a background against which proofs and learning techniques can be taught. The Practice Problems range in difficulty from fairly routine problems to quite a few problems which are really challenging.

2. The style of presentation throughout the book alternates from the formal (using mathematical symbols such as existential and universal quantifiers) to the informal, where only a concise use of the English language is required; it is the view of this author that learning mathematics after the Calculus, both approaches are valuable: learning to write a correct mathematical statement by using the written language in a skilled and succinct way and only making a minimum use of mathematical symbols; or on the other hand, learning how to construct proofs that rely on heavier use of mathematical symbols and less on the extensive use of English words.

3. This text is self-contained, building up from the axiomatic levels and providing the student with the various connecting bits and pieces of information as she/he moves through the material. A notable exception to this axiomatic upwards construction, is Chapter 1 (Relations and Functions) that uses certain familiar notions and basic results that the student knows (or should know) from the basic algebra that underlies precalculus and (first) calculus courses.

4. This text has many rare or hard to find (in other texts) examples and Practice Problems. so in this regard, there is an esoteric novelty about the book. In addition, in Chapter 1, the student will find many Venn Diagrams (about forty or so).

5. Among the seven Chapters in this book, there are three that constitute its focal point. These are Chapter 1, Chapter 2, and Chapter 4; they are the backbone of this text. Chapter 1 (Functions and Relations) deals with functions and relations; it offers an extensive treatment of these two fundamental concepts in mathematics. In Chapter 1, the student will discover many of the familiar (from Calculus) properties of functions, as well as new, unfamiliar properties of functions and relations. Chapter 2 (Rings and Fields) deals with rings and fields, which are of central importance in the world of mathematics. The treatment there (even though on an introductory level) is quite comprehensive and extensive; the student will encounter both the familiar properties of rings and fields, but also some properties that are usually not found in introductory texts on this subject. Thirdly, Chapter 4 (Integers and their Basic Properties) can in effect be used to teach a mini (i.e. half a semester) course as an introduction to elementary number theory.

6. The entire book contains twice as much material that would be needed in order to teach a full semester course that say, meets, three times a week (with each class period being one class hour-fifty regular minutes). This offers the instructor of the course a great deal of flexibility. One could teach such a course by using only a smattering of material from each Chapter. Or instead, one could design the entire semester course on the basis of three Chapters: Chapters 1, 2, and 4. These three chapters in particular, contain a great variety of examples, Practice Problems, and Exercises that the instructor of such a course can draw form.

7. Originally, the author of this text had considered including (what will be) two volumes of this material into one large book. However, practical considerations have advised against such a decision. Thus, this first book is Volume 1 of a two-Volume series project. While Volume 1 deals with relations, functions, and numbers; the main focus of Volume 2 will be equations.

Here is a tentative outline of the Chapters in Volume 2:

Chapter 1: Complex Numbers

Chapter 2: Linear Functions

Chapter 3: Quadratic Functions

Chapter 4: Polynomial Functions

Chapter 5: Cubic and Quartic Equations

Chapter 6: Equations and Systems of Equations in General

Chapter 7: Linear Systems, the Row-Reduction Method, Pivots

Chapter 8: Diophantine Equations and some Linear Diophantine Systems

Chapter 9: Finite Sums

It is my earnest hope that for the independent reader, this book will serve as a vehicle for an exciting ride into the world of mathematics, into the world of algebra to be specific. For the student and instructor of a course using this book as a text, I would like to think that this book will play the rôle of a kaleidoscopic background against which a rewarding learning and teaching experience maybe had.

In closing, I would like to express my gratitude and warmest thanks to Ferna Hartman and Nancy Watson, both staff members with the Mathematics Department of Carnegie Mellon University, for an outstanding technical typing job. Nancy and Ferna, by using state of the art software technologies have succeeded in bringing to fruition the transition of a hand-written manuscript to a typed unpublished manuscript, that is as close to a book format as possible.

Thoughtfully,
Konstantine Dabmian Zelator, Ph.D.
Pittsburgh, Pennsylvania
May 2001

Contents

Chapter 1

Relations and Functions

1.1 Relations

(Binary) Relations from a non-empty set A to a non-empty set B.

The notation $x \in A$ means "x is a member of the subset of A". Also, the notation $A \subset B$ stands for "A is a subset of B". To define a relation from a non-empty set A to a non-empty set B, we need the notion or concept of the cartesian product $A \times B$.

Definition 1: $A \times B$ is the set of all pairs (x, y) such that the first component or "coordinate" or element x belongs to the set A, while the second component or "coordinate" or element y belongs to the set B. We can also say that $A \times B$ is the set of all **ordered** pairs (x, y), with $x \in A$ and $y \in B$.

Definition 2: Let A and B be given or fixed non-empty sets. We call **predicate formula or condition**, of two variables x and y, denoted by $p(x, y)$ and read "p of x and y", every expression containing the two variables x and y and which becomes a specific (meaningful) statement whenever the variables x and y are substituted for specific elements x_0 and y_0 from the sets A and B respectively.

Definition 3: We call relation from a (non-empty) set A to a non-empty set B, any subset of the cartesian product $A \times B$. The set A is called **the starting set** of the relation; B is the **landing or arriving set** of the relation.

Remark 1: A relation from A to B may be denoted by R. When one writes $x_0 R y_0$, it is meant that "x_0 relates to y_0; equivalently $x_0 R y_0$ means exactly that the ordered pair (x_0, y_0) is a member of R which is a subset of $A \times B$.

Remark 2: It is possible that a relation R maybe the empty set, $R = \emptyset$, since $\emptyset$ is a subset of any set, in particular of $A \times B$.

Remark 3: Usually, but not always, a relation may be described or defined by a predicate formula or condition $p(x, y)$, $x \in A$ and $y \in B$. Sometimes, though, a relation R is listed explicitly as a set, with all of its elements (ordered pairs) listed. A relation can also be illustrated with the use of a Venn diagram as it is shown in the examples below.

Definition 4: If the sets A and B are equal or identical, $A = B$, then any relation from A to itself is simply called a relation within the set A. The cartesian product

$A \times A$ can also be denoted by A^2.

Example 1: Consider the sets $A = \{a, b, c, d\}$, where a, b, c, d each denotes a college student from a certain college and $B = \{x \mid x$ foreign language: Spanish, German, or French$\}$, and the condition $p(x, y) =$ "student$x \in A$ is taking a course in $y \in B$, this semester". Also, let us suppose that student a takes Spanish and French, student b takes no foreign language, student c takes German and French, and student d takes Spanish only. Then, the relation that the given condition describes or defines is the set $R = \{(a, \text{Spanish}), (a, \text{French}), (c, \text{French}), (c, \text{German}), (d, \text{Spanish})\}$. The relation R can be illustrated by the following Venn diagram.

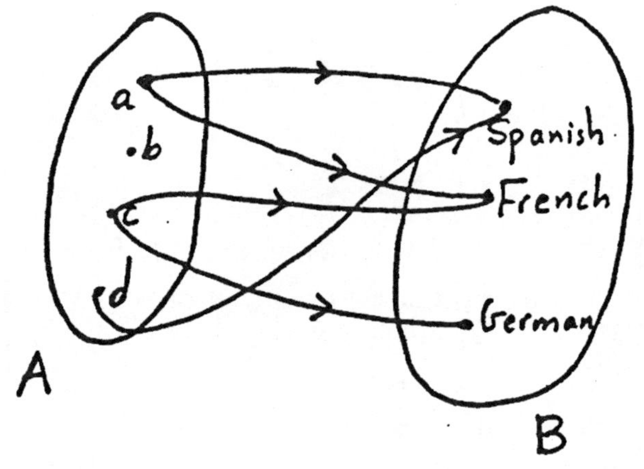

Figure 1

Obviously, precise arrows are needed whenever we use Venn diagrams to represent a relation; if $x_0 R y_0$, for some $x_0 \in A$ and $y_0 \in B$, then an arrow must emanate from x_0 in A and land at the element y_0 in B.

Example 2: Let C, D, E be the three vertices of a given triangle and c, d, and e the three sides of a triangle, as illustrated by the picture below. Let A and B be the sets of the vertices and the sides respectively. $A = \{C, D, E\}$, $B = \{c, d, e\}$

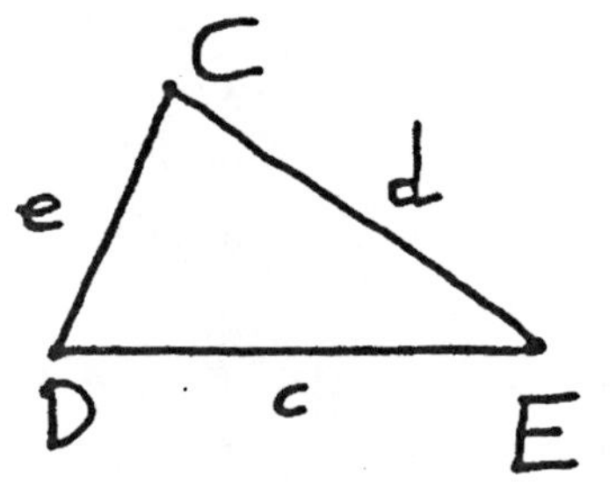

Figure 2

Let $p(x, y) =$ "the (straight) line $x \in A$ passes through $y \in B$".

The predicated formula or condition $p(x, y)$ defines the following relation:

$$R = \{(c, D), (c, E), (d, C), (d, E), (e, C), (e, D)\},$$

a subset of $A \times B$ which contains nine ordered pairs. Even though we can, of course, draw a Venn diagram for this relation, we can also illustrate it by using a rectangular type of graph:

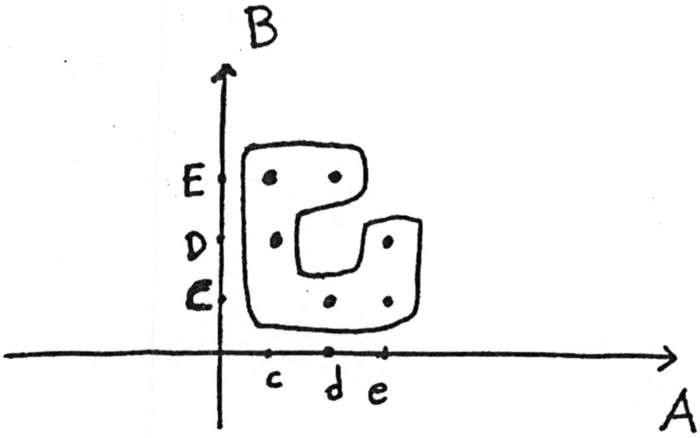

R consists of the six dots falling inside the drawn frame to the left.

Figure 3

Example 3: Let $A = \{1, 2, 3, 4\}$ and $B = \{6, 8, 12\}$, and $p(x, y) =$ "$x \in A$ divides $y \in B$", defines a relation R from A to B given by

$$R = \{(1,6), (1,8), (1,12), (2,6), (2,8), (2,12), (3,6), (3,12), (4,8), (4,12)\}.$$

However, the condition $p(x, y) =$ "$x \in B$ divides $y \in A$", defines a relation from B to A, which is the empty set $\emptyset$, since there is ordered pair in $B \times A$ that has that property.

Remark 4: (i) It is entirely possible for the same condition $p(x, y)$, when pertaining to two different cartesian products, to define two different relations. For example, if we take $A = \{2, 3, 4, 5, 6\}$ and $B = \{3, 4, 6, 7, 8, 10\}$, and the condition

$$p(x, y) = \text{"}x \text{ is twice } y\text{"},$$

then this condition defines, in $A \times B$, the relation $R_1 = \{(6, 3)\}$; but in $B \times A$ it defines the relation $R_2 = \{(4, 2), (6, 3), (8, 4), (10, 5)\}$.

(ii) Two different conditions may define the same relation. For example, let $A = \{1, 2, 4\}$, $B = \{4, 7\}$, and

$$p_1(x, y) = \text{"}x \in A \text{ is less than } y \in B\text{"},$$

$$p_2(x, y) = \text{"}x \in A \text{ satisfies } x \neq y,\ y \in B.\text{"}$$

These two conditions define the relations R_1 and R_2 in $A \times B$ given by

$$R_1 = \{(1,4),(1,7),(2,4),(2,7),(4,7)\},$$

$$R_2 = \{(1,4),(1,7),(2,4),(2,7),(4,7)\}$$

Thus, $R_1 = R_2$. Such conditions are called equivalent.

Definition 5: Two conditions $p_1(x,y)$, $p_2(x,y)$, pertaining to the cartesian product $A \times B \neq \emptyset$ (i.e., both A and B are non-empty), (with the variables x and y "taking values" from A and B respectively), are called equivalent if, and only if, they define the same relation in the cartesian product $A \times B$.

Example 4: Let $A = \{a,b,c,d,e,f,g,h,i,j,k\}$ be a set consisting of girls and boys and $p(x,y) =$ person $x \in A$ has sister person $y \in A$, and assume that the given condition defines a relation R in the set A, whose Venn diagram is given below.

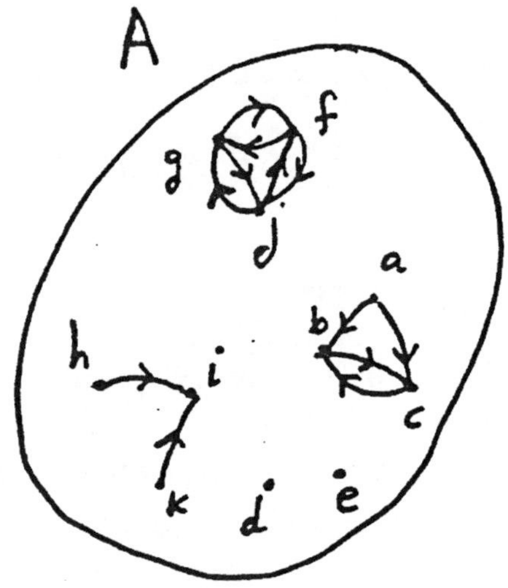

Figure 4

$$R = \{(a,b),(a,c),(b,c),(c,b),(g,f),(f,g),(f,j),(j,f),(g,j),(j,g)(h,i),k,i)\}$$

We can draw the following inferences:

(i) The persons a, h, and k must be boys, since in the Venn diagram there are only arrows emanating from these three elements, but not arrows arriving at a, h, or k. Clearly, if a person is a girl with a sister, then that girl must be a sister of that sister, and hence an arrow must be directed at that girl.

(ii) The persons b, c, f, g, and j must be girls, since each of them is the starting point of an arrow, but also the arriving point of an arrow.

(iii) The person i must be a girl since it receives at least one arrow.

(iv) The persons d, e have no sisters among the remaining persons, nor is d or e sister of any of the remaining persons; nor is d sister of e or vice-versa. The sexes of

d and e remain undetermined.

Example 5: Let $A = \{0,1,2,3,4,5,6,7\}$ and $p(x,y)$ the condition given by, $p(x,y) = x \in A$ is a divisor of $y \in A$. This condition in A defines a relation R in A given by,

$$\begin{aligned} R &= \{(0,0),(1,0),(2,0),(4,0),(5,0),(6,0),(7,0),(1,1),(1,2),(1,3),(1,4), \\ &\quad (1,5),(1,6),(1,7),(2,2),(2,4),(2,6),(3,6),(4,4),(5,5),(6,6),(7,7)\}. \end{aligned}$$

This relation contains 22 elements; R is a subset of the cartesian product $A \times A$ which contains $7^2 = 49$ elements.

Remark 5: If a set A contains n elements, the cartesian product $A \times A$ must contain n^2 elements or ordered pairs.

The Venn diagram of the previous example is illustrated below.

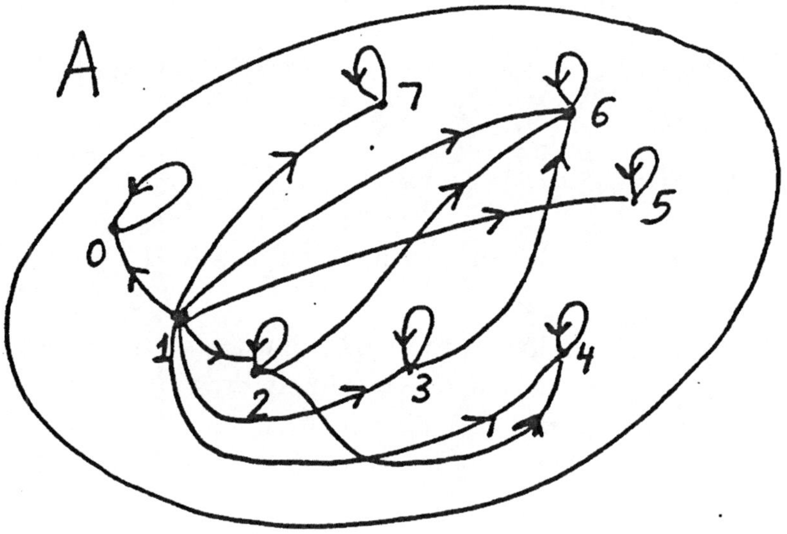

Figure 5

We now introduce the concept of the inverse relation.

Definition 6: Let R be a relation from a non-empty set A to a non-empty set B. We call the inverse relation of R, denoted by R^{-1}, the relation from B to A, which consists of the pairs which are inverse or reverse correspondingly to the pairs of R. We write: For every $(x,y) \in A \times B : (x,y) \in R \Leftrightarrow (y,x) \in R^{-1}$. Naturally, R^{-1} is the inverse relation of R if, and only if, R is also the inverse relation of R^{-1}; $(R^{-1})^{-1} = R$.

Example 6: Let $A = \{1,2,3,4,5\}$ and consider the two conditions,

$$p_1(x,y) = x \in A \text{ is less than } y \in A,$$

$$p_2(x,y) = x \in A \text{ is less than or equal to } y \in A.$$

These two conditions define two relations R_1 and R_2 in A:

$$R_1 = \{(1,2),(1,3),(1,4),(1,5),(2,3),(2,4),(2,5),(3,4),(3,5),(4,5)\}$$

$$\begin{aligned} R_2 = \{&(1,1),(1,2),(1,3),(1,4),(1,5),(2,2),(2,3),(2,4),(2,5),\\ &(3,3),(3,4),(3,5),(4,4),(4,5),(5,5)\}\,. \end{aligned}$$

The inverse R_1^{-1} of the first relation R_1 is given by,

$$R_1^{-1} = \{(2,1),(3,1),(4,1),(5,1),(3,2),(4,2),(5,2),(4,3),(5,3),(5,4)\}$$

Clearly R_1^{-1} may be defined by the condition,

$$p_1^{-1}(x,y) = x \in A \text{ is greater than } y \in A.$$

The inverse R_2^{-1} of the second relation R_2 is given by,

$$\begin{aligned} R_2^{-1} = \{&1,1),(2,1),(3,1),(4,1),(5,1),(2,2),(3,2),(4,2),\\ &(5,2),(3,3),(4,3),(5,3),(4,4),(5,4),(5,5)\} \end{aligned}$$

This relation R_2^{-1} may be defined by the condition,

$$p_2^{-1}(x,y) = x \in A \text{ is greater or equal to } y \in A.$$

Remark 5: If a relation R from set A to set B is illustrated by a Venn diagram, then we can illustrate the inverse relation R^{-1} by simply drawing the reverse arrows from B to A. Also, if $R = \emptyset$, then $R^{-1} = \emptyset$ as well.

1.2 Important Relations

Reflexive relations, anti-reflexive relations, symmetric relations, anti-symmetric relations, and transitive relations

Definition 7: A relation R in a non-empty set A is called reflexive if it contains every pair of the form (x, x), with $x \in A$. In other words (the relation R if reflexive) $\Leftrightarrow$ (for every $x \in A: \ (x, x) \in R$).

Remark 6: Another way to express the meaning of Definition 7, is to say that R is reflexive exactly when D_A is a subset of R; where $D_A = \{(x, y) \in A \times A \ : \ x = y\}$; the set D_A is called the **diagonal** of the cartesian product $A \times A$. So, a relation R in A is reflexive if, and only if, R is a superset of D_A or equivalently, D_A is a subset of R. If $R = D_A$, then R or D_A is also called the **identity** relation in A.

Example 7: Let $A = \mathbb{Z} =$ the set of ordinary integers and $p(x, y) = x \in A$ divides $y \in A$. This condition defines a relation R in $\mathbb{Z}$ that is reflexive since every integer divides itself. Of course here, both A and R are infinite sets, but nevertheless R contains all the pairs of the form (x, x), with $x \in \mathbb{Z}$.

Definition 8: Any relation R in a non-empty set A that does not contain any pair of the form $(x, x), x \in A$, is called **anti-reflexive**. In other words, R is anti-reflexive if, and only if, the intersection set of R with D_A is the empty set; we write $R \cap D_A = \emptyset$.

Example 8: The relation of example 4 is anti-reflexive. In the Venn diagram of an anti-reflexive relation there is no arrow emanating from an element and directed at the same element. Here is another example: let $A = \mathbb{Z}$ and $p(x, y) = x \in A$ is greater than $y \in A$. The relation R in A that this condition defines is antireflexive since there is no integer greater than itself.

Definition 9: A relation R in a non-empty set A is called **symmetric** if R is equal to (or coincides with) its inverse relation R^{-1}. In other words (the relation $R \subset A^2$ is symmetric) $\Leftrightarrow R = R^{-1}$. This is equivalent to saying that R in A is symmetric if, and only if the following statement is true: for every $(x, y) \in A^2 : (x, y) \in R \Rightarrow (y, x) \in R$.

Remark 7: To recognize a symmetric relation R in A from its Venn diagram, one simply checks to see whether every row has a reverse or inverse arrow directed backwards; if a single arrow fails to have a reverse one, then the relation R is not symmetric. Also note that the diagonal relation D_A is, rather trivially, symmetric.

Example 9: Let A be the set of positive integers also denoted by $\mathbb{Z}^+$ and also by P; $A = \{1, 2, 3, 4, \ldots\}$. Let $p(x, y)$ be the condition, $p(x, y) =$ "$x \in A, \ y \in A$, with $\mid x - y \mid = 1$", where $\mid \ \mid$ denotes the real absolute value of real numbers taken for granted here. This condition defines a symmetric relation R in A. To see this consider any pair (x_0, y_0) in R: we must have $\mid x_0 - y_0 \mid = 1 \Rightarrow \mid -(y_0 - x_0) \mid = 1$; this implication is valid because according to the rules of arithmetic in the set of integers $\mathbb{Z}$, addition is a commutative operation: $x_0 - y_0 = x_0 + (-y_0) = -y_0 + x_0 = (-1) \cdot (y_0 - x_0)$, since according to those rules we can factor the number -1; but the absolute values of opposite integers are equal: $\mid (-1) \cdot (y_0 - x_0) \mid = \mid y_0 - x_0 \mid$; hence $\mid -(y_0 - x_0) \mid = 1 \Rightarrow \mid y_0 - x_0 \mid = 1$ which proves that (y_0, x_0) also belongs to the relation R; R is symmetric. It may have seemed pedantic to go over the usual operations of arithmetic in $\mathbb{Z}$ in a detailed fashion; however we do this, to only remind the student of certain operations that are taken for granted. We will see in Chapter 2 that the addition and multiplication structures combined $*$, make $\mathbb{Z}$ into an example of a **ring**. Also note that the relation R in this example is anti-reflexive, there is no pair of the form (x, x) that can satisfy the given condition.

Here we can make a connection with precalculus/calculus: if we use the cartesian coordinate system, we can illustrate the graph of this relation, by illustrating a few points on its graph.

$$R = \{(1,2),(2,1),(2,3),(3,2),(3,4),(4,3),\dots\}$$

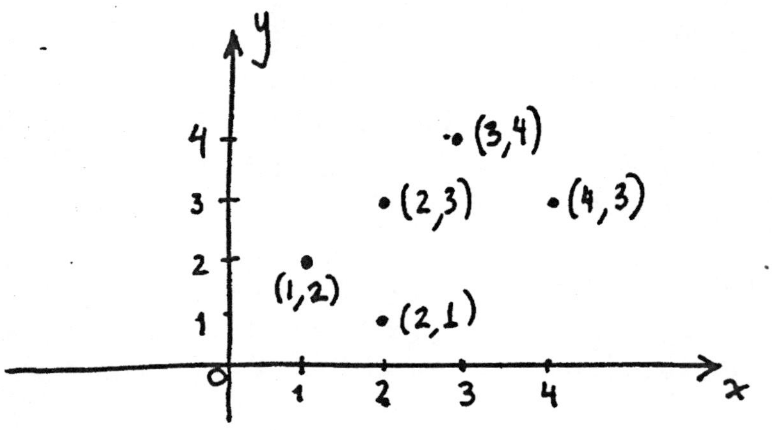

Figure 6

Note that the graph of a symmetric relation, when graphed in a standard coordinate system, is always symmetric with respect to the straight line with equation $y = x$ (the line that is the diagonal of the first and third quadrants).

Example 10: Let $A = \mathbb{Z}$, the set of integers and the condition, $p(x, y) = x \in A$ and $y \in A$ are such that $x + y$ is a multiple of 3. Since $y + x$ must also be a multiple of 3 if $x + y$ is, it follows that the relation R in A that the given condition defines, must be a symmetric relation. Note that R is neither reflexive (for example, it does not contain $(1, 1)$ nor anti-reflexive (for example, it contains $(3, 3)$).

Example 11: Suppose that in a certain family, the four children denoted by the initials of their first names a, b, e, and j, are all female. Let $A = \{a, b, e, j\}$. The condition $p(x, y) =$ person $x \in A$ is a sister of person $y \in A$, defines a symmetric relation R in A, since if x is a sister of y, then y must also be a sister of x, in view of the fact every member of A is a female. Note that R is anti-reflexive since no person can be a sister of herself. Thus $R = \{$(a,b),(b,a),(a,e),(e,a),(a,j),(j,a),(b,e),(e,b),(b,j),(j,b),(e,j),(j,e)$\}$.

Definition 10: A relation R in a non-empty set A is called anti-symmetric if, and onlyif, for every $(x, y) \in A^2$, with $x \neq y$, the following inference or implication is true:

$$(x, y) \in R \Rightarrow (y, x) \notin R, \text{ (i.e., } (y, x) \text{ does not belong to } R).$$

Example 12: Let A be a non-empty set consisting of positive integers only. Define $p(x, y) = x \in A,\ y \in A,\ x \neq y$, and x is a divisor of y. The relation R that $p(x, y)$ defines is anti-symmetric.

Example 13: Let A be the set of positive integers, $p(x, y) = x \in A,\ y \in A$, and $x < y$; where "$<$" has the usual meaning, "x is less than y"; $p(x, y)$ defines an anti-symmetric relation R in A.

Example 14: Let k be a fixed positive integer greater than 1, and A the set of integers; that is $A = \mathbb{Z}$. Define $p(x, y) = x \in A,\ y \in A$, and x is a k-multiple of y; here "k-multiple" simply means $x = k \cdot y$; the relation R that $p(x, y)$ defines in A, is anti-symmetric.

Definition 11: A relation R in an non-empty set A is called transitive if, and only if, for any elements x, y, z in A, distinct or not, the following inference holds true:

$$((x, y) \in R \text{ and } (y, z) \in R) \Rightarrow (x, z) \in R.$$

Remark 8: If the pairs (a, b), (b, c), and (a, c) are members of a transitive relation R in a non-empty set A, there are five possibilities depending on whether the three elements a, b, c are distinct, only two of them equal, or all three equal. Graphically, in a Venn diagram, we can represent these cases as follows:

(i) All three distinct: $a \neq b \neq c \neq a$:

(ii) $a = b$ and $b \neq c$:

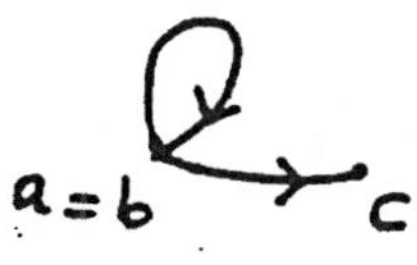

(iii) $a \neq b$ and $b = c$:

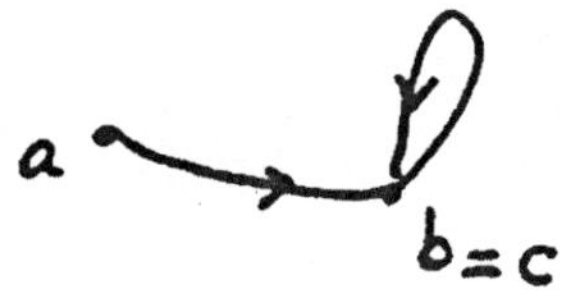

(iv) $a = c$ and $a \neq b$:

(v) $a = b = c$

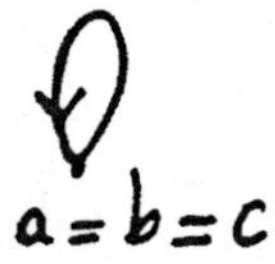

Figure 7

Example 15: Let A be a non-empty subset of the set of integers $\mathbb{Z}$, and $p(x, y) = x \in A,\ y \in A$, and x is a divisor of y.

This condition defines a transitive relation R in A. To see why, suppose that a, b, and c are in A such that a divides b; and b divides c. We will show that a must also divide c. Indeed, since a is a divisor of b, there exists an integer d_1 such that $b = a \cdot d_1$. Likewise, $c = b \cdot d_2$, for some integer d_2. Combining the two equations we obtain $c = (a \cdot d_1) \cdot d_2;\ \ c = a \cdot (d_1 \cdot d_2)$, which proves that a is a divisor of c; thus, R is a transitive relation in A. Also note that the diagonal D_A of A, must be a subset of R. Indeed, since for each $a \in A$, a is a divisor of itself; hence, $(a, a) \in R$, for every element of $a \in A$.

Example 16: Let A be the set of positive integers P, and the two conditions

$$p_1(x, y) = x \in A,\ y \in A,\ \text{and } x < y;$$

$$p_2(x, y) = x \in A,\ y \in A,\ \text{and } x \leq y;$$

where the symbols "<" and "≤" have the usual meanings "less than" and " less than or equal to" respectively. The two conditions define two relations R_1, R_2, in A, both of them transitive; the second condition R_2 is also reflexive; and both of these relations are anti-symmetric.

1.3 Equivalence relations

Definition 12: (i) A relation R in (or on) a non-empty set A is called an **equivalence relation**, if R is reflexive, symmetric, and transitive.

(ii) If $x \in A$, $y \in A$, and $(x, y) \in R$, where R is an equivalence relation in A we say that the element x is equivalent to the element y.

Example 17: Let $A = \{0, 2, 3, 5, 7, 12, 13, 14\}$ and $p(x, y) = x \in A$, when divided with 7, leaves the same remainder as y does when divided by 7. The given condition defined the following equivalence relation R in A:

$$\begin{aligned} R \; = \; & \{(0,0), (2,2), (3,3), (5,5), (7,7), (12,12), (13,13), (14,14) \\ & (0,7), (7,0), (0,14), (14,0), (7,14), (14,7), (5,12), (12,5)\} \end{aligned}$$

Example 18: Let n be a fixed positive integer and $A = \mathbb{Z}$. Define $p(x, y) = x \in A$ leaves the same remainder (when divided by n) as $y \in A$ does. This condition defines an equivalence relation R in A. We know from elementary number theory that given a positive integer n, there are n possible remainders when an integer is divided by n. These remainders are the numbers $0, 1, 2, \ldots, n-1$. If k is any of these n remainder numbers, then all the integers that leave remainder k when divided by n, are equivalent with each other. Any such integer must be of the form $n \cdot q + k$, where q is the quotient of the division; every time we assign a new integer value to q, we obtain a new integer which when divided by n, it leaves remainder k. Assuming the fundamental fact from elementary number theory that any integer which when divided by n, leaves remainder k, with $0 \leq k \leq n-1$; must be of the form $n \cdot q + k$, for some integer k, it is easy to see that the above defined relation R is reflexive, symmetric, and transitive.

Example 19: Let m and n be fixed positive integers and let A be a set containing m integers $a_1, \ldots, a_m$; $A = \{a_1, \ldots, a_m\}$, with $a_i \in \mathbb{Z}$, $i = 1, \ldots, m$. Define $p(x, y) = x \in A$ leaves the same remainder, when divided by n, as $y \in A$ does. Then $p(x, y)$ defines an equivalence relation R in A. This example is a generalization of Example 17.

Example 20: Let A be a non-empty subset of $\mathbb{Z}$, and n a fixed positive integer. The condition $p(x, y) = x \in A$, when divided by n, leaves the same remainder as $y \in A$ does, defines an equivalence relation R in A.

Definition 13: Let R be an equivalence in a non-empty set A, and suppose a is an element of A. The subset E_a of A defined by $E_a = \{x \in A \mid xRa\}$ is called the **equivalence class of a with respect to the equivalence relation R.**

Theorem 1: Let R be an equivalence class in a non-empty set A.

(i) If a and b are nonequivalent (i.e., $(a, b) \notin R$) with respect to R then their corresponding equivalence classes E_a and E_b are disjoint; that is, they have no elements in common. We write $E_a \cap E_b = \emptyset$.

(ii) If a and b are equivalent elements of A (i.e., $(a, b) \in R$) with respect to R, then, their corresponding equivalence classes E_a and E_b are identical or equal: $E_a = E_b$.

(iii) The union of all the equivalence classes is equal to the set A. We write $\cup_{a \in A} E_a = A$.

Remark 9: (i) Note that because R is an equivalence relation, the statement "a is not equivalent to b" does imply "b is not equivalent to b". This is clear since R is a symmetric relation, so one cannot have aRb without at the same time having bRa; or vice-versa. Thus, the statement "if a and b are nonequivalent elements of A with respect to $R\ldots$", has an exact and unambiguous meaning.

(ii) An equivalence relation R in A, always gives rise to what is called **a partition of** A; that happens when A is written as a union of disjoint subsets of A.

Proof: (i) Let a and b be elements of A that are not equivalent. We argue by contradiction, if $E_a \cap E_b \neq \emptyset$, then let c be an element that belongs to both sets E_a and E_b. According to Definition 13, the element c must be equivalent to a since $c \in E_a$. By the same token, c must also be equivalent to b. Therefore, we have cRa and cRb, but R is symmetric $\Rightarrow aRc$. We have $(aRc$ and $cRb) \Rightarrow aRb$ since R is also a transitive relation, but this is a contradiction, by virtue of the fact that a and b have been assumed to be nonequivalent elements. Thus, the assumption that $E_a \cap E_b \neq \emptyset$ must be false since it led to a contradiciton. We conclude that $E_a \cap E_b = \emptyset$.

We may remark here that proving the contrapositive statement of the statement we wish to prove, would be an alternative way of proceeding. In basic propositional calculus (basic logic) one learns from truth tables that the implications $(p \Rightarrow q)$ and $(\rceil q \Rightarrow \rceil p)$, are logically equivalent. Here $\rceil q$ stands for the negation of q; $\rceil q$ reads "non q"; the same, of course, for $\rceil p$: "non p". So, if one proves the truth of $(\rceil q \Rightarrow \rceil p)$, one has proved the truth of $(p \Rightarrow q)$ and vice-versa. The statement $(\rceil q \Rightarrow \rceil p)$ is called the contrapositive of the statement $(p \Rightarrow q)$. In our case above, the contrapositive statement reads "If $E_a \cap E_b \neq \emptyset$, then a and b are equivalent elements", under the assumption R is an equivalence relation.

(ii) Since aRb it follows, by the definition E_b, that $a \in E_b$. Let c be any element of E_a; by definition, cRa and aRb, it follows by the transitivity of R, that cRb. This proves $c \in E_b$. Since every element c of E_a also belongs to E_b, it follows that E_a is a subset of E_b.

Similarly, by starting with bRa, an analogous argument can prove that E_b is a subset of E_a. In conclusion, since the sets E_a and E_b are subsets of each other, they must be identical, $E_a = E_b$.

(iii) Since each equivalence class E_a is a subset of A, it follows that the union $\cup_{a \in A} E_a$ must be a subset of A, since the union consists only of subsets of A. Conversely, A is a subset of the union $\cup_{a \in A} E_a$, simply because each a in A is an element of E_a. Hence, $\cup_{a \in A} E_a$ and A are subsets of each other, hence they are equal. End of proof. ■

Example 21: In Example 17, the equivalence classes are the following sets:

$$\begin{aligned} E_0 &= E_{17} = E_{14} = \{0, 7, 14\}, \\ E_5 &= E_{12} = \{5, 12\}, \\ E_3 &= \{3\}, \text{ and } E_{13} = \{13\} \end{aligned}$$

Example 22: In Example 18, there are exactly n equivalent classes $E_0, E_1, E_2, \ldots, E_{n-1}$. Each one of them contains all the integers which when divided by n leave remainder $0, 1, \ldots, n-1$, respectively. These are infinite set; we write,

$$\begin{aligned}
E_0 &= \{\dots,-4n,-3n,-2n,-n,0,n,2n,3n,4n,\dots\}\\
E_1 &= \{\dots,-4n+1,-3n+1,-2n+1,-n+1,1,n+1,\\
&\quad 2n+1,3n+1,4n+1,\dots\}\\
&\vdots\\
E_k &= \{\dots,-4n+k,-3n+k,-2n+k,-n+k,k,n+k,\\
&\quad 2n+k,3n+k,4n+k,\dots\}\\
&\vdots\\
E_{n-1} &= \{\dots,-4n+(n-1),-3n+(n-1),-2n+(n-1),\\
&\quad -n+(n-1),n-1,n+(n-1),2n+(n-1),\dots\}
\end{aligned}$$

Definition 14: We call Δ a partition of a non-empty set A, a set each of whose member is a non-empty subset of A; and with the property that the union of all its members is the set A and all the members of Δ are disjoint sets.

The following theorem shows that every partition of a non-empty set gives rise to an equivalance relation.

Theorem 2: Let Δ be a partition of a non-empty set A.

(i) The condition $p(x,y) = x \in A,\ y \in A,\ (x,y) \in S \times S$, where S is some member of Δ; defines an equivalence relation R in A.

(ii) If Δ is a finite set, $\Delta = \{S_1,\dots,S_n\}$, with $S_1 \cup \dots \cup S_n = \cup_{i=1}^{n} S_i = A$, then the condition defined by, $p(x,y) = x \in A,\ y \in A,\ (x,y) \in S_i \times S_i$, for some integer i, $1 \le i \le n$; defines an equivalence relation in A.

Remark 10: Sometimes Δ is called the quotient set of A with respect to R, we write $\Delta = A/R$.

Proof: First observe that (ii) is just a special case of (i). So we need only to prove (i). To show that R is reflexive, let $a \in A$; since Δ is a partition of A, it follows that a belongs in one and only one set S, a subset of A and member of Δ. But then $(a,a) \in S \times S$ and so according to the condition $p(x,y)$, aRa; or $(a,a) \in R$. Next, let us prove that R is symmetric. Suppose that $(a,b) \in R$; by the definition of R, that means that $(a,b) \in S \times S$, for some member S of the partition Δ. But $(a,b) \in S \times S \Rightarrow (b,a) \in S \times S$, according to the definition of the cartesian product; and by condition $p(x,y)$, it is now clear that $(b,a) \in R$. The relation is symmetric.

Suppose that $(a,b) \in R$ and $(b,c) \in R$. Then $(a,b) \in S_1 \times S_1$ and $(b,c) \in S_2 \times S_2$, where S_1 and S_2 are members of Δ. By definition of the cartesian products $S_1 \times S_1$ and $S_2 \times S_2$, we know that both elements a and b belong to S_1; and both elements b and c are in S_2. But S_1 and S_2 are disjoing sets, $S_1 \cap S_2 = \emptyset$; unless they are equal, that is, unless they are one and the same set; this is according to the definition of Δ. Since $b \in S_1$ and $b \in S_2$, it follows $S_1 \cap S_2 \neq \emptyset \Rightarrow S_1 = S_2$. The pair (a,c), therefore, is a member of the set $S_1 \times S_1 = S_2 \times S_2$, which proves according to the condition $p(x,y)$ that $(a,c) \in R$. The relation is transitive; we are done. ∎

Remark 11: There is a useful notation for an equivalence class that can be employed; if x is an element of the set A and R an equivalence relation on A, then (xR) denotes the equivalence class of x; we have $A = \cup_{x \in A}(xR)$; and as we have already seen, $(xR) \cap (yR) \neq \emptyset$, if x and y are not equivalent elements of A; that is, if $(x, y) \notin R$. Otherwise, if $(x, y) \in R$, we have $(xR) = (yR)$.

1.4 Ordering or order relations

Let A be a non-empty set.

Definition 15: A relation R in A is called a **a partial ordering or partial order in the narrow sense,** if it satisfies all three requirements below:

(i) R is anti-reflexive,

(ii) R is anti-symmetric, and

(iii) R is transitive.

Remark 12: A partial order R in the narrow sense in a set A, is usually denoted by the symbol $\prec$.

Example 23: Let $A = \{a, b, d, e, h, i\}$ and let $p(x, y) = x \in A,\ y \in A$, and x precedes y alphabetically by at least two spaces.

One can easily see that the relation R that $p(x, y)$ defines, is anti-reflexive, anti-symmetric, and transitive. Note though, that because of the requirement that x precede y by at least two spaces in the alphabet, the letter a does not relate to b; that is, $(a, b) \notin R$. Likewise $(d, e) \notin R$, and $(h, i) \notin R$. Using symbol $\prec$ we see that,

$$a \prec d \prec h,\quad a \prec d \prec i,\quad a \prec e \prec h,\quad a \prec e \prec i,$$

$$b \prec d \prec h,\quad b \prec d \prec i,\quad b \prec e \prec h,\quad b \prec e \prec i,$$

$$d \prec h \qquad d \prec i, \qquad e \prec h, \qquad e \prec i$$

We can also employ the following *Hasse* diagram:

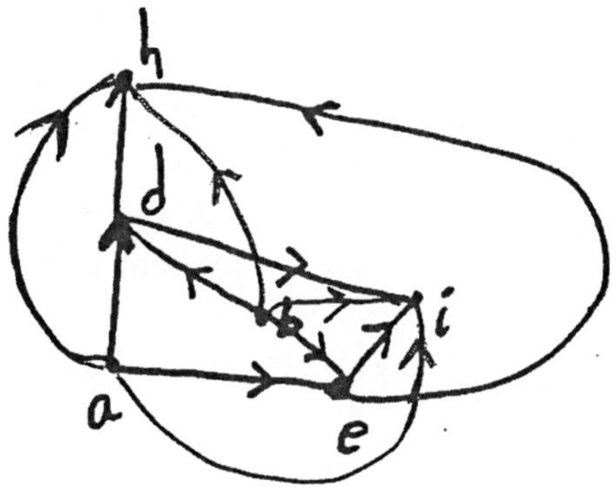

Figure 8

$$R = \{(a, d), (a, e), (a, h), (a, i), (b, d), (b, e), (b, h), (b, i), (d, h), (d, i), (e, h), (e, i)\}$$

Remark 13: Note that since $(a, b) \notin R$, $(d, e) \notin R$, $(h, i) \notin R$, the elements a and b are called **noncomparable**; likewise for d and e and h and i. In contrast, any pair in R consists of two **comparable** elements, for instance, a and d are comparable since $a \prec d$.

Definition 16: A relation R in a non-empty set A is called a **total ordering or total order in the narrow sense**, if the following four requirements are satisfied:

(i) R is anti-reflexive,

(ii) R is anti-symmetric,

(iii) R is transitive, and

(iv) for every pair (x, y) in A^2 with $x \neq y$, either $(x, y) \in R$ or $(y, x) \in R$.

We see from Definition 16, that any two elements of A are comparable with respect to a total order $\prec$; this is illustrated by the following two examples.

Example 24: Let A be the set of the previous example, $A = \{a, b, d, e, h, i\}$ and $p(x, y) = x \in A,\ y \in A$, and x alphabetically precedes y; the relation R that $p(x, y)$ defines is a total order in the narrow sense and we have, $a \prec b \prec d \prec e \prec h \prec i$; the Hasse diagram being much simpler than that of Example 23.

$$R = \{(a, b), (a, d), (a, e), (a, h), (a, i), (b, d), (b, e), (b, h), (b, i), (d, e), (d, h), (d, i), (e, h), (e, i), (h, i)\}$$

Example 25: Let $A = \mathbb{Z}$ and $p(x, y) = x \in A,\ y \in A$, and x is less than y. We may write $x < y$ using the common notation for "less than". The relation R that $p(x, y)$ defines is a total order in the narrow sense. We may use a Hasse diagram:

$$\ldots \rightarrow -6 \rightarrow -5 \rightarrow -4 \rightarrow -3 \rightarrow -2 \rightarrow -1 \rightarrow 0 \rightarrow 1 \rightarrow 2 \rightarrow 3 \rightarrow 4 \rightarrow 5 \rightarrow 6 \rightarrow \ldots$$

$$\ldots < -6 < -5 < -4 < -3 < -2 < -1 < 0 < 1 < 2 < 3 < 4 < 5 < 6 \leftarrow \ldots$$

Definition 17: A relation R in a non-empty set A is called a **partial ordering or partial order in the wide sense**, usually denoted by $\preceq$ if it satisfies the following three requirements:

(i) R is reflexive,

(ii) R is anti-symmetric, and

(iii) R is transitive.

Example 26: (i) Let $A = \mathbb{Z}^+ = P$, the set of positive integers. Consider the condition $p(x, y) = x \in A,\ y \in A$, and x is a divisor of y. Since every positive integer is a divisor of itself, it follows that the relation R, that the given condition defines, must be a reflexive relation. We have $x = 1 \cdot x$ for every positive integer x. To prove that R is anti-symmetric we could argue in two different ways. We can show that if $(x, y) \in R$ and $x \neq y$, then the pair (y, x) does not belong to R. Indeed, suppose that $(x, y) \in R$ with $x \neq y$; that means that x is a divisor of y and $x \neq y$. We have $y = x \cdot k$, k some positive integer. Note that $y = x \cdot k \geq x$, since $k \geq 1$. Hence $(y \geq x \text{ and } y \neq x) \Rightarrow y > x$; since y is greater than x, it follows that y cannot be a divisor of x, because both y and x are positive integers.

Another argument that proves that R is anti-symmetric goes as follows: suppose that both (x, y) and (y, x) are in R. We will show that this assumption implies $x = y$. Indeed, since x is a divisor of y and y is a divisor of x we must have, $x = k_1 \cdot y$ and $y = k_2 \cdot x$, for some positive integers k_1 and k_2. Thus, we have $y = k_2 \cdot k_1 \cdot y$, and since $y \neq 0$, we conclude $1 = k_1 k_2$, which implies $k_1 = k_2 = 1$, since both k_1 and k_2 are positive integers. Therefore $x = y$. To show that R is transitive suppose that x divides y; and y divides z. We have $y = m \cdot x$, $z = n \cdot y$, for positive integers m and

n. Substituting for y in $z = n \cdot y$ we obtain $z = n \cdot (m \cdot x)$; $z = (n \cdot m) \cdot x$, which proves that x is a divisor of z. Hence, R is transitive.

(ii) If we were to take $A = \{1, 2, 3, 4, 5, 6, 7, 8, 9, 10, 11, 12\}$ under the same divisibility condition of part (i), we obtain the following diagram for this partial order in the wide sense:

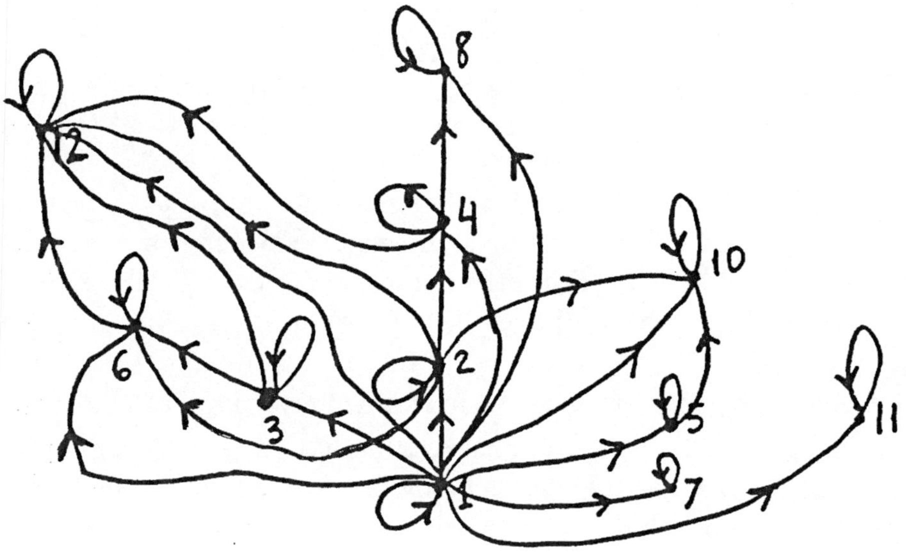

Figure 9

Remark 14: If we were to take $A = \mathbb{Z}$ instead of $\mathbb{Z}^+$ in the previous example, then under the same condition of divisibility, we see that the relation R that is defined by the divisibility condition, fails to be a partial order; because it is no longer anti-symmetric. If x is any integer then its opposite integer, $-x$, is a divisor of $x : x = (-1) \cdot (-x)$; and x is a divisor of $-x : -x = (-1) \cdot x$. Hence, both pairs $(x, -x)$ and $(-x, x)$ are members of R, for every integer x.

Definition 18: A relation R in a non-empty A is called a **total ordering or total order in the wide sense**, if it satisfies the following four requirements:

(i) R is reflexive,

(ii) R is anti-symmetric,

(iii) R is transitive, and

(iv) for every pair $(x, y) \in A^2$ with $x \neq y$, either $(x, y) \in A^2$ or $(y, x) \in A^2$.

Example 27: The relation "$\leq$" (less than or equal to) in the set $A = P = \mathbb{Z}^+$, is a total order on that set, the set of positive integers. The same is true for the set of integers $\mathbb{Z}$, the set of rational numbers $\mathbb{Q}$, and the set of real numbers $\boldsymbol{R}$, which will be introduced in Chapter 3.

Example 28: This example requires some familiarity with the set of complex numbers $\mathbb{C}$ and the basic operations in that set. Each complex number is a number of the form $x + iy$, where x and y are real numbers. In the set $A = \mathbb{C}$, we define the condition $p(z_1, z_2) = z_1 \in A,\ z_2 \in A,\ z_1 = x_1 + iy_1,\ z_2 = x_2 + iy_2$; and either $x_1 < x_2$ or $x_1 = x_2$ and $y_1 \leq y_2$. (Here "$\leq$" has the usual meaning "less than or equal to", and "$<$" also has the usual meaning of "less".) We denote this relation by $\preceq$; we can write, $z_1 \preceq z_2$; $x_1 + iy_1 \preceq x_2 + iy_2 \Leftrightarrow (x_1 < x_2 \text{ or } (x_1 = x_2 \text{ and } y_1 \leq y_2))$. Clearly, this relation is reflexive since if $z_1 = z_2$, then by the definition of equality of complex numbers, $x_1 = x_2$ and $y_1 = y_2$; so $z_1 \preceq z_2$. This relation is anti-symmetric, because if both $z_1 \preceq z_2$ and $z_2 \preceq z_1$ are satisfied, we see that $x_1 = x_2$ (cannot have $x_1 < x_2$ and $x_2 < x_1$). We also see that we must have $y_1 \leq y_2$ and $y_2 \leq y_1$; and in view of the fact that y_1, y_2 are real numbers, we must have $y_1 = y_2$. Thus, $x_1 = x_2$ and $y_1 = y_2$; $z_1 = z_2$. The relation is anti-symmetric. The relation is transitive as well. Suppose $z_1 \preceq z_2$ and $z_2 \preceq z_3$. From the definition $z_1 \preceq z_2$, it follows that either $x_2 < x_2$ or ($x_1 = x_2$ and $y_1 \leq y_2$); and from $z_2 \preceq z_3$, we have $x_2 < x_2$ or $x_2 = x_3$ and $y_2 \leq y_3$.

If $x_1 < x_2$ and $x_2 < x_3$, then $x_1 < x_3$, which immediately implies $z_1 \preceq z_3$.

If $x_1 < x_2$ and $x_2 = x_3$, then again $x_1 < x_3$, so that $z_1 \preceq z_3$. Suppose that $x_1 = x_2$ and $y_1 \leq y_2$; if $x_2 < x_3$, then $x_1 < x_3$ and so $z_1 \preceq z_3$. If, on the other hand, $x_2 = x_3$ and $y_2 \leq y_3$, we conclude $x_1 = x_3$ and $y_1 \leq y_3$. Thus, once more $z_1 \preceq z_3$. The transitivity of the relation is proven. Finally we show that requirement (iv) of Definition 18 is also satisfied . Indeed, let z_1, z_2 be any two complex numbers. From the axiomatic foundation of real numbers (Chapter 3), we know that exactly one of three possibilities must occur: $x_1 < x_2$, $x_1 = x_2$, or $x_2 < x_1$. If the first or third possibility occurs, then by definition of the relation $\preceq$, we conclude $z_1 \preceq z_2$ or $z_2 \preceq z_1$ respectively. If the second possibility occurs, $x_1 = x_2$, then we must have $y_1 \leq y_2$ or $y_2 \leq y_1$ (this is a nonexclusive "or", since y_1 and y_2 may be equal). Thus, again from the definition $\preceq$, we see that $z_1 \preceq z_2$ or $z_2 \preceq z_1$, respectively.

Hence, the relation $\preceq$ is a total order in the wide sense on the set $\mathbb{C}$.

Example 29: See practice problem P 21.

Definition 19: Let A be a non-empty set. We say that the set A is **totally ordered with respect to an order** $\prec$ ($\preceq$), if the order $\prec$ ($\preceq$) is a total order in the narrow sense (wide sense) on A.

Theorem 3: Let R be a relation in a non-empty set A, and R^{-1} the inverse relation of R.

(i) If R is reflexive, so is R^{-1}.

(ii) If R is anti-reflexive, so is R^{-1}.

(iii) If R is symmetric, so is R^{-1}.

(iv) If R is anti-symmetric, so is R^{-1}.

(v) If R is transitive, so is R^{-1}.

(vi) If R satisfies the condition: "for every $x \in A$, $y \in A$, at least one of the pairs (x, y), (y, x) is a member of R"; then R^{-1} satisfies the same condition (where R is replaced by R^{-1}).

Proof: (i) Immediate from Definition 6 (with $A = B$) and Defintion 7.

(ii) Likewise, it follows immediately from Definitions 6 and 8.

(iii) Immediate from definitions 6 and 9.

(iv) Lnmediate from Defintions 6 and 10.

(v) Let (x, y) and (y, z) be members of R^{-1}. By Definition 6, it follows that (y, x) and (z, y) must be members of R. But R is transitive; which means that $((z, y) \in R$ and $(y, x) \in R)$ implies $(z, x) \in R \Rightarrow (x, z) \in R^{-1}$. We have proven the implication $((x, y) \in R^{-1}$ and $(y, z) \in R^{-1}) \Rightarrow (x, z) \in R^{-1}$; R^{-1} is transitive.

(vi) Let $x \in A,\ y \in A$. Since R satisfies the stated conditions, either (x, y) or (y, x) or both belong to R. If $(x, y) \in R$, then by Defintion 6, $(y, x) \in R^{-1}$. If $(y, x) \in R$, then by the same definition, $(x, y) \in R^{-1}$. Hence, R^{-1} satisfies the same condition.

The following theorem is an immediate consequence of Theorem 3 and Definitions 15, 16, 17 and 18.

Theorem 4: Let R be a relation in a non-empty set A.

(i) If R is a partial order in A in the narrow sense, then so is the inverse relation R^{-1}.

(ii) If R is a partial order in A in the wide sense, then so is the inverse relation R^{-1}.

(iii) If R is a total order on A in the narrow sense, then so is the inverse relation R^{-1}.

(iv) If R is a total order on A in the wide sense, then so is the inverse relation R^{-1}.

Definition 20: Let $\prec$ ($\preceq$) be a partial order in a non-empty set A, also denoted by R; let m, M denote elements of A. Consider the sets $S_m = \{x \in A \mid (m, x) \in R\}$ and $S_M = \{x \in A \mid (x, M) \in R\}$.

(i) The element m is called **a minimal element with respect to order** $\preceq$ ($\preceq$) (or simply **minimal**), if the set S_m is non-empty, (which means $m \prec x (m \preceq x)$ for every $x \in S_m$), and there is no element $k \in A$ such that $k \neq m$, and $k \prec m (k \preceq m)$

(ii) The element M is called **a maximal element with respect to the order** $\prec$ ($\preceq$) (or simply **maximal**), if the set S_M is non-empty, (which means $x \prec M (x \preceq M)$ for every $x \in S_M$), and there is no element $k \in A$ such that $k \neq M$ and $M \prec k (M \preceq k)$.

Remark 15: (i) Note that in the above definition, there is no need to include a special mention of total orders, since every total order is automatically a partial order.

(ii) The conditions $k \neq m,\ k \neq M$ in the above definition are superfluous in the case of $\prec$, since such an order is anti-reflexive.

Example 30: In Example 23, the elements a and b are the only minimal elements in the set A; note that $S_a = \{a, e, h, i\}$, $S_b = \{d, e, h, i\}$; the elements h and i are both maximal elements, the only maximal elements; $S_h = \{a, b, d, e\}$, $S_i = \{a, b, d, e\}$.

In Example 24, a is the only minimal element and i is the only maximal element, and $S_a = \{b, d, e, h, i\}$, $S_i = \{a, b, d, e, h\}$. In Example 26, the numbers 7,8,9,10,11,12 are all maximal elements; 1 is the only minimal element. We have $S_1 = \{1,2,3,4,5,6,7,8,9,10,11,12\}$, $S_7 = \{1,7\}$, $S_8 = \{1,2,4,8\}$, $S_9 = \{1,3,9\}$, $S_{10} = \{1,2,5,10\}$, $S_{11} = \{1,11\}$, $S_{12} = \{1,2,3,4,6,12\}$.

Remark 16: Note that in the case of a totally ordered set A, whether under $\prec$ or $\preceq$, there can only be one minimal element and one maximal element. In other words, if a minimal element exists, it must be unique; likewise for a maximal element. This is easy to see because if say m is a minimal element of a totally ordered set A under $\prec$ ($\preceq$), then if x is any element of A with $x \neq m$, then according to Definitions 16 and 18, we must have $x \prec m$ ($x \preceq m$) or $m \prec x$ ($m \preceq x$). If A is under $\prec$, then since m is minimal, it follows that $m \prec x$; if A is under $\prec$, then again since m is minimal it follows that $x \preceq m$; the alternative, $m \preceq x$, can only hold if $m = x$, otherwise it would violate the minimalness of m. Thus, we see that there is room for only one minimal element, since the above argument clearly shows that if m' is any other minimal element, we must then have $m = m'$; a similar argument holds for the case of a maximal element. This leads to the definition below.

Definition 21: Let A be a totally ordered set under $\prec$ ($\preceq$). The unique minimal (maximal) element of A, if it exists, is called the **minimum (maximum)** element of A; denoted by $\min A(\max A)$. If m is the minimum of A, then $m \prec x$, for every $x \in A$, $x \neq m$ ($m \preceq x$, for every $x \in A$). If M is the maximum of A, then $x \prec M$ for every $x \in A$, $x \neq M$ ($x \preceq M$, for every $x \in A$).

Example 31: (i) Under both orders $<$ ("less than") and $\leq$ ("less than or equal to"), the set of positive integers has 1 as its minimum element, but not maximum element.

(ii) Any finite set of integers has a minimum element and a maximum element under both $<$ and $\leq$.

(iii) Here we use a notation from Calculus: "let a and b be distinct real numbers". The interval $[a, b]$ has a as its minimum and b its maximum, under both orders $<$ and $\leq$. The interval $[a, b)$ has only a minimum, namely a, but no maximum. The interval $(a, b]$ under both $<$ and $\leq$ has b as its maximum, but no minimum. The interval $(a, b]$ has neither maximum or minimum under $<$ or under $\leq$.

(iv) The set of certain rational numbers defined as $S = \{\frac{1}{n} \mid n \in \mathbb{Z}^+\}$ is the set of all the reciprocals of the positive integers. This set has only a maximum, the number 1; but no minimum under $<$ or $\leq$.

(v) Any subset of the set of integers $\mathbb{Z}$, that contains infinitely many negative integers cannot have a minimum element under $<$ or under $\leq$.

(vi) The set of complex numbers $\mathbb{C}$, under the total order $\preceq$ of Example 28 has no minimum element, since if $z_1 = x_1 + iy_1$, one can always find a "smaller" element than z_1, distinct from z_1, by taking $z_2 = x_2 + iy_2$, where x_2 is a real number less than x_1.

Definition 22: Let $\prec$ ($\preceq$) be a total order on a non-empty set A. The set A is called **well ordered** under the order $\prec$ ($\preceq$), if, and only if, every non-empty subset of A has a minimum element.

Remark 17: It follows from the definition of a totally ordered set, that if A is totally ordered under $\prec$ ($\preceq$), then every non-empty subset X of A must also be totally ordered under $\prec$ ($\preceq$). So, it is meaningful to talk about a minimum element of a subset of A.

Example 32: (i) The set of positive integers $\mathbb{Z}^+$ is a well ordered set, under the usual order of $<$ ($\leq$); the same is true for the sets of the form $S_a = \{x \mid x \in \mathbb{Z}$ and $a \leq x\}$, where a is some fixed non-positive integer (i.e., $a = 0$ or a is a negative integer).

(ii) None of the sets $\mathbb{Z}$, $\mathbb{Q}^+$ (positive rationals), $\mathbb{R}$, $\mathbb{R}^+$ (positive reals) is well ordered under $<$ ($\leq$). Another example of a set that is not well ordered is the following subset of complex numbers: $S_{z_0} = \{z \mid z \in \mathbb{C},\ z \neq z_0$, and $z_0 \preceq z\}$, where $z_0 = x_0 + iy_0$ is some fixed complex number and $\preceq$ is the total order of Example 28. One can see that in fact S_{z_0} itself has no minimum element, so according to Definition 22, S_{z_0} is not well ordered. Indeed, if $z_1 \neq z_0$ is an element of S_{z_0}, then by definition $z_0 \preceq z_1$; $x_0 + iy_0 \preceq x_1 + iy_1$, which means, according to the definition of $\preceq$, that either $x_0 < x_1$ or otherwise ($x_0 = x_1$ and $y_0 \leq y_1$). Note, however, that if the latter possibility holds we must have $y_o < y_1$ in view of the fact that $z_0 \neq z_1$. Let $z_2 = \frac{x_0+x_1}{2} + \left(\frac{y_0+y_1}{2}\right) \cdot i$; then if $x_0 < x_1$ holds, it follows that $\frac{x_0+x_1}{2} < x_1$; and if $y_0 < y_1$ holds, we must have $\frac{y_0+y_1}{2} < y_1$. In either case, we see that $z_2 \preceq z_1$ and $z_2 \neq z_1$. Moreover, from either $x_0 < x_1$, or from $y_0 < y_1$, it follows respectively that $x_0 < \frac{x_0+x_1}{2}$; or $y_0 < \frac{y_0+y_1}{2}$; so that $z_0 \preceq z_2$ and $z_0 \neq z_2$; hence z_2 is a member of the set S_{z_0}; but the element z_2 is "smaller" than z_1 and it is distinct from z_1. So whenever one picks an element member of S_{z_0}, one can find a "smaller element" in S_{z_0}, and distinct from the element initially chosen. This proves that S_{z_0}, has no minimum element; S_{z_0} is not well ordered under $\preceq$.

(iii) The set A of Example 24 is well ordered, as it is easily seen.

Definition 23: Let A be a non-empty, totally ordered set under $\prec$ ($\preceq$).

(i) A non-empty subset X of A is said to be **bounded above** if there exists an element a of A such that $x \prec a(x \preceq a)$ for every $x \in X$. The element a is called an **upper bound** of the set X.

(ii) A non-empty subset X of A is said to be **bounded below**, if there exists and element b of A such that $b \prec x(b \preceq x)$ for every $x \in X$. The element b is called a **lower bound** of the set X.

Remark 18: Observe that if X is a non-empty subset of a totally ordered set A, the element m is the minimum of X if, and only if, m is an lower bound of X and $m \in X$; and the element M is the maximum of X if and only if M is an upper bound of X and M is a member of X. Also observe that if a is an upper(lower) bound of X, then so is another element of A "greater"("less") than a.

Definition 24: Let X be a non-empty subset of a totally ordered set A.

(i) If X is bounded above, then the minimum of the set of all upper bounds of X, if that minimum exists, is called the **supremum** of X and it is denoted by $\sup S$.

(ii) If X is bounded below, then the maximum of the set of all lower bounds of X, if that maximum exits, is called the **infimum** of X and it is denoted by $\inf X$

Example 33: The set $A = \mathbb{R}$ is a totally ordered set under both $<$ ("less than") and $\leq$ ("less than or equal to"). The subset $X = (a, b)$ of A,where a and b are fixed real numbers with $a < b$, is familiar from calculus. It consists of all reals greater than a but less than b. The set $\bigcup$ of all upper bounds of X consists of all real numbers greater than or equal to b. In calculus notation, $\bigcup = [b, \infty)$; and the set L of all lower bounds of X is the interval $L = (-\infty,\ a]$. The minimum element of $\bigcup$ is b; thus b is the supremum of X; and the maximum element of L is a; that a is the infimum of X.

(ii) Let $A = \mathbb{Q}$ be the set of rational numbers and X be the subset of A defined as $X = \{\frac{1}{n} \mid n \text{ a positive integer}\}$. Under the usual orders $<$ ($\leq$) the set of all upper bounds of X is the set of all rational numbers which are greater than or equal to 1. Thus, that set has a minimum, namely 1; the number 1 is the supremum of X. On the other hand, the set of all lower bounds of X consists all rational numbers which are less than or equal to zero. The maximum of that set is the number zero. Hence, zero is the infimum of X.

(iii) Let $A = \mathbb{Q}$, $X = \{2 - \dfrac{1}{n} \mid n \text{ a positive integer}\}$. The infimum of X is the number 1 and supremum of X is the number 2.

Remark 19: Unlike the set of rational numbers $\mathbb{Q}$, the set of real numbers $\mathbb{R}$ (see Chapter 3) has the property that every bounded above (below) subset of $\mathbb{R}$ possesses a supremum (infimum) under the usual order $<$ ($\leq$) of $\mathbb{R}$.

1.5 Functions and Mappings

In what is to follow, it is convenient to sometimes use certain symbols from symbolic logic or propositional calculus. Other than notations like $x \in A$ ("x is a member of the set A") which we have extensively used thus far. For the introduction of the concept of function, the following symbols come in handy:

The symbol $\exists$ means "there exists".
The symbol $\exists \mid$ means "there exists at most one".
The symbol $\exists *$ means "there exists exactly one".
The symbol $\forall$ means "for every".

And (as we have seen), the symbol $\subset$ means "subset".

Definition 25: Let A and B be non-empty sets.

(i) **A function from the set A to the set B** is a relation f from A to B with the property that, for every element $x \in A$ there is at most one element $y \in B$ such that $(x, y) \in f$. In symbolic notation this statement can be expressed as follows:

(f is a function from A to B) $\Leftrightarrow$ $(f : A \to B) \Leftrightarrow (f \subset A \times B : \forall x \in A \exists \mid y \in B$ with $(x, y) \in f)$.

(ii) If a pair (x_0, y_0) belongs to or is a member of a function f, then y_0 is the unique element of B which is **assigned to** the element x_0 of A. We then say that the function f is **defined** for or at $x = x_0$; and the element y_0 which is also denoted by $f(x_0)$ (it reads "f of x") is called the **image of** x_0, as well as the **"value" of the function f for or at** $x = x_0$. The element x_0 is called the **pre-image** of $y = f(x_0)$. In general , when non-specific elements are involved, we write $f : A \to B,\ y = f(x)$.

(iii) The set $D_f = \{x \in A : \exists * y \in B \text{ with } y = f(x)\}$, which is a subset of A, is called the **domain** of the function f. This definition can be read as follows: the domain D_f of the function f is the set of all elements x of A, such that, for each of them there exists only one element y in B with $y = f(x)$.

(iv) The set B is sometimes called the **codomain** of the function f. The set $R_f = f(D_f) = \{y \in B : \exists x \in A \text{ with } y = f(x)\}$, a subset of B, is called the **range** of the function f.

This can also be read as follows:

The range of the function f, is the set of all elements y in B, such that, for each one of them, there exists at least one element x of A with $y = f(x)$.

(v) **A mapping from (or of) the set A to the set B** is a function from the set A to the set B, such that $D_f = A$. In other words, a mapping from A to B is a function whose domain is the entire set A.

Remark 19: Note that if f is a function from a set A to a set B, then for every $x \in A$, there do not exist two distinct elements y_1 and y_2 in B such that $(x, y_1) \in f$ and $(x, y_2) \in f$.

Definition 26: (i) (The case $A = B$) A function f from a non-empty set A to itself, is called a **function within the set A or simply a function in A**.

(ii) If a function f in a non-empty set A is a mapping (i.e., domain $D_f = A$), such a function is also called a **transformation of the set A**.

Definition 27: Let f be a function from a non-empty set A to a non-empty set B.

(i) The function f is called **surjective or onto**, if $R_f = B$. That is, if the range of f is the entire set of B. If, in addition, f is a mapping, then f is called an **onto or surjective mapping of A onto B**

(ii) The function f is called **injective or one-to-one**, if no two distinct elements x_1, x_2 in the domain D_f of f have the same image.
Symbolically:
$\forall x_1, x_2 \in D_f : x_1 \neq x_2 \Rightarrow f(x_1) \neq f(x_2)$
Or equivalently: $\forall x_1, x_2 \in D_f : f(x_1) = f(x_2) \Rightarrow x_1 = x_2$
Or: $\forall x_1, x_2 \in D_f : ((x_1, y) \in f$ and $(x_2, y) \in f) \Rightarrow x_1 = x_2$; if an injective function is a mapping, then it is called an **injective mapping of A into B**.

Definition 28: A mapping f from a non-empty set A to a non-empty set B is called a **bijection from A onto B**, if f is both a surjective and an injective mapping.

Definition 29: A transformation of a non-empty set A is called a **permutation of the set A**, if it is a bijection from A onto itself.

Definition 30: Two special mappings: let A and B be non-empty sets.

(i) The **identity mapping (or transformation) of A**, denoted by I_A, is the bijection from A onto itself that satisfies $I_A(x) = x$, for every $x \in A$.

(ii) A mapping f from A to B is called **constant**, if R_f is a **singleton** set; that is, $R_f = \{b\}$, where b is some element of B; in other words, $f(x) = b$, $\forall x \in A$.

Theorem 5: Let f be a function from a non-empty set A to a non-empty set B.

(i) The inverse relation f^{-1} of f is a function from B to A if, and only if, f is injective. If that is the case, $D_{f-1} = R_f$ and $R_{f-1} = D_f$, and f^{-1} will be an injective function.

(ii) The inverse relation f^{-1} is a mapping from B to A if, and only if, f is both injective and surjective; and in fact, f^{-1} will be an injective mapping.

(iii) If f is a mapping, then the inverse relation f^{-1} is a mapping if, and only if, f is a bijection from A onto B.

(iv) The function f is a bijection from A onto B if, and only if, the inverse relation f^{-1} is a bijection from B onto A.

Proof: (i) If the inverse relation f^{-1} is a function that means, $\forall y \in B \exists \mid x \in A$ such that $f^{-1}(y) = x$. In other words, for every element y of B there is at most one element of x of A with $f^{-1}(y) = x$. Thus there can be no two pairs (x_1, y) and (x_2, y) in f, with $x_1 \neq x_2$ (because this would imply that (y, x_1), (y_2, x_2) would both be members of the inverse relation f^{-1}, with $x_1 \neq x_2$, violating the assumption that f^{-1} is function). Since no two distinct elements in A can have the same image in B, it follows that f must one-to-one or injective. Now, let us prove the converse: assume f to be injective, $(\forall x_1, x_2 \in D_f$, with $x_1 \neq x_2) \Rightarrow f(x_1) = y_1 \neq y_2 = f(x_2)$. Suppose the pairs (y, x_1) and (y, x_2) are in the inverse relation f^{-1}, it follows that $x_1 = x_2$. For if x_1 were distinct from x_2, then since by definition (x_1, y) and (x_2, y) are members of f, this would violate the assumption that f is injective. We have shown that, $\forall y \in B \exists \mid x \in A$ with $f^{-1}(y) = x$. In other words, we have proved that for every element y of B, there is at most one element x of A such that $(y, x) \in f^{-1}$. Hence, the inverse relation f^{-1} is a function from B to A.

Clearly, from the above, we see that $D_{f^{-1}} = R_f$ and $D_f = R_{f^{-1}}$.

(ii) Suppose that the inverse relation f^{-1} is a mapping from B to A. Since f^{-1} is a mapping from B to A it follows, by definition, that $B = D_{f^{-1}}$; that is, the domain for f^{-1} is the entire set B. Hence, $\forall y \in B$, $\exists x \in A$ such that $f^{-1}(y) = x$. In other words, every element y of B has an image x in A under f^{-1}, but then $(y, x) \in f^{-1} \Rightarrow$ (by definition) $(x, y) \in f$; $f(x) = y$ which proves that every element y in B is the image of some $x \in A$. Thus, $R_f = B$, the function f is surjective. To prove that f is also an injective function observe that there can exist no two pairs (x_1, y) and (x_2, y), both members of f, and with $x_1 \neq x_2$; since this would imply (by definition of the inverse relation) that both (y, x_1) and (y, x_2) would be members of f^{-1}, with $x_1 \neq x_2$; but that cannot happen since f^{-1} has been assumed to be a mapping and hence, in particular, a function. Clearly, from the above, we see that f^{-1} is injective. Now, we prove the converse statement: If f is both injective and surjective, then f^{-1} is a mapping, an injective one. Because f is injective it follows that $\forall y \in B \exists \mid x \in A$ with $f(x) = y$. In other words, every element y of B is the image of at most one element x of A. Hence, it is easy to see that the inverse relation f^{-1} must be a function from B to A. Moreover, since f is surjective, $R_f = B \Rightarrow D_{f^{-1}} = B$, which proves that f^{-1} is a mapping; a one-to-one mapping, as it is readily seen.

(iii) Since f is a mapping, it is, in particular, a function. Thus, according to part (ii), the inverse relation f^{-1} is a mapping if, and only if, f is both injective and surjective. Since, by assmption, f is a mapping, it follows from the definition of bijection that f^{-1} is a mapping if, and only if, f is a bijection.

(iv) This follows immediately from parts (ii) and (iii), since by (ii), f^{-1} must be injective; and since f is a mapping, f^{-1} will also be surjective.

Examples of functions and mappings

Example 34: Consider the relation R in the set

$$A = \{a, b, c, d, e, f, g, h, i, j, k, l, m\},$$

whose Venn diagram is illustrated below.

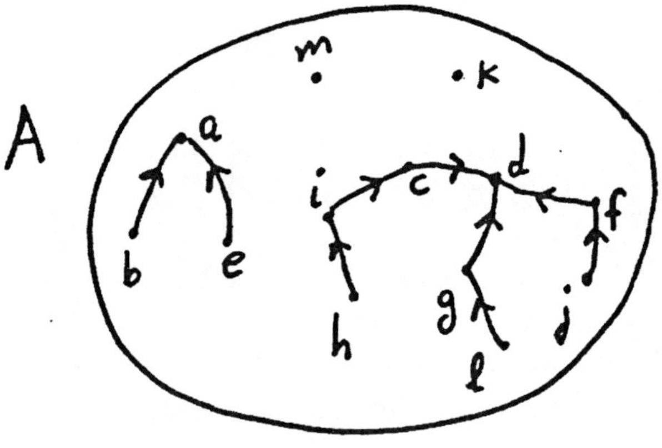

Figure 10

Also, suppose that this relation is defined by the condition, $p(x, y) =$ person $x \in A$ has a mother person $y \in A$

$$R = \{(b, a), (e, a), (h, i), (i, c), (c, d), (f, d), (g, d), (\ell, g), (j, f)\}$$

We see that R is a function f in the set A. The domain of this function f is the set $D_f = \{b, e, h, i, c, g, f, j, \ell\}$. The range is the set $R_f = \{a, c, d, f, g, i\}$.

We can represent or illustrate this function by using the following table:

x	b	c	e	f	g	h	i	j	ℓ
$f(x)$	a	d	a	d	d	i	c	f	g

Note that this function is not a transformation of the set A. That is, this function is not a mapping from A to itself, since the domain D_f is not the entire set A, but only a proper set of A. Also, this function is neither surjective nor injective, as it is easily seen. Hence, in particular, the inverse relation R^{-1} is not a function. From the Venn diagram or the table describing f we can make the following observations:

(i) The mothers of the persons k and m are not members of the set A.

(ii) The persons b and e have the same mother, namely a.

(iii) The persons c, f, and g have the same mother, the person d.

(iv) The person c is the grandmother of person h. Likewise, the person d is the grandmother of persons ℓ and j.

(v) The person d is the great-grandmother of person h.

Example 35: Let p be the coordinate plane we are familiar with from calculus and A the set of all circles that lie in the plane p. $A = \{x \mid x \text{ is a circle in } p\}$, and B is the subset of the set of real numbers, $B = \mathbb{R}^+ \cup \{0\}$. B is the set of non-negative real numbers. The condition $p(x, y) = x \in A$ has radius $y \in B$, defines a mapping f from the set A to the set B; the mapping f is actually surjective since for a given radius $y \geq 0$, there is always at least one circle in the plane p with that radius y. In fact, infinitely many such circles. But, clearly f is not a one-to-one mapping, since there is more than a circle with a given radius.

Example 36: Let C be a given circle in the coordinate plane p; A the set of all chords of the circle C. B is the set of all points in the coordinate plane p. The condition defined by $p(x, y) = x \in A$ has $y \in B$ as its midpoint, defines a mapping f from A to B. Note that this mapping is not injective, since for example, any two diameters (which are chords, of course) share the same midpoint, namely the center of the circle C. The mapping f is not surjective either, since any point on the circle, or outside the disk that the circle defines, is not the mid-point of a chord. That is, such points are not in the range of this mapping. Also note that if we exclude all the diameters from the set A; that is, if we take $A' = \{x \mid x \text{ is a chord of the circle } C, \text{ but not a diameter}\}$, then the same condition above defines a mapping g from the set A' to the set B, which is injective (draw a picture to see this).

Example 37: Let k be a fixed positive integer, $k \geq 2$; and $A = P$, the set of positive integers. The condition, $p(x, y) = x \in A$ and $y \in A$ with $y = k \cdot x$, defines mapping f in the set of positive integers or a transformation of the set of postiive integers, we write $f(x) = k \cdot x$. It is easily seen that this mapping is one-to-one but not onto, since every positive integer that is not a multiple of k, cannot be in the range of this mapping.

Example 38: Let $A = \{2, 3, 5, 7, 11, 13, 17, 19, 23, 29, 31\}$ and the condition $p(x, y)$ $= x \in A$ is a divisor of $y \in A$, defines a relation that is actually a transformation f of the set A, since any of the numbers in the set A, can only divide itself. That is because each member of A is a prime number. We have $f(x) = x$, $\forall x \in A$. In other words, f is none other that the identity mapping I_A of the set A.

Example 39: Let $A = \mathbb{Z}^+$, the set of postiive integers and $p(x, y) = x \in A$ is a divisor of every $y \in A$. This condition defines a relation R in the set A; but the only positive integer that divides every positive integer is the integer $x = 1$. We have $R = \{(1, 1),\ (2, 1),\ (1, 3),\ (1, 4),\ \ldots,\ (n, 1),\ \ldots\}$. Obviously, this relation is not a function. However, the inverse relation $R^{-1} = \{(1, 1),\ (2, 1),\ (3, 1),\ (4, 1),\ \ldots,\ (1, n),\ \ldots\}$ is a function in the set A, described by the formula $F(x) = 1$, $\forall x \in A$. It is a constant function, actually a constant mapping of the set A.

Example 40: Let m and b be fixed rational numbers, $m \neq 0$. Consider the function $f : \mathbb{Q} \to \mathbb{Q}$, defined by $f(x) = mx + b$. It is clearly a function since for each $x \in \mathbb{Q}$, there corresponds only one image $y \in \mathbb{Q}$, namely the function is actually a mapping in the set of rationals. The function is injective: indeed if $f(x_1) = f(x_2)$, for $x_1, x_2 \in \mathbb{Q}$, then $mx_1 + b = mx_2 + b \Rightarrow mx_1 = mx_2 \Rightarrow mx_1 - mx_2 = 0 \Rightarrow$ $m \cdot (x_1 - x_2) = 0$; and since $m \neq 0$, we conclude $x_1 - x_2 = 0 \Rightarrow x_1 = x_2$. The mapping f is also surjective. To show that the range R_f is the entire set $\mathbb{Q}$, pick any $y \in \mathbb{Q}$; let $x = \frac{y-b}{m}$. Note that this choice of x is to obtain simply solving for x in terms of y (solve for the pre-image(s) in term of the image) in the initial formula $y = mx + b$. We have $f(x) = f\left(\frac{y-b}{m}\right) = m \cdot \left(\frac{y-b}{m}\right) + b = y - b + b = y$. This shows that for any $y \in \mathbb{Q}$, there is a number in $\mathbb{Q}$, namely $\frac{y-b}{m}$, whose image under the given mapping is the number y. Clearly this proves that the mapping f is onto or surjective. Hence, since f is also one-to-one, we conclude that f is a bijection from $\mathbb{Q}$ onto itself, that is, a permutation of the set $\mathbb{Q}$. Also note that the graph of this mapping, in the standard calculus coordinate system, is part of the (straight) line with equation $y = mx + b$; but not the entire line. Only those points whose x coordinate is a rational number (and subsequently their y-coordinate as well $-mx + b$), are in the graph of f.

Example 41: In Example 37, we saw that the mapping f from the set of positive integers P to itself, defined by $f(x) = k \cdot x$, k a fixed positive integer ≥ 2, is one-to-one or injective but not surjective. The inverse relation f^{-1} is a function from P to itself, according to Theorem 5(i). However, since f is not onto, the inverse function is not a mapping since its domain is not the entire set P. $D_{f^{-1}} = \{x \mid x \in P \text{ and } x = k \cdot t, \text{ for some } t \in P\}$, a proper subset of P. We can describe the inverse function f^{-1} by $f^{-1}(x) = \frac{1}{k} \cdot x$.

Example 42: Let $A = P \times P$, the cartesian product of the set of positive integers with itself; and $B = Q^+$ the set of positive rationals. Define the mapping $f : A \to B$ by $f((m,n)) = m + \frac{1}{n}$, for $(m,n) \in P \times P$. This mapping is injective, though this is not obvious at a glance. To show this, suppose $f((m_1,n_1)) = f((m_2,n_2))$; $m_1 + \frac{1}{n_1} = m_2 + \frac{1}{n_2}$ for some positive integers m_1, n_1, m_2 and n_2. We have $m_1 - m_2 = \frac{1}{n_2} - \frac{1}{n_1} \Leftrightarrow m_1 - m_2 = \frac{n_1 - n_2}{n_1 n_2} \Leftrightarrow n_1 n_2 \cdot (m_1 - m_2) = n_1 - n_2$; $n_1 n_2 \cdot (m_1 - m_2) + n_2 = n_1$; $[n_1 \cdot (m_1 - m_2) + 1] \cdot n_2 = n_1$. The number $n_1 \cdot (m_1 - m_2) + 1$ is an integer, in fact, a positive one, since both n_2 and n_1 are positive integers. The key observation here is that n_2 divides n_1. Similarly, from $n_1 n_2 \cdot (m_1 - m_2) = n_1 - n_2$, if we apply a similar algebraic step as before, we obtain $n_2 = n_1 \cdot [n_2 \cdot (m_2 - m_1) + 1]$, which proves that n_1 divides n_2. Since the two positive integers divide each other, they must be equal. Going back to $m_1 - m_2 = \frac{1}{n_2} - \frac{1}{n}$, we see that $n_1 = n_2$ proves that f is an *a* one-to-one mapping. However, this mapping is not surjective. If r is a positive rational in lowest term, then $r = \frac{k}{\ell}$, where the positive integers k, ℓ have no positive divisors in common, except 1.

If $\frac{k}{\ell} = m + \frac{1}{n}$, for some $m, n \in P$ then $\frac{k}{\ell} = \frac{mn+1}{n}$; and since the integers $mn + 1$ and n have no positive divisors in common we see that we must have $k = mn + 1$ and $\ell = n$, so that $k = m\ell + 1$. We conclude that the positive rational will be in the range R_f of this mapping if, and only if, k leaves remainder 1 when divided by ℓ: $R_f = \left\{x \in \mathbb{Q}^+ \mid x = \frac{k}{\ell},\ k \in P,\ \ell \in P, \text{ and } k = m\ell + 1, \text{ for some } \ell \in P\right\}$.

Example 43: Let $A = P \cup \{0\}$; A is the set of all nonnegative integers. The function $f : A \to A$ defined as,

$$f(x) = \begin{cases} x + 1, & \text{if } x \text{ is an even non} - \text{negative integer} \\ \\ x - 1, & \text{if } x \text{ is an odd non} - \text{negative integer,} \end{cases}$$

is a permutation of the set A.

We can create the following table:

x	0	1	2	3	4	5	6	7	8	9	...
$y = f(x)$	1	0	3	2	5	4	7	6	9	8	...

Example 44: In other mathematics courses, such as discrete mathematics, students learn that if a finite set A contains n elements, then there are exactly $n!$ permutations of the set A, $n!$ is called 'n factorial', something that the reader of this book is most likely familiar with from calculus, $n! = 1 \cdot 2 \ldots (n-1) \cdot n$, the product of the first n positive integers. For instance, if $n = 3$, and say, $A = \{a_1, a_2, a_3\}$, there are exactly $3! = 1 \cdot 2 \cdot 3 = 6$ permutations of the set A. One of them is the identity mapping I_A, of course. We illustrate them below by using Venn diagrams (or, one can use tables, or a cross between Venn diagrams and tables, e.g. $\begin{array}{ccc} a_1 & a_2 & a_3 \\ \downarrow & \downarrow & \downarrow \\ a_2 & a_1 & a_3 \end{array}$)

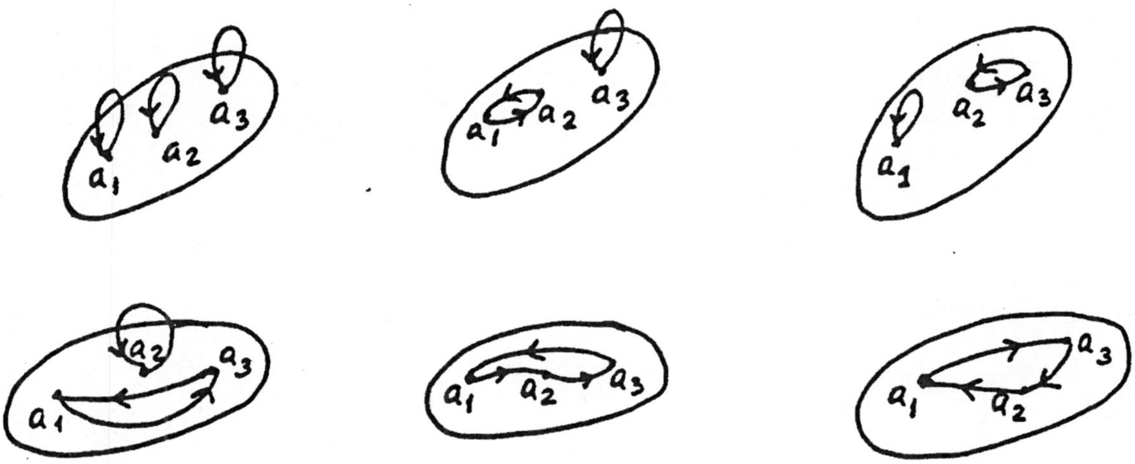

Figure 11

Example 45: Let a, b, c, d be fixed or given integers, and with not both c and d zero (i.e., assume at least one of c or d is nonzero). Consider the two functions, described by the same formula,

$$f: \mathbb{Q} \to \mathbb{Q}, \quad f(x) = \frac{ax+b}{cx+d}$$

$$g: \mathbb{Z} \to \mathbb{Q}, \quad g(x) = \frac{ax+b}{cx+d}$$

Note that these are different functions since their starting sets are different; $\mathbb{Q}$ and $\mathbb{Z}$. However, the fact that they are described by the same formula and the set of integers $\mathbb{Z}$ is a subset of the set of rational numbers $\mathbb{Q}$, creates a phenomenon that is described as one function being an extension of the other, f is an extension of g. It also creates the phenomenon of one function being a restriction of the other; g is a restriction of f. We will define these two concepts in the next section.

(i) If $c = 0$, the domain of f, D_f consists of all rationals, $D_f = \mathbb{Q}$. However, if $c \neq 0$, D_f consists of all rationals except the rational number $-d/c$ because the image of $-d/c$ is undefined under the given formula, $f(-d/c)$ is undefined as a rational number. So, for $c \neq 0$, $D_f = \{x \mid x \in \mathbb{Q},\ x \neq -d/c\}$. On the other hand, the domain of g, D_g is all of $\mathbb{Z}$; $D_g = \mathbb{Z}$, unless c is a divisor of d. If c is a divisor of d, say $d = k \cdot c$, for some integer k, then, assuming $k \neq 0$, we see that the value $f(-k)$ is undefined since $c \cdot (-k) + d = 0$. On the other hand, if $d = 0$, then according to assumption $c \neq 0$, $f(0)$ is undefined in that case. In summary, we have

α) $D_g = \mathbb{Z}$, if c is not a divisor of d; or

β) $D_g = \{x \mid x \in \mathbb{Z},\ x \neq -d/c\}$, if $d \neq 0$ and c is a divisor of d; d/c being the integer k; or

γ) $D_g = \{x \mid x \in \mathbb{Z},\ x \neq 0\}$, if $d = 0$.

(ii) Let $D = ad - bc$. D is an integer since a, d, b and c are integers. We will show if $D \neq 0$, and only if $D \neq 0$, the function f is injective. The same holds true for the function g. To see this, suppose that $f(x_1) = f(x_2)$, for some x_1, x_2 in the domain D_f.

$$\frac{ax_1 + b}{cx_1 + d} = \frac{ax_2 + b}{cx_2 + d} \Rightarrow$$

(multiply both sides by the product $(cx_1 + d) \cdot (cx_2 + d)$)

$$\Rightarrow (ax_1 + b) \cdot (cx_2 + d) = (cx_1 + d) \cdot (ax_2 + b)$$

$$\Rightarrow acx_1x_2 + adx_1 + bcx_2 + bd = cax_1x_2 + cbx_1 + dax_2 + db$$

$$\Rightarrow adx_1 - adx_2 + bcx_2 - bcx_1 = 0;$$

$$\Rightarrow ad \cdot (x_1 - x_2) - bc(x_1 - x_2) = 0;$$

$$\Rightarrow (ad - bc) \cdot (x_1 - x_2) = 0.$$

Thus, if $D = ad - bc \neq 0$, the last product implies $x_1 - x_2 = 0$; $x_1 = x_2$, which proves that f is an injective function. Conversely, we can show that if f is an injective function, then $D \neq 0$. It suffices to prove the contrapositive: if $D = 0$, then f is not a one-to-one function. Indeed, if $D = 0$, then starting with any x_1, x_2 in the domain D_f, we have $(ad - bc) \cdot (x_1 - x_2) = 0$. Since all the previous steps are reversible, we can trace back to $\frac{ax_1+b}{cx_1+d} = \frac{ax_2+b}{cx_2+d}$ holding true for any x_1, x_2; $x_1 \neq x_2$ in D_f. Hence, if f is not injective, we are done. Actually there is more: if $D = 0$, f is a constant function. To see this, suppose first that $c \neq 0$. We can write $f(x) = \frac{ax+b}{cx+d} = \frac{a}{c} \cdot \left(\frac{acx+bc}{acx+ad}\right) = \frac{a}{c} \cdot \left(\frac{acx+bc-ad+ad}{acx+ad}\right) = \frac{a}{c} \cdot \left(\frac{acx+ad}{acx+ad} + \frac{bc-ad}{acx+ad}\right) = \frac{a}{c} \cdot \left(1 + \frac{bc-ad}{acx+ad}\right)$, $\forall x \in D_f$. Since $D = ad - bc = 0 \Rightarrow bc - ad = 0$. Thus, $f(x) = \frac{a}{c} \cdot \left(1 + \frac{0}{acx+ad}\right)$; $f(x) = \frac{a}{c}$, for every $x \in D_f$.

Going back, if $c = 0$, then since $D = 0 = ad - bc$, it follows $ad = 0$; but not both d and c can be zero. Thus, $a = 0$ and we arrive at $f(x) = \frac{b}{d}$.

In summary, we see that if $D = 0$ and $c \neq 0$, then $f(x) = a/c$, $\forall x \in D_f$. While if

$D=0$ and $c=0$, $f(x)=\frac{b}{d}$, $\forall x \in D_f$; f a constant function.

(iii) In this part we will take a look at the ranges R_f and R_g of the functions f and g. We also derive a formula for the inverse relation f^{-1} of f. Below we see that actually, for $D \neq 0$, f^{-1} is a function. So we start with the function f. In part (ii) we saw that if $D=0$ and $c \neq 0$, then $f(x)=\frac{a}{c}$, $\forall x \in D_f$; thus, $R_f=\{a/c\}$ in that case and if $D=0=c$, then $f(x)=\frac{b}{d}$, $\forall x \in D_f$, which means $R_f=\{b/d\}$ when $D=c=0$.

Now suppose that $D \neq 0$. We have already shown that f must be injective when this happens. Let $y \in R_f$; then $y=\frac{ax+b}{cx+d}$, for a unique $x \in \mathbb{Q}$, with $x \neq -d/c$.

We solve for x in terms of y: $y \cdot (cx+d)=ax+b \Leftrightarrow y \cdot c \cdot x + y \cdot d = ax+b \Leftrightarrow d \cdot y - b = (a - c\cdot y)\cdot x \Leftrightarrow x = \frac{d \cdot y - b}{-c \cdot y + a}$. Clearly, the rational number y must be distinct from a/c. Otherwise, the denominator of the last fraction would be zero; all this assuming $c \neq 0$. If $c=0$, then y can be any rational. Now let any $y \in \mathbb{Q}$, $y \neq a/c$, $c \neq 0$. If $x=\frac{dy-b}{-cy+a}$, then clearly $x \in \mathbb{Q}$, since d, y, b, c, a are all in $\mathbb{Q}$; we show that $f(x)=y$. Indeed,

$$
\begin{aligned}
f(x) &= f\left(\frac{dy-b}{-cy+a}\right) = \frac{a\left(\frac{dy-b}{-cy+a}\right)+b}{c\cdot\left(\frac{dy-b}{-cy+a}\right)+d} \\
&= \frac{\left[a\cdot\left(\frac{dy-b}{-cy+a}\right)+b\right]\cdot(-cy+a)}{\left[c\cdot\left(\frac{dy-b}{-cy+a}\right)+d\right]\cdot(-cy+a)}, \\
&= \frac{a(dy-b)+b\cdot(-cy+a)}{c\cdot(dy-b)+d\cdot(-cy+a)} \\
&= \frac{ady-ab-bcy+ab}{cdy-cb-dcy+ad} \\
&= \frac{(ad-bc)y}{ad-bc} = y
\end{aligned}
$$

since $D=ad-bc \neq 0$.

This proves that $R_f=\mathbb{Q}-\{a/c\}=$ all rationals except a/c.

Now, if $c=0$, then from $x=\frac{dy-b}{-cy+a}$, we obtain $x=\frac{dy-b}{a}$; in this case $R_f=\mathbb{Q}$. It is now clear that under the assumption $D \neq 0$, since f is injective, and from the above expression of x in terms of y the following is true:

(1) If $c \neq 0$, then the inverse relation f^{-1} is a function given by

$$f^{-1}(x) = \frac{d \cdot x - b}{-c \cdot y + a}; \; D_{f^{-1}} = R_f \;=\; \mathbb{Q} - \{a/c\},$$

$$\text{and } R_{f^{-1}} = D_f \;=\; \mathbb{Q} - \{-d/c\}$$

(2) If $c = 0$, then the inverse relation f^{-1} is a function given by

$$f^{-1}(x) = \frac{d}{a} \cdot x - \frac{b}{a}; \; D_{f^{-1}} = R_f = \mathbb{Q}$$

$$\text{and } R_{f^{-1}} = D_f = \mathbb{Q}$$

(Note that $a \neq 0$, because we have assumed $D = ad - bc \neq 0$)

Now let us look at the range R_g of the function g.

As in the case of the function f, g is injective if and only if $D \neq 0$. Specifically, if $D \neq 0$ the inverse relation g^{-1} is a function from $\mathbb{Q}$ to $\mathbb{Z}$, defined by $g^{-1}(x) = \frac{dx-b}{-cx+a}$, if $c \neq 0$. Otherwise, if $c = 0$, $g^{-1}(x) = \frac{d}{a}x - \frac{b}{a}$. As we know from Theorem 5(i), $R_g = D_{g^{-1}}$ and $R_{g^{-1}} = D_g$. Under the assumption $D \neq 0$ we now determine the range R_g. If the rational number $r = \frac{\rho}{\rho'}$, ρ, $\rho' \in \mathbb{Z}$, $\rho' \neq 0$, (with ρ and ρ' having no positive divisors in common other than 1), is in R_g; then $\frac{ax+b}{cx+d} = \frac{\rho}{\rho'}$, for some $x \in \mathbb{Z}$. Since $ax + b$ and $cx + d$ are both integers, we must have $ax + b = m \cdot \rho$ and $cx + d = m\rho'$ (since $\frac{\rho}{\rho'}$ is in lowest terms) for some integer m; solving for x we have $x = \frac{m\rho-b}{a} = \frac{m\rho'-d}{c}$, assuming that neither a nor c is zero. From $\frac{m\rho-b}{a} = \frac{m\rho'-d}{c}$, we obtain $m(c \cdot \rho - a\rho') = bc - ad$. The last statement says that $c \cdot \rho - a\rho'$ is a divisor of $-D = bc - ad$, therefore, also a divisor of D. Converely, if $c\rho - a\rho'$ is a divisor of minus the determinant, that is, a divisor of $bc - ad$, then we must have $m \cdot (c\rho - a\rho') = bc - ad$, for some $m \in \mathbb{Z}$; this because $c\rho - a\rho'$ is a divisor of $bc - ad$. From that we obtain $\frac{m\rho-b}{a} = \frac{m\rho'-d}{c}$; and so by taking $x = \frac{m\rho-b}{a} = \frac{m\rho'-d}{c}$, we arrive at $\frac{ax+b}{cx+d} = \frac{\rho}{\rho'}$, which shows that $\frac{\rho}{\rho'}$ is in the range R_g. Thus, we see that when $D \neq 0$, $a \neq 0$ and $c \neq 0$, $R_g = \left\{\frac{\rho}{\rho'} \mid \rho, \rho' \in \mathbb{Z}, \rho' \neq 0, \text{ and } c \cdot \rho - a\rho' \text{ is a divisor of } bc - ad\right\}$.

Note that such a divisor of $bc - ad$ must be divisible by the greatest common divisor of a and c. As it turns out, when one studies linear diophantine equations in two unknowns and if we pick any divisor δ of $D = ad - bc$, which is also divisible by the greatest common divisorof a and c, then there are infinitely many integers ρ and ρ' such that $c \cdot \rho - a\rho' = \delta$ (assuming $a \neq 0$, $c \neq 0$).

What happens when one of a, c is zero? If $a = 0$ and $c \neq 0$, we may simply state $R_g = \{\frac{b}{cx+d} \mid x \in \mathbb{Z},\; x \neq -d/c\}$. If $a \neq 0$ and $c = 0$, then $R_g = \{\frac{a}{d} \cdot x + \frac{b}{d} \mid x \in \mathbb{Z}\}$. Under $D = ad - bc =\neq 0$, obviously not both a and c can be zero.

What about $D = ad - bc = 0$? If $D = 0$, then as we saw in the case of the function f (by using the same technique,-go to part (ii)), we see that g must be a constant function:

$$g(x) = \frac{a}{c}, \; \forall x \in Dg, \text{ if } D = 0 \text{ and } c \neq 0$$

Otherwise $g(x) = \frac{b}{d}$, $\forall x \in Dg$, if $D = 0$ and $c = 0$ (which implies $a = 0$ as well).

Definition 31: Let A, B, C, D be non-empty sets and f, g two functions; f a function from the set A to the set B and g a function from the set C to the set D, $f : A \to B$ and $g : C \to D$. We say the at the two functions are **equal** if, and only if, $D_f = D_g = D$, and $f(x) = g(x)$, $\forall x \in D$. (As usual, D_f stands for the domain of f; and D_g for the domain of g.)

Remark 20: Note that unless D is the empty set, in which case the definition will hold vacuously, it follows from the definition that if f and g are equal, then the common domain D will be a non-empty subset of the intersection set $A \cap C$; and the common range $R_g = R_f$ will be a non-empty subset of the intersection set $B \cap D$.

Example 46: (i) Consider the following functions. $f : \{0,1\} \to \{1,3\}$, defined by $f(x) = 2x^2 + 1$, with domain $D_f = \{0,1\}$; and the function $g : \{0,1\} \to \{1,3\}$ defined by $g(x) = x^3 + x^2 + 1$ and with domain $D_g = \{0,1\}$. These two functions are equal since $D_f = D_g = \{0,1\}$ and $f(0) = 1 = g(0)$, $f(1) = 3 = g(1)$.

(ii) The function $f : \mathbb{Z} \to \{0,1,2\}$, with domain $D_f = \mathbb{Z}$, defined by $f(x) =$ remainder of the division of the integer x by 3, is equal to the function $g : \mathbb{Z} \to \{0,1,2\}$ with domain $D_g = \mathbb{Z}$, and defined by $g(x) =$ remainder of the division of the integer $(x^3 + 3x)$ by 3.

To see why they are equal, note that if r is the remainder of the division of x by r, then by definition $r = 0, 1$, or 2; and $x = 3t + r$, t being the quotient of the division. Thus, we see that $f(x) = r = 0, 1$, or 2 according to whether $r = 0, 1$, or 2 respectively. Now consider $g(x) = x^3 + 3x$; or, $g(x) = (3t + r)^3 + 3(3t + r)$. Even without expanding the cube $(3t + r)^3$ we see that all the terms on the right-hand side of the last equation, except possibly for r^3, will be multiples of 3. In other words, $g(x) =$ remainder of the integer (r^3 + (multiple of 3)) when divided by 3; $g(x) =$ remainder of r^3 when divided by 3; but $r^3 = 0^3, 1^3$ or $2^3 = 0, 1$, or 2. We now see that $f(0) = 0 = g(0)$; $f(1) = 1 = g(1)$; and $f(2) = 2$, while $g(2) =$ remainder of 8 when divided by $2 = 2$. Thus, $g(2) = 2 = f(2)$. The two functions are equal.

(iii) The functions f and g of Example 45 are not equal, even though they are defined by the same formula, because they have different domains.

Definition 32: Let f be a mapping from a non-empty set A to a non-empty set B; and A_1 a non-empty subset of A; and A a subset of the set A_2. Assume $A_1 \neq A \neq A_2$.

(i) The mapping f_1 from the set A_1 to the set B, defined by $f_1(x) = f(x)$, $\forall x \in A_1$, is called **a restriction of the mapping f on the set A_1**.

(ii) If f_2 is a function from the set A_2 to the set B such that $f_2(x) = f(x)$, $\forall x \in A$, then f_2 is called **an extension of the mapping f to the set A_2**,

Example 47: (i) From calculus, consider the mapping

$$f : \mathbb{R} \to \mathbb{R},\ f(x) = \begin{cases} \frac{1}{x}, & \text{if } x \neq 0 \\ 0, & \text{if } x = 0 \end{cases},$$

and the mapping

$$f_1 : \mathbb{R} - \{0\} \to \mathbb{R},\ f_1(x) = \frac{1}{x}.$$

The mapping f_1 is a restriction of f on the set $\mathbb{R} - \{0\}$ and the mapping f is an extension of f_1 to the set $\mathbb{R}$.

(ii) Again from calculus the three mappings
$f_1 : A_1 = (-\infty, -1) \to \mathbb{R},\ f_1(x) = -x;$
$f : A = (-\infty, 1) \to \mathbb{R}$, defined by $f(x) = \begin{cases} -x, & \text{if } x < -1 \\ 1, & \text{if } -1 \leq x \leq 1 \end{cases}$

$f_2 : A_2 = (-\infty, \infty) = \mathbb{R} \to \mathbb{R}$, defined by $f_2 = \begin{cases} -x, & \text{if } x < -1 \\ 1, & \text{if } -1 \leq x \leq 1 \\ \frac{x}{2} + \frac{1}{2}, & \text{if } 1 < x, \end{cases}$

satisfy Definition 32. The mapping f_1 is a restriction of the mapping f on the set $A_1 = (= \infty, -1)$ and the mapping f_2 is an extension of the mapping f to the set $A_2 = \mathbb{R}$.

We finish this section by establishing a theorem which says that every mapping gives rise to an equivalence relation.

Theorem 6: Let $f : A \to B$ be a mapping from a non-empty set A to a non-empty set B. Define a relation R in the set A by the following condition:

$$p(x, y) = x \in A,\ y \in A \text{ and } f(x) = f(y).$$

Then, relation R is an equivalence relation.

Proof: The relation R is reflexive since $f(x) = f(x)$, for every x in A. The relation R is symmetric because if $(x, y) \in R$, then $f(x) = f(y) \Rightarrow f(y) = f(x) \Rightarrow (y, x) \in R$; and the relation R is transitive because if $(x, y) \in R$ and $(y, z) \in R$, then $f(x) = f(y)$ and $f(y) = f(z)$ implies $f(x) = f(z)$ and so $(x, z) \in R$.

Remark 21: Each equivalence class of this relation consists of all the elements in A that have the same image in B under the mapping f. So, if a is an element of A, its equivalence class is given by $E_a = \{x \in A \mid f(x) = f(a)\}$.

Example 48: This example is from calculus.

(i) Let $f : \mathbb{R} \to \mathbb{R}$ be the mapping defined by $f(x) = x^2$. Then, if a is a non-zero real number, its equivalence class is given by $E_a = \{-a, a\}$, since $-a$ and a are the only two real numbers whose square is equal to a^2. The class of zero, E_0, contains only one element, namely zero; $E_0 = \{0\}$.

(ii) Let $g : \mathbb{R} \to \mathbb{R}$ be the mapping piecewise defined by,

$$g(x) \begin{cases} 1, & \text{if } x < -2 \\ 0, & \text{if } -2 \leq x \leq 3 \\ 4, & \text{if } x > 3 \end{cases}$$

There are only three equivalence classes of the relation that this mapping gives rise to according to Theorem 6. The three classes are the three intervals $(-\infty, -2)$; $[-2, 3]$, and $(3, +\infty)$. We can write,

$$E_a = (-\infty, -2) = \{x \mid x \in \mathbb{R} \text{ and } x < -2\}, \text{ for any } a \text{ less than } -2.$$
$$E_a = [-2, 3] = \{x \mid x \in \mathbb{R} \text{ and } -2 \leq x \leq 3\}, \text{ for any } a \text{ between } -2 \text{ and } 3.$$
$$E_a = (3, \infty) = \{x \mid x \in \mathbb{R} \text{ and } 3 < x\}, \text{ for any } a \text{ greater than } 3.$$

1.6 Composition of functions and relations in general

In this section we will first define the composition of functions since the student is undoubtedly familiar with such a concept from calculus. We will state three theorems and their proofs. After that, we will define the composition of relations, state the theorem that says that the composition of relations is an associative operation, but not a commutative one. We will also prove two more theorems pertaining to the composition of a transitive relation with itself.

Definition 32: Let A, B, and C be three non-empty sets and $g : A \to B$, $f : B \to C$, be functions from A to B and from B to C respectively. Furthermore, assume that the range of the function g and the domain of the function f have a non-empty intersection: $R_g \cap D_f \neq \emptyset$. The relation, which is actually a function (see remark below), from the set A to the set C; symbolized by $f \circ g$, and defined by

$$f \circ g = \{(x, z) \mid x \in A,\ z \in C \text{ and } \exists y \in B : y = g(x) \text{ and } z = f(y);\ z = f(g(x))\},$$

is called the **composite of the functions g and f**; or the **composite of the function g (first) with the function f (second).**

Remark 22: (i) It is easy to see from Defintion 32 that the composite relation $f \circ g$, under the assumptions of Defintion 32, is actually a function called the composite function of g and f. This is clear since for every $x \in A$ there is at most one z in C such that $z = f(g(x))$. That is because both g and f have been assumed to be functions. If x is the domain D_g and if the range value y in R_g that corresponds to x falls inside the domain D_f of f, then x will go to a unique value z in c since y goes to that unique value. This is illustrated in the following Venn diagram:

$y = g(x)$, $\qquad$ $z = f(y)$, $\qquad$ $z = f(g(x))$

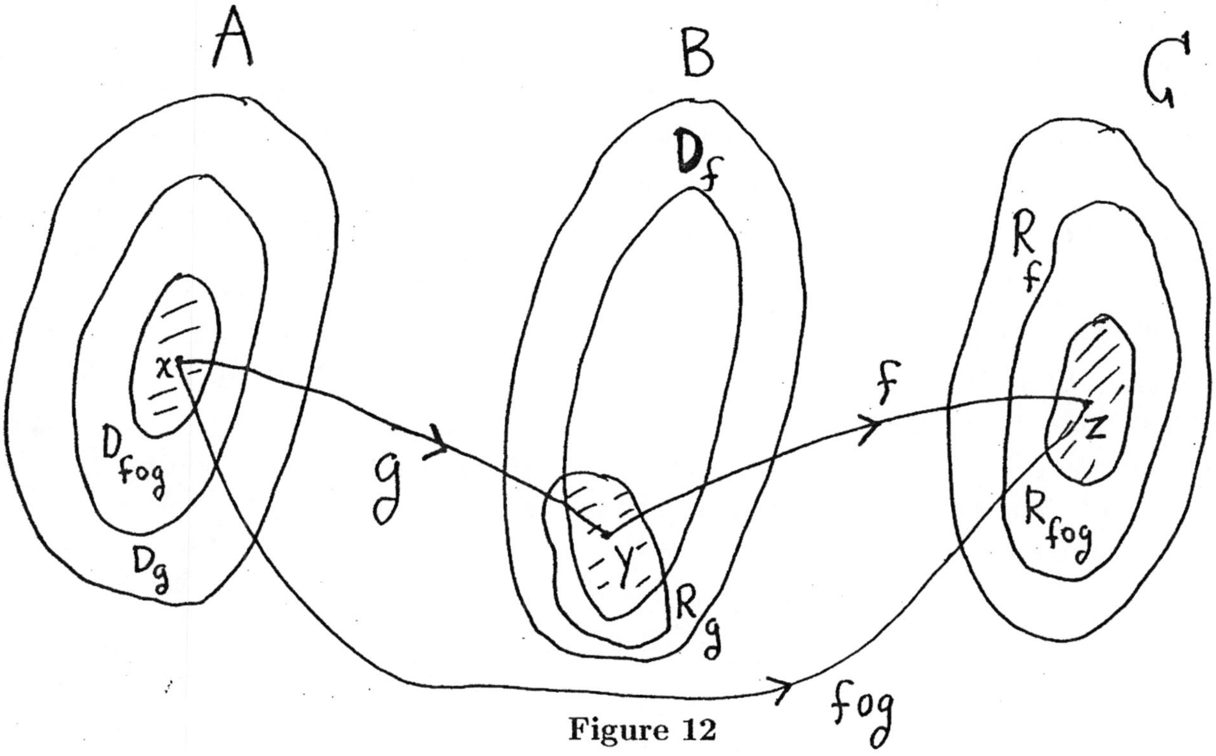

Figure 12

It should be clear that the domain D_{fog} is a subset of the domain D_g and it consists only of those elements x in D_g for which their range values y fall inside the domain D_f of f. Symbolically,

$$D_{fog} = \{x \in D_g \mid g(x) \in D_f\}$$

and the range R_{fog} is a subset of the range R_f:

$$R_{fog} = \{z \mid z \in R_f \text{ and } z = f(y), \text{ for some } y \in R_g \cap D_f\}$$

(ii) Because of the condition $R_g \cap D_f \neq \emptyset$, according to the way the composite function $f \circ g$ is defined in Definition 32, it can never be the empty relation $\emptyset$. Obviously, if one was to waive the requirement $R_g \cap D_f \neq \emptyset$, then it would be possible to have $f \circ g$ being the empty relation. Examples from calculus abound.

(iii) It is easy to see that, in general and in fact, most of the time, the two composite function $f \circ g$ and $g \circ f$ are not equal, even if the three sets A, B, C are assumed to be equal; and even if the three sets D_g, R_g, D_f, and R_f are assumed to be one and the same set. Still, far more frequently than not, $f \circ g$ will be different from the composite $g \circ f$. The following is at the crux of the matter. Consider the functions $g : A \to B$; $g = \{(a, b)\}$; $g(a) = b$ and $f : B \to C$; $f = \{(b, c)\}$; $f(b) = c$. Then the composite function $fog : A \to C$ is $fog = \{(a, c)\}$, $(fog)(a) = f(g(a)) = c$. Clearly, unless $c = a$, the composite function $g \circ f$ will not be defined; and in the case where $a = c$, $f \circ g$ and $g \circ f$ will be distinct (for $a \neq b$) and actually inverse functions.

Generally speaking, it is a rare phenomenon to have $f \circ g = g \circ f$. Other than the phenomenon of inverse functions, it is not easy to come up with examples where $f \circ g = g \circ f$. Consider the following example from calculus.

Example 49: Let a, b, c, d be real constants, i.e., fixed real numbers. Consider the linear functions $f : \mathbb{R} \to \mathbb{R}$, $f(x) = ax + b$, and $g : \mathbb{R} \to \mathbb{R}$, $g(x) = cx + d$. What condition must the four constants satisfy in order that the two composite functions $f \circ g$ and $g \circ f$ be equal? To answer this question, we start with the obvious, $(f \circ g)(x) = f(g(x)) = a \cdot (cx + d) + b = acx + ad + b, \forall x \in \mathbb{R}$ and $(g \circ f)(x) = g(f(x)) = c \cdot (ax + b) + d = cax + cb + d$, $\forall x \in \mathbb{R}$.

Since $ac = ca$, it is now clear that we will have $f \circ g = g \circ f$ (obviously both of these two composites have domain $\mathbb{R}$) if, and only if, the four constants a, b, c, d satisfy the condition $ad + b = cb + d \Leftrightarrow \boxed{d \cdot (a - 1) = b \cdot (c - 1)}$. We analyze the situation further by considering five cases:

(1) If all four real numbers $d, a - 1, b$, and $c - 1$ are non-zero, that is, if $d \neq 0, a \neq 1, b \neq 0$, and $c \neq 1$, then the above general condition is tantamount to $\frac{b}{d} = \frac{a-1}{c-1}$. What this says, from a calculus perspective, is that the ratio of the two constant terms or y-intercept of the two straight lines (both f and g are linear functions, their graphs are straight lines) must be equal to the ration of their respective slopes each reduced by 1. An example of this is $a = 4, b = 3, c = 3, d = 2$; $f(x) = 4x+3$, $g(x) = 3x+2$; $\frac{3}{2} = \frac{4-1}{3-1}$, and

$$(f \circ g)(x) = f(g(x)) = 4(3x + 2) + 3 = 12x + 11$$

$$(g \circ f)(x) = g(f(x)) = 3(4x + 3) + 2 = 12x + 11$$

(2) Suppose $d = 0$ and $b = 0$. Then the above condition is satisfied; $f(x) = ax$, $g(x) = cx$; note that both graphs of f and g would be lines through the origin in the standard xy coordinate system; $(f \circ g)(x) = acx = cax = (g \circ f)(x)$. This clearly states that the composite of two straight lines through the origin is also a straight line through the origin, and it does not matter which of the two composites we take. They are both equal.

(3) The next possibility is $d = 0$ and $c = 1$. We have $f(x) = ax+b$ and $g(x) = x+1$

$$(f \circ g)(x) = f(g(x)) = ax + b + 1$$

$$(g \circ f)(x) = g(f(x)) = ax + b + 1$$

(4) The fourth possibility is $a = 1$, $b = 0$. We have $f(x) = x$ and $g(x) = cx + d$. Since f is the identity transformation on $\mathbb{R}$, it is not surprising that in this case $(f \circ g)(x) = cx + d = (g \circ f)(x)$.

(5) The fifth case is $a = 1$, $c = 1$; $f(x) = x + b$, $g(x) = x + d$. Both functions have graphs that are straight lines of slope 1; $(f \circ g)(x) = x + b + d = (g \circ f)(x)$.

Theorem 7: Let $f : A \to B$ be a mapping from a (non-empty) set A to a (non-empty) set B; and I_A, I_B be the identity transformations on the sets A and B

respectively.

(i) $f \circ I_A = f = I_B \circ f$

(ii) If f is injective, then the inverse relation f^{-1} is an injective function and, $f^{-1} \circ f = I_A$.

(iii) If f is a bijection, then so is the inverse relation f^{-1} and, $f^{-1} \circ f = I_A$ and $f \circ f^{-1} = I_B$.

Proof: (i) Since f is a mapping, the domain D_f is equal to A, $D_f = A$; for every $x \in A$, $I_A(x) = x \Rightarrow (f \circ I_A)(x) = f(I_A(x)) = f(x)$. Since $D_{f \circ I_A} = D_f = A$ and $(f \circ I_A)(x) = f(x)$, $\forall x \in A$, it follows by Definition 31 that the functions $f \circ I_A$ and f are equal.

The proof for $f = I_B \circ f$ is similar. Clearly, $A = D_f = D_{I_B \circ f}$, since the range of f, R_f, is a subset of the domain $D_{I_B} = B$. Furthermore, $(I_B \circ f)(x) = I_B(f(x)) = f(x)$, $\forall x \in A$. The last step, $I_B(f(x)) = f(x)$, is obviously true since I_B is the identity function on B.

(ii) First observe that the inverse relation f^{-1} is an injective function according to Theorem 5(i). Also, by the definition of the inverse relation, (Definition 6), the image $(f^{-1} \circ f)(x)$ will be defined or exist for every $x \in A$, which shows that $D_f = A = D_{f^{-1} \circ f}$; and for every $x \in A$, $(f^{-1} \circ f)(x) = f^{-1}(f(x)) = x = I_A(x)$. The step $f^{-1}(f(x)) = x$ being justified by the fact that f^{-1} is the inverse relation of f. Hence, $f^{-1} \circ f = I_A$.

(iii) Since f is a bijection, it is in particular an injection, so by part (ii) we have $f^{-1} \circ f = I_A$. But f is a bijection, so by Theorem 5(iii), f^{-1} must be an (injective) mapping from B to A; and since by definition, $(f^{-1})^{-1} = f$ (this is true for any relation); by applying part (ii) again we conclude $(f^{-1})^{-1} \circ f^{-1} = I_B$; $f \circ f^{-1} = I_B$.
■

The next theorem specifies the conditions two functions must satisfy so their composition or composite function is injective; and for their composite function to be surjective.

Theorem 8: Let A, B, C be non-empty sets and $g : A \to B$, a function from the set A to the set B; and $f : B \to C$, a function from the set B to the set C. Suppose that $R_g \cap D_f \neq \emptyset$. Then,

(i) The composite function $f \circ g$; $f \circ g : A \to C$, will be injective if, and only if, **both** conditions listed below are satisfied.

Condition 1: The restriction of the function g to the set $S = \{x \in D_g \mid g(x) \in D_f\}$, is one-to-one or injective. Condition 2: The restriction of the function f on the set $R_g \cap D_f$, is injective.

(Note that since ${}_g \cap D_f \neq \emptyset$, the set S the domain of $f \circ g$ is also non-empty; $S = D_{f \circ g}$).

(ii) The compostite function $f \circ g$ is surjective if, and only if, the restriction of the function f on the set $R_g \cap D_f$ is surjective.

Proof: (i) First we prove that if the two conditions are satisfied then $f \circ g$ is injective. To do this, let $x_1, x_2 \in D_g$, $x_1 \neq x_2$ and such that both images $g(x_1), g(x_2)$ fall inside D_f, which is to say, $g(x_1)$ and $g(x_2)$ are in $R_g \cap D_f$. By Condition 1 $\Rightarrow$

(since $x_1 \neq x_2$) $g(x_1) \neq g(x_2)$ and by Condition 2, we infer that since f's restriction on $R_g \cap D_f$ is one-to-one, we must have $(f \circ g)(x_1) = f(g(x_1)) \neq f(g(x_2)) = (f \circ g)(x_2)$. Thus we have proven that $f \circ g$ is one-to-one. Now we prove the converse. Namely, that if $f \circ g$ is injective, then both Conditions 1 and 2 must be satisfied. To do this, we may prove the contrapositive statement which is logically equivalent. If either of the two conditions fail, then $f \circ g$ cannot be one-to-one. Indeed, suppose that Condition 1 fails; then, there exist $x_1, x_2 \in S$, $x_1 \neq x_2$ and with $g(x_1) = g(x_2)$; but then, obviously, $f(g(x_1)) = f(g(x_2))$; $(f \circ g)(x_1) = (f \circ g)(x_2)$ with $x_1 \neq x_2$, which shows that $f \circ g$ fails to be one-to-one. Next, assume that the second condition fails. Then there exists $y_1, y_2 \in R_g \cap D_f$, with $y_1 \neq y_2$, and such that $f(y_1) = f(y_2)$. Now, pick a pre-image x_1 of y_1, which must exist since $y_1 \in R_g \cap D_f$ and so $y_1 \in R_g$, in particular.

We have $g(x_1) = y_1$; similarly, let x_2 be a pre-image of y_2 so that $g(x_2) = y_2$. The crucial observation to make now is that x_1 must be distinct from x_2 (no matter which pre-images of y_1 and y_2 we choose). Why? Because, if it were $x = x_1 = x_2$, then we would have $g(x) = y_1 \neq y_2 = g(x)$. In other words, x in D_g would be assigned to two distinct y values in the range R_g, which cannot happen since g is, by hypothesis, a special type of relation, namely a function. Hence, $x_1 \neq x_2$; but we also have $g(x_1) = y_1$, $g(x_2) = y_2$ and $f(y_1) = f(y_2)$, so that $f(g(x_1)) = f(g(x_2))$; $(f \circ g)(x_1) = (f \circ g)(x_2)$ with $x_1 \neq x_2$, which proves that $f \circ g$ is not one-to-one. End of proof. ■

(ii) Suppose that $f \circ g$ is surjective, which means, since $f \circ g$ is a function from A to C, that the range of $f \circ g$ is all of C; $R_{f \circ g} = C$. We will prove that the restriction of f to the set $R_g \cap D_f$ is also surjective. Pick any $z \in C$, since $C = R_{f \circ g} \Rightarrow z = (f \circ g)(x)$ for some $x \in D_g$. Thus, $z = f(g(x))$, but the element $y = g(x)$ is in $R_g \cap D_f$. We have shown that for every $z \in C$, there exists y in $R_g \cap D_f$ such that $f(y) = z$, which proves that the restriction of f to the set $R_g \cap D_f$ is a surjective function. Now, let us prove the converse. Suppose that the restriction of f to the set $R_g \cap D_f$ is surjective. That means for every $z \in C$, there exists $y \in R_g \cap D_f$, such that $f(y) = z$; but, $y \in R_g$, in particular, so that $y = g(x)$, for some $x \in D_g$. Hence, $f(y) = z \Rightarrow f(g(x)) = z$; $(f \circ g)(x) = z$, for some $x \in D_g$, which proves that $f \circ g$ is an onto function. The proof is complete. ■

Remark 23: (i) The Venn diagram below illustrates Theorem 8(i)

$S = D_{f \circ g}$
$y = g(x), \quad z = f(y).$

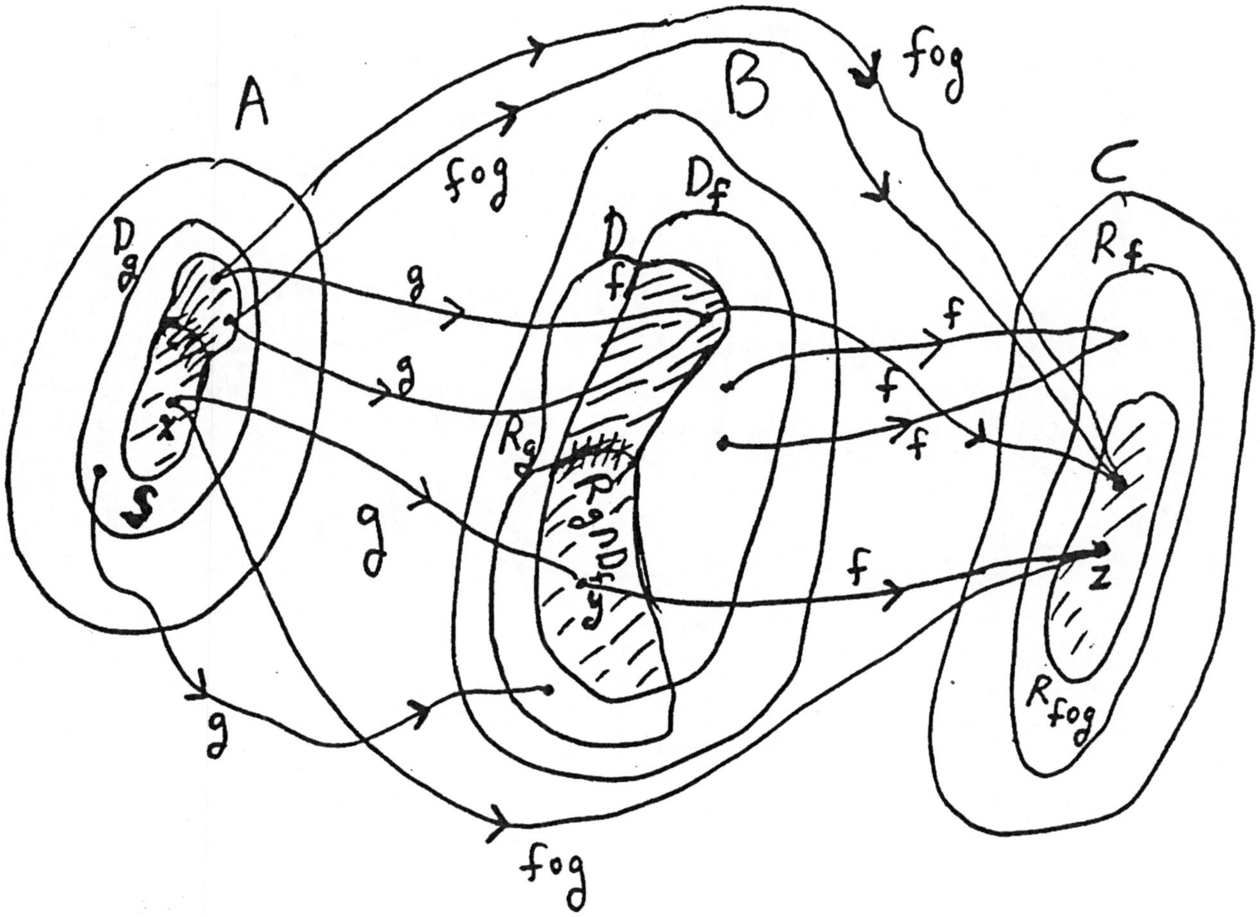

Figure 13

For $f \circ g$to be one-to-one, it is necessary and sufficient that the restriction of g to the set $S = \{x \in D_g \mid g(x) \in D_f\} = D_{f \circ g}$ is injective and the restriction of f to the set $R_g \cap D_f$ is also injective. Neither of the functions g or f has to be injective.

(ii) The Venn diagram below illustrates Theorem 8(ii)

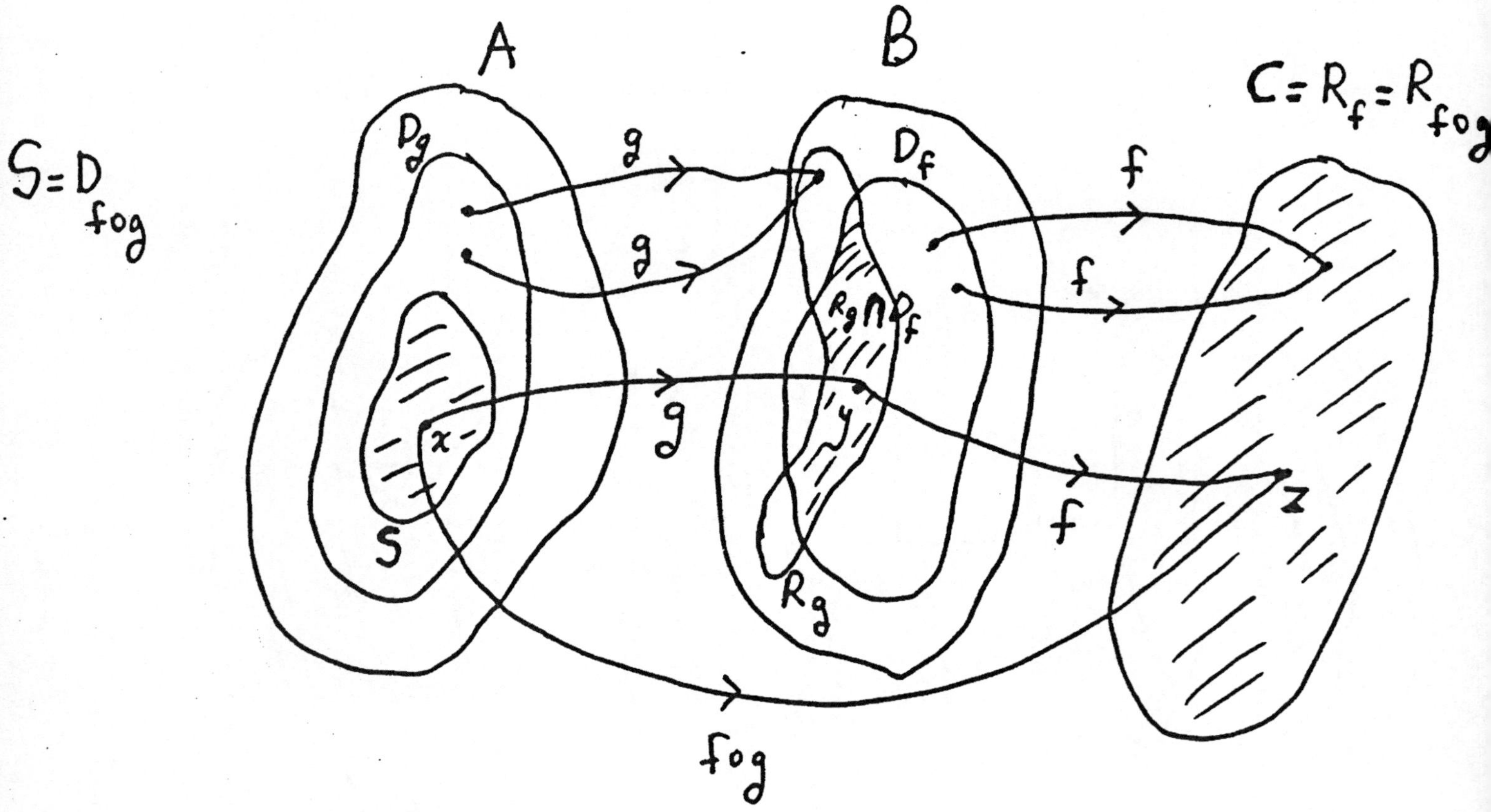

Figure 14

The following theorem comes as no surprise. It pertains to the composite function of two bijections.

Theorem 9: Let A, B, and C be non-empty sets, the function $g : A \to B$ being a mapping from the set A to the set B; and the function $f : B \to C$, being a mapping from the set B to the set C. Then,

(i) If both mappings g and f are injective, then their composite function $f \circ g$ is an injective mapping from the set A to the set C. Futhermore, the inverse relations $f^{-1} : C \to B$ and $g^{-1} : B \to A$, are both injective functions and, $(f \circ g)^{-1} = g^{-1} \circ f^{-1}$.

(ii) If both mappings g and f are surjective, so is their composite function $f \circ g$; a surjective mapping from A onto C.

(iii) If both g and f are bijections, then so is $f \circ g$.

Proof: (i) Since both g and f are injective functions, they satisfy the two conditions in Theorem 8(i), so it follows immediately that $f \circ g$ is an injective function. Also, $D_g = A$, since g is a mapping and the domain of $f \circ g$, $D_{f \circ g} = S$ is simply the entire set A because g is a mapping; $A = D_g = S = D_{f \circ g}$. Hence, $f \circ g$ is a mapping from the set A to the set C and since it is also an injective function, we conclude that $f \circ g$ is an injective mapping. That the inverse relations f^{-1} and g^{-1} are both injective functions follows immediately from Theorem 5(i). To complete the proof, we must show $(f \circ g)^{-1} = g^{-1} \circ f^{-1}$. To prove these two functions equal, we first show that they have equal domains. By Theorem 5(i) (we have already shown that $f \circ g$ is injective), the range of $f \circ g$ is exactly equal to the domain of $(f \circ g)^{-1}$; $R_{f \circ g} = D_{(f \circ g)^{-1}}$.

Let $z \in D_{(f \circ g)^{-1}} \Leftrightarrow z \in R_{f \circ g}$, $z = f(y)$ and $y = g(x)$, for some $x \in D_g$. By definition, $z = f(y) \Rightarrow y = f^{-1}(z)$ and $y = g(x) \Rightarrow x = g^{-1}(y)$. Thus, $x = g^{-1}(f^{-1}(z))$; $x = (g^{-1} \circ f^{-1})(z)$, which proves that $z \in D_{g^{-1} \circ f^{-1}}$. We have just shown that every element z in $D_{(f \circ g)^{-1}}$ also belongs in the domain of $D_{g^{-1} \circ f^{-1}}$. Hence, $D_{(f \circ g)^{-1}}$ is a subset of $D_{g^{-1} \circ f^{-1}}$; $D_{(f \circ g)^{-1}} \subset D_{g^{-1} \circ f^{-1}}$.

Now pick any element w of $D_{g^{-1} \circ f^{-1}}$. Since w is in the domain of $g^{-1} \circ f^{-1}$, then $t = f^{-1}(w) \in R_g \cap D_f$, and $g^{-1}(t) = u \in R_{g^{-1} \circ f^{-1}}$, a subset of D_g. We have $f(t) = w$ and $t = g(u)$; $f(g(u)) = w \Rightarrow (f \circ g)(u) = w \Rightarrow w \in R_{f \circ g}$, but $R_{f \circ g} = D_{(f \circ g)^{-1}}$ since $f \circ g$ and $(f \circ g)^{-1}$ are inverse functions. Hence, $w \in D_{(f \circ g)^{-1}}$, which proves that $D_{g^{-1} \circ f^{-1}}$ is a subset of $D_{(f \circ g)^{-1}}$; $D_{g^{-1} \circ f^{-1}} \subset D_{(f \circ g)^{-1}}$. Hence $D_{(f \circ g)^{-1}} = D_{g^{-1} \circ f^{-1}} = D$. Let $z \in D$. Then $z = f(y)$, for some $y \in R_g \cap D_f$, and $y = g(x)$, for some in $x \in D_g$. Altogether, $z = f(g(x))$; $z = (f \circ g)(x)$; but also from $z = f(y) \Rightarrow y = f^{-1}(z)$; and from $y = g(x) \Rightarrow x = g^{-1}(y)$. We have, $x = g^{-1}(f^{-1}(z))$; $x = (g^{-1} \circ f^{-1})(z)$. Moreover, from $z = (f \circ g)(x) \Rightarrow x = (f \circ g)^{-1}(z)$. Thus we have proven that $(x = (g^{-1} \circ f^{-1})(z)$ and $x = (f \circ g)^{-1}(z)) \Rightarrow (g^{-1} \circ f^{-1})(z) = (f \circ g)(z)$; $\forall z \in D$.

The two functions $g^{-1} \circ f^{-1})(z) = (f \circ g)^{-1}$ are equal.

(ii) Since g is mapping from A to B, we have $D_g = A$; f a function from B to C, so the composite $f \circ g$ is a mapping from A to C. Since the function f is surjective, it follows by Theorem 5(ii), that the compositive function $f \circ g$ is surjective, but we have already proved it to be a mapping. Therefore $f \circ g$ is a surjective mapping from the set A to the set C.

(iii) Since a bijection is a surjective and injective mapping, this part is an immediate consequence of parts (ii) and (iii).

Before we give the definition of the composite of two relations, we define the domain and range of a relation, something we have already done for functions, which are a special type of a relation. We have

Definition 33: Let A and B be non-empty sets and R a relation from A to B.

(i) The domain D_R of the relation R is defined to be the set

$$D_R = \{x \mid x \in A \text{ such that with } (x, y) \in R\}$$

(ii) The range R_R of the relation f is defined to be the set

$$R_R = \{y \mid y \in B \text{ such that } \exists x \in A \text{ with } (x, y) \in R\}.$$

We are now in a position to define the composite of two relations.

Definition 34: Let A, B, and C be non-empty sets, g a relation from A to B and f a relation from B to C. Assume that the range R_g and the domain D_f have a non-empty intersection; $R_g \cap D_f \neq \emptyset$.

The relation from the set A to the set C defined by

$$f \circ g = \{(x, z) \mid x \in A,\ z \in C, \text{ and } \exists y \in B : (x, y) \in g \text{ and } (y, z) \in f\},$$

is called the **composite of the relations g and f** or the **composite of the function g (first) with the function f (second).**

Remark 24: As in the case of the composite of two functions, the composite of two relations, g and f, will have a domain D_{fog} which is a subset of D_g;

$$D_{fog} = \{x \in D_g \mid \exists y \in D_f \text{ such that } (x, y) \in g\}$$

and the range R_{fog} will be a subset of R_f

$$R_{fog} = \{z \in R_f \mid \exists y \in R_g \cap D_f \text{ such that } (y, z) \in f\}$$

Example 50: Let $A = B = C = \mathbb{Z}$, the set of integers. Consider the relation g in $\mathbb{Z}$, defined by the condition $p(x, y) = x \in \mathbb{Z},\ y \in \mathbb{Z},\ x = y^4$.

$$g = \{(0,0), (1,1), (1,-1), (16,2), (16,-2), \ldots (n^4, n), (n^4, -n), \ldots\}$$

The domain of g is the set $D_g = \{x \mid x \in \mathbb{Z}, \text{ and } x \text{ is a fourth integer power}\} = \{0, 1, 16, 81, 256, 625, 1296, \ldots\}$. The range of g is the set $R_g = \mathbb{Z}$. Next consider the relation f in $\mathbb{Z}$ defined by the condition, $p(y, z) = y \in \mathbb{Z},\ z \in \mathbb{Z}, y = z^2$.

$$f = \{(0,0), (1,1), (1,-1), (9,3), (9,-3), \ldots, (m^2, m), (m^2, -m) \ldots\}$$

The domain of f is the set $D_f = \{x \mid x \in \mathbb{Z}, \text{ and } x \text{ is an integer square}\} = \{0, 1, 4, 9, 16, 25, 36, 49, 64, 81, 100, 121, 144, \ldots\}$ The range of f is the set $R_f = \mathbb{Z}$.

The domain of the composite relation $f \circ g$ set $D_{fog} = \{x \mid x \in \mathbb{Z}, \text{ and } x \text{ is an eighth integer power}\}\ \{x \mid x \in \mathbb{Z}, \text{ and } x = k^8,\ k \in \mathbb{Z}\}$. The range of $f \circ g$ is the set $R_{fog} = \mathbb{Z}$. We have,

$$f \circ g = \{(x, z) \mid x \in \mathbb{Z}, z \in \mathbb{Z}, x = k^8, z = k \text{ or } z = -k, \text{ for some } k \in \mathbb{Z}\},$$

$$f \circ g = \{(0,0), (1,1), (1,-1), (2^8, 2), (2^8, -2), \ldots (k^8, k), (k^8, -k), \ldots\}$$

It is left to the student to show that in this particular example, the other composite function $g \circ f$ is in fact equal to $f \circ g$. As we have seen in the case of functions, the two composite functions $f \circ g$ and $g \circ f$ are, in general, different from one another.

Example 51: This example will be familiar to students from Calculus. Let p and q in the set $A = [0, +\infty)$ (the closed interval consisting of all nonnegative real numbers). Relation g is defined by the condition, $p_1(x, y) = x \in A, y \in A$, and $x^p = y$. Relation f is defined by the condition,

$$p_2(y, z) = y \in A,\ z \in A, \text{ and } y^q = z.$$

As it is well known from calculus $D_g = R_g = A = D_f = R_f$, and, in fact, both relations g and f are functions. Furthermore, since $(x^p)^q = x^{pq} = (x^q)^p$, it follows that the two composite functions $f \circ g$ and $g \circ f$ are in fact equal; they both equal the function $h : A \to A$, defined by $h(x) = x^{pq}, \forall x \in A$.

Example 52: Let $A = \{1, 2, 3, 4, 5\}$, $B = \{4, 7, 8, 11\}$, $C = \{12, 14, 15, 16, 17, 18\}$ and consider the relation g from the set A to the set B defined by the condition, $p_1(x, y) = x \in A, y \in B$ and $x + y$ is a multiple of 3. Also, consider the relation f from the set B to the set C defined by the condition, $p_2(y, z) = y \in B$, $z \in C$, and $x + y$ is a multiple of 7.
$D_g = \{1, 2, 4, 5\}$,
$R_g = \{4, 7, 8, 11\}$,
$g = \{(1, 8), (1, 11), (2, 4), (2, 7), (4, 8), (5, 4), (5, 7)\}$

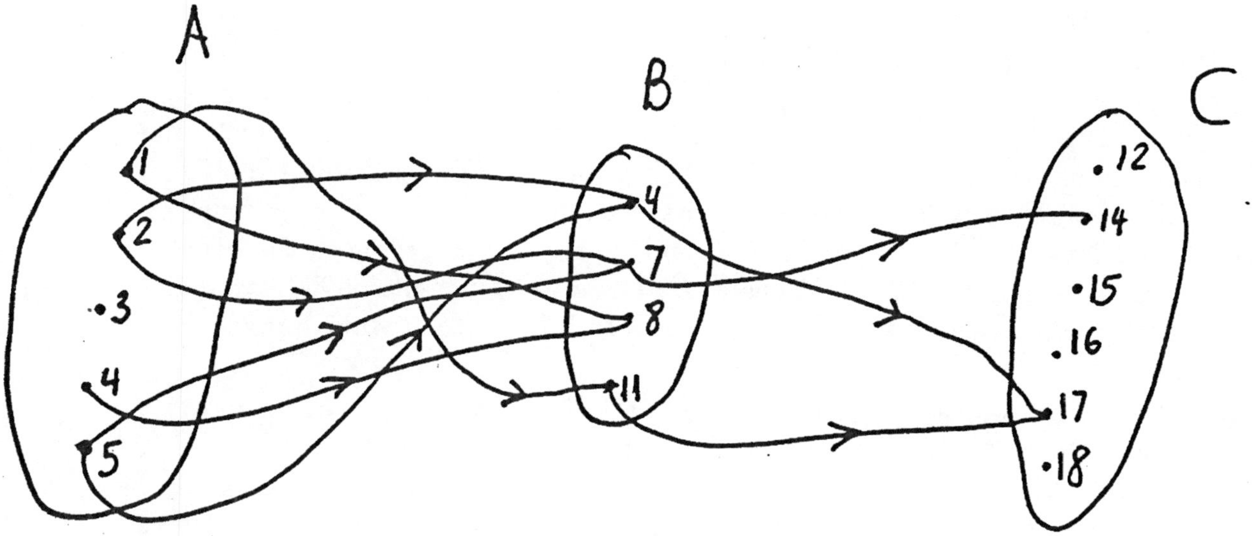

Figure 15

$$D_f = \{4, 7, 11\}, \quad R_f = \{14, 17\}, \text{ and } f = \{(4, 17), (7, 14), (11, 17)\}$$

The domain of the composite function $f \circ g$ is $D_{f \circ g} = \{1, 2, 5\}$, and the range the set $R_{f \circ g} = \{14, 17\}$. The composite function $f \circ g$ is the set, $f \circ g = \{$ (1,17), (2,17), (2,14), (5,17) $\}$.

Even though the operation of composing two relations is not commutative, it is associative (as the reader is probably familiar from calculus, when relations are functions). Also, the inverse of the composite of two relations is equal to the composite of the inverse relations in reverse order. Without proof, we state the following two theorems.

Theorem 10: Let A, B, C, and D be non-empty sets, h a relation from the set A to the set B, g a relation from the set B to the set C, and f a relation from the set C to the set D. Moreover, assume that $R_h \cap D_g \neq \emptyset$ and $I \cap D_f \neq \emptyset$, where I stands for the image of the set $R_h \cap D_g$ under the function g. Then both composite relations $f \circ (g \circ h)$ and $(f \circ g) \circ h$ are defined and they are in fact equal: $f \circ (g \circ h) = (f \circ g) \circ h$

The student or reader of this book is strongly encouraged to try to prove Theorem 10 and 11 by drawing the appropriate Ven diagram if necessary.

Theorem 11: Let A, B, and C be non-empty sets, g a relation from A to B and f a relation from B to C; g^{-1} and f^{-1} the inverse relations of g and f respectively. Also, assume that $R_g \cap D_f \neq \emptyset$. Under these assumptions, $(f \circ g)^{-1} = g^{-1} \circ f^{-1}$.

We finish this section by listing and proving two theorems pertaining to transitive and symmetric relations.

Theorem 12: Let R be a relation in a non-empty set A.

(i) R is transitive if, and only if, the composite relation of R with itself, $R \circ R$, is a subset of R, that is, $R \circ R \subset R$.

(ii) If R is both transitive and reflexive, then $R \circ R = R$ (the converse of this statement is false, see remark below.)

Proof: (i) Suppose that R is transitive. To be able to compose R with itself we must, whenever we pick a pair $(x, y) \in R$, pick another pair in R of the form (y, z), so that $(x, z) \in R \circ R$; but since R is transitive, it follows at once, that (x, z) must also be a member of R. This proves that if R is transitive the composite $R \circ R$ must be a subset of R. Now let us prove the converse: if $R \circ R \subset R$, then R must be a transitive relation. Indeed, let $(x, y) \in R$ and $(y, z) \in R$; according to the definition of the composite function $R \circ R$, the pair (x, z) must be a member of $R \circ R$; but $R \circ R$ is a subset of R by assumption. Hence, (x, z) also belongs to R which proves that R is transitive.

(ii) Since R is transitive it follows, from part (i), that $R \circ R \subset R$, we must also show that $R \subset R \circ R$. To do this, pick any pair (x, y) of R. Since $x \in A$, $y \in A$, and R is reflexive, the pair (y, y) is also a member of R. The pair (x, y) can clearly be composed with the pair (y, y), producing the pair (x, y) again which them, it must, by definition, be in $R \circ R$. We have thus shown that if $(x, y) \in R$, then $(x, y) \in R \circ R$. This proves that $R \subset R \circ R$ and since $R \circ R \subset R$ as well, we conclude $R = R \circ R$.

Remark 25: The converse of Theorem 12(ii) if false: a transitive relation R may satisfy $R \circ R = R$, without being reflexive. Indeed, consider the set $A = \{a, b\}$, a and b distinct elements and the relation $R = \{(a, a), (a, b)\}$. We have $R \circ R = \{(a, a), (a, b)\} = R$, but R is not reflexive since (b, b) is not in R; $(b, b) \notin R$.

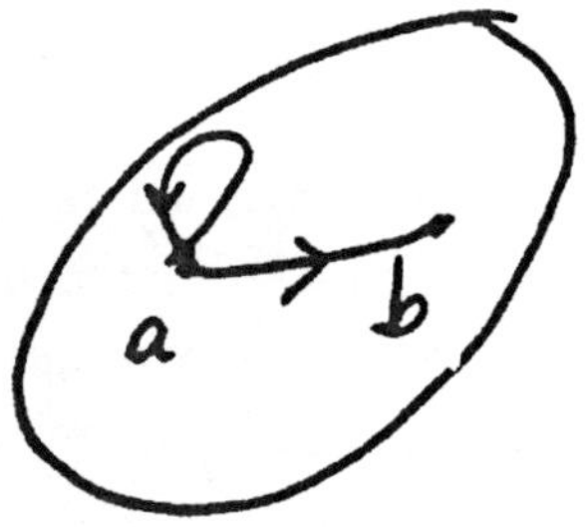

Figure 16

Theorem 13: Let R and S be symmetric relations in a non-empty set A. Then the composite relation $R \circ S$ is symmetric if, and only if, $R \circ S = S \circ R$.

Proof: "$\Rightarrow$" First we will show that if $R \circ S$ is symmetric, then $R \circ S = S \circ R$. To do this, we will prove that $R \circ S$ is a subset of $R \circ R$ and vice-versa. Pick a member of $R \circ S$; a pair (x, z) with $x, z \in A$.

Since $(x, z) \in R \circ S$ and $R \circ S$ is assumed to be symmetric, it follows that $(z, x) \in R \circ S$. By definition of the composite $R \circ S$ we conclude that there exists an element $y \in A$ such that $(z, y) \in S$ and $(y, x) \in R$ (so that $(z, x) \in R \circ S$) But both S and R are symmetric by the hypothesis of the theorem. Hence, $((x, y) \in R$ and $(y, z) \in S) \Rightarrow (x, z) \in S \circ R$. This proves that $R \circ S$ is a subset of $S \circ R$; in a similar manner one can prove that $S \circ R$ is a subset of $R \circ S$, so that $R \circ S = S \circ R$. ■

"$\Leftarrow$" Suppose that $R \circ S = S \circ R$ and let (x, z) be a member of $R \circ S$; then by definition there exists $y \in A$ such that $(x, y) \in S$ and $(y, z) \in R$. Again, since S and R are both symmetric, we must have $(z, x) \in S \circ R$. However, $S \circ R = R \circ S \Rightarrow (z, x) \in R \circ S$. We have proved that whenever $(x, z) \in R \circ S$, (z, x) must also be a member of $R \circ S$. Hence, $R \circ S$ is a symmetric relation. ■

1.7 Practice Problems

P1. Let $A = \{a, b, c, d, e\}$ be a set of five straight lines on a coordinate plane. The Venn diagram below is supposed to describe a relation R in A, defined by the condition $p(x, y) = x \in A$ is perpendicular to $y \in A$.

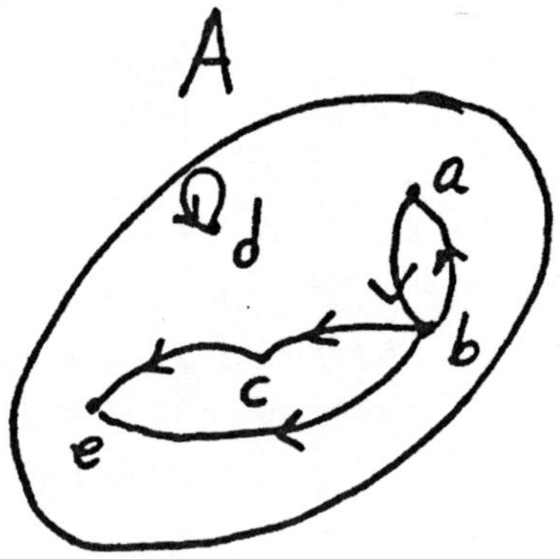

Figure 17

However, this diagram is erroneous, it cannot describe any relation in A defined by the above condition. Explain why and draw another Venn diagram that could describe a relation in A defined by the above condition.

Solution: The first error is that d relates to itself, which is impossible under the above conditions. A line cannot be perpendicular to itself. Also, such a relation would have to be symmetric. If x is perpendicular to y, then y must also be perpendicular to x. So $b \perp c$ also implies $c \perp b$ and $c \perp e \Rightarrow e \perp c$. Finally, b could not possibly be perpendicular to e, if at the same time b is perpendicular to c and c is perpendicular to e; b would be parallel to e. The following Venn diagram would be consistent with the given condition:

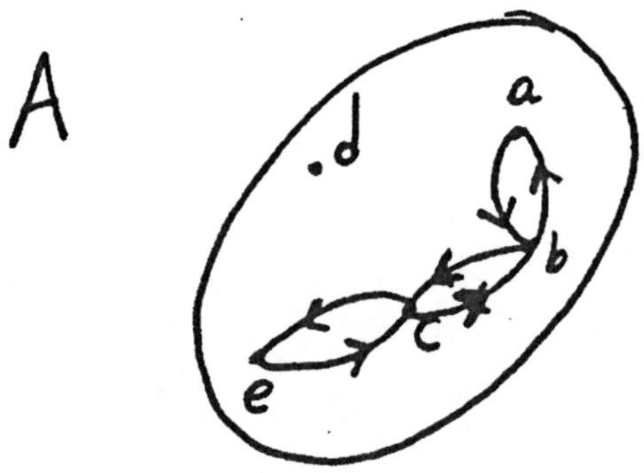

Figure 18

P2. Let $A = \{-3, -2, -1, 0, 1, 2, 3\}$, and the condition, $p(x, y) = x \in A$, $y \in A$, and $| x - y | = 2$. List all the pairs that are members of the relation R that the given condition defines. Draw a Venn diagram for this relation and determine whether R is reflexive or symmetric.

Solution: Note that the condition $| x-y | = 2$ is equivalent to $(x-y = 2$ or $x-y = -2) \Leftrightarrow (y = x - 2$ or $y = x + 2)$

$$R = \{(-1,-3),(0,-2),(1,-1),(2,0),(3,1),(-3,-1),(-2,0),$$
$$(-1,1),(0,2),(1,3)\}$$

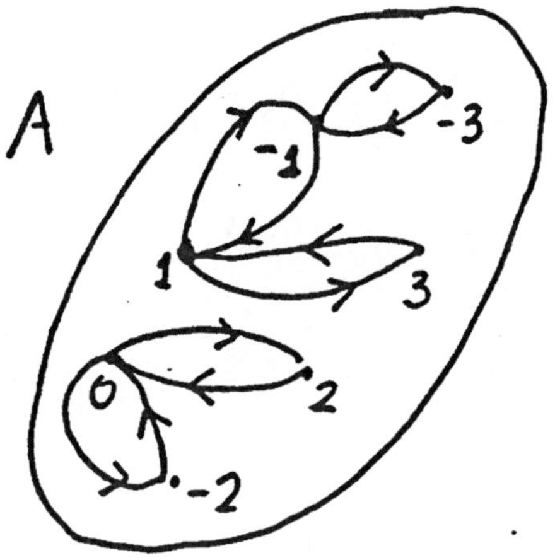

Figure 19

Clearly, this relations is symmetric since $| x - y | = 2 \Rightarrow | -(y - x) | = 2;\ | y - x | = 2$. It is not reflexive since $| x - x | = 0 \neq 2$. In fact, no pair of the form (x, x) is a member of R, which means that R is a anti-reflexive relation.

P3. Determine whether the given Venn diagram below describes a relation that is reflexive, anti-reflexive, symmetric, anti-symmetric, transitive, or none of the above.

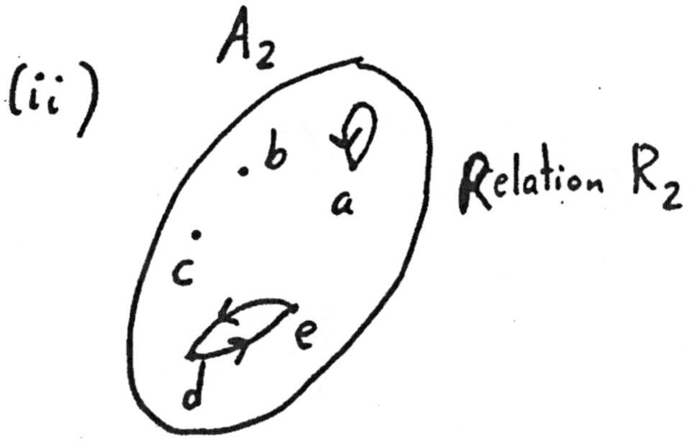

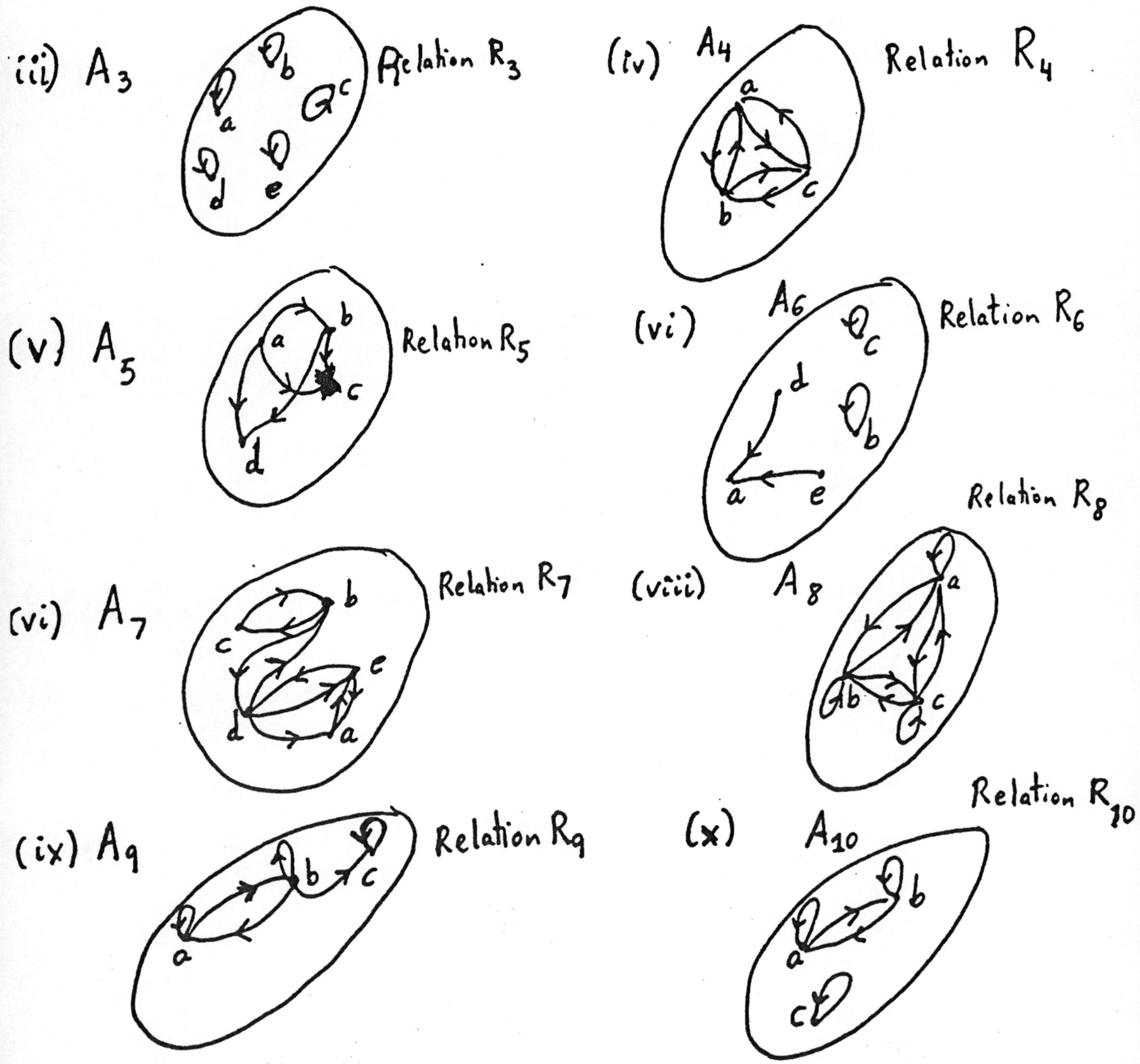
(iii) A3
Relation R3
(iv) A4
Relation R4
(v) A5
Relation R5
(vi) A6
Relation R6
(vi) A7
Relation R7
(viii) A8
Relation R8
(ix) A9
Relation R9
(x) A10
Relation R10

Figure 20

Solution: (i) Relation R_1 is antireflexive only.

(ii) Relation R_2 is symmetric only.

(iii) Relation R_3 is reflexive, symmetric, antisymmetric, and transitive. It is, in fact, the identity relation on the set A_3.

(iv) Relation R_4 is symmetric and transitive.

(v) Relation R_5 is antireflexive and transitive.

(vi) Relation R_6 is antisymmetric only.

(vii) Relation R_7 is antisymmetric only.

(viii) Relation R_8 is reflexive, symmetric, and transitive. That is, it is an equivalence in the set A_8.

(ix) Relation R_9 is reflexive only.

(x) Relation 10 is reflexive, symmetric, and transitive. That is, it is an equivalence relation in the set A_{10}.

(xi) Relation 11 is transitive only.

P4. Consider the set $P(S)$ where its members are the subsets of the set $S = \{a, b, c\}$ and the condition, $p(x, y) = x \in P(S),\ y \in P(S)$, and with $x = y^c$, where y^c stands for the complement of the set y. List all the pairs in the relation R that the given condition defines, and determine whether R is reflexive, antireflexive, symmetric, antisymmetric, transitive, or none of the above.

Solution: We have $P(S) = \{\emptyset, \{a\}, \{b\}, \{c\}, \{a, b\}, \{a, c\}, \{b, c\}\{a, b, c\}\}$. The compliment of a subset y of the set S will consist of all the elements in S that do not belong to the set y. We have,

$$\begin{aligned} R &= \{(\emptyset, S), (\{a\}, \{b, c\},), (\{b\}, \{a, c\}), (\{c\}, \{a, b\}) \\ &\quad (\{a, b\}, \{c\}), (\{a, c\}, \{b\}), (\{b, c\}, \{a\}), (S, \emptyset)\} \end{aligned}$$

If we label the eight subsets of S, S_1 through S_8, starting with $S_1 = \emptyset,\ S_2 = \{a\},\ S_3 = \{b\},\ S_4 = \{c\},\ S_5 = \{a, b\},\ S_6 = \{a, c\},\ S_7 = \{b, c\}$, and $S_8 = S = \{a, b, c\}$, we have the following Venn diagram.

The relation R in $P(S)$ is symmetric and antireflexive.

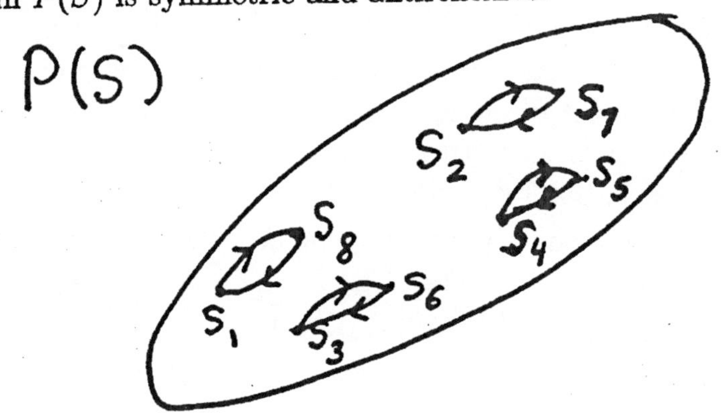

Figure 21

P5. If R is a relation in the set A, and I_A is the identity transformation or relation in A, is the relation $R \cup I_A$ reflexive, regardless of what R is?

Solution: Yes, $R \cup I_A$ will always be reflexive for every relation R in A since $R \cup I_A$ will contain all pairs of the form (x, x), with $x \in A$.

P6. In the set of rational numbers $\mathbb{Q}$ we define the following relation R defined by the condition,

$$p(x, y) = x \in \mathbb{Q},\ y \in \mathbb{Q} \text{ and } x \cdot y = 1.$$

Examine whether the relation R is reflexive, antireflexive, symmetris, antisymmetric, transitive, or none of the above.

Solution: (i) R fails to be reflexive since if $x^2 = 1$, then $x = 1$ or -1. Thus only the pairs $(1, 1)$ and $(-1, -1)$ are in R.

(ii) R also fails to be antireflexive, since at least one pair of the form (x, x) (in fact, two of them) $(1, 1)$ and $(-1, -1)$ are members of R.

(iii) R is symmetric because $x \cdot y = 1 \Rightarrow y \cdot x = 1$.

(iv) R is not antisymmetric since, with the exception of $(1, 1)$ and $(-1, -1)$, all the other pairs (x, y) in the relation satisfy $x \neq y$; and whenever (x, y) is in R, so is (y, x).

(v) R is not transitive because if $x \cdot y = 1$ and $y \cdot z = 1$ then $y = \frac{1}{x} = \frac{1}{z} \Rightarrow x = z$; but then $x \cdot y = 1$ and $y \cdot x = 1$ would imply, if the relation were transitive, that $x \cdot x = 1$, which only happens for $x = 1$ and $x = -1$.

Note that R is none other than the set $R = \{(x, \frac{1}{x}) \mid x \neq 0,\ x \in \mathbb{Q}\}$. The Venn

P7. Let $A = \{1, 6, 12, 18, 24, 36\}$ and R the relation in A defined by $p(x, y) = x \in A$, $y \in A$, and the greatest common divisor of x and y is 6. List all the elements of R and examine whetheter R is antireflexive and/or symmetric.

Solution:

$$\begin{aligned} R = \{&(6,6), (6,12), (12,6), (6,18), (18,6), (6,24), (24,6), (6,36), (36,6), \\ &(12,18), (18,12), (18,24), (24,18)\}. \end{aligned}$$

This relation is clearly symmetric (without even listing the elements of R we know this by the definition of the greatest common divisor of two numbers), but not antireflexive since it contains the pair $(6, 6)$. Note, however, that R is not reflexive (easy to see). It is not transitive either (even though it is close to being transitive) because for instance, both $(12, 18)$ and $(18, 12)$ are in R, but $(12, 12) \notin R$.

P8. Determine whether the given statement below is true or false.

A is a non-empty set.

(i) If R is symmetric, then so is the inverse relation R^{-1}.

(ii) If R is antisymmetric, then so is the inverse relation R^{-1}.

Solution: (i) True by the definition of a symmetric relation and the inverse relation.

(ii) True, because if R is antisymmetric, then it never contains two pairs (x, y) and (y, x), with $x \neq y$. Then R^{-1} cannot contain two such pairs either; for if it did, R would have to contain them as well; by simply switching x with y.

P9. Let R be a relation in a non-empty set A. Decide whether the given statement below is true or false.

(i) If R is reflexive, then $R \cap R^{-1} \neq \emptyset$, where R^{-1} is the inverse relation.

(ii) If R is symmetric, then $R \cap R^{-1} \neq \emptyset$.

(i) is true, in fact, $D_A \subset R \cap R^{-1}$, where D_A is the diagonal of the cartesian product A. Recall that $D_A = \{(x, x) \mid x \in A\}$. Since R is reflexive, D_A is a subset of R. Thus, it is also a subset of R^{-1} by the definition of the inverse relation R^{-1}, but $\emptyset \neq D_A$, since A is non-empty; consequently, $\emptyset \neq D_A \subset R \cap R^{-1}$.

(ii) is false since a symmetric relation R may be the empty set. If you read Definition 9 carefully, you will see that it is possible that R may be the empty set. In which case, $R^{-1} = \emptyset$ as well (refer to Remark 5). If, on the other hand, R is assumed to be non-empty, then the modified element, "if R is a non-empty symmetric relation in a non-empty set A, then $R \cap R^{-1} \neq \emptyset$", would be true. Note that if $R \neq \emptyset$, then, since $R = R^{-1}$ (because R is symmetric), we would, in fact, have $R \cap R^{-1} = R \neq \emptyset$.

P10. True or false: (assume R_1 and R_2 to be relations in a non-empty set A)

(i) If both R_1 and R_2 are antisymmetric, so is their union $R_1 \cup R_2$.

(ii) If both R_1 and R_2 are antisymmetric, so is their intersection $R_1 \cap R_2$.

(iii) If both R_1 and R_2 are transitive, so is their union $R_1 \cup R_2$.

(iv) If both R_1 and R_2 are transitive, so is their intersection $R_1 \cap R_2$.

Solution: (i) True, because the union $R_1 \cup R_2$ cannot contain any pair of the form (x, x). If it did, such a pair would have to belong in R_1 or to R_2 (since any element in the union belongs in one set or the other); but this could not happen, since by assumption, both R_1 and R_2 are antisymmetric. Neither of them contains elements of the form (x, x).

(ii) True, it is a mere consequence of the fact that any subset of an antisymmetric relation must also be antisymmetric, and trivially $R_1 \cap R_2 \subset R_1$. So, since $R_1 \cap R_2$ is a subset of R_1, $R_1 \cap R_2$ must be antisymmetric.

(iii) False, this can easily be seen in any number of counterexamples. For instance, let A be the set, $A = \{a, b, c, d, e, f\}$; a set of six distinct elements. Consider the relations in A, $R = \{(a, b), (b, c), (a, c), (a, d)\}$ and $R' = \{(d, e), (e, f), (d, f)\}$.

Both R and R' are transitive, but the union,

$$R \cup R' = \{(a, b), (b, c), (a, c), (a, d), (d, e), (e, f), (d, f)\},$$

since $(a, e) \notin R \cup R'$, while both (a, d) and (d, e) are members of $R \cup R'$.

(iv) True, it merely follows from the fact that any subset of a transitive relation will itself be transitive. In our case, $R_1 \cap R_2 \subset R_1$, and since R_1 is transitive, $R_1 \cap R_2$ will also be transitive.

P11. In each case below a condition that defines a relation R in a given set A, is given. Determine whether the relation R is reflexive, antireflexive, symmetric, antisymmetric, transitive, or none of the above.

(a) $A = R,\ R : p(x,y) = x \in A,\ y \in A,\ x \leq y$

(b) $A = P - \{1\}$ = set of postiive integers greater than 1, $R : p(x,y) = x \in A,\ y \in A$, and the greatest common divisor of x and y is greater than 1.

(c) $A = P,\ R : p(x,y) = x \in A,\ y \in A$, and x is divisible by y.

(d) $A = \mathbb{Z},\ R : p(x,y) = x \in A,\ y \in A$, and $x < y$.

(e) $A = P,\ R : p(x,y) = x \in A,\ y \in A$, and $k < x - y$, where k is a fixed positive integer.

In $(f) - (k)$, the given set is $A = \{x \mid x \in P,\ x \leq 12\}$ =the set of all positive integers not exceeding 12.

(f) $R : p(x,y) = x \in A,\ y \in A$, and $x \cdot y = 24$.

(g) $R : p(x,y) = x \in A,\ y \in A$, and $x \cdot y = 12$.

(h) $R : p(x,y) = x \in A,\ y \in A$, and $x^2 + y^2 \leq 100$.

(i) $R : p(x,y) = x \in A,\ y \in A$, and $x^2 - 4y^2 = 0$.

(j) $R : p(x,y) = x \in A,\ y \in A$, and $x - 2y = 0$.

(k) $R : p(x,y) = x \in A,\ y \in A$, and $x + 3y = 1$.

Solution: (a) R is reflexive, transitive, and antisymmetric.

(b) R is clearly symmetric by the definition of the greatest common divisor. Also, $(x,x) \in R$ for every $x \in A$, since the greatest common divisor of x with itself, is itself; and since $x > 1$, the pair (x,x) must be in R, for every $x \in A$. Hence, R is also reflexive. But it is not transitive since the greatest common divisors of x and y, and y and z, may both be greater than 1. But the greatest common divisor of x and z may still equal 1. For example, $x = 35,\ y = 10,\ z = 6$. The greatest common divisor of x and y is 5; of y and z is 2; but the greatest common divisor of x and z is 1.

(c) R is reflexive and transitive. Obviously, it is not symmetric or antireflexive. Also, it is not antisymmetric either, since for example, any two opposite integers must divide each other (but if they have the same sign only one can divide the other, but not both, unless, of course, they are equal).

(d) This is an antireflexive, antisymmetric, and transitive relation.

(e) R cannot contain any pairs of the form (x,x), since this would require $k < x - x$; $k < 0$, not possible since k is a positive integer. Hence R is antireflexive. Moreover, R cannot contain two pairs (a,b) and (b,a), for this would imply $(k < a - b$ and $k < b - a) \Rightarrow 2k < a - b + b - a \Rightarrow 2k < 0$, again contradicting the assumption that k is a positive integer. Thus R is also antisymmetric. Is R transitive? Suppose that $(x,y) \in R$ and $(y,z) \in R : k < x - y$ and $k < y - z$. Adding the two inequalities we obtain $2k < x - z$; but $k \leq 2k$. Since, in fact, k is a positive integer; thus $k < x - z \Rightarrow (x,z) \in R$. Yes, R is also transitive.

(f) $R = \{(2,12),(12,2),(3,8),(8,3),(4,6),(6,4)\}$

Upon inspection we see that R is a symmetric and an antireflexive relation, but not transitive, reflexive, or antisymmetric.

(g) $R = \{(1,12),(12,1),(2,6),(6,2),(3,4),(4,3)\}$ Same answer as in part (f). This is only a symmetric and antireflexive relation.

(h) Clearly R is symmetric, since $x^2 + y^2 \leq 100$ implies $y^2 + x^2 \leq 100$. But R is not reflexive, since, in fact, if $x \geq 8$, then $x^2 + x^2 \geq 128 > 100$. So none of the pairs $(8,8), (9,9), (10,10), (11,11), (12,12)$ lies in R (of course the absence of only one such pair is sufficient to make the relation nonreflexive). Clearly, it is not antisymmetric, since it is symmetric and it is not a subset of the identity relation or transformation and it is not antireflexive, since it contains pairs of the form (x,x); for example, $(1,1)$. Is it transitive? No, it is not, since ($x^2+y^2 \leq 100$ and $y^2+z^2 \leq 100$) does not imply $x^2 + z^2 \leq 100$; y could be small in comparison to both x and z. For example, $7^2+1^2 \leq 100$ and $1^2+8^2 \leq 100$; but $7^2+8^2 = 49+64 = 113 > 100$. Another example is $(7,6)$ and $(6,8)$. Both these pairs are in R, but $(7,8)$ is not a member of R. Thus R is symmetric only.

(i) Note that the condition $x^2 - 4y^2 = 0$ does not imply $y^2 - 4x^2 = 0$; for example $(6,3) \in R$ but $(3,6) \notin R$; hence R is not symmetric. It is not reflexive either since $(1,1)$, for example, is not in R. Actually, R is antireflexive: no pair (x,x) can be R, for this would imply $x^2 - 4x^2 = 0$; $-3x^2 = 0$; $x = 0$ which cannot happen since x is a positive integer between 1 and 12. R is antisymmetric. To see this, assume (x,y) and $(y,x) \in R$, with $\neq y : x^2-4y^2 = 0$ and $y^2-4x^2 = 0 \Leftrightarrow -3\cdot(x^2+y^2) = 0,\ x^2+y^2 = 0$, an impossibility, since both x and y are positive integers. This proves (by contradiction), that R is antisymmetric. Is R transitive? Assume $x^2 - 4y^2 = 0$ and $y^2 - 4z^2 = 0$. Solving for $y^2 = 4z^2$ and substituting for y^2 in the first equation we obtain,

$$x^2 - 16z^2 = 0.$$

If R were transitive, we would also have $x^2 - 4z^2 = 0$, which when combined with $x^2 - 16z^2 = 0$ implies $x = z = 0$, again impossible. Hence R is antireflexive and antisymmetric.

$$R = \{(2,1), (4,2), (6,3), (8,4), (10,5), (12,6)\}$$

An alternative solution would be to realize that the condition $x^2 - 4y^2 = 0$ is equivalent to $(x-2y)(x+2y) = 0$, but $x + 2y > 0$, since both x and y are positive integers. Hence, $x - 2y = 0,\ x = 2y$.

(j) This relation is identical to that of the previous part (i), since the condition $x^2 - 4y^2 = 0$ is equivalent to $x - 2y = 0$ in view of the fact that x and y are positive integers (and so the possibility $x = -2y$ is excluded).

$$R = \{(2,1), (4,2), (6,3), (8,4), (10,5), (12,6)\}$$

As we have already seen, this is an antireflexive and antisymmetric relation.

(k) R is the empty relation. No pair (x,y) of positive integers can satisfy $x+3y = 1$. Technically speaking, the empty relation satisfies (in a vacuous sense) the definitions of a antireflexive, symmetric, antisymmetric, and transitive relation.

P12. Consider the set $A = \mathbb{Z} \times (\mathbb{Z} - \{0\}) = \{(t,u) \mid t \in \mathbb{Z}, u \in \mathbb{Z} - \{0\}\} : p(x,y) = x \in A,\ y \in A,\ x = \frac{m}{n},\ y = \frac{k}{\ell}$, and $m\ell = n \cdot k$, with m, n, k, and $\ell \in \mathbb{Z}$; and $n \cdot \ell \neq 0$. Prove that the given condition defines an equivalence relation R in the set A.

Solution: We must prove that R is reflexive, symmetric, and transitive. If $x = (m,n) \in A$, then clearly $(x,x) \in R$, since obviously $m \cdot n = n \cdot m$. Next suppose that $(x,y) \in R$, with $n \cdot \ell \neq 0$, $x = (m,n)$ and $y = (k,\ell)$, $z = (r,t)$ and with $n \cdot \ell \neq 0$, $\ell \cdot t \neq 0$, $m \cdot \ell = n \cdot k$, and $k \cdot t = \ell \cdot r$. Multiplying the last two equations we obtain, $(m \cdot \ell) \cdot (k \cdot t) = (n \cdot k) \cdot (\ell \cdot r)$; $(m \cdot t) \cdot (k \cdot \ell) = (n \cdot r) \cdot (k \cdot \ell) \Rightarrow$ (since $\ell \neq 0$) $(m \cdot t) \cdot k = (n \cdot r) \cdot k$. So, if $k \neq 0$, then the last equation implies $m \cdot t = n \cdot r$. What if $k = 0$? If $k = 0$, then from the previous two equations $m \cdot \ell = n \cdot k$ and $k \cdot t = \ell \cdot r$, and by virtue of the fact that $n \cdot \ell \neq 0$ and $\ell \cdot t \neq 0$ (which means that all three n, ℓ and t are non-zero), we conclude that $m = 0 = r$. Hence we see that the equality $m \cdot t = n \cdot r$, with $n \cdot t \neq 0$ still holds, which implies $(x,z) \in R \Rightarrow R$ is transitive. Hence R is an equivalence relation.

P13. A relation R in a non-empty set A is called **cyclic** if, and only if, for any three elements x, y, z of A, the following inference holds true:

$$((x,y) \in R \text{ and } (y,z) \in R) \Rightarrow (z,x) \in R$$

According to this definition, prove that R is an equivalence relation in A if, and only if, R is both cyclic and reflexive.

Solution: "$\Rightarrow$" Suppose that R is an equivalence relation in A; we will show that R is cyclic. Let x, y, z any three elements of A, (not necessarily distinct) and such that $(x,y) \in R$ and $(y,z) \in R$. Since R is transitive,

$$((x,y) \in R \text{ and } (y,z) \in R) \Rightarrow (x,z) \in R.$$

But R is also symmetric, which means that $(x,z) \in R \Rightarrow (z,x) \in R$. This proves that R is cyclic. Note that in this part of the proof we have only made use of the fact that R is symmetric and transitive. So, clearly, any relation symmetric and transitive must be cyclic. But R is also reflexive (since it is an equivalence relation). We are done. ∎

"$\Leftarrow$ " Suppose that R is cyclic and reflexive; we prove that R is also symmetric and transitive. Indeed, suppose that $(x,y) \in R$ (x and y may not be distinct). Since R is reflexive $\Rightarrow (y,y) \in R$. Now apply the assumption that R is cyclic, by taking x, y, and $z = y$:

$$((x,y) \in R \text{ and } (y,y) \in R \Rightarrow (y,x) \in R,$$

which proves that R is symmetric. Now let x, y, z (not necessarily distinct) be any three elements of A; and such that $(x,y) \in R$ and $(y,z) \in R$. Since R is cyclic $\Rightarrow (z,x) \in R \Rightarrow$ (since R is symmetric) $(x,z) \in R$, which proves that R is transitive. End of proof. ∎

P14. (i) Draw the Venn diagram of a relation R in a non-empty set A, that is both symmetric and transitive, but not reflexive.

(ii) Let R be a symmetric and transitive relation in a non-empty set A. Prove that if R is non-empty, then it is reflexive in the set D_R (domain of R); and since R

is also symmetric and transitive in A, and thus in D_R, R is an equivalence relation in its domain D_R.

Solution:

(i) $R = \{(a,b),(b,a),(a,a),(b,b)\}$

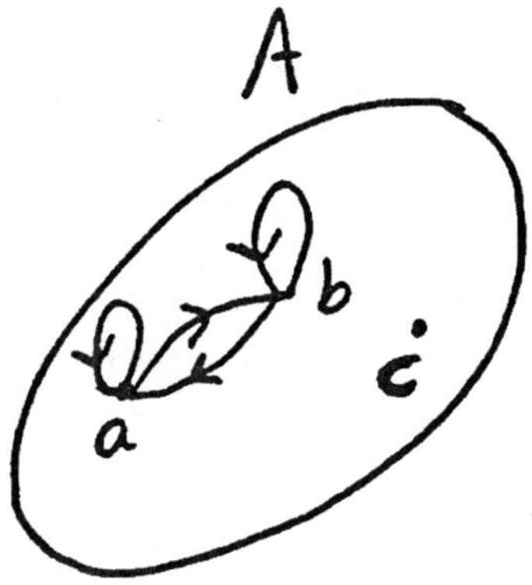

Figure 22

(ii) By definition $D_R = \{x \mid x \in$ and $(x,y) \in R$, for some $y \in A\}$. Pick any $x \in D_R$; then $\exists y \in A$, such that $(x,y) \in R$. Since R is symmetric $\Rightarrow (y,x) \in R$, but R is also transitive which means that $((x,y) \in R$ and $(y,z) \in R) \Rightarrow (x,z) \in R$; in our case $z = y : ((x,y) \in R$ and $(y,x) \in R) \Rightarrow (x,x) \in R$ which proves that R is reflexive in its own domain D_R, and hence an equivalence relation in D_R. Clearly, when $D_R = A$, R will be an equivalence relation in A, but not otherwise. That is not, when D_R is a proper subset of A, as the Venn diagram of part (i) illustrates.

P15. Let $A = \mathbb{Z}^+ \times \mathbb{Z}^+ = \{(t,u) \mid t \in \mathbb{Z}^+, u \in \mathbb{Z}^+\}$, the cartesian product of the set of positive integers with itself. Consider the condition $p(x,y) = x \in A,\ y \in A,\ x = (m,n),\ y = (k,\ell)$, and with $m + \ell = n + k$. Let R be the relation this condition defines.

(i) Prove that R is an equivalence relation in the set A.

(ii) List the equivalent class of the element $(4,7)$.

(iii) Describe the equivalent classes.

Solution: (i) R is reflexive, since for any $x = (m,n) \in A$, obviously $m + n = n + m$. R is also symmetric since if $(x,y) \in R$, with $x = (m,n)$ and $y = (k,\ell)$ we have $m + \ell = n + k \Rightarrow k + n = m + \ell$. To prove the transitivity of R, let $x = (m,n)$, $y = (k,\ell)$, and $z = (r,t)$, and suppose $(x,y) \in R$ and $(y,z) \in R$. We have $m + \ell = n + k$ and $k + t = \ell + r$. Adding these two equations memberwise we obtain $(m+\ell)+(k+t) = (n+k)+(\ell+r) \Rightarrow m + t = n + r$, which proves that $(x,z) = ((m,n),(r,t))$ is a member of R. Hence R is transitive.

(ii) All the elements (k,ℓ) equivalent to $(4,7)$ must satisfy $4+\ell = 7+k \Leftrightarrow \ell = k+3$. That the equivalence class $E_{(4,7)}$ of $(4,7)$ consists of all pairs (k,ℓ); $k,\ell \in \mathbb{Z}^+$, such that ℓ exceeds k by 3. That is, all pairs of the form $(k, k+3)$, ℓ is a positive integer.

$$E_{(4,7)} = \{(k, k+3,) \mid \ell \in \mathbb{Z}^+\}$$

(iii) Following part (ii), we see that for given positive integers m and n, the equivalence class of the element (m, n) is the set $E_{(m,n)} = \{(m - n + \ell, \ell) \mid \ell \in \mathbb{Z}^+\}$.

P16 The following condition in the set of postive integers $\mathbb{Z}^+$; $p(x, y) = x \in \mathbb{Z}^+, y \in \mathbb{Z}^+,\ y = n \cdot x$, where n is a fixed positive integer; $n \geq 2$ defines a relation R in $\mathbb{Z}$.

Is R partial order in the narrow sense? If it is a partial order, is it partial order in the narrow or wide sense?

Solution: R is antireflexive, since we cannot have $x = n \cdot x$, in virtue of the fact that $n \geq 2$. Also, R is antisymmetric, since if $(x, y) \in R$, then $y = n \cdot x$. We could not have $x = n \cdot y$ at the same time (which would be the case if $(y, x) \in R$), since the two equalities would imply $y = n \cdot (n \cdot y) \Leftrightarrow y = n^2 \cdot y$. By cancelling out the nonzero factor y from both sides, we get $n^2 = 1$ which is impossible since $n \geq 2$. Thus R is antisymmetric. However, R is not transitive. Indeed, if $(x, y) \in R$ and $(y, z) \in R$ we must have $y = n \cdot x$ and $z = n \cdot y$. Combining these two we obtain $z = n \cdot (n \cdot x)$; $z = n^2 \cdot x$. But then we could not have at the same time $z = n \cdot x$ for this would imply $n^2 \cdot x = n \cdot x \Rightarrow$ (since $x \in \mathbb{Z}^+$) $n^2 = n \Rightarrow$ (since $n \in \mathbb{Z}^+$) $n = 1$; contrary to the hypothesis that $n \geq 2$. Thus, $(x, z) \notin R$, which proves that R is not a transitive relation.

Conclusion: R is antireflexive, antisymmetric, and nontransitive (note, as we have shown above, that in fact, no combination of pairs $(x, y) \in R$ and $(y, z) \in R$ ever implies $(x, z) \in R$) because R fails to be transitive, we conclude that R is not a partial order.

P17. Consider the sets $A = \{x \mid x \geq 0,\ x \in \mathbb{Z},\ \text{and } x = 2t,\ \text{for some } t \in \mathbb{Z}\}$; and $B = \{x \mid x \geq 0, x \in \mathbb{Z},\ \text{and } x = 3t,\ \text{for some } t \in \mathbb{Z}\}$; A is the set of all nonnegative integers and B is the set of all nonnegative integers which are multiples of 3,

$$A = \{0, 2, 4, 6, 8, 10, \dots\},\quad B = \{0, 3, 6, 9, 12, 15, 18, \dots\}$$

Let $p(x, y) = x \in A,\ y \in B$ and $y = 3x$. The condition $p(x, y)$ defines a relation R from the set A to the set B.

(i) Is R a function from B to A?

(ii) Describe the domain D_R of R.

(iii) Is R a mapping from B to A?

(iv) If R is a function, is it one-to-one or injective? If it is injective, describe the inverse function R^{-1} (if R is not injective, there is still the inverse relation R^{-1}).

(v) If R is a function, is it surjective? At any rate, whether R is a function or not, describe the range R_R of the relation R.

Solution: (i) Yes, R is a function, since for every element x of A, there is at most one element y in B assigned to it; in fact, exactly one as the formula $y = 3x$ clearly demonstrates.

(ii) By inspection $D_R = A$.

(iii) Yes, R is a mapping from A to B, since $D_R = A$.

(iv) R is an injective function since, obviously, if $x_1 \neq x_2$ are both integers in the set A, we have $3x_1 \neq 3x_2$. From $y = 3x$, by solving for x we obtain $x = \frac{1}{3} \cdot y$; so $R^{-1} : B \to A$, defined by $R^{-1}(y) = \frac{y}{3}$, for every $y \in D_{R^{-1}} = R_R$. The domain of the inverse function is, as we know, equal to the range R_R of R. The range R_R of R consists of all the nonnegative integers which are multiples of 6 : $D_{R^{-1}} = R_R = \{x \mid x \geq 0, x \in \mathbb{Z}$ and $x = 6t,$ for some $t \in \mathbb{Z}\}$. Note that the inverse function R^{-1} is not a mapping from B to A, since its domain is not the entire set B, but only a proper subset of it.

(v) R is not surjective. As we have already seen in part (iv), the range R_R of R consists all the nonnegative integers which are multiples of 6.

P18. In each case below a function from a non-empty set A to a non-empty set B is given. Determine the domain of each function.

(i) $f : A = \mathbb{R} \to B = \mathbb{R};\ \ f(x) = \frac{3}{x^2-1}$

(ii) $\varphi : A = \mathbb{R} \to B = \mathbb{R};\ \ \varphi(x) = \frac{x^2+1}{\sqrt{(x-1)(x+3)}}$

(iii) $\tau : A = \mathbb{R} \to B = \mathbb{R};\ \ \tau(x) = \sqrt{1-x^2}$

Solution: (i) $D_f = \mathbb{R} - \{1, -1\}$ = all reals except 1 and -1 = (in Calculus interval notation) $(-\infty, -1) \cup (1, 1) \cup (1, \infty)$.

(ii) $D_\varphi = \{x \mid x \in \mathbb{R}$ and $(x-1)(x+3) > 0\} = (-\infty, -3) \cup (1, \infty)$

(iii) $D_\tau = \{x \mid x \in \mathbb{R}$ and $1 - x^2 \geq 0;\ (1-x)(1+x) \geq 0\} = [-1, 1]$ = all reals between -1 and 1.

P19. (i) $\sigma : \mathbb{R} \to \mathbb{R};\ \ \sigma(x) = \frac{\sqrt{(|x|+1)(x+2)}}{|x|-1}$

(ii) $g : \mathbb{R} \to \mathbb{R};\ \ g(x) = \frac{\sqrt{(x+1)(x+2)}}{\sqrt{|x|-x}}$

(iii) $h : \mathbb{R} \to \mathbb{R};\ \ h(x) = \frac{\sqrt{4-x}}{|x-1|-1}$

Solution: (i) The following must be true: $x \in D_\sigma \Leftrightarrow ((|x|+1)(x+2) \geq 0$ and $|x| - 1 \neq 0)$. Solving $(|x|+1)(x+2) \geq 0$ we find $x + 2 \geq 0;\ x \geq -2$ (since $|x| + 1 \geq 1 > 0$ for all $x \in \mathbb{R}$).

Also, the condition $|x| - 1 \neq 0$ is equivalent to $x \neq 1, -1$.

Conclusion:

$$D_\sigma = \{x \mid x \in \mathbb{R} \text{ with } x \geq -2,\ x \neq -1, \text{ and } x \neq 1\};$$

$$D_\sigma = [-2, -1) \cup (-1, 1) \cup (1, +\infty).$$

(ii) $x \in D_g \Leftrightarrow ((x+1) \cdot (x+2) \geq 0$ and $|x| - x > 0)$.
Solving $(x+1)(x+2) \geq 0$, we obtain $x \geq -1$ or $x \leq -2$. The condition $|x| - x > 0$ is equivalent to $x < 0$ (since for $x \geq 0,\ |x| - x = 0$). Thus,

$$D_g = \{x \mid x \in \mathbb{R}, \text{ with } x \geq -1 \text{ or } x \leq -2; \text{ and with } x < 0\};$$

$$D_g = \{x \mid x \in \mathbb{R}, \text{ with } x \leq -2; \text{ or with } -1 \leq x < 0\}$$

$$D_g = (-\infty, -2] \cup [-1, 0).$$

(iii) $x \in D_h \Leftrightarrow (4 - x \geq 0$ and $|x-1| - 1 \neq 0)$. The condition $|x-1|-1 \neq 0 \Leftrightarrow |x-1| \neq 1$ (it is easier to solve $|x-1| = 1$; $x - 1 = 1$ or $x - 1 = -1$; $x = 2$ or $x = 0$), is equivalent to $x \neq 0, 2$. We conclude that, $D_h = \{x \mid x \in \mathbb{R}$ with $4 - x \geq 0,\ x \neq 0$, and $x \neq 2\} = \{x \mid x \in \mathbb{R}$ with $x \leq 4,\ x \neq 0$, and $x \neq 2\} = (-\infty, 0) \cup (0, 2) \cup (2, 4]$.

P20. Consider the function $f : \mathbb{Q} \to \mathbb{Q}$, $f(x) = \frac{3x}{x^2 - 2x}$, from the set of rationals to itself. Is this function a mapping from $\mathbb{Q}$ to itself?

Solution: If $D_f = \mathbb{Q}$, then f will be a mapping from $\mathbb{Q}$ to itself. But $x^2 - 2x = 0 = x(x-2)$, for $x = 0$ and for $x = 2$. So neither $f(0)$ not $f(2)$ is defined via the given rule that defines the function, $D_f = \mathbb{Q} - \{0, 2\}$; f is not a mapping from $\mathbb{Q}$ to $\mathbb{Q}$, just a function.

P21. Explain why the function $f : \mathbb{Q} \to \mathbb{Q}$; $f(x) = \frac{x}{x^2 - 2}$, from the set of rationals to itself, is actually a mapping from $\mathbb{Q}$ to itself. While the extension g of f to the set of reals $\mathbb{R}$ defined by $g : \mathbb{R} \to \mathbb{R}$, $g(x) = \frac{x}{x^2-2}$, is not a mapping from $\mathbb{R}$ to itself.

Solution: The denominator $x^2 - 2$ can never vanish when x is a rational number. Since both $\sqrt{2}$ and $-\sqrt{2}$ are irrational numbers, as it is well known. Thus, $D_f = \mathbb{Q}$; f is a mapping from $\mathbb{Q}$ to itself. However, $D_g = \mathbb{R} - \{\sqrt{2}, -\sqrt{2}\} = (-\infty, -\sqrt{2}) \cup (-\sqrt{2}, \sqrt{2}) \cup (\sqrt{2}, +\infty)$, the function g is not a mapping from $\mathbb{R}$ to itself, since D_g is not equal to $\mathbb{R}$, but it is just a proper subset of $\mathbb{R}$.

P22. According to Theorem 6, every mapping from a non-empty set A to a non-empty set B, defines an equivalence relation in the set A. Apply Theorem 6 and Renark 21 to describe the equivalence classes of the equivalence relation that arises from the mapping $f : \mathbb{Z} \to \mathbb{Z}$; $f(x) = x^2 + n^2$, where n is a fixed integer.

Solution: Let $a \in \mathbb{Z}$. We will describe the equivalence class E_a of the element a : $E_a = \{x \mid x \in \mathbb{Z}$ and with $f(x) = f(a)\}$. We must find those integers x such that $x^2 + n^2 = a^2 + n^2 \Leftrightarrow x^2 = a^2$ which gives $x = a$ or $x = -a$. Hence $E_a = \{a, -a\}$, if $a \neq 0$. If $a = 0$, $E_0 = \{0\}$.

P23. Consider the mapping f from $\mathbb{Z}$ to itself defined by,

$$f : \mathbb{Z} \to \mathbb{Z};\ f(x) = \begin{cases} \dfrac{x}{2}, & \text{if } x \text{ is even} \\[2ex] \dfrac{x+1}{2}, & \text{if } x \text{ is odd} \end{cases}.$$

(i) Is this mapping f injective? If it is, describe the inverse function f^{-1}.

(ii) Is f surjective? Describe the range R_f of f.

Solution: (i) We must examine whether it is possible to have $f(x_1) = f(x_2)$, with $x_1 \neq x_2$. There are four possibilities:

(a) $f(x_1) = f(x_2)$, with x_1 and x_2 both even.
(b) $f(x_1) = f(x_2)$, with x_1 and x_2 both odd.
(c) $f(x_1) = f(x_2)$, with x_1 odd and x_2 even; or vice-versa.

If (a) occurs $\frac{x_1}{2} = \frac{x_2}{2} \Rightarrow x_1 = x_2$. So the first possibility can only occur when the two x values are equal, $x_1 = x_2$.
If (b) occurs, $\frac{x_1+1}{2} = \frac{x_2+1}{2} \Rightarrow x_1 = x_2$; same situation as in (a).
If (c) occurs, say with x_1 even and x_2 odd, we would have

$$\frac{x_1}{2} = \frac{x_2+1}{2} \Rightarrow x_1 = x_2 + 1.$$

Conclusion: f is not an injective function, in fact $f(k) = f(k+1)$ is true for any integer k. So there is no inverse function f^{-1}, just the inverse relation f^{-1}.

(ii) f is clearly surjective. You only have to look at the formula $\frac{x}{2}$ to see this. If n is any given integer, then $f(2n) = \frac{2n}{n} = n$, which proves that f is an onto mapping; range $R_f = \mathbb{Z}$.

P24. Let $f : \mathbb{R} - \{0\} \to \mathbb{R}$; $f(x) = x + \frac{1}{x}$. As it is easily seen this funciton is a mapping from the set of nonzero real numbers to the set of real numbers.

(i) Determine whether f is an injective mapping; and (ii) Describe the range of f and decide whether f is surjective.

Solution: (i) Suppose that $f(x_1) = f(x_2)$, for $x_1, x_2 \in \mathbb{R} - \{0\}$. We have $x_1 + \frac{1}{x_1} = x_2 + \frac{1}{x_2} \Rightarrow x_1 - x_2 + \frac{1}{x_1} - \frac{1}{x_2} = 0 \Rightarrow \frac{(x_1-x_2)\cdot x_1x_2 - (x_1-x_2)}{x_1x_2} = 0 \Rightarrow (x_1-x_2)\cdot x_1x_2 - (x_1-x_2) = 0 \Rightarrow (x_1 - x_2)\cdot(x_1x_2 - 1) = 0$. We see that there are two possibilities:

$$x_1 - x_2 = 0 \quad \text{or} \quad x_1x_2 - 1 = 0;$$

$$x_1 = x_2 \qquad \text{or} \quad x_1x_2 = 1$$

Thus we see that f is not one-to-one, since if we take $x_2 = \frac{1}{x_1}$ in the domain $\mathbb{R} - \{0\}$ of f; we have $f(x_2) = f(\frac{1}{x_1}) = \frac{1}{x_1} + \frac{1}{\frac{1}{x_1}} = \frac{1}{x_1} + x_1 = x_1 + \frac{1}{x_1} = f(x_1)$. So any two reciprocal numbers will have the same image under the mapping f; e.g., the numbers $\frac{1}{2}$ and 2 both have image the rational number $\frac{5}{2}$. Thus f is not injective.

(ii) For this question, the method of calculus would be handy indeed (note that f is an odd function, as the term is being used in calculus: $f(-x) = -f(x)$, $\forall x \in D_f = \mathbb{R} - \{0\}$. The graph of f is symmetric with respect to the origin $(0,0)$). Without the use of calculus, one may observe that $x + \frac{1}{x} \geq 2$, for every positive value of x; and $x + \frac{1}{x} \leq -2$, for every negative value of x (these two statements could be put into one: $\mid x + \frac{1}{x} \mid \geq 2$). Let us first prove that $x + \frac{1}{x} \geq 2$; for $x > 0$.

We have $x + \frac{1}{x} \geq 2 \Leftrightarrow \frac{x^2+1}{x} \geq 2$. Under the assumption $x > 0$, the inequality $\frac{x^2+1}{x} \geq 2$ is equivalent to $x \cdot (\frac{x^2+1}{x}) \geq x \cdot 2 \Leftrightarrow x^2 + 1 \geq 2x \Leftrightarrow x^2 - 2x + 1 \geq 0 \Leftrightarrow (x-1)^2 \geq 0$, which is true for all real values of x; in particular, for $x > 0$. Similarly, under the assumption $x < 0$, on can prove $\frac{x^2+1}{x} \leq -2$. Multiply both sides by $-x > 0 : -x\cdot(\frac{x^2+1}{x}) \leq (-x)\cdot(-2) \Leftrightarrow -(x^2+1) \leq 2x \Leftrightarrow 0 \leq x^2+1+2x$; $0 \leq (x+1)^2$, which is true.

Next we prove that the range R_f is the set $(-\infty,-2]\cup[2,+\infty)$. Note that $f(1) = 1+\frac{1}{1} = 2$; and $f(-1) = -1+\frac{1}{-1} = -2$. Pick any real number a in the set $(-\infty,-2]\cup[2,+\infty)$. Suppose $a \in [2,+\infty)$; we show that there exists at least one nonzero real number x such that $f(x) = a \Leftrightarrow x+\frac{1}{x} = a$; $x\cdot(x+\frac{1}{x}) = x\cdot a \Leftrightarrow x^2+1-ax = 0$; $x^2-ax+1 = 0$. According to the quadratic formula the two roots of this quadratic equation are given by,

$$r_1 = \frac{a+\sqrt{a^2-4}}{2} \text{ and } r_2 = \frac{a-\sqrt{a^2-4}}{2}.$$

To complete the proof, we must show that these are real roots, and not genuine complex roots (by that we mean complex numbers $a+bi$, $a-bi$, with $b\neq 0$). Indeed, the discriminant of the quadratic equation is equal to a^2-4, but $a^2-4 = (a-2)(a+2)\geq 0$, since $a\in[2,\infty)$ and so $a\geq 2$. Thus, both r_1, r_2 are real numbers (equal when $a=2$). The proof for the case $a\in(-\infty,-2]$ is essentially the same. We omit it. We have proved that the range R_f of f is the set, $R_f = (-\infty,-2]\cup[2,\infty)$. Hence R_f is not surjective.

P25. Every rational number r can be in form, $r = \frac{m}{n}$, where $m\in\mathbb{Z}$ and $n\in\mathbb{Z}^+$; and m and n are in lowest terms, meaning that their greatest common divisor is equal to 1. Using this simple fact, determine whether the mapping, $f:\mathbb{Q}\to\mathbb{Z}\times\mathbb{Z}^+$; $f\left(\frac{m}{n}\right) = (m,n)$ (with m an integer, n a positive integer and the greatest common divisor of m and n equal to 1), is an injective one.

Solution: It is injective, since if $f\left(\frac{m}{n}\right) = f\left(\frac{k}{\ell}\right)$; $(m,n) = (k,\ell) \Leftrightarrow m = k$ and $n = \ell$.

P26. Prove that if f is a mapping from a non-empty set A to a non-empty set B, and S_1, S_2 are any two subsets of the set A, then $lm(S_1\cup S_2) = lmS_1\cup lmS_2$, where $lm(S_1\cup S_2)$ stands for the image of the union set $S_1\cup S_2$ under the mapping f; lmS_1, lmS_2 are the images of S_1 and S_2 respectively under f(Generally, $lmS = \{y\in B \mid y = f(x),$ for some $x\in S\}$.

Solution: We will prove that $lm(S_1\cup S_2)$ is a subset of $lmS_1\cup lmS_2$ and vice-versa. Let y be an element of $lm(S_1\cup S_2)$, then by definition (of $lm(S_1\cup S_2)$), there exists $x\in S_1\cup S_2$ such that $f(x) = y$. But since $x\in S_1\cup S_2 \Rightarrow x\in S_1$ or $x\in S_2$ ("or" is not to be interpreted as an exclusive one. It means $x\in S_1$ or $x\in S_2$ or x belongs to both sets. As mentioned in the glossary of symbols exclusive "or" in this text is always expressed as "either ... or ... "). If $x\in S_1$, then since $y = f(x)$, it follows that $y\in lmS_1$. Likewise, if $x\in S_2$; $y = f(x)$ implies $y\in lmS_2$. Thus, ($y\in lmS_1$ or $y\in lmS_2$) $\Rightarrow y\in lmS_1\cup lmS_2$, which proves that $\forall y\in lm(S_1\cup S_2)\Rightarrow y\in lmS_1\cup lmS_2$. In other words, $lm(S_1\cup S_2)$ is a subset of $lmS_1\cup lmS_2$. Let us now prove that converse: that $lmS_1\cup lmS_2$ is a subset of $lm(S_1\cup S_2)$. Indeed, pick any $y\in lmS_1\cup lmS_2$ then $y\in lmS_1$ or $y\in lmS_2$; if $y\in lmS_1\Rightarrow y = f(x)$, for some $x\in S_1$. Likewise, if $y\in lmS_2\Rightarrow y = f(x)$, for some $x\in S_2$. In either case we see that $y = f(x)$, for some $x\in S_1\cup S_2$, which proves that $y\in lm(S_1\cup S_2)$. We have shown that $\forall y\in lmS_1\cup lmS_2\Rightarrow y\in lm(S_1\cup S_2)$. Hence $lmS_1\cup lmS_2$ is a subset of $lm(S_1\cup S_2)$. Therefore, $(lm(S_1\cup S_2)\subset lmS_1\cup lmS_2$ and $lmS_1\cup lmS_2\subset$

$lm(S_1 \cup S_2)) \Rightarrow lm(S_1 \cup S_2) = lmS_1 \cup lmS_2$. The proof is complete. ■

P27. Let a, b, and c be fixed rational numbers such that $a \neq -1$ and $b \neq c$. Consider the two functions,

$$f: \; \mathbb{Q} - \{1\} \text{ (all rationals except 1)} \to \mathbb{Q} - \{1\}; \; f(x) = \frac{x+a}{x-1}; \text{ and}$$

$$g: \; \mathbb{Q} - \{-c\} = \text{(all rationals except } -c) \to \mathbb{Q} - \{1\}; \; g(x) = \frac{x+b}{x+c}.$$

(i) Show that g is a bijection from the set $\mathbb{Q} - \{-c\}$ onto the set $\mathbb{Q} - \{1\}$; that f is a permutation of the set $\mathbb{Q} - \{1\}$; that is, f is a bijection from the set $\mathbb{Q} - \{1\}$ onto itself.

(ii) Apply Theorem 9(iii) to conclude that $f \circ g$ is bijection from $\mathbb{Q} - \{-c\}$ onto $\mathbb{Q} - \{1\}$. Find a formula for the composite function $f \circ g$.

Solution: (i) We need to show that f is injective and that the range R_f is equal to $\mathbb{Q} - \{1\}$. That is, f is also a surjective function from $\mathbb{Q} - \{1\}$ to itself. Injective: Let $x_1 x_2 \in \mathbb{Q} - \{1\}$ and such that $f(x_1) = f(x_2)$. Then $\frac{x_1+a}{x_1-1} = \frac{x_2+a}{x_2-1} \Rightarrow (x_1+a)(x_2-1) = (x_1-1)(x_2+a) \Rightarrow x_1x_2 - x_1 + a \cdot x_2 - a = x_1x_2 + a \cdot x_1 - x_2 - a \Rightarrow -x_1 + ax_2 - ax_1 + x_2 = 0 \Rightarrow -x_1 \cdot (a+1) + (a+1) \cdot x_2 = 0 \Rightarrow (a+1) \cdot (-x_1 + x_2) = 0$; but $a \neq -1$, and so $a + 1 \neq 0$. Hence $-x_1 + x_2 = 0 \Rightarrow x_1 = x_2$, which proves that f is injective.

Next, we prove that f is surjective: Pick $z \in \mathbb{Q} - \{1\}$. We must show that there exists $x \in \mathbb{Q} - \{1\}$ such that $f(x) = z$. To do this, set $z = \frac{x+a}{x-1}$ and solve for $x : z \cdot (x-1) = x + a \Rightarrow x \cdot (z-1) = z + a$; $x = \frac{z+a}{z-1}$ (incidentally, this shows that f equals its own inverse function). Now, for any given $z \in \mathbb{Q} - \{1\}$, the rational number $\frac{z+a}{z-1}$ is also in $\mathbb{Q} - \{1\}$ (because if it were $\frac{z+a}{z-1} = 1$; this would imply $a = -1$ contrary to the hypothesis $a \neq -1$). Hence the rational number $\frac{z+a}{z-1}$ lies in the domain of f, which is the set $\mathbb{Q} - \{1\}$.

Now apply the function

$$f : f\left(\frac{z+a}{z-1}\right) = \frac{\frac{z+a}{z-1} + a}{\frac{z+a}{z-1} - 1} = \frac{z + a + (z-1) \cdot a}{z + a - (z-1)}$$

$$= \frac{z + a + z \cdot a - a}{z + a - z + 1} = \frac{z \cdot (a+1)}{a+1} = z.$$

We have proved that for every $z \in \mathbb{Q} - \{1\}$ there exists $x \in \mathbb{Q} - \{1\}$, namely $x = \frac{z+a}{z-1}$, such that $f(x) = z$; f is a surjective function. Since f is both injective and surjective from the set $\mathbb{Q} - \{1\}$ to itself, if f is a bijection from $\mathbb{Q} - \{1\}$ to itself. The proof that the other function, g, is a bijection from the set $\mathbb{Q} - \{-c\}$ to $\mathbb{Q} - \{1\}$ is similar and it is left to the reader.

(ii) According to Theorem 9(iii), the composite of two bijections is also a bijection. In our case a bijection from $\mathbb{Q} - \{-c\}$ to $\mathbb{Q} - \{1\}$. We have $(f \circ g)(x) = f(g(x)) = \frac{\frac{x+b}{x+c} + a}{\frac{x+b}{x+c} - 1}$, $\forall x \in \mathbb{Q} - \{-c\}$. So that $(f \circ g)(x) = f(g(x)) = \left(\frac{\frac{x+b}{x+c} + a}{\frac{x+b}{x+c} - 1}\right) = \frac{x + b + a \cdot (x+c)}{x + b - (x+c)}$; $(f \circ$

$g)(x) = \frac{(a+1)\cdot x+b+ac}{b-c}$; $(f \circ g)(x) = \left(\frac{a+1}{b-c}\right) \cdot x + \frac{b+ac}{b-c}$. It is interesting to note here that the resulting composite function $f \circ g$ is a linear function with domain the set of all rationals except $-c$. From calculus, we know that any linear function has a graph that is a straight line. In our case, the above composite function $f \circ g$ has a graph that is part of the graph of the straight line with equation $y = \left(\frac{a+1}{b-c}\right) \cdot x + \frac{b+ac}{b-c}$. The graph of $f \circ g$ contains only those points on that straight line that have rational x-coordinates (and consequently, since a, b, c are all rational, also rational y-coordinates) and with the point $(-c, 1)$ removed.

P28. In each case below a set A (subset of the reals) which is totally ordered under the usual total order "$\leq$" of the reals, is specified. In addition, a subset X of A is also specified. Find $\inf X$(infimum of X), $\min X$(minimum of X), $\sup X$(supremum of X), and $\max X$ (maximum of X), if any of these four elements in A exists.

(i) $A = \mathbb{Q}$ set of all rationals, $X = \{x \in A \mid x = \frac{1}{n},\ n \text{ a positive integer}\} = \{1, \frac{1}{2}, \frac{1}{3}, \frac{1}{4}, \dots\}$ = the set of all the reciprocals of the positive integers.

(ii) $A = \mathbb{Q}^+ \cup \{0\}$, $X = \{x \in A \mid x = at + b$, and with $0 \leq t < n$ and $t \in \mathbb{Q}\}$, where a, b are fixed positive rationals; n a fixed positive integer.

Solution: (i) $\text{Inf}\, X$ is, according to Definition 24, the greatest of all the lower bounds for the set X. Clearly all the negative rationals and zero are all lower bounds of the set X, but no positive rational can be a lower bound for X. Since the members of the set X can get arbitrarily close to 0 (in this particular situation, and in the language of calculus we would write $\lim\limits_{n \to +\infty} \frac{1}{n} = 0$). Hence, zero is the greatest of all lower bounds for X; $\underline{\text{Inf}\, X = 0}$. However, $\min X$ does not exist, since 0 is not a member of the set X. Next, by inspection we see that $\sup X = \frac{1}{1} = 1 = \max X$.

(ii) Since a is positive, it is easy to see that $at_2 + b > at_1 + b$, whenever $t_1 < t_2$, $0 \leq t_1, t_2 < n$, and $t_1, t_2 \in \mathbb{Q}$. This shows that $\text{Inf}\, X = a \cdot 0 + b = b = \min X$. Also $\sup X = an + b$, since members of the set X get arbitrarily close to the rational number $an + b$, but $an + b$ is not a member of X. Hence, $\max X$ does not exist.

P29. Consider the functions, $\sigma \mathbb{R} \to \mathbb{R}$; $\sigma(x) = \begin{cases} -x, & \text{if } x < -1 \\ 0, & \text{if } x = 0 \\ x, & \text{if } x > 1 \end{cases}$, and $\varphi : \mathbb{R} \to \mathbb{R}$; $\varphi(x) = x^2$.

Find formulas for the two composite functions $\sigma \circ \varphi$ and $\varphi \circ \sigma$; give their domains and ranges.

Solution: By inspection, we see that the range of σ is the set $R_\sigma = (-\infty, -1) \cup \{0\} \cup (1, +\infty)$; and the range of φ is the set $R_\varphi = [0, +\infty)$. According to Definition 32, in order for $\sigma \circ \varphi$ to be defined (that is, to be non-empty), we must have $R_\varphi \cap D_\sigma \neq \emptyset$. The domain D_σ of σ is the set $D_\sigma = (-\infty, -1) \cup \{0\} \cup (1, +\infty)$. Thus, $R_\varphi \cap D_\sigma = \{0\} \cup (1, +\infty)$.

To determine the domain $D_{\sigma \circ \varphi}$, note that the values of x^2 must be either 0 or greater than 1. If the real number x is to be a member of $D_{\sigma \circ \varphi}$: solving $x^2 > 1 \Leftrightarrow (x-1)(x+1) > 0 \Leftrightarrow (x < -1 \text{ or } x > 1)$; and solving $x^2 = 0$ produces $x = 0$. Thus, $D_{\sigma \circ \varphi} = (-\infty, -1) \cup \{0\} \cup (1, +\infty)$, a proper subset of $D_\varphi = \mathbb{R}$. We have,

$$(\sigma \circ \varphi)(x) = \sigma(\varphi(x)) = \begin{cases} x^2, & \text{if } x < -1 \\ 0, & \text{if } x = 0 \\ x^2, & \text{if } x > 1 \end{cases}$$

Next consider the composite $\varphi \circ \sigma$; we have $R_\sigma = (-\infty, -1) \cup \{0\} \cup (1, +\infty)$ and

$$\begin{aligned} R_\sigma \cap D_\varphi &= (\{0\} \cup (1, +\infty)) \cap \mathbb{R} \\ &= \{0\} \cup (1, +\infty). \end{aligned}$$

We see that $R_\sigma \cap D_\varphi = R_\sigma$. In other words R_σ is a subset of D_φ. Hence, $D_{\varphi \circ \sigma} = D_\sigma = (-\infty, -1) \cup \{0\} \cup (1, +\infty)$.

We have

$$(\varphi \circ \sigma)(x) = \varphi(\sigma(x)) = \begin{cases} (-x)^2 = x^2, & \text{if } x < -1 \\ 0^2 = 0, & \text{if } x = 0 \\ x^2, & \text{if } x > 1. \end{cases}$$

Interestingly the two composites $\sigma \circ \varphi$ and $\varphi \circ \sigma$ are equal functions! Their range is the set $R_{\sigma \circ \varphi} = R_{\varphi \circ \sigma} = \{0\} \cup (1, +\infty)$.

P30. Consider the mappings,

$$f : \mathbb{Z}^+ \to \mathbb{Q}; \quad f(x) = \frac{1}{x+1}, \text{ and}$$

$$g : \mathbb{Z}^+ \to \mathbb{Z}^+; \quad g(x) = x^2.$$

(i) Which of the composite functions $f \circ f$, $g \circ g$, $f \circ g, g \circ f$ are defined according to Definition 32? In other words, which of them are non-empty?

(ii) For those that are defined, find formulas that describe them.

(iii) Of those compositive functions, of part (i), that are defined, which are mappings?

Solution: (i) We have

$$\begin{aligned} D_f = \mathbb{Z}^+, \quad R_f &= \{y \mid y = \tfrac{1}{n}, \text{ with } n \in \mathbb{Z}^+ \text{ and } n \geq 2\}; \\ R_f &= \{\tfrac{1}{2}, \tfrac{1}{3}, \tfrac{1}{4}, \tfrac{1}{5}, \ldots\} \end{aligned}$$

We also have $D_g = \mathbb{Z}^+$ and $\begin{aligned} R_g &= \{z \mid z = n^2, \text{ with } n \in \mathbb{Z}^+\}; \\ R_g &= \{1, 4, 9, 16, 25, \ldots\} \end{aligned}$.

(a) For $f \circ f$ to be defined, we must have $R_f \cap D_f \neq \emptyset$. We have $R_f \cap D_f = \emptyset$, so $f \circ f$ is not defined; $f \circ f = \emptyset$.

(b)]For $g \circ g$ to be defined, we must have $R_g \cap D_g \neq \emptyset$. We have $R_g \cap D_g = R_g = \{z \mid z = n^2 \text{ with } n \in \mathbb{Z}^+\}$, since R_g is a subset of $D_g = \mathbb{Z}^+$; yes, $g \circ g$ is defined.

(c) We have $R_f \cap D_g = \emptyset$; so $g \circ f$ is undefined; $g \circ f = \emptyset$

(ii) Of the four composite functions, only $g \circ g$ is defined or non-empty. We have $(g \circ g)(x) = g(g(x)) = (x^2)^2 = x^4$, with domain $D_{gog} = \mathbb{Z}^+$, and range $R_{gog} = \{y \mid y = n^4, \text{ with } n \in \mathbb{Z}^+\}$.

(iii) Clearly $g \circ g$ is a mapping from $\mathbb{Z}^+$ to itself, since $D_{gog} = \mathbb{Z}^+$. (Or one can say, since g is a mapping, $g \circ g$ must also be a mapping.)

P31. Let z, b, c, be fixed positive integers. Consider the relation R in the set of positive integers $\mathbb{Z}^+$, defined by the condition, $p(x, y) = x \in \mathbb{Z}^+,\ y \in \mathbb{Z}^+$, and $ax + by = c$. What condition must the constants a, b, and c must satisfy so that R is a symmetric relation?

Solution: A few observations may be helpful. It is possible that R may be the empty relation (e.g., take $a = 2, b = 1, c = 1$). Regardless of the positive integer values of a, b, and c, relation R is always a function from $\mathbb{Z}^+$ to itself, since for every $x \in \mathbb{Z}^+$, there is at most value $y \in \mathbb{Z}^+$ such that $ax + by = c$. (For example, if we take $a = 2, b = 1, c = 5$, then $R = \{(2, 1), (1, 3)\}$).

Furthermore, R is an injective function, no matter what the vales of the positive integers a, b, c are. This can be proved easily (the reader should do so). Because R is an injective function, R will be symmetric if, and only if, R coincides with its inverse function R^{-1} (since R is one-to-one, the inverse relation will also be a one-to-one functions). At any rate, if R is required to be symmetric, we have

$$\left.\begin{array}{r} ax + by = c \\ \text{and} \qquad\qquad \\ ay + bx = c \end{array}\right\} .$$

for every pair $(x, y) \in R$, because both (x, y) and (y, x) must be in R.

Subtracting the second question from the first we obtain

$$a \cdot (x - y) - b \cdot (x - y) = 0 \Leftrightarrow (a - b) \cdot (x - y) = 0$$

.

Since R is supposed to be symmetric, then, unless it is a subset of the diagonal of the cartesian product $\mathbb{Z}^+ \times \mathbb{Z}^+$ (the set of all pairs of the form (n, n), $n \in \mathbb{Z}^+$) it will contain at least one pair (x, y), with $x \neq y$. Hence, for $(a - b) \cdot (x - y) = 0$ to be satisfied, we must have $a - b = 0$; $a = b$. Thus, the condition $a = b$ is necessary if we assume that R contains at least one pair (x, y), with $x \neq y$. Furthermore, if $a = b$, then $ax + by = c$ implies $ax + ay = c \Rightarrow a \cdot (x + y) = c$. So a must be a divisor of c.

Finally, what if R contains only pairs of the form (x, x), $x \in \mathbb{Z}^+$? Then, $ax + by = c$ implies $ax + bx = c$; $(a + b) \cdot x = c$, from which we obtain $x = \frac{c}{a+b}$. In conclusion, R will be symmetric if, and only if, either of the following two conditions occur:

(1) $a = b$ and a is a divisor of c.

(2) $a + b$ is a divisor of c and there are no positive integers x and y, with $x \neq y$, satisfying $ax + by = c$.

If condition (1) occurs, then $R = \{(x, y) \mid x \in \mathbb{Z}^+,\ y \in \mathbb{Z}^+, \text{ and with } x + y = \frac{c}{a}\}$. (Note that $\frac{c}{a}$ is a positive integer since a is a divisor of c.)

If condition (2) occurs, then $R = \{(\frac{c}{a+b}, \frac{c}{a+b})\}$. Note in condition (2), it is note enough to just say that $a + b$ is a divisor of c. For example, consider $2x + y = 6$; $a = 2, b = 1, c = 6$; $a + b = 3$ is a divisor of 6; but R in this case also contains other pairs (x, y), with $x \neq y$, specifically $R = \{(2,2), (1,4)\}$. However, if we consider $3x + 5y = 8$; $a = 3, b = 5, c = 8$, we see that $a + b = 8$ is a divisor of $c = 8$ and $R = \{(1,1)\}$. Another such example is $3x + 5y = 16$; $a = 3, b = 5, c = 6$; $a + b = 8$ is a divisor of $c = 16$; and $R = \{(2,2)\}$.

Finally, an example pertaining to condition (1) would be, $5x + 5y = 20$; $a = 5 = b$; $c = 20$; 5 is a divisor of 20. We have

$$R = \{(x,y) \mid x \in \mathbb{Z}^+, y \in \mathbb{Z}^+, \text{ and with } x + y = \frac{20}{4} = 4\}$$

$$R = \{(1,3), (3,1), (2,2)\}$$

P32. Let H be the set of half (or semi) straight lines or rays that emanate from the origin $0(0,0)$ of the xy coordinate plane we are accustomed to in calculus. The condition, $p(x,y) = x \in H$ is perpendicular to $y \in H$, defines a relation R in the set H. Explain why this relation is not a function. If R symmetric? Is it transitive, reflexive, antisymmetric, or antireflexive? Also, describe the composite $R \circ R$. Is it a function or not?

Solution: R is not a function, since if $(x,y) \in R$, x is perpendicular to y, but then $(x, y') \in R$, where y' is the ray oppositive from y; and obviously $y \neq y'$. So for every $x \in H$, there corresponds, in fact, two distinct rays y and y' in H, both perpendicular to x. Relation R is obviously symmetric since if x is perpendicular to y, then y must be perpendicular to x. R is not transitive because if x is perpendicular to y, and y perpendicular to z, it does not follow that x must be perpendicular to z. In fact, z would have to be either x itself, or x', the ray oppositive from x. Obviously R is not antisymmetric and it is not reflexive. R is antireflexive since no $x \in H$ can be perpendicular to itself. In summary, R is only an antireflexive and symmetric relation and not a function.

Next consider the composite $R \circ R$. If $(x,y) \in R$, then y is one of the two rays perpendicular to x. To compose (x,y) with another pair in R, according the Definition 33, the pair (x,y) can only be composed with a pair of the form $(y,z) \in R$. So that $(x,z) \in R \circ R$, since $x \in H$ is perpendicular to $y \in H$, and $y \in H$ is perpendicular to $z \in H$; it follows that z must necessarily be either the ray opposite x or x itself. Therefore,

$$\begin{aligned} R \circ R &= \{(x,z) \mid \text{ either } x \text{ and } z \text{ areopposite rays in } H \text{ or } x = z\} \\ &= \{(x,x) \mid x \in H\} \cup \{(x,t) \mid x \in H, t \in H, \text{ and } x \text{ and } t\text{are} \\ &\qquad \text{opposite rays}\} \end{aligned}$$

Clearly $R \circ R$ is not a function since for each $x \in H$, there correspond two distinct elements in H, via this relation: x itself and the ray opposite x; both domain and the range of H are equal to the set H.

P33. Consider the following two mappings from the set of real numbers to itself: $f : \mathbb{R} \to \mathbb{R};\ \ f(x) = \mid x-1 \mid + \mid x+2 \mid$ and,

$g : \mathbb{R} \to \mathbb{R};\ \ g(x) = \frac{|x|}{2} - \mid x-3 \mid$

(i) Give piecewise expressions (without absolute values) for each function, draw the two graphs, and state the ranges R_f and R_g.

(ii) Find formulas for the composite function $f \circ g$ (which is also a mapping from $\mathbb{R}$ to itself, since clearly $R_g \subset D_f = \mathbb{R}$), sketch its graph, and state its range $R_{f \circ g}$.

Solution: (i) We make use of the fact (from Calculus or Precalculus) that if $t \in \mathbb{R}$, then $\mid t \mid = \begin{cases} -t, & \text{if } t < 0 \\ t, & \text{if } t \geq 0 \end{cases}$. Apply this to $\mid x-1 \mid$ and we obtain $\mid x-1 \mid =$ $\begin{cases} 1-x, & \text{if } x-1 < 0 \\ x-1, & \text{if } x-1 \geq 0 \end{cases}$; $\mid x-1 \mid = \begin{cases} 1-x, & \text{if } x < 1 \\ x-1, & \text{if } x \geq 1 \end{cases}$

Likewise, $\mid x+2 \mid = \begin{cases} -x-2, & \text{if } x < -2 \\ x+2, & \text{if } x \geq -2 \end{cases}$; combining the piecewise expressions for $\mid x-1 \mid$ and $\mid x+2 \mid$ we arrive at,

$$f(x) = \begin{cases} -2x-1, & \text{if } x < -2 \\ 3, & \text{if } -2 \leq x < 1 \\ 2x+1, & \text{if } x \geq 1 \end{cases}$$

Working similarly for the second function g we find that,

$$\frac{\mid x \mid}{2} = \begin{cases} -\frac{x}{2}, & \text{if } x < 0 \\ \frac{x}{2}, & \text{if } x \geq 0 \end{cases} \quad \text{and} \quad \mid x-3 \mid = \begin{cases} 3-x, & \text{if } x < 3 \\ x-3, & \text{if } x \geq 3 \end{cases}$$

Combining these two we obtain,

$$g(x) = \frac{\mid x \mid}{2} - \mid x-3 \mid = \begin{cases} \frac{x}{2} - 3, & \text{if } x < 0 \\ \frac{3}{2} \cdot x - 3, & \text{if } 0 \leq x < 3 \\ -\frac{x}{2} + 3, & \text{if } x \geq 3 \end{cases}$$

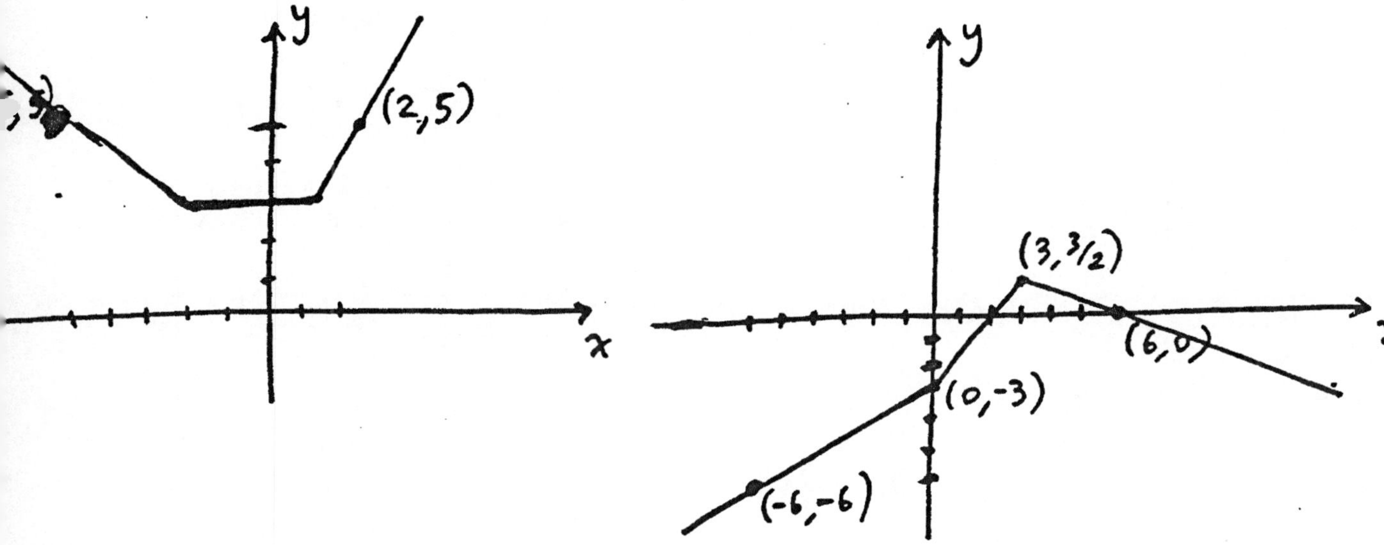

Figure 23

Even though both ranges R_f and R_g can be found algebraically through the use of the piecewise formulas for the two functions we can simply read off the ranges from the two graphs:

$$R_f = [3, +\infty) \text{ and } R_g = (-\infty, \frac{3}{2}]$$

(ii) Let $x \in \mathbb{R}$, $y \in \mathbb{R}$, $z \in \mathbb{R}$ and such that $y = g(x)$ and $z = f(y)$.

(a) If $x < 0$, then $y = g(x) = \frac{x}{2} - 3$. We must find the range of the y values: $x < 0 \Rightarrow \frac{x}{2} < 0 \Rightarrow \frac{x}{2} - 3 < -3$; so $y < -3$. Looking at the formulas for f, we see that for $y < -3$, $z = f(y) = -2y - 1$. Thus, $z = f(g(x)) = -2 \cdot (\frac{x}{2} - 3) - 1 = -x + 5$.

Conclusion: $(f \circ g)(x) = -x + 5$, if $x < 0$.

(b) Next suppose $0 \le x < 3$, then $y = g(x) = \frac{3}{2}x - 3$. Moreover, $0 \le x < 3 \Rightarrow 0 \le \frac{3}{2}x < \frac{9}{2} \Rightarrow -3 \le \frac{3}{2}x - 3 < \frac{9}{2} - 3$; $\frac{3}{2}$ (not inclusively). But, if we look at the formulas for f, we see that we need to consider three cases:

(b_1) If $-2 \le y < 1$, then $z = f(y) = 3$.

(b_2) While for $1 \le y < \frac{3}{2}$, $z = f(y) = 2y + 1 = 2 \cdot (\frac{3}{2} \cdot x - 3) + 1 = 3x - 6 + 1 = 3x - 5$

(b_3) If $-3 \le y < -2 \Rightarrow z = f(y) = -2y - 1 = -2(\frac{3}{2}x - 3) - 1 = -3x + 5$. Note that when $-2 \le y < 1$, we have $-2 \le \frac{3}{2}x - 3 < 1 \Leftrightarrow -2 + 3 \le \frac{3}{2} \cdot x - 3 + 3 < 1 + 3 \Leftrightarrow 1 \le \frac{3}{2}x < 4 \Leftrightarrow$ (multiply by $\frac{2}{3}$), $\frac{2}{3} \le x < \frac{8}{3}$.

Also, note that when $1 \le y < \frac{3}{2}$, we have $1 \le \frac{3}{2}x - 3 < \frac{3}{2} \Leftrightarrow 4 \le \frac{3}{2} \cdot x < \frac{9}{2} \Leftrightarrow$(multiply by $\frac{2}{3}$) $\frac{8}{3} \le x < 3$.

Finally, when $-3 \le y < -2$; $-3 \le \frac{3}{2}x - 3 < -2$; $0 \le \frac{3}{2}x < 1$; $0 \le x < \frac{2}{3}$.

Conclusion:
$$\begin{array}{ll} (f \circ g)(x) = 3, & \text{if } \frac{2}{3} \le x < \frac{8}{3}, \text{ and} \\ (f \circ g)(x) = 3x - 5, & \text{if } \frac{8}{3} \le x < 3; \text{ and} \\ (f \circ g)(x) = -3x + 5, & \text{if } 0 \le x < \frac{2}{3} \end{array}$$

(c) Finally, suppose $x \geq 3$; then $y = g(x) = -\frac{x}{2}+3$. Now $x \geq 3$ implies $-\frac{x}{2} \leq -\frac{3}{2}$. Thus, $-\frac{x}{2}+3 \leq -\frac{3}{2}+3$; $-\frac{x}{2}+3 \leq \frac{3}{2}$. We see that $y \leq \frac{3}{2}$. To determine the range of $z = f(y)$ for $y < \frac{3}{2}$, we must consider three cases: $1 \leq h \leq \frac{3}{2}$, $-2 \leq y < 1$, and $y < -2$.

(c_1) If $1 \leq y \leq \frac{3}{2}$, then $z = f(y) = 2y+1 = 2(-\frac{x}{2}+3)+1 = -x+6+1 = -x+7$. Also, $1 \leq y \leq \frac{3}{2} \Leftrightarrow 1 \leq -\frac{x}{2}+3 < \frac{3}{2}$; $1-3 \leq -\frac{x}{2} < \frac{3}{2}-3 \Leftrightarrow -2 \leq -\frac{x}{2} \leq -\frac{3}{2} \Leftrightarrow$ (multiply by -2)$4 \geq x \geq 3$. So, for $3 \leq x \leq 4$, $(f \circ g)(x) = -x+7$.

(c_2) If $-2 \leq y < 1$, then $z = f(y) = 3$. Also, $-2 \leq y < 1 \Leftrightarrow -2 \leq -\frac{x}{2}+3 < 1 \Leftrightarrow -2-3 \leq -\frac{x}{2} < 1-3 \Leftrightarrow -5 \leq -\frac{x}{2} < -2 \Leftrightarrow$ (multiply by -2) $10 \geq x > 4$. Thus, for $4 < x \leq 10$, $(f \circ g)(x) = 3$.

(c_3) If $y < -2$, then $z = f(y) = -2y-1 = -2(-\frac{x}{2}+3)-1 = x-7$. Also, $y < -1 \Leftrightarrow -\frac{x}{2}+3 < -2 \Leftrightarrow -\frac{x}{2} < -5 \Leftrightarrow x > 10$. Therefore, for $x > 10$ we have $(f \circ g)(x) = x-7$.

Conclusion:

$$\begin{aligned}
(f \circ g)(x) &= -x+7, && \text{if } 3 \leq x \leq 4; \text{ and}\\
(f \circ g)(x) &= 3, && \text{if } 4 < x \leq 10; \text{ and}\\
(f \circ g)(x) &= x-7, && \text{if } 10 < x.
\end{aligned}$$

Putting everything together we obtain the following piecewise description of the composite function $f \circ g$:

$$(f \circ g)(x) = \begin{cases} -x+5, & \text{if } x < 0 \\ -3x+5, & \text{if } 0 \leq x < \frac{2}{3} \\ 3, & \text{if } \frac{2}{3} \leq x < \frac{8}{3} \\ 3x-5, & \text{if } \frac{8}{3} \leq x < 3 \\ -x+7, & \text{if } 3 \leq x \leq 4 \\ 3, & \text{if } 4 < x \leq 10 \\ x-7, & \text{if } x > 10 \end{cases}.$$

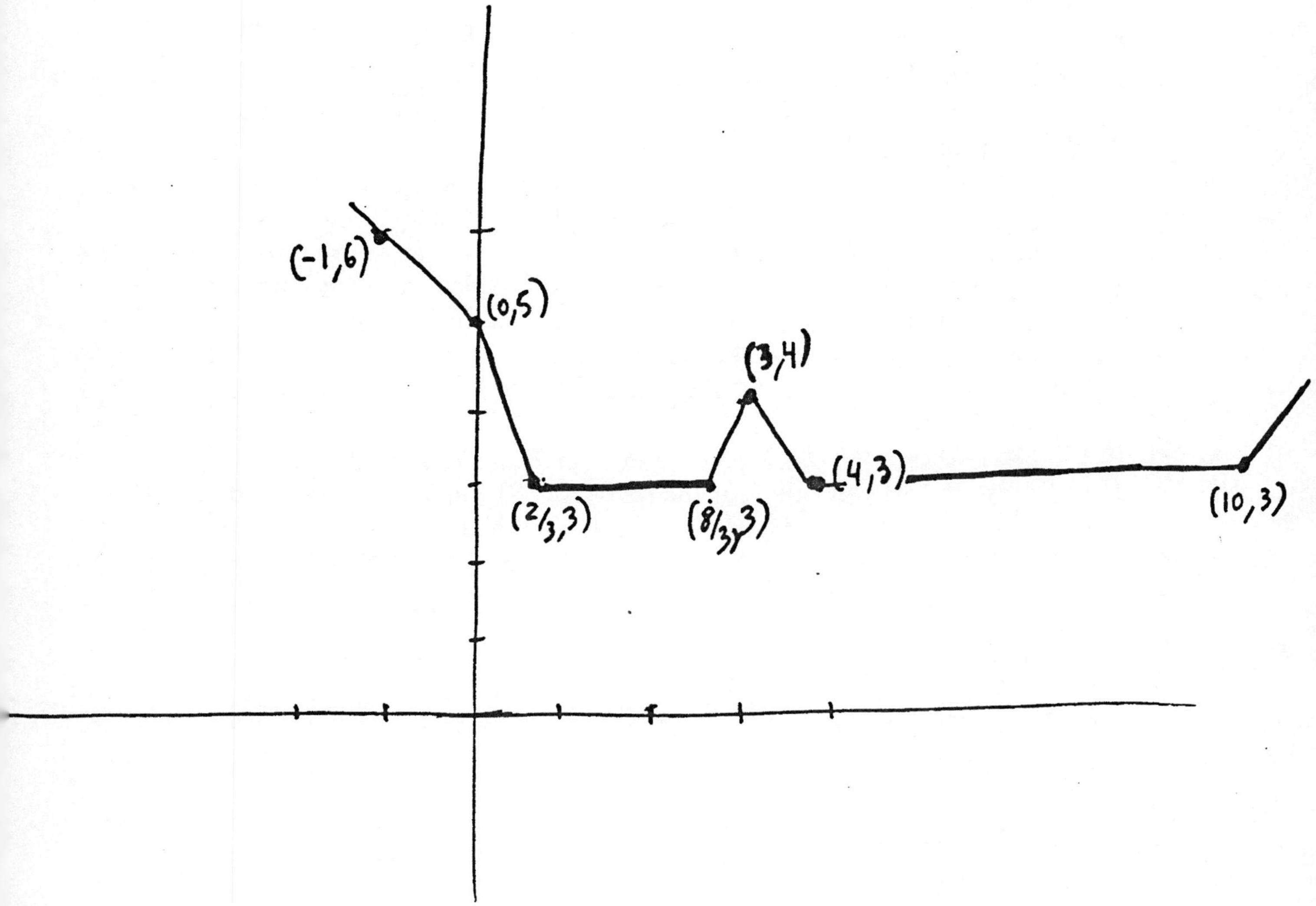

Figure 24: The graph of $f \circ g$

The range of $f \circ g$ is the set $R_{f \circ g} = [3, +\infty)$. Note the range of $f \circ g$ could easily be found from the graphs of f and g alone, before any formulas for $f \circ g$ have been obtained and before its graph is drawn. Indeed, if we go back, we know that the range of g is the set $(-\infty, \frac{3}{2}]$. So if $y \in (-\infty, \frac{3}{2}]$, by scanning the graph of f, we see that as y takes all possible values in $(-\infty, \frac{3}{2}]$; $z = f(y)$ takes all possible values in $[3, +\infty)$ (again look at the graph of f).

P34. Let A and B be non-empty sets. Prove the following statement: if f is a surjective mapping of A onto B, and S and subset of A, then $B - f(S)$ must be a subset of the set $f(A-S)$. Is the converse of this statement true? Prove it or disprove it by giving a counter-example or by explaining what can go wrong in the general situation.

Note: (1)]If T is a subset of A, then $f(T)$ stands for the image set of T under $f: f(T) = \{y \mid y \in B \text{ and } y = f(x) \text{ for some } x \in A\}$.

(2) $A - T$ stands for the **difference** set, the set that contains those elements of A, which do not belong to T.

Solution: Suppose that $y \in B - f(S)$, then y is an element of B but not of $f(S)$, meaning that there is no element s in S such that $f(s) = y$. Nevertheless, since $y \in B$ and f is a surjective mapping from A onto B, there must exist an element $a \in A$ such that $f(a) = y$. But, the element a, for reasons already explained, is not a member of S. Thus, since $a \in A$ and $a \notin S$, it follows by definition that a is an element of the difference set $A - S$; $a \in A - S$ and since $f(a) \in f(A-S)$; $y \in f(A-S)$. We have proved, $(\forall y \in B - f(S)) \Rightarrow y \in f(A-S)$. Hence, the set $B - f(S)$ is a subset of the set $f(A-S)$. We are done. ∎

The converse statement is also true. If for a mapping f, and for every subset S of A, the set $B - f(S)$ is a subset of the set $f(A-S)$, then f must be surjective. Indeed, just take $S = \emptyset$, the empty set: then both S and $f(S)$ are the empty set. Hence $B - f(S) = B$ and $f(A-S) = f(A)$. Thus, according to the given condition we have,

$$B \subset f(A).$$

But by definition, $f(A) \subset B$, since $f(A)$ is none other than the range of f. Hence $B \subset f(A)$ and $f(A) \subset B$, which implies $f(A) = B$: f is surjective. End of proof. ∎

P35. Let a, b, c, d, fixed positive integers and the function $f : \mathbb{Q}^+ \cup \{0\} \to \mathbb{Q}^+ \cup \{0\}$; $f(x) = \frac{ax+b}{cx+d}$. This function is a mapping from the set of the nonnegative rational numbers to itself, since, obviously, its domain is the entire set $\mathbb{Q}^+ \cup \{0\}$. The set $\mathbb{Q}^+ \cup \{0\}$ is totally ordered under the usual order "$\leq$" of the real numbers. Let R_f be the range of the function f. Show the following:

(i) If $ad - bc < 0$, show that both the supremum and maximum are equal to $\frac{b}{d}$, the infimum is $\frac{a}{c}$, and the minimum does not exist.

(ii) If $ad - bc = 0$, show that all four, infimum, minimum, supremum, and maximum, are equal to the rational number $\frac{a}{c}$ because in this case the function f is constant.

(iii) If the integer $ad - bc$ is positive, the the infimum of the set R_f is equal to $\frac{b}{d}$; the minimum is also $\frac{b}{d}$, the supremum is equal to $\frac{a}{c}$, and the maximum of R_f does not exist.

Solution: Note that since all a, b, c, d are positive, that means that a and c are, in particular, positive (and hence non-zero). We can write

$$\begin{aligned} f(x) &= \frac{ax+b}{cx+d} = \frac{a}{c}\cdot\frac{c}{a}\cdot\left(\frac{ax+b}{cx+d}\right) = \frac{a}{c}\cdot\left(\frac{cax+cb}{acx+ad}\right) \\ &= \frac{a}{c}\cdot\left[\frac{cax+ad+cb-ad}{acx+ad}\right] = \frac{a}{c}\cdot\left[\frac{cax+ad}{acx+ad}+\frac{cb-ad}{acx+ad}\right] \\ &= \frac{a}{c}\cdot\left[1-\frac{(ad-bc)}{acx+ad}\right] \end{aligned}$$

thus $f(x) = \frac{a}{c} - \frac{a}{c}\cdot\frac{(ad-bc)}{acx+ad} \Rightarrow f(x) = \frac{a}{c} - \frac{a}{c}\cdot\frac{(ad-bc)}{a\cdot(cx+d)}$; $f(x) = \frac{a}{c} - \frac{(ad-bc)}{c\cdot(cx+d)}$. The last formula or expression is valid for every $x \in \mathbb{Q}^+ \cup \{0\}$, since the denominator can never vanish, in virtue of the fact that c and d are both positive.

(i) If $ad - bc < 0$, then $-(ad - bc) > 0$, and the fraction $\frac{-(ad-bc)}{c\cdot(cx+d)}$ will always be a positive rational number, for every value of $x \in \mathbb{Q}^+ \cup \{0\}$. Furthermore, whenever $x_1, x_2 \in \mathbb{Q}^+ \cup \{0\}$ and with $x_1 < x_2$, we must have $\frac{-(ad-bc)}{c\cdot(cx_2+d)} < \frac{-(ad-bc)}{c\cdot(cx_1+d)}$. In other words, the fraction $\frac{-(ad-bc)}{c\cdot(cx+d)}$ decreases, as $x \in \mathbb{Q}^+ \cup \{0\}$ increases in value. Hence the given function, $f(x) = \frac{a}{c} - \frac{(ad-bc)}{c\cdot(cx+d)}$ is a decreasing function over its domain $\mathbb{Q}^+ \cup \{0\}$. Thus, the smallest value in the domain, namely 0, will produce the highest value in the range: $f(0) = \frac{b}{d}$ is both the maximum and supremum of the range R_f. On the other hand, for sufficiently large values of x, the fraction $\frac{-(ad-bc)}{c\cdot(cx+d)}$ gets arbitrarily close to zero in value. In the language of calculus we specifically have $\lim_{x\to+\infty} f(x) = \lim_{x\to+\infty}\left[\frac{a}{c} - \frac{(ad-bc)}{c\cdot(cx+d)}\right] = \frac{a}{c} - 0 = \frac{a}{c}$. We see that the infimum of the range R_f is indeed $\frac{a}{c}$. There is no minimum element in R_f, since this value $\frac{a}{c}$, is never attained by the function f for any value of x in $\mathbb{Q}^+ \cup \{0\}$.

(ii) Obviously, if $ad - bc = 0$; $f(x) = \frac{a}{c}$, $\forall x \in \mathbb{Q}^+ \cup \{0\}$. And so, trivially, all four: $\inf R_f$, $\min R_f$, $\sup R_f$, and $\max R_f$ are equal to $\frac{a}{c}$.

(iii) Because $ad - bc > 0$; $-(ad - bc) < 0$, the roles are reversed (see part (i)). The details are left to the reader. ■

Note: The function $g : \mathbb{R}^+ \cup \{0\} \to \mathbb{R}^+ \cup \{0\}$; (from the set of nonnegative reals to itself) $g(x) = \frac{ax+b}{cx+d}$, is an extension of the function f to the set $\mathbb{R}^+ \cup \{0\}$. From calculus, we are familiar with the graph of such a positive function (the graph is confined in the first quadrant and y-axis of the standard xy coordinate system.

(1)$ad - bc < 0$, which implies, since a, d, b, c are all positive, $\frac{a}{c} < \frac{b}{d}$

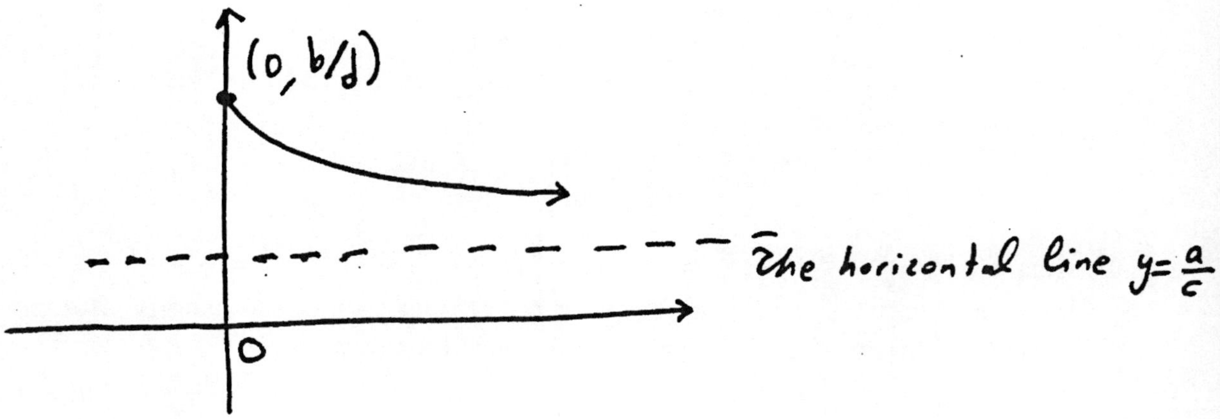

(2) $ad - bc = 0$ $\frac{a}{c} = \frac{b}{d} = r$

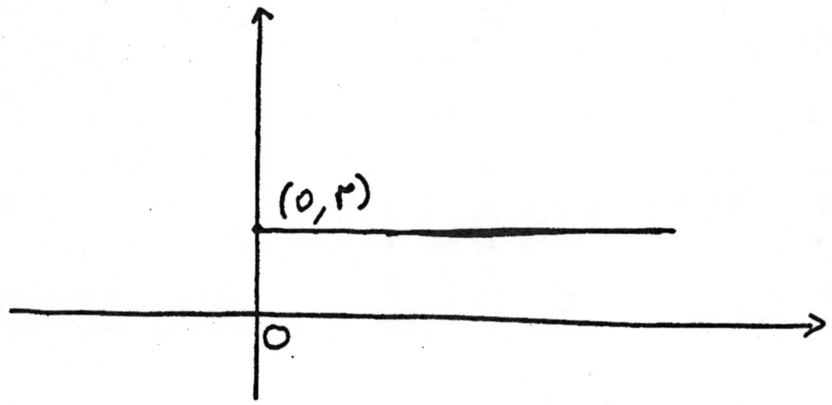

(3) $ad - bc > 0$; $\frac{a}{c} > \frac{b}{d}$

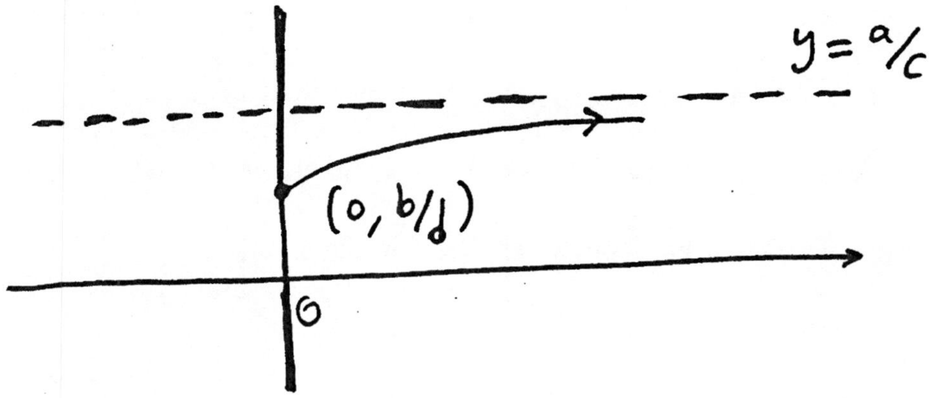

Figure 25

In all three cases, the horizontal line $y = \frac{a}{c}$, is a horizontal asymptote for the graph of the function $g : \mathbb{R}^+ \cup \{0\} \to \mathbb{R}^+ \cup \{0\}$; $g(x) = \frac{ax+b}{cx+d}$.

1.8 Exercise Set

1.(a) When is a relation R in a non-empty set A not reflexive?

(b) When is a relation R in a non-empty set A not antireflexive?

(c) When is a relation R in a non-empty set A not symmetric?

(d) When is a relation R in a non-empty set A not antisymmetric?

(e) When is a relation R in a non-empty set A not transitive?

2. Let $A = \{1, 2, 3, 4, 6, 9, 12\}$ and the two conditions, $p_1(x, y) = x \in A,\ y \in A,$ and $x = \frac{1}{3} \cdot y$; $p_2(x, y) = x \in A,\ y \in A,$ and $x = 3y$. Sketch the two Venn diagrams of the two relations R_1 and R_2 that the given conditions define.

3.

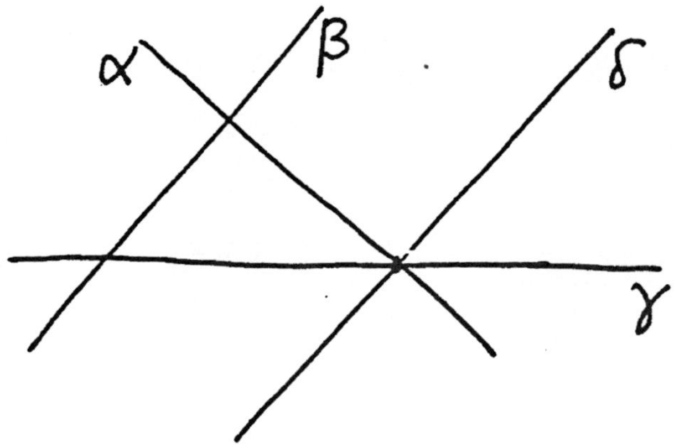

Figure 26

Let A be the set, $A = \{\alpha, \beta, \gamma, \delta\}$, where $\alpha, \beta, \gamma,$ and δ are the four straight lines in Figure 23, under the assumption that δ is parallel to a. Consider $p_1(x, y) = x \in A,\ y \in A$, and either x is parallel to y or x is coincident to y; and $p_2(x, y) = x \in A,\ y \in A$, and neither x is parallel to y nor is it coincident to y. Draw the two Venn diagrams of the two relations R_1 and R_2 that the two conditions define.

4. Let $A = \{2, 6, 8, 10, 12\}$ and $B = \{5, 9, 10, 11, 15\}$ and the condition $p(x, y) = x \in A\ y \in A$, and $y = x + 3$. This condition defines a relation R from A to B. Find the relation R by listing all the pairs that are members of R. Also find the inverse relation R^{-1} by listing its members and give a condition $p^{-1}(x, y)$ that defines R^{-1}.

5. Below your are given four Venn diagrams, each representing a relation R_i in a set A_i; $i = 1, 2, 3, 4$. Which of these relations are reflexive and which are symmetric?

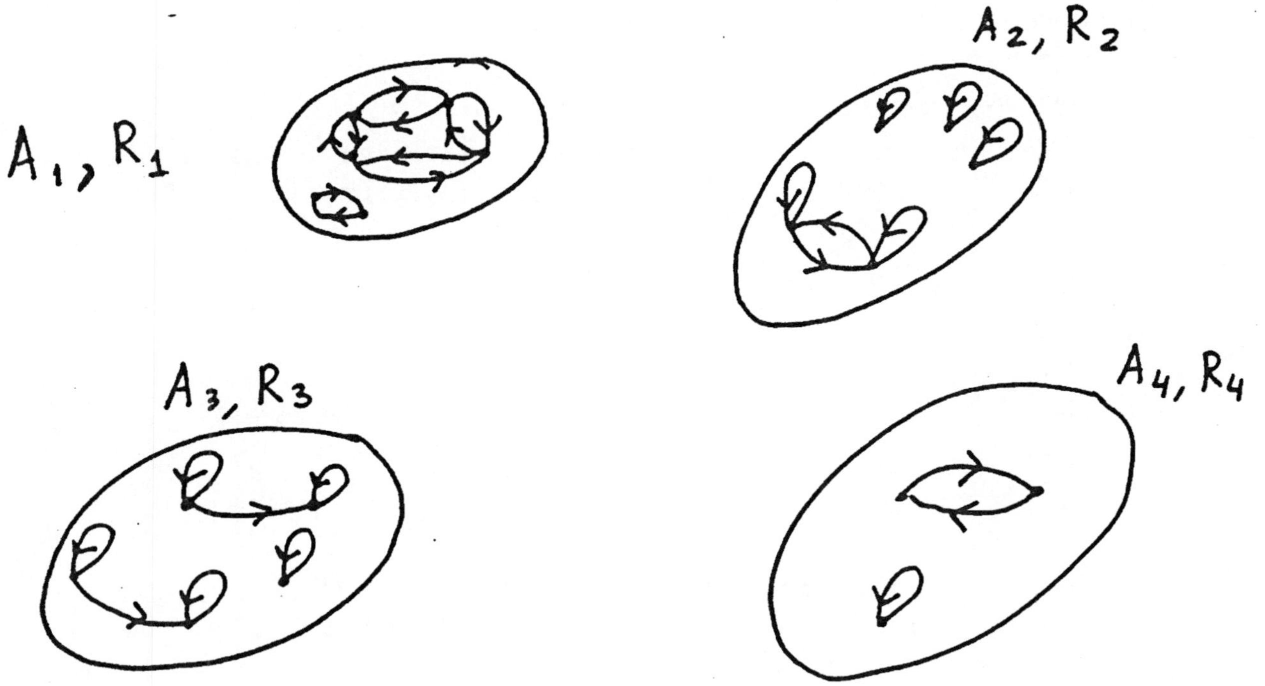

Figure 27

6. The identity relation or transformation I_A in a non-empty set A, is a subset of every reflexive relation R in A. Explain why.

7. Let $A = \{0, 1, 2, 3, 4, 5, 6, 7, 8, 9\}$ and $p(x, y) = x \in A,\ y \in A$, and such that $x + y - xy = 1$. Is the relation R defined by the give condition symmetric? List the member pairs of R. Note that all pairs of the form $(1, n)$ and $(n, 1)$, where n is a nonnegative integer between 0 and 9, are members of R. Are there any other pairs?

8. List all the member pairs of R, a relation defined in the set $A = \{0, 1, 2, 3, 4\}$, by the condition $p(x, y) = x \in A,\ y \in A$, and such that $y = x + 2$, $y = x$, or $y = x - 2$. If R a symmetric relation?

9. A student reasons that if a non-empty relation R (in a non-empty set A) is both symmetric and transitive, it must also be reflexive by using the following

argument. Let $(x, y) \in R$; since R is symmetric, (y, x) is also a member of R. By transitivity, the student argues that

$$((x, y) \in R \text{ and } (y, x) \in R) \Rightarrow (x, x) \in R.$$

The student concludes that R must also be reflexive. However, there is a subtle flaw in the student's reasoning. Find that flaw. What condition must be added to the assumption that R is both symmetric and transitive, to be able to draw the conclusion that R must also be reflexive?

Draw Venn diagrams, if necessary, to illustrate a case where R is symmetric, transitive, but not reflexive; and a case in which R is symmetric, transitive and reflexive.

10. Let R_1 and R_2 be two relations in a non-empty set A. Decide whether the given statement below is true or false. Justify your answer.

(a) If at least one of R_1, R_2 is reflexive, then the relation $R_1 \cup R_2$ must be reflexive.

(b) If both R_1 and R_2 are reflexive, then the relation $R_1 \cap R_2$ is reflexive.

11. A relation R is a non-empty set A, maybe reflexive, antireflexive, symmetric, antisymmetric, transitive, or none of the above. In each case below, give an example by using a Venn diagram, if necessary, of a relation R in a set A that satisfies the given condition.

(a) R is only reflexive and symmetric.

(b) R is only antireflexive and symmetric.

(c) R is only reflexive and antisymmetric.

(d) R is only antireflexive and antisymmetric.

(e) R is only reflexive and transitive.

(f) R is only antireflexive and transitive.

(g) R is only symmetric and transitive.

(h) R is only antisymmetric and transitive.

(i) R is reflexive, symmetric and transitive.

(j) R is reflexive, symmetric, and antisymmetric.

(k) R is reflexive, antisymmetric, and transitive.

(l) R is antireflexive, antisymmetric, and transitive.

(m) R is antireflexive only.

(n) R is none of the above.

12. Consider the relations R_1 and R_2 in the set of real numbers $\mathbb{R}$:

$$R_1 = \{(x,y)) \in \mathbb{R} \times \mathbb{R} : 2x - 3y = 1\}, \ R_2 = \{(x,y) \in \mathbb{R} \times \mathbb{R} : 3x + 2y = 2\}$$

Find the relation $R_1 \cap R_2$. Is $R_1 \cap R_2$ an antireflexive, antisymmetric, and transitive relation?

13. Let $A = \mathbb{Z}$, and $p(x,y) = x \in A,\ y \in A$, and such that $x - y$ is a multiple of 3. Prove that the relation R that this condition defines is an equivalence relation. Give the equivalence classes of R.

14. Let $P(A)$ be the power set, that is, the set of all subsets of the set $A = \{a,b,c\}$, and $p(x,y) = x \in P(A),\ y \in P(A)$ and with $x \cap \{a\} = y \cap \{a\}$. Show that the relation R that this condition defines is an equivalence relation, and draw a Venn diagram for R.

15. Show that if R_1 and R_2 are equivalence relations in a non-empty set A, then the relation $R_1 \cap R_2$ is also an equivalence relation in R.

16. If $\mathbb{R}^3$, the three-dimensional Euclidean space ($\mathbb{R}^3$ is the set of all triples (x,y,z): $x, y, z \in \mathbb{R}$), and E the set of all points of a specific plane in $\mathbb{R}^3$, E^c the complement of E with respect to $\mathbb{R}$ that is, $E^c = \mathbb{R}^3 - E$, consider the relation $R = \{(x,y) \in E^c \times E^c :$ the line segment xy has no points in common with $E\}$. Show that R is an equivalence relation in E, and find the equivalence classes of R.

17. Let $\mathbb{R}^2$ be the dimensional Euclidean space ($\mathbb{R}^2$ is the set of all pairs $(x,y) : x \in \mathbb{R},\ y \in \mathbb{R}$), F a fixed point of $\mathbb{R}^2$, and $A = \{F\}^c = \mathbb{R}^2 - \{F\}$, the complement of F with respect to $\mathbb{R}^2$: the set of all points in $\mathbb{R}^2$ except F. The condition $p(x,y) = x \in A,\ y \in A$, and the straight line through x and y contains the point F, defines a relation R in the set A. Show that R is an equivalence relation and describe the equivalence classes of R.

18. Consider the set $U = \{A,B,C,D,E\}$, where A,B,C,D, and E are sets, as illustrated in the Venn diagram below.

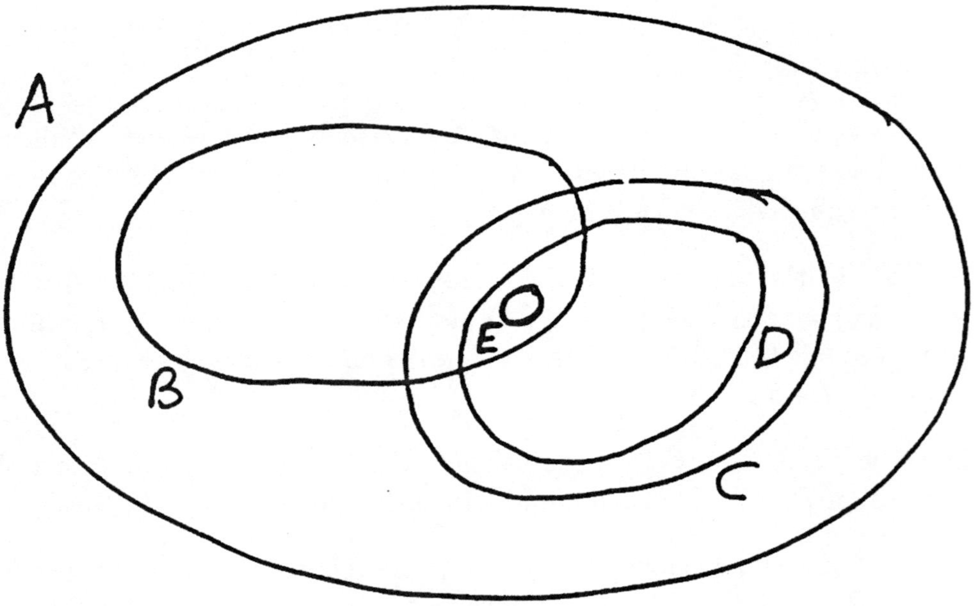

Figure 28

Let $p(x, y) = x \in U$, $y \in U$, and $x \subset y$; and let R be the equivalence relation that this condition defines. Show that R is a partial ordering in the wide sense. Is R actually a total ordering in the wide sense? Justify your answer. Draw a Venn diagram for R.

19. Let A be the set of all fractions in lowest terms, between 0 and 1: $A = \{\frac{m}{n} \mid m, n \in \mathbb{Z}^+, \frac{m}{n} < 1$, and the greatest common divisor of m and n is $1\}$. Consider the relation "$\preceq$" which is defined as follows:

$$\frac{m_1}{n_1} \preceq \frac{m_2}{n_2} \Leftrightarrow [(m_1 + n_1 < m_2 + n_2) \text{ or } (m_1 + n_1 = m_2 + n_2 \text{ and } m_1 \leq m_2)].$$

(i) Show that this relation is a total order in the wide sense in the set A.

(ii) Find all the fraction members $\frac{m}{n}$ of A, such that $m + n = 20$, then order these fractions according to the given ordering "$\preceq$".

20. Let A, B non-empty sets, and suppose that "$\preceq$" is an order in the set A; and "$\trianglelefteq$" is an order in the set B. Consider the condition $p(x, y)$, pertaining to the cartesian product $A \times B$, defined as follows:

$$p(x, y) : x = (a_1, b_1) \in A \times B,\ y = (a_2, b_2) \in A \times B,$$

and such that

$$a_1 \preceq a_2 \text{ and } b_1 \overset{\prec}{=} b_2.$$

(i) If both orderings $\preceq$ and $\overset{\prec}{=}$ are partial orderings in the narrow sense, so is the relation R.

(ii) If both orders $\preceq$ and $\overset{\prec}{=}$ are partial orders in the wide sense, so is the relation R.

(iii) If both orderings $\preceq$ and $\overset{\prec}{=}$ are total orderings in the narrow(wide) sense, so is the relation R.

21. Prove that if R_1 and R_2 are two partial orderings in the wide sense in a non-empty set A, so is the relation $R_1 \cap R_2$. Conclude that the same is true for R_1 and R_2 being total orders in the wide sense. Moreover, show that if at least one of R_1, R_2 is a partial ordering in the narrow sense in A, then $R_1 \cap R_2$ is a partial ordering in the narrow sense. Conclude the same for total orders in the narrow sense, where at least one of R_1, R_2 is a total order in the narrow sense.

22. Let $A = \{1, 2, 3, 4, 5, 6, 7, 8, 9, 10, 11, 12\}$ and consider the relations

$$R_1 \;:\; (x, y) \in A \times A, \text{ and with } y = 2x$$

$$R_2 \;:\; (x, y) \in A \times A, \text{ and with } y = 3x$$

(i) List all the element pairs of R_1, and list all the pairs that are members of R_2.

(ii) Show that both R_1 and R_2 are antireflexive antisymmetric, but not transitive relations, thus they fail to be partial orderings in the narrow sense.

23. Let A, B be sets, each consisting of persons. Which of the given conditions below define functions and which define relations that may not be functions? Explain.

(i) $p_1(x, y) = x \in A$ has father $y \in B$.

(ii) $p_2(x, y) = x \in A$ is the mother of $y \in B$.

(iii) $p_3(x, y) = x \in A$ has grandfather from mother $y \in B$.

(iv) $p_4(x, y) = x \in A$ has brother $y \in B$.

24. Let $A = \{0, 2, 4, 6, 8, 10, 12\}$, $B = \{0, 3, 6, 9, 12, 15, 18, 21, 24\}$, $p(x, y) = x \in A,\ y \in B$, and $y = 3x$, R the relation from A to B, that this condition defines. Is R a mapping from A to B? Give the domain D_R. Sketch a Venn diagram for R and find the inverse relation R^{-1} by listing its member pairs.

25. Find the domain of each of the following functions:

(i) $f : \mathbb{R} \to \mathbb{R};\ f(x) = \frac{3}{x^2-1} + \frac{1}{x+2}$

(ii) $g : \mathbb{R} \to \mathbb{R};\ f(x) = \frac{x^2}{\sqrt{(x-1)\cdot(x+3)}}$.

(iii) $h : \mathbb{R} \to \mathbb{R};\ h(x) = \sqrt{c^2 - x^2}$, c a positive constant.

26. Let $f : \mathbb{Z}^+ \to \mathbb{Z}^+;\ f(x) = k \cdot x$, k a fixed positive integer, be a mapping from $\mathbb{Z}^+$ to itself.

 (i) Is f injective? Explain.
 (ii) Is f surjective? Explain.

 Is there a value of k that makes f a permutation of $\mathbb{Z}^+$?

27. Show that if m and b are fixed real numbers, m nonzero, then the function $\varphi : \mathbb{R} \to \mathbb{R}$, $\varphi(x) = mx + b$ is an injective and surjective mapping onto itself. In other words, φ is a bijection of $\mathbb{R}$ onto itself; a permutation of $\mathbb{R}$.

28. Let φ be the function of Exercise 27.

 (i) Find the inverse function φ^{-1}.
 (ii) For which values of the constants m and b does φ equal φ^{-1}? Sketch a graph in that case. What do you observe?

29. List all the transformations from the set $S = \{a, b\}$ to itself; that is, all mappings from S to itself.

30. Prove that if A is a finite set with elements, $A = \{a_1, a_2, \ldots, a_n\}$ and f a mapping from A to itself; prove that the following three statements are equivalen:

 (i) f is surjective.
 (ii) f is injective.
 (iii) f is a bijection.

31. Let $f : \mathbb{Z}^+ \to \mathbb{Z}^+;\ f(x) = (x-3)^2$. Describe all the pairs (x_1, x_2), with $x_1, x_2 \in \mathbb{Z}^+$, and $x_1 \neq x_2$, such that $f(x_1) = f(x_2)$.

32. The condition $y = \frac{x^2}{x+2}$ defines a mapping f of the set (open interval) $A = (-2, +\infty)$ to the set of real numbers $\mathbb{R}$; and a function g from $\mathbb{Z}$ to $\mathbb{R}$. Give both domains D_f and D_g. Why is g not a mapping from $\mathbb{Z}$ to $\mathbb{R}$?

33. According to Theorem 6, the function $f : A \to \mathbb{Z}$, where $A = \{x \in \mathbb{Z} :\mid x \mid\leq 5\}$, defined by $f(x) = x^4 - x^2$, gives rise to the equivalence relation $R = \{(x, y) \mid f(x) = f(y)\}$. Find R by listing its member pairs, and determine the equivalence classes of R.

34. Show that if f is a mapping from A to B, then:

(i) f is surjective if, and only if, $\forall y \in B$, the equation $y = f(x)$ has at least one solution.

(ii) f is injective, if and only if, $\forall y \in B$, the equation $y = f(x)$ has at most one solution.

(iii) f is a bijection if, and only if, $\forall y \in B$, the equation $y = f(x)$ has exactly one solution.

35. Show that the function $f : \mathbb{R} \to \mathbb{R}$; $f(x) = x \cdot \sqrt{|\, x\,|}$, is a bijection from $\mathbb{R}$ to itself; that is a permutation of $\mathbb{R}$.

36. Let $f : \mathbb{Z} \to \mathbb{Z}$, be a function piecewise defined,

$$f(x) = \begin{cases} \frac{x}{2}, & \text{if } x \text{ is even} \\ \frac{x+1}{2}, & \text{if } x \text{ is odd.} \end{cases}$$

(i) Is f injective? Justify your answer.

(ii) Is f surjective? Justify your answer.

37. Consider the function $f : \mathbb{R} \to \mathbb{R}$; $f(x) =|\, x^2 - 1\,|$. Find the restriction g of the function f to the set $[-1, 1]$ (closed interval). Give an absolute value-free formula that describes g.

38. The Venn diagram below describes a relation R in the set $A = \{1,2,3,4,5,6,7,8,9,10\}$. As it can easily be seen from this diagram, R is reflexive, antisymmetric, and transitive. Hence, it is a partial order in the set A, in the wide sense. It is not a total order though, since there are pairs $(x, y) \in A \times A$ such that neither (x, y) nor (y, x) is a member of R. Identify, all the maximal and minimal elements in A, under the given partial order R.

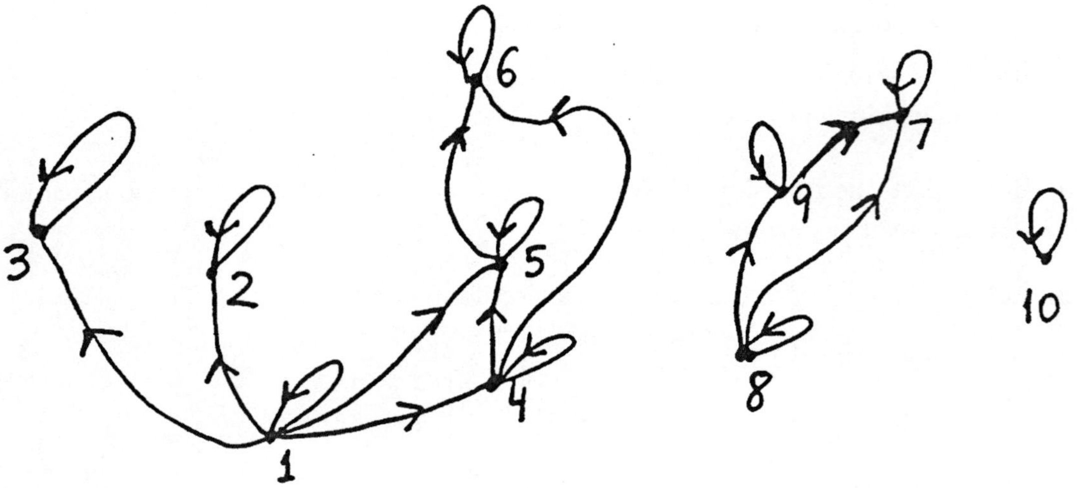

Figure 29

39. Consider the sets $A = \{-4,-3,-2,-1,0,1,2,3,4\}$ and $B = \{-1,0,2,3,9,14\}$ and $f : A \to B$, be the mapping from A to B defined by

$$f(x) = \begin{cases} x^2, & \text{if } x \text{ is of the form, } x = 3k, k \in \mathbb{Z} \\ x^2 - 1, & \text{if } x \text{ is of the form, } x = 3k+1, k \in \mathbb{Z} \\ x^2 - 2, & \text{if } x \text{ is of the form, } x = 3k+2, k \in \mathbb{Z} \end{cases}$$

(i) If f injective? Justify your answer.

(ii) Is f surjective? Justify your answer.

(iii) If the notation $f(S)$ stands for the image of set S under the mapping f, find the following images: $f(A), f(S), f(A) - f(S)$, and $f(S - A)$, where S is the set, $S = \{-1, 0, 1, 2\}$ and the notation $C - D$ stands for the difference sets. It consists of the elements that are in C but not in D.

40. If f is a mapping from the non-empty set A to the non-empty set B, prove the following:

(i) $f(A_1 \cap A_2) \subset f(A_1) \cap f(A_2)$, for any subsets A_1, A_2 of A.

(ii) Show by offering a counterexample or by explaining the general situation that the equality $f(A_1 \cap A_2) = f(A_1) \cap f(A_2)$ does not hold for every mapping f from A to B.

(iii) Conclude from (i) and (ii) that the statement,

$$f(A_1) \cap f(A_2) \subset f(A_1 \cap A_2),$$

does not hold true for every mapping f from A to B.

(iv) Show that if f is injective, then it is true that $f(A_1) \cap f(A_2) \subset f(A_1 \cap A_2)$ (and consequently by (ii) that $f(A_1) \cap f(A_2) = f(A_1 \cap A_2)$).

(v) Find the necessary and sufficient condition that a mapping from f from A to B must satisfy in order that,

$$f(A_1) \cap f(A_2) = f(A_1 \cap A_2)$$

Hint for (iii): It is possible to have a mapping $f : A \to B$ such that $f(a_1) = f(a_2)$, where $a_1 \in A_1 - A_2$ and $a_2 \in A_2 - A_1$.

41. If f is a mapping from a nonempty set A to a nonempty set B prove the following:

(i) For any subsets A_1, A_2 of A, $f(A_1) - f(A_2) \subset f(A_1 - A_2)$.

(ii) Show by offering a counterexample, or by explaining the general situation that the equality $f(A_1) - f(A_2) = f(A_1 - A_2)$.

(iii) Conclude that the statement $f(A_1 - A_2) \subset f(A_1) - f(A_2)$ fails for some mappings f.

(iv) Show that if f is injective, then $f(A_1 - A_2) = f(A_1) - f(A_2)$.

(v) Find the necessary and sufficient condition that a mapping f from A to B must satisfy in order that $f(A_1 - A_2) = f(A_1) - f(A_2)$.

42. Let f be a mapping from a nonempty set A to a nonempty set B. Show each of the following:

(i) $f(A) - f(S) \subset f(A - S)$, for every subset S of A.

(ii) If f is surjective, then $B - f(S) \subset f(A - S)$, for every subset S of A.

(iii) The equality $B - f(S) = f(A - S)$, $\forall S$ subset of A, holds true if, and only if, f is a bijection from the set A onto the set B.

Hint: Use Exercises 40 and 41.

43. Let U; $U \neq \emptyset$, be a reference set, a universe if you like, and S a subset of U. We define the characteristic function of the set S as follows:

$$x_x(x) = \begin{cases} 1, & \text{if } x \in S \\ 0, & \text{if } x \in S \end{cases}$$

Prove the following:

(i) $A = B \Leftrightarrow \forall x \in U : X_A(x) = X_B(x)$

(ii) If $A^c = U - A$ is the complement of A with respect to U, then $X_A(x) + X_{A^c}(x) = 1, \ \forall x \in U$.

(iii) The functions $X_{A \cap B}$ and $X_A \cdot X_B$ are equal.

(iv) $X_{A \cup B} = X_A + X_B - X_A \cdot X_B$.

44. Use Exercise 43 to show that if A, B, and C are nonempty sets of a universe $U \neq \emptyset$, then if both equalities $A \cap B = A \cap C$ and $A \cup B = A \cup C$ hold true, the sets B and C must be equal.

45. In a set A, consisting of persons, the functions f, g, and h are defined by the following conditions:

$f : x$ has father y;

$g : x$ has mother y;

and $h : x$ has brother y

(It is assumed that any person in A, can have at most one brother in A)

Verbally describe the conditions that define the following composite functions:

$$f \circ g, \ g \circ f, \ f \circ f, \ g \circ g, \text{ and } h \circ g.$$

Question: Is it true that $h \circ h$ must be a subset of the diagonal $D_A = \{(x, x) \mid s \in A\}$? The assumptions here (as stated above) is that every person in A has at most one brother in A.

46. Let A be a set of persons and R_1, R_2 be two relations in A defined by the conditions $p_1(x, y)$ and $p_2(x, y)$ respectively:

$p_1(x, y) = x \in A, \ y \in A$, and x is a son of y.

$p_2(x, y) = x \in A, \ y \in A$, and x is a brother of y.

Verbally describe the conditions that define the inverse relations R_1^{-1} and R_2^{-1}. Furthermore, verbally describe the conditions that define the composite relations $R_1 \circ R_2, \ R_2 \circ R_1, \ R_1^{-1} \circ R_2^{-1}, \ R_2^{-1} \circ R_1^{-1}, \ R_1^{-1} \circ R_2, \ R_2 \circ R_1^{-1}$.

47. Consider the mapping $f : \mathbb{Z}^+ \cup \{0\} \to \mathbb{Z}^+ \cup \{0\}$, from the set to the nonnegative integers to itself defined as,

$$f(x) = \begin{cases} x + 1, & \text{if } x \text{ is even} \\ \\ x - 1, & \text{if } x \text{ is odd} \end{cases}$$

Prove that f is a permutation of the set $\mathbb{Z}^+ \cup \{0\}$ and find the composite mapping $f \circ f$. What do you observe? Is it true that f is equal to its inverse function f^{-1}?

48. Consider the mappings $g : \mathbb{R} \to \mathbb{R}$; $g(x) = 7x - 2$, and $h : \mathbb{R} \to \mathbb{R}$; $h(x) = 2x + 3$. Show that these two mappings are permutations of the set $\mathbb{R}$. Then, so must be the inverse functions g^{-1} and h^{-1}. Find formulas for g^{-1} and h^{-1}. Use the formulas so that you can also find formulas for the composite permutations $g \circ h$, $h \circ g$, $g^{-1} \circ h^{-1}, g \circ h^{-1}$, and $g^{-1} \circ h$. Verify that $(g \circ h)^{-1} = h^{-1} \circ g^{-1}$ and $(h \circ g)^{-1} = g^{-1} \circ h^{-1}$.

49. Consider the function $F : \mathbb{R} \to \mathbb{R}$, defined as $F(x) = \frac{2x-1}{x-3}$.

(i) Find the domain D_F of the function F.

(ii) Is F a mapping from the set $\mathbb{R}$ to itself? Explain.

(iii) Find the range R_F of the function and show that F is a bijection from D_F onto R_F. Verify that,

$F^{-1} \circ F = I_D$ (identity mapping D_F) and

$F \circ F^{-1} = I_{R_F}$ (identity mapping on R_F)

50. Consider the two mappings,

$\varphi : \mathbb{R} - \{\frac{1}{2}\} \to \mathbb{R}$; $\varphi(x) = \frac{1}{2x-1}$; and,

$f : \mathbb{R} \to \mathbb{R}$; $f(x) = \frac{x}{2} + 1$

(i) Find the ranges R_φ and R_f and show that both φ and f are injective; but only f is surjective (and thus a bijection).

(ii) Find formulas for the composite functions $\varphi \circ f$ and $f \circ \varphi$, both injective mappings (since each of them is the composite of two injective mappings). Also find the values $(f \circ \varphi)(1)$ and $(\varphi \circ f)(4)$.

(iii) Give the domains $D_{\varphi \circ f}$, and $D_{f \circ \varphi}$ the ranges $R_{\varphi \circ f}$, $R_{f \circ \varphi}$.

(iv) Find $(f \circ \varphi)(\mathbb{R} - \{\frac{1}{2}\})$ that is, the image of of the set $\mathbb{R} - \{\frac{1}{2}\}$ under the composite mapping $f \circ \varphi$.

51. Let $f : \mathbb{R} \to \mathbb{R}$; $f(x) = 3x - 1$ and $g : \mathbb{R} \to \mathbb{R}$; $g(x) = x^2 - 1$. Find formulas for the composite functions $f \circ g$ and $g \circ f$; determine the domain and range of each.

52. Let $f : \mathbb{R} \to \mathbb{R}$; $f(x) = 6x + 2$, $g : \mathbb{R} \to \mathbb{R}$; $g(x) = x^2$, and $h : \mathbb{R} \to \mathbb{R}$; $h(x) = \sin(2x)$ (this is the sine function from calculus). Find formulas for the following:

(a) $f \circ g$

(b) $g \circ f$

(c) $f \circ h$

(d) $h \circ f$

(e) $g \circ h$

(f) $h \circ g$

(g) $(f \circ g) \circ (g \circ f)$

(h) $(g \circ h) \circ ((h \circ g)$

(i) $(f \circ h) \circ (h \circ f)$

(j) $(f \circ g) \circ h$

(k) $f \circ (g \circ h)$

(All of the eleven composites listed above are mappings from $\mathbb{R}$ to itself). Also verify that, $(f \circ g) \circ h = f \circ (g \circ h)$.

53. Let b be a fixed real number; r a fixed positive real. Consider the functions (both are mappings).
$F : \mathbb{R} \to \mathbb{R}$; $F(x) = -x + b$; and $G : [0, r] \to [0, r]$; $G(x) = \sqrt{r^2 - x^2}$

(i) Prove that both F and G are bijections (and so F is a permutation of $\mathbb{R}$ while G is a permutation of the interval $[0, r]$.

(ii) Show that $F = F^{-1}$ and $G = G^{-1}$; that is, each of these two functions, equals to its own inverse.

(iii) For $r = b = 1$, find a formula for the composite function $G \circ F$; give its domain and range.

54. Consider the sets,
$A = \{1, 2, 3, 4, 5\}$, $B = \{\alpha, \beta, \gamma, \delta\}$, $C = \{x, y, z, w, \varphi, t\}$; R a relation fromt he set A to the set B; and S a relation from the set B to the set C:

$$R = \{(1, \beta),\ (1, \gamma),\ (2, \alpha),\ (4, \gamma),\ (5, \alpha)\}$$

$$S = \{(\alpha, z),\ (\alpha, t),\ (\beta, w),\ (\gamma, x),\ (\gamma, z),\ (\delta, x)\}$$

Find the relations $S \circ R, R^{-1}, S^{-1}, R^{-1} \circ S^{-1}$, and verify that $(S \circ R)^{-1} = R^{-1} \circ S^{-1}$.

55. In the set of positive integers $\mathbb{Z}^+$, consider the relation R defined by the condition:

$$p(x, y) = x \in \mathbb{Z}^+,\ y \in \mathbb{Z}^+, \text{ and } 3x + 2y = 20$$

Find the relations R, R^{-1}, $R^{-1} \circ R$, and $R \circ R^{-1}$, by listing the member pairs of each.

56. Let $A = \{0, 1, 2, 3, 4, 5, 6, 7\}$ and consider the relation, $R = \{(x, y) \mid (x, y) \in A \times A \text{ and } (x = y \text{ or } x + y = 7)\}$.

(i) Find R by listing all of its elements.

(ii) Show, from the way R is defined, that R is an equivalence relation.

(iii) Find and list all the equivalence classes.

57. Show that a relation R, in a nonempty set A, is an equivalence relation if, and only if, the following three conditions are satisfied:

(a) $D_{A^2} \subset R$,

(b) $R^{-1} = R$,

(c) $R \circ R = R$.

(Here, the notation D_{A^2} has the usual meaning: it stands for the diagonal of the set A; that is the subset of the cartesian product $A \times A$, that consists of all pairs of form (x, x), $x \in A$).

58. In the set of integers $\mathbb{Z}$, we consider two relations defined respectively by the conditions:

$R_1 : p_1(x, y) = x \in \mathbb{Z},\ y \in \mathbb{Z}$, and $x - y$ is multiple of both 2 and 3.

$R_2 : p_2(x, y) = x \in \mathbb{Z},\ y \in \mathbb{Z}$, and $x - y$ is multiple of 2 or of 3.

(i) Is the relation R_1 an equivalence relation in $\mathbb{Z}$? If your answer is in the affirmative, describe the equivalence classes.

(ii) Is R_2 an equivalence relation? If not, which of the three requirements fails? It is true that R_2 is both reflexive and symmetric but not transitive? Explain.

59. Let $A = \mathbb{Z}^+ \times \mathbb{Z}^+$ and R a relation in A defined by

$$R = \{((x_1, y_1), (x_2, y_2)) \mid (x_1, y_1) \in \mathbb{Z}^+, (x_2, y_2) \in \mathbb{Z}^+, \text{ and } x_1 + y_2 = x_2 + y_1\}$$

Prove that R is an equivalence relation in R. How would you describe the equivalence classes in general? Describe the equivalence class of the element $(3, 4)$.

60. Consider the relation R in the set $A = \mathbb{Z}^+ \times \mathbb{Z}^+$, defined by the condition $p(t, w) = (t = (x_1, y_1) \in \mathbb{Z}^+,\ w = (x_2, y_2) \in \mathbb{Z}^+$, and $x_1 + y_1 = x_2 + y_2)$. Prove that R is an equivalence relation in the set A. Describe the equivalence class of the elemenbt $(3, 6)$.

61. In the set $A = \mathbb{Q} \times \mathbb{Q}$, consider the relation R defined as follows: $(x_1, y_1)R(x_2, y_2) \Leftrightarrow (x_1 \leq x_2 \text{ and } y_1 \leq y_2)$

(i) Is R a partial order int he wide sense? Explain.

(ii) Is R actually a total order in the wide sense? Explain.

62. In the set $A = \mathbb{Q}\times\mathbb{Q}$, consider the relation R defined by the following condition:

$p(t,w) = (t = (x_1, y_1) \in \mathbb{Q},\ w = (x_2, y_2) \in \mathbb{Q}$ and either $x_1 < x_2$; or otherwise $x_1 = x_2$ and $y_1 \leq y_2)$.

Prove that this relation is a total ordering in the wide sense.

63. Suppose R is a partial order (whether in the narrow or wide sense) in a nonempty set A. Show that R will be a total order if, and only if, $R \cup R^{-1} = A \times A$.

64. Let $A = \mathbb{N}_0 \times \mathbb{N}_0$, where $\mathbb{N}_0$ is the set of nonnegative integers. Consider the three conditions:

(a) $p_1(x,y) : x \in A,\ y \in A$, and $y = 2x - 6$.

(b) $p_2(x,y) : x \in A,\ y \in A$, and $y = 3x - 12$.

(c) $p_3(x,y) : x \in A,\ y \in A$, and $y = 5x - 25$.

Explain why the three relations R_1, R_2, and R_3, that these three conditions define, are not mappings from A to itself, but only functions from A to itself. Specify the domains D_{R_1}, D_{R_2}, and D_{R_3}.

65. (i) Consider the mapping $f : \mathbb{Z} \to \mathbb{Z};\ f(x) = 3x - 1$.
Is this mapping injective? Is it surjective? Is it both (thus a bijection)? Is it neither?

(ii) Next, consider the mapping $g : \mathbb{Q} \to \mathbb{Q};\ g(x) = 3x - 1$, which is simply an extension of the mapping f to the set $\mathbb{Q}$. Answer the same (four) questions.

66. Let $A = \{a, b\},\ B = \{b, c\},\ C = \{a, b, c\},\ D = \{b, c, d\}$

(i) Find all bijections from the set A onto the set B. Draw the corresponding Venn diagrams.

(ii) Find all bijections from the set C onto the set D. Draw the corresponding Venn diagrams.

67. In a set A consisting of persons, consider the relations:

R_1: female person x is the daugher of male person y.

R_2: male person y is the child of male person z (that is, male person z is the father of male person y).

R_3: male person z is the son of female person w (that is, female person w is the mother of male person z).

(i) Verbally describe the composite relation $R_3 \circ R_2 \circ R_1$

(ii) Verbally describe the relation $(R_3 \circ R_2 \circ R_1)^{-1}$

(iii) Verbally describe the composite relation $R_2 \circ R_1 \circ R_3$.

68. Let $A = \{1, 2, 3, 4, 5\}$ and R be the relation in A defined by the condition, $p(x, y) = x \in A,\ y \in A$, and $x > y$.

(i) Draw a Venn diagram for R.

(ii) Draw a Venn diagram for the composite $R \circ R$; also denoted by R^2.

(iii) Since the operation of composing relations is an associative one, we have $R \circ (R \circ R) = (R \circ R) \circ R$; also denoted by R^3. Verify this and draw a Venn diagram for R^3.

69. Consider the relation $R = \{(a, b), (a, c), (a, d), (b, c), (b, d), (c, d)\}$ in the set $A = \{a, b, c, d\}$. Find the relations R^{-1}, $R \circ R^{-1}$, $R^{-1} \circ R$, and $(R^{-1} \circ R)^{-1}$. Are the second and fourth of these relations equal according to a certain theorem? Which theorem is it?

70. Consider the function $f : \mathbb{R} \to \mathbb{R};\ f(x) = \begin{cases} x, & \text{if } x \text{ is rational} \\ 0, & \text{if } x \text{ is irrational} \end{cases}$. Is f injective, surjective, bijective, or none of the above? Justify your answer.

71. Let $f : \mathbb{R} \to \mathbb{R};\ f(x) = 7x + 3$, and $g : \mathbb{R} \to \mathbb{R};\ g(x) = -2x + 4$.

(i) Show that both f and g are bijections and find formulas for the inverse bijections f^{-1} and g^{-1}.

(ii) Use your answer in part (i) to find formulas for the following functions:

(a) $f \circ g$

(b) $g \circ f$

(c) $f^{-1} \circ g^{-1}$

(d) $g^{-1} \circ f^{-1}$

(e) $f^{-1} \circ g$

(f) $(f \circ g)^{-1}$

72. Of the two mappings $f : \mathbb{R} \to \mathbb{R};\ f(x) = 2x - 4$, and $g : \mathbb{R} \to \mathbb{R};\ g(x) = 3x^2 - 1$, only one is a bijection. Which one is it? Justify your answer.

73. Let $A = [0, +\infty) =$ set of nonnegative reals; $B = [-1, +\infty)$, and $C = [-6, +\infty)$. Prove that the function $f : A \to B;\ f(x) = 3x^2 - 1$, is a bijection from the set A onto B; also prove that the function $g : B \to C;\ g(x) = 2x - 4$, is a bijection from B onto C. Hence, according to Theorem 9, the composite $g \circ f$ is a bijection from A onto C. Verify this and find a formula for $g \circ f$.

74. According to Theorem 6, each function f from a nonempty set A to a nonempty set B gives rise to an equivalence relation $R : xRy \Leftrightarrow f(x) = f(y)$. In each case below, describe the equivalence classes of R:

(a) $t : A = \mathbb{Z} \rightarrow B = \mathbb{Z};\ t(x) = x^2$

(b) $\varphi : A = \mathbb{Z} \rightarrow B = \mathbb{Z};\ \varphi(x) = 2x^2 - 5$

(c) $h : A = \mathbb{Z} \rightarrow B = \mathbb{Z}^+;\ h(x) = x^2 + 2x + 5$

(d) $g : A = \mathbb{Z} \rightarrow B = \{0, 1, 2, 3\};\ g(x) =$ remainder of the division of x with 4.

75. Let $A = \{1, 2, 3, 4, 5, 6, 7\}$ and $\mathbb{N}$ = set of natural numbers or positive integers. Consider the mapping, natural numbers or positive integers. Consider the mapping, $f : A \rightarrow \mathbb{N};\ f(x) = 2x + 3$.

(i) Is this mapping injective? Is this mapping surjective? Is this mapping a bijection?

(ii) Find the images $f(B), f(C)$, of the sets B and C respectively; where $B = \{1, 2, 3, 4\}$ and $C = \{3, 4, 5\}$.

(iii) Find the sets $f(B \cup C)$ and $f(B) \cup f(C)$. If they are equal, is this always the cases with any mapping, or just particular to this problem?

(iv) Find the sets $f(B \cap C)$ and $f(B) \cap f(C)$.

76. Let $A \neq B$, A and B fixed nonempty subsets of a set U. In the set U we define two relations R_1 and R_2 as follows:

$R_1 : (x, y) \in R_1$ if, and only if, $x \in A$ and $y \in B$

$R_2 : (x, y) \in R_2$ if, and only if, $x \in A$ or $y \in B$

Prove each of the following:

(i) R_1 is a transitive relation, but it is neither reflexive nor symmetric no matter what nonempty subsets A, B, of U are chosen; under $A \neq B$.

(ii) R_1 is antireflexive if, and only if, $A \cap B = \emptyset$.

(iii) R_1 is antisymmetric if, and only if, either $A \cap B = \emptyset$; or $A \cup B = \{u\}$; $u \in U$ (that is, the intersection of A with B contains only one element, it is a singleton set).

(iv) R_2 is transitive.

(v) R_2 is reflexive if, and only if, $A \cup B = U$

(vi) R_2 is symmetric if, and only if, either A is a subset of B or B is a subset of A (keep in mind A and B are not equal sets, by hypothesis).

(vii) Conclude from parts (v) and (vi) that R_2 is an equivalence relation in U if, and only if, exactly one of the sets A, B is equal to U: that is, either $A = U$ or $B = U$.

(viii) R_2 is never antireflexive, no matter how the sets A, B are chosen (under $A \neq B$ and $A \neq \emptyset$, $B \neq \emptyset$)

(ix) R_2 is antisymmetric if, and only if, $A \cap B = \emptyset$.

77. Consider the functions: $f : \mathbb{Z} \to \mathbb{Z};\ f(x) = 3x + 1$, and $g : \mathbb{Q} \to \mathbb{Q};\ g(x) = 3x + 1$.

(i) Is f a restriction of g, to what set? And g an extension to f, to what set? Which definition would you refer to?

(ii) Show that f is an injective mapping, but not surjective, hence not a bijection.

(iii) Show that the range of f, R_f, consists of all integers which when divided by 3, leave remainder 1.

(iv) Show that g is a bijection.

(v) Since both f and g are injective mappings, both composite functions $f \circ g$ and $g \circ f$ would have to be injective functions (as long as neither $f \circ g$ nor $g \circ f$ is empty - refer to definition). Which Theorem guarantees this?

(vi) Show that $D_{f \circ g} = \{\frac{n}{3} \mid n \in \mathbb{Z}\} = \mathbb{Z} \cup \{\frac{m}{3} \mid m \in \mathbb{Z} \text{ and } m \text{ is not divisible by } 3\}$. This proves that $f \circ g$ is not a mapping from $\mathbb{Q}$ to $\mathbb{Z}$; but just a function from $\mathbb{Q}$ to $\mathbb{Z}$. Why?

(vii) Find a formula for $f \circ g$ and describe its range $R_{f \circ g}$.

(viii) Find a formula for $g \circ f$, and describe its range $R_{g \circ f}$.

(ix) Observe that even though $D_{f \circ g} = \mathbb{Z}$ is a proper subset of $D_{f \circ g}$, the range $R_{f \circ g}$ is a superset of $R_{g \circ f}$ (that is $R_{g \circ f}$ is a subset of $R_{f \circ g}$).

78. Let $\varphi : \mathbb{R} \to \mathbb{R};\ y(x) = x - 1$, and $g : \mathbb{R} \to \mathbb{R};\ g(x) = \frac{1}{1+x^2}$.
For each of the functions listed below, do the following:

(i) Determine whether the function is injective, surjective, both, or neither.

(ii) Give the domain and the range of the function.

(a) φ

(b) g

(c) $\varphi \circ g$

(d) $g \circ \varphi$

79. Consider the functions f and g, from the set of real numbers $\mathbb{R}$ to itself; defined by $f(x) = \frac{x-1}{x+1}$ and $g(x) = \frac{x+1}{x-1}$, respectively.

(a) Find a formula for the composite $g \circ f$ and determine its domain $D_{g \circ f}$

(b) Find a formula for the composite $f \circ g$ and determine its domain $D_{f \circ g}$.

80. Let $f : \mathbb{N}_0 \to \mathbb{N}_0$ (set of nonnegative integers); $f(x) = 2x$, and

$$g : \mathbb{N}_0 \to \mathbb{N}_0;\ g(x) = \begin{cases} \frac{x}{2}, & \text{if } x \text{ is an even nonnegative integer} \\ 0, & \text{if } x \text{ is an odd nonnegative integer} \end{cases}$$

(i) Determine whether f is injective, surjective, or a bijection.

(ii) Repeat part (i) for the function g.

(iii) Find formulas for the composite mappings $f \circ g$ and $g \circ f$.

Note: By inspection one can see that both f and g are mappings from $\mathbb{N}_0$ to itself; hence $f \circ g$, $g \circ f$ will also be mappings from $\mathbb{N}_0$ to $\mathbb{N}$.

81. Below you are given three functions from a given set A to a given set B. In each case,

(i) Show that the given function is an injection from A onto B.

(ii) Specify which of the three functions are actually bijections from A onto B.

(iii) Find a formula for the inverse of each function and specify its domain. (**Note:** If you need to use the square root function, then use its meaning as you have known it from calculus or precalculus).

(a) $f : A = \mathbb{R} - \{2\} \to B = \mathbb{R} - \{1\};\ f(x) = \frac{x+1}{x-2}$.

(b) $\varphi : A = \mathbb{R}^+ \to B = \mathbb{R};\ \varphi(x) = 2 - x^2$; $\mathbb{R}_0^+$ stands for the set of nonnegative reals; that is, $\mathbb{R}_0^+ = [0, \infty]$.

(c) $g : A = \mathbb{R}_0^+ \to B = (-\infty, -1) \cup [0, +\infty);\ g(x) = \frac{x^2}{1-x^2}$.

82. Prove that the bijection $f : A = \mathbb{R} - \{3\} \to D = R - \{2\}$, defined by $f(x) = \frac{2x+5}{x-3}$ is the composite of the following three bijections:

$f_1 : A = \mathbb{R} - \{3\} \to B = \mathbb{R} - \{0\};\ f_1(x) = x - 3$,

$f_2 : B = \mathbb{R} - \{0\} \to C = \mathbb{R} - \{0\};\ f_2(x) = \frac{11}{x}$,

$f_3 : C = \mathbb{R} - \{0\} \to D = \mathbb{R} - \{2\};\ f_3(x) = x + 2$.

In other words, prove that $f = f_3 \circ f_2 \circ f_1$.

83. Let a, b, c, d be real numbers, with $c \neq 0$. Consider the mappings:

$f_1 : \mathbb{R} - \{-\frac{d}{c}\} \to \mathbb{R} - \{0\};\ f_1(x) = x + \frac{d}{c}$,

$f_2 : \mathbb{R} - \{0\} \to \mathbb{R};\ f_2(x) = \frac{k}{x}$, where $k = \frac{1}{c} \cdot \left(\frac{c \cdot b - ad}{c}\right)$.

$f_3 : \mathbb{R} - \{0\} \to \mathbb{R} - \{\frac{a}{c}\};\ f_3(x) = x + \frac{a}{c}$.

Prove that f_1, f_2, and f_3, are all bijections; and that $f = f_3 \circ f_2 \circ f_1$.

84. If f is a injective mapping from a nonempty set A to a nonempty set B, show that for every subset S of A the following statement is true:

$$f(A) - f(S) \subset f(A - S) \subset B - f(S)$$

Is the converse true? In other words, if the above statement is true for every subset S of A, does it follow that the mapping f must be injective? Why or why not?

85. Prove that if f is a bijection from a nonempty set A to a nonempty set B, then for every subset S of A, the following statement is true:

$$B - f(S) = f(A - S)$$

Is the converse of this statement true? In other words if the equality $B - f(S) = f(A-S)$ holds for every subset S of A, does it follow that f must be a bijection? Why or why not?

86. Show that if f is a mapping from a nonempty set A to a nonempty set B then the following statements are true:

(a) $f^{-1}(S_1 \cup S_2) = f^{-1}(S_1) \cup f_2^{-1}(S_2)$, for any subsets S_1 and S_2 of B.

(b) $f^{-1}(S_1 \cap S_2) = f^{-1}(S_1) \cap f^{-1}(S_2)$, for any subsets S_1 and S_2 of B.

(c) $f^{-1}(S_1) \cup f^{-1}(S_2) = A$, where S_1, S_2 are subsets of B, such that $S_1 \cup S_2 = B$ and $S_1 \cap S_2 = \emptyset$.

87. Prove that if f is a mapping from a nonempty set A to a nonempty set B the following statement holds true: $S \subset f^{-1}[f(S)]$, for every subset S of A.

88. Show that if f is a bijection from the set A to the set B then,

$$S = f^{-1}[f(S)],$$

for every subset S of A. Is the converse true? Why, or why not?

89. Let A, B be nonempty sets, f a mapping from A to B. Prove,

$$f[f^{-1}(S)] \subset S, \text{ for every subset } S \text{ of } B.$$

90. Show that if a mapping f from a set A to a set B is such that the statement,

$$f[f^{-1}(S)] = S, \text{ for every subset } S \text{ of } B,$$

holds true, then f must be a surjective mapping.

Is the converse of this statement true? That is, if $f[f^{-1}(S)] = S$ holds true for every subset S of B, does it follow that f must be surjective?

91. Let $f : \mathbb{N}_0 \times \mathbb{N}_0 \to \mathbb{N}_0$ (set of nonnegative integers) be the mapping defined as, $f((m,n)) = n,\ m \in \mathbb{N}_0,\ n \in \mathbb{N}_0$.

According to practive problem P40, if A_1 and A_2 are any subsets of $\mathbb{N}_0 \times \mathbb{N}_0$, we must have $f(A_1 \cap A_2) \subset f(A_1) \cap f(A_2)$.

(i) Prove that $f(A_1 \cap A_2)$ will be a proper subset of $f(A_1) \cap f(A_2)$ (that is, $f(A_1 \cap A_2) \neq f(A_1) \cap f(A_2)$), if, and only if, these exist (at least) two pairs $(m_1, n_1) \in A_1$ and $(m_2, n_2) \in A_2$, such that $m_1 \neq m_2$, $n_1 = n_2$, $(m_1, n_1) \neq A_2$, and $(m_2, n_2) \notin A_1$.

(ii) Show that the mapping f is surjective but not injective.

92. Below, you are given six mappings. In each case, find the values of the infimum, minimum, supremum, and maximum of the range R_f of the given mapping; if, of course, any of these four values exists as a real number; otherwise, say that it does not exist.

(i) $f : \mathbb{R}_0^+ = [0, +\infty) \to \mathbb{R}_0^+$; $f(x) = \frac{3x+2}{5x+3}$, where $\mathbb{R}_0^+$ is the set of nonnegative real numbers.

(ii) $g : (-2, 4) \to \mathbb{R}$; $g(x) = -3x + 7$, where $(-2, 4)$ stands for the usual open interval consisting of all reals strictly between -2 and 4.

(iii) $h : \mathbb{R}_0^+ \to \mathbb{R}_0^+$; $h(x) = \frac{4x+1}{3x+2}$

(iv) $u : [-1, 2] \to \mathbb{R}$; $u(x) = 2x + 6$, where $[-1, 2]$ stands for the usual closed interval consisting of all real numbers between -1 and 2.

(v) $t : \mathbb{R}_0^+ \to \mathbb{R}_0^+$; $t(x) = \frac{6x+3}{4x+2}$

(vi) $\varphi : [0, 4) \to \mathbb{R}$; $\varphi(x) = -\frac{1}{3}x + 6$; where $[0, 4) = \{x \in \mathbb{R} \mid 0 \leq x < 4\}$.

93. Let a, b, and d be fixed positive integers; c a fixed nonnegative integer. Consider the mapping, $f : \mathbb{Z}^+ \to \mathbb{Q}^+$ (from the set of positive integers to the set of positive rationals) defined as $f(x) = \frac{ax+b}{cx+d}$

(i) Suppose that $d = 1$. Under this assumption, show that range R_f of f is a subset of $\mathbb{Z}^+$ is, and only if, $c = 0$.

Hint: If $c \neq 0$, then, as we know from calculus, the values of the function f will be arbitrarily close to the rational number $\frac{a}{c}$; in the language of calculus we specifically have $\lim_{x \to +\infty} \frac{ax+b}{cx+d} = \frac{a}{c}$.

(ii) Suppose that $c = 0$. Under this assumption, prove that the range R_f of f will be a subset of $\mathbb{Z}^+$ if, and only if, d is a common divisor of both a and b.

(iii) Suppose $c \neq 0$. Let $D = bc - ad$. Prove that if $D = 0$, the range R_f will be a subset of $\mathbb{Z}^+$ if, and only if, c is a divisor of a. In that case, $R_f = \{\frac{a}{c}\}$.

(iv) Suppose $c \neq 0$ and $D = bc - ad \neq 0$. Let $i = f(x)$; i being the image of some $x \in \mathbb{Z}^+$, under the mapping f. Prove that the two conditions below are the necessary and sufficient conditions for i to be a (positive) integer.

(a) $a + \ell = c \cdot i$ and

(b) $\frac{D}{\ell} - ad = c \cdot x$.

where ℓ is some positive integer divisor of D.

Conclude that there can only be finitely many (if at all) values of i in $\mathbb{Z}^+$

Hint: We can write $\frac{ax+b}{cx+d} = \frac{a}{c} + \frac{cb-ad}{c\cdot(cx+d)}$ (verify this).

Chapter 2

Rings and Fields

2.1 Axioms, Definitions, and Important Examples

To be able to precisely define the concept of a ring R, we must first give a concise definition of the concept of a binary operation in or over a nonempty set S.

Definition 1: (i) Let S be a nonempty set. Any mapping of the Cartesian product $S \times S$ (to some other set S') is called a **binary operation on the set** S.

(ii) If the range of a binary operation (defined in (i)) on S is a subset of S itself, then we say that the **set** S **is closed under the given binary operation.**

It is noteworthy to remark here that the familiar operations of addition and multiplication on familiar sets such as $\mathbb{Z}$ $\mathbb{Q}$, and $\mathbb{R}$, are binary operations on these sets such that the said sets are closed under them. We are now in a position to define the concept of a ring R.

Definition 2: A **ring** is a nonempty set R together or equipped with two binary operations on R, denoted by $\oplus$ and $\odot$ ($\oplus$ is usually called "addition" and $\odot$ is usually called "multiplication"), that satisfy the following eight principles or axioms:

Axiom 1. The set R is closed under the operations $\oplus$. In other words, for any two elements x_1 and x_2 of R, the element $x_1 \oplus x_2$ is a member of R, symbolically: $(\forall x_1, x_2 \in R) \Rightarrow (x_1 \oplus x_2 \in R)$.

Axiom 2. For all $x_1, x_2, x_3 \in R$ the following statement holds true:

$$x_1 \oplus (x_2 \oplus x_3) = (x_1 \oplus x_2) \oplus x_3.$$

(So the sum $x_1 \oplus x_2 \oplus x_3$ is unambiguous.) This principle or axiom is called **the associative property of addition.**

Axiom 3. For any $x_1, x_2 \in R$ the following statement holds true: $x_1 \oplus x_2 = x_2 \oplus x_1$. This principle or axiom is called the **commutative property of addition.**

Axiom 4. There exists an element of R, denoted by 0_R, that satisfies $x \oplus 0_R = x = 0_R \oplus x$, for every element x of R. Such an element 0_R in R is called a **neutral element** of R, or a "**zero**" element of R.

Axiom 5. For every element x of R, there exists an element x' of R such that

$x \oplus x' = 0_R$ where 0_R is a neutral element of R. The element x' is said to be an element **opposite** or "**negative**" of the element X.

Note: By commutativity of the addition operation $\oplus$, obviously $x' \oplus x = 0_R$ as well.

Axiom 6. The set R is closed under the operation $\odot$. In other words, for any two elements x_1 and x_2 of R, the element $x_1 \odot x_2$ is a member of R; symbolically, $(\forall x_1, x_2 \in R) \Rightarrow (x_1 \odot x_2 \in R)$.

Axiom 7. For all $x_1, x_2, x_3 \in R$, the following statement holds true:

$$x_1 \odot (x_2 \odot x_3) = (x_1 \odot x_2) \odot x_3$$

(So the product $x_1 \odot x_2 \odot x_3$ is unambiguous.) This principle or axiom is called the **associative property of multiplication.**

Axiom 8: For all $x_1, x_2, x_3 \in R$, the following statement holds true:

$$x_1 \odot (x_2 \oplus x_3) = (x_1 \odot x_2) \oplus (x_1 \odot x_3) \text{ and } (x_1 \oplus x_2) \odot x_3 = (x_1 \odot x_3) \oplus (x_2 \odot x_3)$$

This principle or axiom is called the **distributive property of multiplication with respect to addition.**

Now that we have stated the eight axioms that a nonempty set R must satisfy in order to be a ring, we must make the following remark below:

Remark 1: (i) A neutral element 0_R in a ring R, is unique. In other words, there can be only one neutral element. So we can call it "the neutral element." This is easy to prove and we will do so in the next section, Section 2, where we will discuss the basic properties of rings.

(ii) Likewise, each element x of R, can only have one opposite or "negative" element x'. That is symbolically we write: $\forall x \in R \exists | x' \in R$ such that $x \oplus x' = 0_R$, where $\exists|$ means "there exists one and only one." We have already used this existential symbol $\exists|$ in Chapter 1. The uniqueness of the opposite element can be easily proved. We will do so in then next section.

Definition 3. A **commutative ring** is a nonempty set R that satisfies two conditions below:

(a) R is a ring and

(b) **Axiom 9:** For all $x_1, x_2 \in R$, the following statement holds true:

$$x_1 \odot x_2 = x_2 \odot x_1$$

This principle or axiom is called **commutative multiplication.**

Definition 4: A ring R is called a **ring with identity** if there exist an element of R, denoted by 1_R, such that

Axiom 10: $x \odot 1_R = 1_R \odot x$. Such an element 1_R is called an *identity* element.

Remark 2: As it can be easily proven, there can only be one identity element 1_R (see next section) in a ring R. The proof of the uniqueness of 1_R is easy indeed for if the elements 1_R and $1'_R$ are both identity elements, then by definition 4, we must have

$1_R \odot 1'_R = 1'_r = 1'_R \odot 1_R$. But also, by the same definition, $1'_R \odot 1_R = 1_R = 1_R \odot 1'_R$. Hence $1_R = 1'_R$.

Throughout this chapter when we write $a + b$ and $a \cdot b$ we will always mean the ordinary addition and multiplication used in $\mathbb{Z}, \mathbb{Q}$, and $\mathbb{R}$. Otherwise when a different kind of multiplication and addition is used, we will be using symbols such as $\odot$ and $\oplus$.

Examples of Rings

Example 1: Any singleton set $R = \{r\}$ is a ring trivially, with neutral element $0_R = r$. Both operations $\oplus$ and $\odot$ are simply defined by $r \oplus r = r$ and $r \odot r = r$. So r is also the identity element by definition $1_R = r$.

Example 2: The familiar sets of integers $\mathbb{Z}$, the rationals $\mathbb{Q}$ and the reals $\mathbb{R}$, are all commutative rings with identity under the operations of ordinary addition and ordinary multiplication of numbers. Because the proof of this fact is found in many introductory mathematics books, we omit the proof. Moreover, all three rings $\mathbb{Z}, \mathbb{Q}$, and $\mathbb{R}$ are commutative rings with identity the usual number 1 and neutral element the usual number zero.

Example 3: Let n be a fixed positive integer $n \geq 2$ and let $R = \{x | x \in \mathbb{Z}$ and $x = n \cdot y$, for some $y \in \mathbb{Z}\}$. In other words, R is the set of all integer multiples of n. Under the ordinary addition and multiplication of integers, R is a commutative ring without identity. To see why, let us look at the axioms. Axiom 1 is satisfied since if $x_1, x_2 \in R$. Then $x_1 = n \cdot y_1, x_2 = n \cdot y_2$, so that $x_1 \oplus x_2 = n \cdot y_1 + n \cdot y_2 = n \cdot (y_1 + y_2) \Rightarrow x_1 \oplus x_2 \in R$, since $y_1 + y_2 \in \mathbb{Z}$ (obviously, since both $y_1, y_2 \in \mathbb{Z}$). Axiom 2 is satisfied by inspection since ordinary addition is associative in $\mathbb{Z}$ and R is a subset of $\mathbb{Z}$, closed under addition. Axiom 3 is also satisfied by inspection, since addition in $\mathbb{Z}$ is commutative. The same is true for Axiom 4 since the zero number integer is a member of $R : 0 = n \cdot 0$. As for Axiom 5, it is clear that the integer $x' = -(n \cdot y)$ is the opposite of the integer $x = n \cdot y$, since $x + x' = n \cdot y - n \cdot y = 0 = x' + x$. Axiom 6 is satisfied because if $x_1, x_2 \in$, then $x_1 = n \cdot y_1, x_2 = n \cdot y_2$, with $y_1, y_2 \in \mathbb{Z}$ so that $x_1 \odot x_2 = (n \cdot y_1) \cdot (n \cdot y_2) = n \cdot (y_1 \cdot n \cdot y_2) \in \mathbb{Z}$. Both Axioms 8 and 7 are satisfied by inspection, since addition in $\mathbb{Z}$ is distributive with respect to multiplication and multiplication in $\mathbb{Z}$ is associative. Axiom 9 is also satisfied because multiplication in $\mathbb{Z}$ is commutative and R is a subset of $\mathbb{Z}$. Finally, the ring R has no identity 1_R. Such an element would satisfy, in particular, $1_R \cdot 1_R = 1_R$. Let $1_R = n \cdot y$, for some $y \in \mathbb{Z}$ we would have $(n \cdot y) \cdot (n \cdot y) = n \cdot y \Leftrightarrow y \cdot n \cdot y = y$. Since n is nonzero and if $y \neq 0$, the last equation would imply $n \cdot y = 1$ which is impossible by virtue of the fact n is integer greater than 1. Hence the above equation, $y \cdot n \cdot y = y$ can be satisfied only by $y = 0$.

Example 4: Consider the set $R = \{x \in \mathbb{Z} | x = ny; \} y \in \mathbb{Z}\}$ of the previous example where n is a fixed positive integer greater than 1. If we take the addition operation $\oplus$ to be the ordinary addition of integers, but we take $\odot$ to be a multiplication operation defined by $x_1 \odot x_2 = \frac{x_1 x_2}{n} = n y_1 y_2$ then the set R becomes a commutative ring with identity, under the operations $\oplus$ and $\odot$. With neutral element $O_R = 0$, and identity element $1_R = n$: indeed for any $x \in R$, $x \odot 1_R = x \odot n = \frac{x \cdot n}{n} = x$. Likewise, $1_R \odot x = n \odot x = \frac{n \cdot x}{n} = x$.

Example 5: This and the next example will bear some familiarity to students and readers who have been exposed to the concent of modular arithmetic. Consider the set $\mathbb{Z}_4 = \{0, 1, 2, 3\}$ and define the operations of addition $\oplus$ and multiplication $\odot$ as follows:

$x_1 \oplus x_2 =$ remainder of the division of the integer $(x_1 + x_2)$ with 4

$x_1 \odot x_2 =$ remainder of the division of the integer $x_1 \cdot x_2$ with 4

We can construct the following two tables, one for each operation.

$\oplus$	0	1	2	3
0	0	1	2	3
1	1	2	3	0
2	2	3	0	1
3	3	0	1	2

$\odot$	0	1	2	3
0	0	0	0	0
1	0	1	2	3
2	0	2	0	2
3	0	3	2	1

Figure 1

We leave it to the reader to show that $\mathbb{Z}_4$ is commutative ring with identity. The identity element being $1_R = 1$ and the neutral element being $0_R = 0$. Note that in this ring, the element 1 is the opposite of the element 3 and vice versa; 2 is the opposite of itself (0_R is always the opposite of itself, trivially).

Example 6: Let $n \geq 2$ be a fixed positive integer. This example is a generalization of the previous one. consider the set $\mathbb{Z}_n = \{0, 1, 2, \ldots, n-1\}$ equipped with the following two operations of addition $\oplus$ and multiplication $\odot$:

$x_1 \oplus x_2 =$ the remainder of the division of the integer $(x_1 + x_2)$ with n.

$x_2 \odot x_2 =$ the remainder of the division of the integer $x_1 \cdot x_2$ with n.

In practice problem P10 will prove that $\mathbb{Z}_n$ is a commutative ring with identity. The identity element being $1_R = 1$ and the neutral element being $0_R = 0$. If we allowed $n = 1$, then $\mathbb{Z}_1$ would simply be the trivial ring $\{0\}$. Insofar, all six examples presented have been examples of commutative rings. The next four examples are examples of **noncommutative rings**. A noncommutative ring is a ring that fails to be commutative. That is, there exist at least two elements x_1 and x_2 in R such that $x_1 \odot x_2 \neq x_2 \odot x_1$.

Example 7: Consider the set $M_2(\mathbb{R})$ of all 2×2 (2 by 2") matrices whose entries are real numbers. Parenthetically we remark here that for given m and n positive integers, an $m \times n$ (read "m by n") matrix, is simply a rectangular array or arrangement containing exactly m rows and n columns, and exactly $m \cdot n$ entries. So, $M_2(\mathbb{R})$ is the set

$$M_2(\mathbb{R}) = \left\{ \begin{pmatrix} x_{11} & x_{12} \\ x_{21} & x_{22} \end{pmatrix} \middle| \, x_{11}, x_{21}, x_{12}, x_{22} \in \mathbb{R} \right\}$$

equipped with the operations $\oplus$ and $\odot$ defined as follows: (the notation x_{ij} simply reminds us that the element x_{ij} occupies the ith row and jth column of the matrix.)

$$\begin{pmatrix} x_{11} & x_{12} \\ x_{21} & x_{22} \end{pmatrix} \oplus \begin{pmatrix} y_{11} & y_{22} \\ y_{21} & y_{22} \end{pmatrix} = \begin{pmatrix} x_{11}+y_{11} & x_{12}+y_{12} \\ x_{21}+y_{21} & x_{22}+y_{22} \end{pmatrix}; \text{ and}$$

$$\begin{pmatrix} x_{11} & x_{12} \\ x_{21} & x_{22} \end{pmatrix} \odot \begin{pmatrix} y_{11} & y_{22} \\ y_{21} & u_{22} \end{pmatrix} = \begin{pmatrix} (x_{11}y_{11}+x_{12}y_{21}) & (x_{11}y_{22}+x_{12}y_{22}) \\ (x_{21}y_{11}+x_{22}y_{21}) & (x_{21}y_{22}+x_{22}y_{22} \end{pmatrix}$$

To students and readers familiar with linear algebra, these two operations are simply the usual addition of matrices and matrix multiplication. (Recall from linear algebra that in order to perform the matrix multiplication $A \odot B$, the matrix A must have size $m \times n$ and the matrix B must have size $n \times k$. In other words, the number of columns of the factor A must match the number of rows of the second factor B). Matrix addition is easy to perform. One simply adds the corresponding entries of matrices of equal size. Matrix multiplication is more complicated. However, in order to produce the number that will occupy the ijth the entry (i.e. ith row and jth column) of the product matrix one must "multiply" the ith row of the first matrix with the jth row of the second matrix. By simply taking the sum of the products of the corresponding entries along the ith row (of the first matrix) and the jth column (of the second matrix). For example, to find the entry c_{12} (first row, second column) of the product $\begin{pmatrix} x_{11} & x_{12} \\ x_{21} & x_{22} \end{pmatrix} \odot \begin{pmatrix} y_{11} & y_{12} \\ y_{21} & y_{22} \end{pmatrix}$, we consider the first row of the first matrix. That row is $[x_{11} \ x_{12}]$ and the second column of the second matrix; that column is $\begin{bmatrix} y_{12} \\ y_{22} \end{bmatrix}$ we have two products: $x_{11} \cdot y_{12}$ (product of first entries) and $x_{12} \cdot y_{22}$ (product of the second entries). The resulting sum is $c_{12} = x_{11} \cdot y_{12} + x_{12} \cdot y_{22}$, which occupies the first row and second column in the product matrix. $M_2\mathbb{R}$ is a noncommutative ring with identity. It is easy to see that it is noncommutative. In fact, if one picks two 2×2 matrices with real number entries at random, chances are, that the product will not commute. For example $\begin{pmatrix} 5 & 2 \\ 1 & 0 \end{pmatrix} \odot \begin{pmatrix} -3 & 0 \\ -2 & 4 \end{pmatrix} = \begin{pmatrix} -19 & 8 \\ -3 & 0 \end{pmatrix}$, but $\begin{pmatrix} -3 & 0 \\ -2 & 4 \end{pmatrix} \odot \begin{pmatrix} 5 & 2 \\ 1 & 0 \end{pmatrix} = \begin{pmatrix} -15 & -6 \\ -6 & -4 \end{pmatrix} \neq \begin{pmatrix} -19 & 8 \\ -3 & 0 \end{pmatrix}$. the identity element of $M_2(\mathbb{R})$ is the matrix $1_R = \begin{pmatrix} 1 & 0 \\ 0 & 1 \end{pmatrix}$; and the neutral element is the matrix $0_R = \begin{pmatrix} 0 & 0 \\ 0 & 0 \end{pmatrix}$.

Example 8: The set $R = \{a, b, c, d\}$ equipped by the two operations $\oplus$ and $\odot$, defined by the two tables below, is a noncommutative ring:

$\oplus$	a	b	c	d
a	a	b	c	d
b	b	a	d	c
c	c	d	a	b
d	d	c	b	a

$\odot$	a	b	c	d
a	a	a	a	a
a	a	a	a	a
c	a	a	c	c
d	a	a	d	d

Figure 2

We urge the reader to verify that the eight axioms for a ring are satisfied even though this may take a while. The neutral element is $0_R = a$. There is no identity element 1_R. d seems to come close to being the identity element since $a \odot d = a = d \odot a$, $b \odot d = b$, $c \odot d = c$, and $d \odot d = d$; but $d \odot b = a \neq b = b \odot d$. Likewise, $d \cdot c = d \neq c = c \odot d$. Also, this ring is noncommutative since, for example, $b \odot c = b \neq c \odot b = a$. Another example would be $d \odot b = a \neq b \odot d = b$.

Example 9: See exercise 17 in the exercise set (at the end of this chapter). In that exercise, the ring of **real quaternions** is presented, an example of a noncommutative ring.

Example 10: Consider the set S of all mappings of $\mathbb{R}$ to itself, equipped with the ordinary addition of functions and ordinary multiplication of functions. R is commutative ring with identity. The ideality element 1_R being the constant function $1_R : \mathbb{R} \to \mathbb{R}$; $1_R(x) = 1$, $\forall x \in \mathbb{R}$. The neutral element 0_R being the constant function $0_R : \mathbb{R} \to \mathbb{R}$; $0_R(x) = 0$, $\forall x \in \mathbb{R}$. We note here that $\mathbb{R}$ may be replaced by any other fixed domain D subset of $\mathbb{R}$. The resulting ring R would consist of all functions from D to $\mathbb{R}$. The neutral element 1_R would then be the identity function I_D on the set D. The neutral element would be the constant zero function with domain D.

We now introduce the concept of subring S of a ring R. If we simply look at the three rings $\mathbb{Z}, \mathbb{Q}$ and $\mathbb{R}$ with the usual operations of addition and multiplication of numbers, we see that $\mathbb{Z}$ is a ring inside a larger ring $\mathbb{Q}$ which in turn is part of a yet larger ring, namely $\mathbb{R}$. We have the following definition.

Definition 5: If both sets S and R are rings under the same operations $\oplus$ and $\odot$, and if S is a subset of R, we say that S is a **subring** of the ring R. A word of caution here is needed. As it will be seen in Theorem 1 stated below, a subring S of a ring R, must have the same neutral element as R. That is $0_S = 0_R$. However, if R has an identity element 1_R, a subring S or R may not have an identity element 1_S. Or even if it has an identity 1_S, it may not equal 1_R. Below theorem 1 and its proof, we offer a number of examples to illustrate this observation.

Theorem 1 Let R be a ring with operations $\oplus$ and $\odot$. A subset S of R will be a subring of S if, and only if, the following necessary and sufficient conditions hold:

(a) S is closed under the addition operation $\oplus$.

(b) S is closed under the multiplication operation $\odot$.

(c) The neutral element 0_R of R lies in S. That is $0_R \in S$ (and $0_S = 0_R$).

(d) S is closed under opposites. That is, for every element $s \in S$, its opposite element s' in R actually lies in $S : s' \in S$.

Remark 3: The usefulness of this Theorem lies in that, given a subset S of a ring R, to actually check whether S is a ring itself under the same operations as R, one does not have to prove all the eight axioms for a ring. One need only check whether S is closed under $\oplus$ and $\odot$. Whether $0_R \in S$ and whether S is closed under opposites.

Proof: If S, a subset of R, satisfies the four conditions (a) through (d) of the hypotheses of the Theorem, it is easy to see that S satisfies all the Axioms 1-8 under $\oplus$ and $\odot$. Axioms 1 and 6 are satisfied by inspection since S satisfies conditions (a) and (b). Axioms 2,3, and 7 are satisfied just because S is a subset of R and R is a ring. Since S satisfies condition (c), it follows that $s \oplus 0_R = 0_R \oplus s = s$, for every element s of S. Thus 0_R is the neutral element of $S; 0_R = 0_S$. Lastly, by virtue of the fact that S satisfies condition (d), it follows at once that S satisfies Axiom 5 because by condition (d) $\forall s \in S, \exists s' \in S$ such that $s + s' = 0_R$. We have already shown that $0_R = 0_S$. This completes the proof that if conditions (a) through (d) are satisfied then S is a ring with operations $\oplus$ and $\odot$; hence by definition, a subring of R. Now let us prove the converse: If S is a subring of R, then conditions (a)-(d) must hold true. Clearly conditions (a) and (b) are satisfied since by definition, S must be closed under $\oplus$ and $\odot$. To show that condition (c) holds true, let 0_S be the neutral element of S; 0_R the neutral element of R. We will prove $0_S = 0_R$. Indeed, we have $0_S \oplus 0_R = 0_S$ because $0_S \in S \subset R$ and 0_R is the neutral element of R and $0_S \oplus O'_S = 0_R$; by Axiom 5, where $0'_S$ is the opposite element of 0_S in the ring R. We have

$$\left.\begin{array}{l} 0_s \oplus 0_R = 0_S \\ 0_S \oplus 0'_S = 0_R \end{array}\right\} \text{(add memberwise)} \Rightarrow \qquad \text{(1 and 2)}$$
$$\Rightarrow \quad (0_S \oplus 0_R) \oplus (0_S \oplus 0'_S) = 0_S \oplus 0_R.$$

Now apply Axiom 4: $0_S \oplus 0_R = 0_S$ so that we altogether obtain

$$0_S \oplus (0_S \oplus 0'_S) = 0_S$$

by Axiom 2 (Associativity)

$$\Rightarrow (0_S \oplus 0_S) \oplus 0'_S = 0_S. \tag{3}$$

By Axiom 4 again, $0_S \oplus 0_S = 0_S$. Therefore equation (3) implies

$$0_S \oplus 0'_S = 0_S \tag{4}$$

Compare (2) and (4). The left-hand sides are equal, which proves that the right-hand sides must also be equal: $0_S = 0_R$. this proves that condition (c) is satisfied. Finally to show that condition (d) is satisfied pick any element $s \in S$, let s' be its opposite element in $R : s \oplus s' = 0_R$. We have already proven that $0_R = 0_S$. Hence $s \oplus s' = 0_S$. If s'' is the opposite element of s in S, then $s \oplus s'' = 0_S$ as well, by Axiom 5. Then we deduce that

$$
\begin{aligned}
& s \oplus s' = s \oplus s'' \Rightarrow s' \oplus (s \oplus s') = s' \oplus (s \oplus s'') \Rightarrow \\
& \Rightarrow \text{ (by Axiom 2) } (s' \oplus s) \oplus s' = (s' \oplus s) \oplus s'' \Rightarrow \\
& \Rightarrow \text{ (by Axiom 3) } (s \oplus s') \oplus s' = (s \oplus s') \oplus s'' \\
& \Rightarrow \text{ (by Axiom 5) } 0_S \oplus s' = 0_S \oplus s'' \Rightarrow \\
& \Rightarrow \text{ (byAxiom 4) } 0_S \oplus s' = s''
\end{aligned}
$$

However $0_S = 0_R$ which means (by Axiom 4) that $0_S \oplus s' = 0_R \oplus s' = s'$; hence $s' = s''$. Therefore s' is actually in S which proves that condition (d) is satisfied. This concludes the proof of the Theorem.

We know at present some examples usually without proof, of subrings. the reader should try to verify in each example that S is a subring of R.

Example 11: As mentioned earlier, under the usual operations of addition and multiplication, $\mathbb{Z}$ is a subring of $\mathbb{Q}$ and $\mathbb{Z}$ is a subring of $\mathbb{R}$; and $\mathbb{Q}$ is a subring of $\mathbb{R}$.

Example 12: Any ring R has at least two distinct subrings; itself and the trivial subring $\{0_R\}$. (Unless R is itself trivial in which case it has only one subring,namely, itself.)

Example 13: the ring of example 3, is a subring of $\mathbb{Z}$. As pointed out in that example, that ring has no identity element.

Example 14: Given a ring R with operations$\oplus$ and $\odot$, the Cartesian product $R \times R$ is a ring with addition $*$ and multiplication $\otimes$ defined as follows: (verify that $R \times R$ is a ring!). For all (x_1, y_1) and $(x_2, y_2) \in R \times R$,

$$
\begin{aligned}
&(x_1, y_1) * (x_2, y_2) = (x_1 \oplus x_2, y_1 \oplus y_2); \text{ and} \\
&(x_1, y_1) \otimes (x_2, y_2) = (x_1 \odot x_2, y_1 \odot y_2).
\end{aligned}
$$

The neutral element of the ring $R \times R$ is the element $(0_R, 0_R)$, where 0_R is the neutral element of R. If R has an identity element 1_R, then $(1_R, 1_R)$ is the identity element of $R \times R$. If S and S' are subrings of R, then $S \times S'$ is a subring of $R \times R$. In particular the Cartesian products (under the operations $*$ and $\otimes$),

$$
\begin{aligned}
R \times \{0_R\} &= \{x \in R \times R | x = (r, 0_R), \text{ with } r \in R\}; \text{ and} \\
\{0_R\} \times R &= \{y \in R \times R | y = (0_R, r), \text{with } r \in R\},
\end{aligned}
$$

are subrings. We will see in Section 3 that both of these subrings of $R \times R$ have essentially the same structure as R does. They are isomorphic (this will be defined in Section 3) to the ring R.

Example 15: Consider $\mathbb{Z}_{10} = \{0, 1, 2, 3, 4, 5, 6, 7, 8, 9\}$, the ring defined in example 6 for $n = 10$. The set $S = \{0, 2, 4, 6, 8\}$, equipped with the operations $\oplus$ and $\odot$ as defined in example 6, is a subring of $\mathbb{Z}_{10}$. Note, however, that whereas $\mathbb{Z}_{10}$ has identity element, the element $1_R = 1$, S has identity element the element $1_S = 6$. So here we have $1_R \neq 1_S$. Below the two operation tables for $\mathbb{Z}_6$ are provided

$\oplus$	0	2	4	6	8
0	0	2	4	6	8
2	2	4	6	8	0
4	4	6	8	0	2
6	6	8	0	2	4
8	8	0	2	4	6

$\odot$	0	2	4	6	8
0	0	0	0	0	0
2	0	4	8	2	6
4	0	8	6	4	2
6	0	2	4	6	8
8	0	6	2	8	4

Figure 3

Example 16: Let n be a positive integer, $n \geq 2$. The ring $\mathbb{Z}_n$ was defined in example 6. Let d be a positive integer, divisor of n. It can be proved that the set

$$S = \left\{x \middle| x \in \mathbb{Z}_n,\ x = d \cdot y,\ y \in \mathbb{Z} \text{ and } 0 \leq y \leq \frac{n}{d} - 1\right\},$$

under $\oplus$ and $\odot$ as defined in example6, is a subring of $\mathbb{Z}_n$ (see practice problem P11).

Example 17: Consider the subset $\left\{x \in \mathbb{Q} \middle| x = \frac{m}{10^n},\ m \in \mathbb{Z},\ n \in \mathbb{Z}^+\right\}$ of the set of rationals $\mathbb{Q}$; S is a subring of $\mathbb{Q}$. Both S and $\mathbb{Q}$ share the same identity element $1_R = 1$.

Example 18: This example is in the same flavor as the previous one. Let p be a fixed prime number and consider the set $S = \left\{x \in \mathbb{Q} \middle| x = \frac{m}{p^n},\ m \in \mathbb{Z},\ n \in \mathbb{Z}^+\right\}$ is a subring of the ring $\mathbb{Q}$.

Example 19: Consider the ring R of all mappings of $\mathbb{R}$ to itself under the usual addition and multiplication of functions, as stated in example 10. From calculus we know that the product and sum of two continuous functions are both continuous functions. The product and sum of two differentiable functions, are both differentiable functions.

Because of this, it is not hard that if S_1 is the set of all continuous mappings of $\mathbb{R}$ to itself, and S_2 is the set of all differentiable mappings of $\mathbb{R}$ to itself, then S_1 is a subring of S_2. Both S_1 and S_2 are subrings of R, the ring of all mappings of $\mathbb{R}$, to itself.

Example 20: Consider the ring $\mathbb{Z}_n$ of example 6. It can be shown that if n is a prime number, the only subrings of $\mathbb{Z}_n$ are the trivial subring $\{0_R\}$ and $\mathbb{Z}_n$ itself whereas when n is a composite number (i.e. n has positive integer divisors other than 1 and n), according to example 16, $\mathbb{Z}_n$ will have other subrings as well.

Example 21: Let R be any ring with operations $\oplus$ and $\odot$ and consider the ring $R \times R$ (according to example 14). Let b be a fixed element of R.

Consider the sets $S_1 = \{z \in R \times R \,|\, z = (x, y) \text{ with } x = b \odot y\}$ and $S_2 = \{t \in R \times R \,|\, t = (x, y)\text{, with } y = b \odot x\}$. Both S_1 and S_2 are subrings of $R \times R$. Also note that the subset S of $R \times R$ defined as $S = \{(x, x) | x \in R\}$ is a subring of $R \times R$ as well (note that if R has an identity element 1_R and we take $b = 1_R$, then $S_1 = S_2 = S$, but if R has no identity, S can still be defined as a subring of $R \times R$).

Example 22: Consider $M_2(\mathbb{R})$ from example 7. The ring of all 2×2 matrices with real number entries and with operations the usual addition of matrices and matrix multiplication. Consider the following subset of $M_2(\mathbb{R}$:

$$S = \left\{ \begin{pmatrix} x & y \\ -y & x \end{pmatrix} | x \in \mathbb{R}, \ y \in \mathbb{R} \right\}.$$

Using Theorem 1, we can show that S is a subring of $M_2(\mathbb{R})$. We show that it satisfies the four conditions (a) through (d) of theorem 1:

(a) Pick any two members of S, $\begin{pmatrix} x & y \\ -y & x \end{pmatrix}$ and $\begin{pmatrix} x' & y' \\ -y' & x' \end{pmatrix}$. We have $\begin{pmatrix} x & y \\ -y & x \end{pmatrix} \oplus \begin{pmatrix} x' & y' \\ -y' & x' \end{pmatrix} = \begin{pmatrix} (x+x') & (y+y') \\ -(y+y') & (x+x') \end{pmatrix} \in S$; which shows that S is closed under addition $\oplus$.

(b) Next multiply $\begin{pmatrix} x & y \\ -y & x \end{pmatrix} \odot \begin{pmatrix} x' & y' \\ -y' & x' \end{pmatrix} = \begin{pmatrix} (x \cdot x' - yy') & (xy' + yx') \\ (-yx' - xy') & (-yy' + xx') \end{pmatrix} = \begin{pmatrix} (x \cdot x' - y \cdot y') & (xy' + yx') \\ -(xy' + yx') & (xx' - yy') \end{pmatrix} \in S$; which shows that S is closed under multiplication $\odot$.

(c) Clearly the neutral element $0_R = \begin{pmatrix} 0 & 0 \\ 0 & 0 \end{pmatrix}$ is in S, since one can take $x = 0 = y$.

(d) The opposite element of $s = \begin{pmatrix} x & y \\ -y & x \end{pmatrix}$ in $M_2(\mathbb{R})$ is the element $s' = \begin{pmatrix} -x & -y \\ -(-y) & -x \end{pmatrix} = \begin{pmatrix} -x & -y \\ y & -x \end{pmatrix}$. As one can see by inspection: $s + s' = 0_R = \begin{pmatrix} 0 & 0 \\ 0 & 0 \end{pmatrix}$. Clearly $s' \in S$, which shows that S is closed under the taking of opposites in $M_2(\mathbb{R})$.

We now introduce two concepts that are of great significance in the theory of rings. That of a zero divisor and the concept of a unit.

Definition 6: Let R be a ring with operations $\oplus$ and $\odot$. An element a of R is said to be a **zero divisor** of the ring R if it satisfies two conditions:

(i) $a \neq 0_R$. That is a is a "nonzero" or nonneutral element in R.

and

(ii) There exists some element b in R with $b \neq 0_R$, such that $a \odot b = 0_R$ or $b \odot a = 0_R$.

What the above definition really says is that if R contains two nonzero elements whose "product" is the zero element 0_R, then we actually have two zero divisors (unless $a = b$). As we will show in the beginning of the next Section, the products $0_R \odot x$ and $x \odot 0_R$ are both equal to 0_R, in any ring R and for every element x of R. This is something familiar from the well known rings $\mathbb{Z}, \mathbb{Q}$, and $\mathbb{R}$. When you multiply zero times any element, you always get zero. What is new though, which does not happen in the rings $\mathbb{Z}, \mathbb{Q}$, and $\mathbb{R}$, is the possibility of the product of two nonzero elements being equal to the zero element.

Example 23: Of all the examples of rings we have seen so far only the following ones have zero divisors:

(a) $\mathbb{Z}_4$ of example 5. It has exactly one zero divisor namely the element $2 : 2 \odot 2 = 0 = 0_R$.

(b) The ring $\mathbb{Z}_n$ of example 6, only if n is not a prime number (that is, n is a composite number); if n is not a prime, then it will have at least two divisors d_1, d_2 such that $n = d_1 \cdot d_2$ and $2 \leq d_1, d_2 \leq n - 1$. Both d_1 and d_2 are zero divisors since $d_1 \odot d_2 =$ remainder of the division of the integer $d_1 \cdot d_2 = n$ with n is equal to $0 = 0_R$. Also, $d_2 \odot d_1 = 0 = 0_R$. If, however, $n = p$ =prime, $\mathbb{Z}_p$ has no zero divisors since the only positive divisors of $n = p$ are 1 and p itself. So the only way we can have $d_1 \odot d_2 = 0_R = 0$ in $\mathbb{Z}_p$ is with at least one of d_1 and d_2 being $0_R = 0$ (Recall that the choices of d_1, d_2 in $\mathbb{Z}_p$ are among the integers $0, 1, 2, \ldots, p - 1$.)

(c) The ring $M_2(\mathbb{R})$ of example 7 has a plethora of zero divisors. For instance, if we take the matrices $\begin{pmatrix} x & 0 \\ w & 0 \end{pmatrix}$ and $\begin{pmatrix} 0 & 0 \\ y & z \end{pmatrix}$ where x, w, y, z may be any real numbers see that $\begin{pmatrix} x & 0 \\ w & 0 \end{pmatrix} \odot \begin{pmatrix} 0 & 0 \\ y & z \end{pmatrix} = \begin{pmatrix} 0 & 0 \\ 0 & 0 \end{pmatrix} = 0_R$. Thus if not both x and w are zero, and if not both y and z are zero, it is then clear that both matrices would be zero divisors in $M_2(\mathbb{R})$. So the sets

$$\mathbb{Z}_1 = \left\{ \begin{pmatrix} x & 0 \\ w & 0 \end{pmatrix} | x \in \mathbb{R},\ w \in \mathbb{R} \text{ not both } x \text{ and } w \text{ are zero} \right\}$$

and

$$\mathbb{Z}_2 = \left\{ \begin{pmatrix} 0 & 0 \\ y & z \end{pmatrix} | y \in \mathbb{R},\ z \in \mathbb{R}, \text{ not both} y \text{ and } z \text{ are zero} \right\},$$

are both subsets of $M_2(\mathbb{R})$. Both of these sets consist entirely of zero divisors of $M_2(\mathbb{R})$.

Also note the sets $Z_1 \cup \left\{ \begin{pmatrix} 0 & 0 \\ 0 & 0 \end{pmatrix} \right\}$ and $Z_2 \cup \left\{ \begin{pmatrix} 0 & 0 \\ 0 & 0 \end{pmatrix} \right\}$ are actually subrings of $M_2(\mathbb{R})$.

(d) The noncommutative ring of example 8 has some zero divisors. In fact, all nonzero elements in that ring are zero divisors. These are the elements b, c and d.

(e) The ring of all mappings of the set of real numbers $\mathbb{R}$ to itself, contains zero divisors. For instance, consider the mappings $f(x) = \begin{cases} c_1, \text{ if } x < 0 \\ 0, \text{ if } x \geq 0 \end{cases}$ and $g(x) = \begin{cases} 0, \text{ if } x < 0 \\ c_2, \text{ if } x \geq 0 \end{cases}$, where c_1 and c_2 are given or fixed nonzero real numbers (constants). One easily sees that the product function is the zero function: $(f \cdot g)(x) = f(x) \cdot g(x) = \begin{cases} c_1 \cdot 0 = 0, \text{ if } x < 0 \\ 0 \cdot c_2 = 0, \text{ if } x \geq 0 \end{cases}$. So that $f(x) \cdot g(x) = 0,\ \forall x \in \mathbb{R}$. Hence $f \circ g = 0_R =$ the zero function even though neither f nor g is the zero function.

(f) In example 14, we saw that if R is a ring under $\oplus$ and $\odot$, then so is $R \times R$ under the operations $*$ and $\otimes$ defined below:

$$\begin{aligned}(x_1, y_z) * (x_2, y_2) &= (x_1 \oplus x_2,\ y_1 \oplus y_2) \text{ and} \\ (x_1, y_1) \otimes (x_2, y_2) &= (x_1 \odot x_2,\ y_1 \odot y_2)\end{aligned}$$

If R contains at least two elements, then the following subsets of $R \times R$, denoted by ZD_1 and ZD_2 consist entirely of zero divisors of $R \times R$:

$$ZD_1 = \{(x, 0_R)\,|x \in R \text{ and } x \neq 0_R\}$$

$$ZD_2 = \{(0_R, y)\,|x \in R \text{ and } y \neq 0_R\}$$

If we multiply an element of ZD_1 with an element of ZD_2 we get the zero element $0_R : (x, 0_R) \otimes (0_R, y) = (x \odot 0_R, 0_R \odot y) = (0_R, 0_R) =$ the zero element of $R \times R$. Here we have assumed the fact that in any ring, the product of any element "times" the zero element is equal to the zero element. The product of the zero element "times" any element, produces the zero element. As we have already stated, this fact will be proved in Section 2. Also note above, that under the assumption $x \odot 0_R = 0_R \odot x = 0_R,\ \forall x in R$. It is also true that $(0_R, y) \otimes (x, 0_R) = 0_R \odot x,\ y \odot 0_R) = (0_R, 0_R)$.

Definition 7: Let R be a ring witn operations $\oplus$ and $\odot$ and with identity 1_R. A nonzero element u in $R(u \neq 0_R)$ is called a **unit** of the ring R, if there exists an element $u' in R,\ u' \neq 0_R$, such that $u \odot u' = 1_R = u' \odot u$. The element u' is called a **multiplicative inverse of** u.

Remark 4: (i) If R has at least two elements (that is, if $R \neq \{0_R\}$), then due to the fact that $x \odot 0_R = 0_R = x \odot 0_R$ (see Theorem 3), it follows immediately from the above definition that a unit cannot be the zero element 0_R(since when R has at least two elements, $1_R \neq 0_R$). So other than the trivial rings $R = \{0_R\}$, the word "nonzero" in definition 7, is redundant.

(ii) The relation, defined in the ring R, by the phrase "u' is a multiplicative inverse of u", is obviously symmetric, by definition: u is also the multiplicative inverse of u'. The same is true for the relation in R defined by "the element x' is opposite of the element x", as it is pretty clear from Axiom 5.

(iii) If u has a multiplicative inverse u', then u' is unique. This is easily shown in (Section 2).

Example 24: Obviously, since $1_R = 1_R \odot 1_R$, each ring R with identity, has at least one unit. Let us take a look at the previous examples of rings to find the units of those rings in our examples that have an identity element. In the ring $\mathbb{Z}$ there are two units: 1 and -1. In the ring $\mathbb{Q}$, every nonzero element is a unit since if $m/n : min\mathbb{Z}, n\ in\mathbb{Z}, n \neq 0$, is a rational number; then the rational number n/mi s the multiplicative inverse of $m/n : m/n \cdot n/m = 1$. The same is true for the set of reals $\mathbb{R}$. Every nonzero real number is a unit.

(a) The ring of example 3 has no units.

(b) The ring of example 4 has only two units: n (identity element) and $-n$.

(c) The ring $\mathbb{Z}_n = \{0, 1, 2, \ldots, n-1\}$ of example 6 is an interesting one. It can be proven that if k (a positive integer between 1 and $n-1$) is relatively prime (i.e. has no positive integer factors in common) to n, then k is a unit. The proof of this requires the knowledge of some number theory. For instance, when $n = 10$, the elements of $\mathbb{Z}_{10} : 1, 3, 7$, and 9 are the units of $\mathbb{Z}_{10}$. Note that $3 \odot 7 = 1_R = 1 = 7 \odot 3$ and $9 \odot 9 = 1_R = 1$ so that 7 is the multiplicative inverse of 3 and 9 is its own multiplicative inverse.

An important special case arises when $n = p =$ prime. In $\mathbb{Z}_p$ all non zero elements: $1, 2, 3, \ldots, p-1$ (that is every member of $\mathbb{Z}_p$ except 0) are units.

(e) To find and understand the units of $M_2(\mathbb{R})$ it is more easily done when the student has some knowledge of the Linear Algebra. In linear algebra one learns that a 2×2 matrix $\begin{pmatrix} x_{11} & x_{12} \\ x_{21} & x_{22} \end{pmatrix}$ is a unit (which means in the language of linear algebra that it has an **inverse** matrix) if, and only if the quantity $x_{11} \cdot x_{22} - x_{21} \cdot x_{12}$ is **nonzero**. That quantity is called the **determinant** of the 2×2 matrix. Below we find that inverse. For simplicity, let $\begin{pmatrix} a & b \\ c & d \end{pmatrix} = U$ be a 2×2 matrix. Assume $ad - bc \neq 0$. Consider the matrix $U' = \begin{pmatrix} \frac{d}{ad-bc} & \frac{-b}{(ad-bc)} \\ \frac{-c}{(ad+bc)} & \frac{a}{(ad-bc)} \end{pmatrix}$. Below we verify that $U \odot U' = 1_R$:

$$\begin{pmatrix} a & b \\ c & d \end{pmatrix} \odot \begin{pmatrix} \dfrac{d}{(ad-bd)} & \dfrac{-b}{(ad-bc)} \\[2ex] \dfrac{-c}{(ad-bc)} & \dfrac{a}{(ad-bd)} \end{pmatrix} = \begin{pmatrix} \dfrac{ad-bc}{(ad-bc)} & \dfrac{-ab+ba}{(ad-bc)} \\[2ex] \dfrac{cd-dc}{(ad-bc)} & \dfrac{-cb+da}{(ad-bc)} \end{pmatrix} = \begin{pmatrix} 1 & 0 \\ 0 & 1 \end{pmatrix} = 1_R.$$

Similarly one can verify that $U' \odot U = 1_R$. Conversely, one can prove that if the matrix $U = \begin{pmatrix} a & b \\ c & d \end{pmatrix}$ is a unit in $M_2(\mathbb{R})$, then $ad - bc \neq 0$ and the multiplicative inverse U' of U is given by

$$U' = \begin{pmatrix} \dfrac{d}{ad-bc} & \dfrac{-b}{(ad-bc)} \\[2ex] \dfrac{-c}{(ad-bc)} & \dfrac{a}{(ad-bc)} \end{pmatrix}.$$

The proof of this though is more difficult.

(f) The ring of example 8 has no units.

(g) In the ring of quaternions (see exercise 17), every non zero element is a unit.

(h) In the ring of all mappings of $\mathbb{R}$ to itself (example 10), every mapping f that does not vanish at any real numbers, is a unit; $f(x) \neq 0, \forall x \in \mathbb{R}$; its multiplicative inverse in the function $1/f$.

Definition 8: (i) A commutative ring R with identity 1_R is said to be an **integral domain**, if it has no zero divisors.

(ii) A ring R with identity 1_R is said to be a **division ring**, if every nonzero element of R is a unit.

(iii) A **field** is a commutative division ring.

Example 25: (a) Among examples 1 through 10, the following rings are integral domains: $\mathbb{Z}, \mathbb{Q}, \mathbb{R}$, the ring of example 3, the ring of example 4, the ring of example 6 only when $n = p$ =prime, and the ring of example 9.

(b) Among the examples 1 through 10, the following rings are division rings: $\mathbb{Q}, \mathbb{R}$, the ring of example 6 for $n = p$ =prime, and the ring of example 9 (a noncommutative one).

(c) Among examples 1 through 10, the following rings are actually fields: $\mathbb{Q}, \mathbb{R}$, and the ring of example 6 when $n = p$ =prime.

Remark 5: By definition 8, it automatically follows that every field is a division ring. A division ring is an integral domain, as we show in Theorem 5.

Example 26: Consider the subring $S = \left\{ \begin{pmatrix} x & y \\ -y & x \end{pmatrix} | x \in \mathbb{R}, y \in \mathbb{R} \right\}$ of $M_2(\mathbb{R})$ in example 22. In that example we showed that S is actually a subring of R; S is actually a field. To see this, first observe that the identity element $1_R = \begin{pmatrix} 1 & 0 \\ 0 & 1 \end{pmatrix}$ of $M_2(\mathbb{R})$, actually lies in S by taking $x = 1$ and $y = 0$. Next we show that S is commutative:

$$\text{Indeed } \begin{pmatrix} a & b \\ -b & a \end{pmatrix} \odot \begin{pmatrix} a' & b' \\ -b' & a' \end{pmatrix} = \begin{pmatrix} (aa' - bb') & (ab' + ba') \\ (-ba' - ab') & (-bb' + aa') \end{pmatrix}$$

$$\text{and } \begin{pmatrix} a' & b' \\ -b' & a' \end{pmatrix} \odot \begin{pmatrix} a & b \\ -b & a \end{pmatrix} = \begin{pmatrix} (a'a - b'b) & (a'b + b'a) \\ (-b'a - a'b) & (-b'b + a'a) \end{pmatrix}$$

By comparing their respective entries we see that the two products are actually equal. Hence S is a commutative ring with identity. According to example 24(e), every 2×2 matrix with non zero determinant, must have a multiplicative inverse. If we take any matrix of the form $\begin{pmatrix} a & b \\ -b & a \end{pmatrix} \neq \begin{pmatrix} 0 & 0 \\ 0 & 0 \end{pmatrix} = 0_R$ (which means that not both real numbers a and b are zero; at least one of them must be non zero). Then, determinant $= a \cdot a - b \cdot (-b) = a^2 + b^2 \neq 0$; since not both a and b are zero. Hence every non zero element $\begin{pmatrix} a & b \\ -b & a \end{pmatrix}$ is a unit, and has multiplicative inverse the matrix $\begin{pmatrix} \frac{a}{a^2+b^2} & \frac{-b}{a^2+b^2} \\ \frac{b}{a^2+b^2} & \frac{a}{a^2+b^2} \end{pmatrix}$. Thus, S is actually a field. The following should come as no surprise.

Theorem 2 Let R be a ring with operations $\oplus$ and $\odot$ and S be a ring with operations $+$ and $\otimes$.

We define two operations (addition)$*$ and (multiplication) $\overline{x}$ as follows:

$$(r, s) * (r', s') = (r \oplus r'.\ s + s'),\ \forall r, r' \in R;\ \text{and}\ \forall s,\ s' \in S$$
$$\text{and}\quad (r, s)\overline{x}(r', s') = (r \odot r',\ s \otimes s'),\ \forall r, r \in R;\ \text{and}\ \forall s,\ s' \in S$$

We have the following:

(i) $R \times S$ equipped with the operations $*$ and $\overline{x}$, is a ring with neutral or zero element, the pair $(0_R, 0_S)$, where 0_R and 0_S are the zero element of R and S, respectively.

(ii) If R has an identity element 1_R, and$S$ has identity 1_S, then the pair $(1_R, 1_S) = 1_{R\times S}$ is the identity element of the ring $R \times S$.

(iii) If R_1 is a subring of R and S_1 a subring of S, then $R_1 \times S_1$ is a subring of $R \times S$.

Proof: Exercise.

Theorem 3 Let R, under the operations $\oplus$ and $\odot$, be any ring with neutral or zero element 0_R. Then, $x \odot 0_R = 0_R = 0_R \odot x$, for every element x of 0_R.

Proof: By Axiom 8, $x \odot (0_R \oplus 0_R) = (x \odot 0_R) \oplus (x \odot 0_R)$. But also, $0_R \oplus 0_R = 0_R$ by Axiom 4. Hence $x \odot 0_R = 0_R = (x \cdot 0_R) \oplus (x \odot 0_R)$. Let y be an opposite element of the element $x \odot 0_R$ (Axiom 5). Adding y to both sides of the above equation, from the right, we obtain

$$(x \odot 0_R) \oplus y = [(x \odot 0_R) \oplus (x \odot 0_R)] \oplus y. \tag{1}$$

By Axiom 5, the left-hand side of equation (1) is equal to 0_R and (the right-hand side),

$$[(x \odot 0_R) \oplus (x \odot 0_R)] \oplus y = (x \odot 0_R) \oplus [(x \odot 0_R) \oplus y] = (x \oplus 0_R) \oplus 0_R = x \oplus 0_R,$$

by Axioms 2, 5, and 4, respectively. Thus we conclude, $0_R = x \oplus 0_R$. In a similar manner one can prove that $0_R = 0_R \oplus x$ (just start with $(0_R + 0_R) \odot x = (0_R \odot x) \oplus (0_R \oplus x)$). The proof is complete. ∎

Next we show that in a ring R with an identity 1_R, no unit can be a zero divisor.

Theorem 4 Let R be a ring, with operations $\oplus$ and $\odot$, and identity element 1_R. Then, (i) every unit u or R is **not** a zero divisor; and, (ii) every zero divisor of R is **not** a unit.

Proof: (i) Let u be a unit in R. Then, according to definition 7, we must have $u \odot u' = 1_R = u' \odot u$, for some element u' of R. Both u and u' are nonzero: $u \neq 0_R$ and $u' \neq 0_R$. To prove that u cannot be a zero divisor of R, we show that the assumption that u is a zero division in conjunction with the hypothesis that u is a unit, will lead to a contradiction. Indeed, suppose that u is a zero divisor. Then according to definition 6, there would exist an element b, with $b \neq 0_R$ and such that,

$$(u \odot b = 0_R \text{ or } b\ \odot u = 0_R)$$

Suppose that

$$u \odot b = 0_R \Rightarrow u' \odot (u \odot b) = u' \odot 0_R \tag{2}$$

By Axiom 7 (associativity of multiplication) we have

$$u' \odot (u \odot b) = (u' \odot u) \odot b = 1_R \odot b = b,$$

by definition 4 (Axiom 10). Also, $u' \odot 0_R = 0_R$ by Theorem 3; consequently from equation (1) we obtain $b = 0_R$ contrary to the fact that $b \neq 0_R$ (since b was assumed to be a zero divisor). The other case, $b \odot u = 0_R$ is similar: $b \odot u = 0_R \Rightarrow (b \odot u) \odot u' = 0_R \odot u' \Rightarrow b \odot (u \odot u') = 0_R \odot u' \Rightarrow b \odot 1_R = 0_R \odot u' \Rightarrow b = 0_R$, a contradiction once more. Since the assumption that u is a zero divisor leads to a contradiction, we conclude that the unit u cannot be a zero divisor. The proof is complete. ■

(ii) This is simply the contrapositive statement of (i). It is logically equivalent to (i). We conclude this Section with the following Theorem:

Theorem 5 (i) Every division ring is an integral domain.

(ii) Every field is an integral domain.

Proof: (i) This is an immediate consequence of definition 8(ii) (which says that every element in a division ring other than the zero element 0_R, is a unit); theorem 4(i) and definition 8(i).

(ii) This is an immediate consequence of part (i) of this Theorem, in conjunction with definition 8(iii).

2.2 Properties of Rings

We start this section by introducing some notation.

Definition 9: (i) In a ring R, an element x' that is an opposite of the element can be denoted by $\ominus x$; we write $x' = \ominus x$. It reads "**minus** x."

(ii) In a ring R with an identity element 1_R, the multiplicative inverse u' of a unit u can also be denoted by u^{-1}. We write $u' = u^{-1}$. It reads "u to the power minus one."

(iii) In a ring R with operations $\oplus$ and $\odot$ the expression $b \ominus a$ simply means the sum of b and the opposite of the element a. We write $b \ominus a = b \oplus (\ominus a)$. So the notation of "minus: reduces the number of symbols involved in an algebraic sum of elements of a ring R. Also $b \ominus a$ reads "**b minus a**," and $\ominus$ is the symbol for **subtraction** in a ring R. The next Theorem lists all the basic properties of rings.

Theorem 6 Let R be a ring with addition and multiplication operations denoted by $\oplus$ and $\odot$, respectively; and $\ominus$ the symbol of subtraction; and 0_R denoting a neutral or zero element for R.

(i) R has a unique neutral element 0_R. If R has identity element 1_R, then 1_R is also unique. Also each element x of R, has a unique opposite element $x' = \ominus x$. Each unit u in R has a unique multiplicative inverse u^{-1}.

(ii) For each $x \in R$, $\quad x \odot 0_R = 0_R = 0_R \odot x$.

(iii) $x_1 \odot (\ominus x_2) = \ominus(x_1 \odot x_2) = (\ominus x_1) \odot x_2, \ \forall x_1, x_2 \in R$.

(iv) $\ominus(\ominus x) = x, \quad \forall x \in R$.

(v) $\ominus(x_1 \oplus x_2) = (\ominus x_1) \oplus (\ominus x_2)$, for all $x_1, x_2 \in R$.

(vi) $\ominus[x_1 \ominus x_2] = \ominus x_1 \oplus x_2$, for all $x_1, x_2 \in R$.

(vii) $(\ominus x_1) \odot (\ominus x_2) = x_1 \odot x_2$, for all $x_1, x_2 \in R$.

(viii) $(\ominus 1_R) \odot x = \ominus x$, for every $x \in R$ (if R has an identity 1_R).

(ix) If $x_1 \oplus x_2 = x_1 \oplus x_3$, for some elements $x_1, x_2, x_3 \in R$, then $x_2 = x_3$.

(x) Let a, b be fixed elements of R. Then the equation $a \oplus x = b$ has a unique solution. Namely, the element $b \ominus a$. In other words, the solution set of this equation is $S = \{b \ominus a\}$.

(xi) If the ring R has an identity 1_R, u and b fixed elements of R and u is a unit, then each of the equations $u \odot x = b$ and $y \odot u = b$ has a unique solution in R. The solution set of the first being the set $\{u^{-1} \odot b\}$, while the solution set of the second equation is the set $\{b \odot u^{-1}\}$.

Remark 6: If one was to simply take the operations $\oplus$ and $\odot$ to be the usual operations of addition and multiplication, in the rings $\mathbb{Z}, \mathbb{Q}$ and R; the subtraction $\ominus$ becomes the usual $-$. All eleven parts of Theorem 6 simply become the familiar properties of high school algebra. For instance, property (vi): $-(x_1 - x_2) = x_1 - x_2$, well known distribution property of the minus sign from precollege algebra.

Proof: (i) The uniqueness of 0_R is immediate if $0'_R$ is also a neutral element. Then by Axiom 4 we have $0_R \oplus 0'_R = 0_R = 0'_R \oplus 0_R$ and $0'_R \oplus 0_R = 0'_R = 0_R \oplus 0'_R$. Hence $0_R = 0'_R$. The uniqueness of 1_R is also immediate. By Axiom 10 (Definition

4), $1'_R \odot 1_R = 1'_R = 1_R \odot 1'_R$; and $1_R \odot 1'_R = 1_R = 1'_R \odot 1_R$. Hence $1_R = 1'_R$. Next let $x \in R$. If x' and x'' are both opposites of x, then $x \oplus x' = 0_R$ and $x \oplus x'' = 0_R$. Thus, $x \oplus x' = x \oplus x'' \Rightarrow x' \oplus (x \oplus x') = x' \oplus (x \oplus x'') \Rightarrow$ (by associativity, Axiom 2) $(x' \oplus x) \oplus x' = (x' \oplus x) \oplus x''$. By Axiom 3, $x' \oplus x = x \oplus x'$. Since $x \oplus x' = 0_R$, we also have $x' \oplus x = 0_R$. Thus we obtain $0_R \oplus x' = 0_R \oplus x'' \Rightarrow$ (by Axiom 4) $x' = x''$.

Suppose u is a unit and u', u'' both multiplicative inverses of u. According to Definition 7, u, u', and u'' are all nonzero and $u \cdot u' = 1_R = u' \cdot u$; and $u \cdot u'' = 1_R = u'' \cdot u$. We therefore have $u \odot u' = u \odot u'' \Rightarrow u' \cdot (u \odot u') = u' \odot (u \odot u'') \Rightarrow$ (by Axiom 7, associativity of multiplication), $(u' \odot u) \odot u' = (u' \odot u) \odot u'' \Rightarrow 1_R \odot u' = 1_R \odot u'' \Rightarrow$ (Axiom 10) $u' = u''$.

(ii) This has already been proven, it is Theorem 3.

(iii) By part (ii), $x_1 \odot 0_R = 0_R$. By Axiom 5, since $\ominus x_2$ is the opposite of x_2, we have $x_2 \oplus (\ominus x_2) = 0_R$. Hence, $x_1 \odot 0_R = 0_R \Rightarrow x_1 \odot [x_2 \oplus (\ominus x_2)] = 0_R \Rightarrow$ (by Axiom 8) $(x_1 \odot x_2) \oplus (x_1 \odot (\ominus x_2)] = 0_R$ which says that the element $x_1 \odot (\ominus x_2)$ is opposite of $x_1 \odot x_2$. Since by part (i) we know that the opposite unique and that $\ominus(x_1 \odot x_2)$ is the opposite of $x_1 \odot x_2$ we conclude that,

$$x_1 \odot (\ominus x_2) = \ominus(x_1 \odot x_2).$$

In a similar fashion, by starting with $0_R \odot x_1 = 0_R$, one proves that $\ominus x_1 \odot x_2 = \ominus(x_1 \odot x_2)$.

(iv) We know that $x \oplus (\ominus x) = 0_R = (\ominus x) \oplus x$, since $\ominus x$ is the opposite of x and addition is commutative. The same statement says that x is the opposite of $\ominus x$. Hence $\ominus(\ominus x) = x$.

(v) If we add the element $(\ominus x_1) \oplus (\ominus x_2)$ to the element $x_1 \oplus x_2$ and use associatively and commutatively of addition, we obtain

$$[(\ominus x_1) \oplus (\ominus x_2)] \oplus (x_1 \oplus x_2) = [((\oplus x_1) \oplus (\ominus x_2)) \oplus x_1] \oplus x_2 =$$

$$= [((\ominus x_2) \oplus (\ominus x_1)) \oplus x_1] \oplus x_2 = [(\ominus x_2) \oplus ((\ominus x_1) \oplus x_1)] \oplus x_2 =$$

$$= (\ominus x_2 \oplus 0_R) \oplus x_2 = \ominus x_2 \oplus x_2 = 0_R.$$

We have proved that $[(\ominus x_1) \oplus (\ominus x_2)] \oplus (x_1 \oplus x_2) = 0_R$ which shows that the element $(\ominus x_1) \oplus (\ominus x_2)$ is opposite of $x_1 \oplus x_2$. Hence $\ominus x_1 \oplus \ominus x_2 = \ominus(x_1 \oplus x_2)$.

(vi) This follows at once from parts (iv) and (v), by replacing x_2 by its opposite, $\ominus x_2$.

(vii) By part (iii) we know that $x_1 \odot (\ominus x_2) = (\ominus x_1) \odot x_2$. In this equation, just replace x_1 by $\ominus x_1$ and we use the fact that $\ominus(\ominus x_1) = x_1$ (part (iv)) we have,

$$(\ominus x_1) \odot (\ominus x_2) = [\ominus(\ominus x_1)] \odot x_2 \Rightarrow (\ominus x_1) \odot (\ominus x_2) = x_1 \odot x_2$$

(viii) Apply the statement of part (vii):

$$(\ominus x_1) \odot (\ominus x_2) = x_1 \ominus x_2,$$

by taking $x_1 = 1_R$ and $x_2 = \ominus x$. We have,

$$(\ominus 1_R) \odot [\ominus(\ominus x)] = 1_R \odot (\ominus x)$$

By part (iv), $\ominus(\ominus x) = x$. Since 1_R is the identity element and we must also have $1_R \odot (\ominus x) = \ominus x$. Combining these with the last equation above we arrive at $\ominus 1_R \odot x = \ominus x$.

(ix) $x_1 \oplus x_2 = x_1 \oplus x_3 \Rightarrow \ominus x_1 \oplus (x_1 \oplus x_2) = \ominus x_1 \oplus (x_1 \oplus x_3) \Rightarrow (\ominus x_1 \oplus x_1) \oplus x_2 = (\ominus x_1 \oplus x_1) \oplus x_3; \Rightarrow 0_R \oplus x_2 = 0_R \oplus x_3 \Rightarrow x_2 = x_3$.

(x) Suppose that $a \oplus x = b$. Then $\ominus a \oplus (a \oplus x) = \ominus a \oplus b \Rightarrow (\ominus a \oplus a) \oplus x = \ominus a \oplus b \Rightarrow 0_R \oplus x = \ominus a \oplus b \Rightarrow x = \ominus a \oplus b$; $x = b \oplus (\ominus a) = b \ominus a$, by commutativity. We verify that $b \oplus a$ is a solution: $a \oplus (b \ominus a) = a \oplus (b \oplus (\ominus a)) = a \oplus ((\ominus a) \oplus b) = (a \oplus (\ominus a)) \oplus b = 0_R \oplus b = b$, which proves that $b \oplus a$ is a solution to the equation $a \oplus x = b$.

To show the uniqueness, suppose that both x_1 and x_2 are solutions to the equation $a \oplus x = b$. Then $a \oplus x_1 = b$ and $a \oplus x_2 = b$. So that $a \oplus x_1 = a \oplus x_2 \Rightarrow$ (by part (ix)) $x_1 = x_2$.

(xi) We have $u \odot (u^{-1} \odot b) = (u \odot u^{-1}) \odot b = 1_R \odot b = b$; and $(b \odot u^{-1}) \odot u = b \odot (u^{-1} \odot u) = b \odot 1_R = b$. This proves that the elements $u^{-1} \odot b$ and $b \odot u^{-1}$ are solutions to the equations $u \odot x = b$ and $y \odot u = b$, respectively. To show uniqueness take the first equation and suppose that both x_1, x_2 are solutions to $u \odot x = b$. We must have $u \odot x_1 = b = u \odot x_2 \Rightarrow u^{-1} \odot (u \odot x_1) = u^{-1} \odot (u \odot x_2) \Rightarrow (u^{-1} \odot u) \odot x_1 = (u^{-1} \odot u) \odot x_2 \Rightarrow 1_R \odot x_1 = 1_R \odot x_2 \Rightarrow x_1 = x_2$. The uniqueness of the solution of the other equation is proved similarly. ■

Theorem 7 Let R be a ring with operations $\oplus$ and $\odot$; and subtraction $\ominus$ as defined in definition 9. Suppose that S is a nonempty subset of R. Assume that S satisfies the two conditions below:

(i) $x_1 \ominus x_2 \in S$, for all $x_1, x_2 \in S$ and

(ii) $x_1 \odot x_2 \in S$, for all $x_1, x_2 \in S$.

Then S is a subring of R.

Note: The converse of this theorem is also true (the reader should try to prove it by using Theorem 1).

Proof: According to Theorem 1, if we prove that S satisfies the four conditions of Theorem 1, then S must be a subring of S.

If we pick any element x_1 of S and take $x_2 = x_1$, then according to hypothesis (i), the element $x_1 \ominus x_1 = 0_R$ is in S. This proves that condition (c) of Theorem 1 is satisfied. Next, apply hypothesis (i) again. This time with $x_1 = 0_R \in S$ and x_2 any element of S: $0_R \ominus x_2 = \ominus x_2$ must also be a member of S, which proves that S is closed under opposites, condition (d) of Theorem 1 is satisfied. Now let x_1 and x_2 be any two elements of S. Since S is closed under opposites, it follows that $\ominus x_2$ is also in S. Now apply condition (i) again: $x_1 \ominus (\ominus x_2)$ must belong is S; but $\ominus(\ominus x_2) = x_2$, according to Theorem 6(iv). Hence we conclude that $x_1 \ominus (\ominus x_2) = x_1 \oplus x_2$ is in S, which proves condition (a) of Theorem1. Finally condition (b) is simply the same as hypothesis (ii). This concludes the proof.

The next Theorem illustrates the cancellation property, an important property used in the solving of equations, as the reader is undoubtedly familiar with the use of this property in $\mathbb{Z}, \mathbb{Q}$, and $\mathbb{R}$.

Theorem 8 Let R be a ring with operations $\oplus$ and $\odot$; $\ominus$ the symbol of subtraction, and neutral element 0_R.

(i) If an element b of R with $b \neq 0_R$ is **not** a zero divisor of R, then the statement $b \odot x_1 = b \odot x_2$ implies $x_1 = x_2$ (Cancellation Law), likewise the statement $x_1 \odot b = x_2 \odot b$ implies $x_1 = x_2$.

(ii) If an element a of R satisfies $a \odot c = c \odot a = 0_R$ then one of three possibilities must occur:

(a) $a = 0_R$, (b) a is a zero divisor and $c \neq 0_R$ or (c) $c = 0_R$. (Note that only possibilities (a) and (b) are mutually exclusive.)

(iii) If R has no zero divisors, then whenever $a \odot b = a \odot c$ occurs, b must equal to c; $b = c$.

(iv) Every integral domain satisfies the Cancellation Law:

$$(\forall a, b, c \in R \text{ and such that } a \odot b = a \odot c) \Rightarrow b = c.$$

Proof: (i) Suppose $b \odot x_1 = b \odot x_2$. This implies:

$$(b \odot x_1) \ominus (b \odot x_2) = (b \odot x_2) \ominus (b \odot x_2) \Rightarrow (b \odot x_1) \oplus [\ominus(b \odot x_2)] = 0_R.$$

By Theorem 6(iii) we know that $\ominus(b \odot x_2) = b \odot (\ominus x_2)$, so the above equation implies,

$$(b \odot x_1) \oplus [b \odot (\ominus x_2)] = 0_R$$

$$\Rightarrow \text{ (by Axiom 8) } \quad b \odot [x_1 \oplus (\ominus x_2)] = 0_R;$$
$$b \odot (x_1 \ominus x_2) = 0_R.$$

Since b is **not** a zero divisor, and $b \neq 0_R$ then by definition 6, it follows that the element $x_1 \ominus x_2$ must be the zero element 0_R; $x_1 \ominus x_2 = 0_R \Rightarrow x_1 = x_2$. The proof that $x_1 \odot b = x_2 \odot b$ implies $x_1 = x_2$ is similar and left to the reader.

(ii) If $a = 0_R$ then possibility (a) occurs. If on the other hand $a \neq 0_R$, then a may be a zero divisor. If $a \neq 0_R$ is actually a zero divisor, then the statement $a \odot c = 0_R$ implies that $c \neq 0_R$ (the same is true if $c \odot a = 0_R$ holds). Finally if $a \neq 0_R$ and a is **not** a zero divisor, then (according to definition 6), the element c must be zero; $c = 0_R$.

(iii) This part is an immediate consequence of part (i) since a, in particular, is assumed not to be a zero divisor.

(iv) In view of the fact that every integral domain contains no zero divisors (definition 8), this part is an immediate consequence of part (iii).

The following Theorem is an interesting one.

Theorem 9 (i) Every **finite** (i.e. containing a finite number of elements) commutative ring with no zero divisors is a field.

(ii) Every **finite** integral domain is a field.

Remark 7: The above Theorem fails if one removes the assumption "finite." For example, $\mathbb{Z}$ is a commutative ring with any infinite number of elements and no zero divisors, but $\mathbb{Z}$ is not a field, the only units are 1 and -1.

Proof: (i) Let $R = \{a_1, \ldots a_{n-1}, 0_R\}$ be a commutative ring with no zero divisors and containing n elements denoted by $a_1, \ldots, a_{n-1}$, $a_n = 0_R$ and n a positive integer with $n \geq 2$ and with $\oplus$ and $\odot$ denoting addition and multiplication in R.

We only need to prove that R possesses an identity element 1_R and that every nonzero element of R is a unit. Let a be one of R's nonzero elements. $a \in R$ and $a \neq 0_R$. Consider the mapping f of R to itself defined as follows: $f : R \to R$; $f(x) = a \odot x$. This is an injective mapping: if $f(x_1) = f(x_2)$ for some $x_1, x_2 \in R$, we would then have $a \odot x_1 = a \odot x_2$, and according to the Cancellation Law for a ring R with no zero divisors (Theorem 8(iii)), it follows that $x_1 = x_2$. Hence whenever $f(x_1) = f(x_2)$, then $x_1 = x_2$. Or equivalently if $x_1 \neq x_2$ then $f(x_1) \neq f(x_2)$. Now, since f is a surjective mapping from R to itself, and R is a finite set, it follows that f must also be surjective or onto; it is easy to see why, for the images $f(a_1), f(a_2), \ldots, f(a_n)$ of the n elements of R, must all be distinct since f is one-to-one. Thus the elements $f(a_1), f(a_2), \ldots, f(a_n)$ are simply some rearrangement of the sequence $a_1, a_2, \ldots, a_n$. The mapping f is onto as well. We have thus shown that each of the $(n-1)$ mappings $f_1, \ldots f_{n-1}$ from R to itself; $f_i : R \to R$; $f_i(x) = a_i \odot x$, for $i = 1, \ldots, (n-1)$ is a bijection (one-to-one and onto) of R onto itself. That is, a permutation of the ring R. Thus, each of the $(n-1)$ equations: $a_1 \odot x = a_1, \ldots, a_{n-1} \odot x = a_{n-1}$, must have a solution in R. The claim is that they all have a unique solution and, in fact, the exact same solution. Indeed if x_1 is a solution to $a_1 \odot x = a_1$ and x_2 is a solution of $a_2 \odot x = a_2$ then,

$$\left.\begin{array}{l} a_1 \odot x_1 = a_1 \\ a_2 \odot x_2 = a_2 \end{array}\right\} \Rightarrow \left.\begin{array}{l} a_2 \odot (a_1 \odot x_1) = a_2 \odot a_1 \\ a_1 \odot (a_2 \odot x_2) = a_1 \odot a_2 \end{array}\right\} \Rightarrow \begin{array}{l} (a_2 \odot a_1) \odot x_1 = a_2 \odot a_1 \\ (a_1 \odot a_2) \odot x_2 = a_1 \odot a_2 \end{array}$$

Multiplication is commutative since R has been assumed to be commutative. Hence $a_1 \odot a_2 = a_2 \odot a_1$ and thus the above system of equations gives
$\left.\begin{array}{l} (a_1 \odot a_2) \odot x_1 = a_1 \odot a_2 \\ (a_1 \odot a_2) \odot x_2 = a_1 \odot a_2 \end{array}\right\} \Rightarrow (a_1 \odot a_2) \odot x_1 = (a_1 \odot a_2) \odot x_2$. However, a_1 and a_2 are among the $(n-1)$ nonzero elements of R. Since a_1 and a_2 are both not zero divisors, their product $a_1 \odot a_2$ cannot be a zero divisor either. (Why? See explanation at the end of this proof.) By applying Theorem 8(i), we conclude from the above equation (by cancelling out the factor $a_1 \odot a_2$) that $x_1 = x_2$.

Following the same reasoning, one can prove that any two of $(n-1)$ equations $a_1 \odot x = a_1, \ldots, a_{n-1} \odot x = a_{n-1}$, have exactly the same solution. Let 1_R be that solution. We have proved that $a_1 \odot 1_R = a_1, \ldots, a_{n-1} \odot 1_R = a_{n-1}$. Since R is commutative, we have $1_R \odot a_1 = a_1, \ldots, 1_R \odot a_{n-1} = a_{n-1}$. Moreover, the same is true for the nth element of R, namely $a_n = 0_R$. Indeed by Theorem 6(iii) (or by

Theorem 3) we have

$$1_R \odot a_n = 1_R \odot 0_R = 0_R = 0_R \odot 1_R = a_n \odot 1_R$$

In effect we have shown that $a \odot 1_R = a = 1_R = a = 1_R \odot a$, $\forall a \in R$, establishing the element 1_R as the identity element of R. To conclude the proof we must also show that every nonzero element of R is a unit. We know that $a_1, \dots, a_{n-1}$ are the nonzero elements of R. We have already established that the $(n-1)$ mapping s $f_1, \dots, f_n$ are surjective. If a is any of the elements $a_1, \dots, a_{n-1}$ there exists a' in R such that $a \odot a' = 1_R$. By commutativity, $1_R = a' \odot a$ also; $a' \neq 0_R$ since otherwise we would have $0_R = 1_R$ which is impossible for a ring of at least two elements. Thus, the definition of unit is satisfied. We are done. ■

(ii) This is immediate since every integral domain has no zero divisors. Part (ii) is the direct consequence of part (i). ■

Note: In any ring R if a_1 and a_2 are both not zero divisors, $a_1 \odot a_2$ and $a_2 \odot a_1$ are not zero divisors either. Suppose $(a_1 \odot a_2) \odot b = 0_R$. Then $a_1 \odot (a_2 \odot b) = 0_R \Rightarrow$ (since a_1 is not a zero divisor) $a_2 \odot b = 0_R \Rightarrow$ (since a_2 is not a zero divisor) $\Rightarrow b = 0_R$. Similarly one proves that $b \odot (a_1 \odot a_2) = 0_R$ implies $b = 0_R$. In a similar way one shows that the element $a_2 \odot a_1$ is not a zero divisor either.

2.3 Homomorphisms and Isomorphisms

Before we introduce the notions of homomorphism and isomorphism, we need to clarify what it means to add an element to itself n times in a ring R; and to raise an element to the nth power.

Definition 10: Let R be a ring with operations $\oplus$ and $\odot$, a any element of R, and n a positive integer.

(i) The notation $\boldsymbol{na}$ stands for $\underbrace{\boldsymbol{a \odot a \odot \ldots \odot a}}_{\textbf{n times}}$; $\boldsymbol{a^n} = \underbrace{\boldsymbol{a \odot a \odot \ldots \odot a}}_{\textbf{n times}}$.

We are now ready to introduce the notions of homomorphism and isomorphism. We hae the following definition.

Definition 11: Let R_1 be a ring with operations $\oplus$ and $\odot$, R_2 a ring with operations (addition) $*$ and multiplication $\otimes$; and f a mapping of R_1 to R_2.

(i) The mapping f is said to have the **additive property** or simply be **additive** if $f(x_1 \oplus x_2) = f(x_1) * f(x_2)$ for **all** x_1, x_2 elements of R_1. Verbally this property can be stated as follows: **The image of the sum of any two elements in R_1, is equal to the sum of the images.**

(ii) The mapping f is said to have **the multiplicative property** or simply be **multiplicativeif**, $f(x_1 \odot x_2) = f(x_1) \otimes f(x_2)$ for **all** x_1, x_2, elements of R_1. Verbally this property can be stated as follows: **The image of the product of any two elements in R_1, is equal to the product of the images.**

(iii) The mapping f is said to be a **homomorphism** from the ring R_1 to the ring R_2, if it is **both** additive and multiplicative.

(iv) The mapping f is said to be an **isomorphism** from the ring R_1 to the ring R_2, if it is **both** a homomorphism and a bijection (that is, one-to-one and onto mapping). We then say that R_1 and R_2 are **isomorphic** and we may use the notation $\boldsymbol{R_1 \simeq R_2}$.

Remark 8: (i) As it will be shown in Theorem 10, under an additive mapping f, the zero element of the ring R_1 is **always** mapped on the zero element the second ring R_2. The proof of this is very short.

(ii) Under a multiplicative mapping f, even if each of the rings R_1 and R_2 possesses an identity, the image of the identity element of R_1 may not equal the identity element of R_2 (see examples 7 and 8). If however, as Theorem 11 shows, the mapping f is a surjective (or onto) homomorphism and both R_1 and R_2 have identity elements, then the image of the identity element of R_1 must equal the identity element of R_2.

(iii) It is important for the student to realize early on that if f is an isomorphism, then R_1 and R_2 are essentially carbon copies of each other. The fact that f is **both** a homorphism and a bijection allows f to **preserve the structure** of R_1 onto R_2. The bijection part allows for a relabeling of all the elements of R_1, to obtain one by one, all the elements of R_2. If R_1 and R_2 are finite isomorphic rings, then their tables of operations are finite. By relabeling all the elements of one, we can simply obtain the elements of the other. The fact that f is a homomorphism, allows us to obtain from the table of one, the table of the other. Any to isomorphic rings must have exactly the same properties.

(iv) According to exercise 50, if f is an isomorphism from R_1 to R_2 then the inverse bijection f^{-1} is an isomorphism from R_2 to R_1. This should come as no surprise.

Below we present twelve examples of additive mappings; multiplicative mappings, homomorphisms, and isomorphisms. We will then state Theorems 10 and 11 and their proofs. After that, we present another 7 examples. We finish the section with Theorem 12.

Example 27: If S is a subring of a ring R, then by taking $R_1 = S$ and $R_2 = R$. Consider the two mappings: $f : S \to R$; $f(x) = 0_R$ (the zero element of R) and $I_S : S \to R$; $I_S(x) = x$. The first mapping is the zero mapping on S (since S and R have the same zero element) and the second mapping is the identity mapping on S. They are both homomorphisms. The second mapping is injective. It is easy to show that both of these two mappings are both additive and multiplicative. We leave this to the reader.

Example 28: The three mappings $f_1 : \mathbb{Z} \to \mathbb{Z}$; $f_1(x) = n \cdot x$, $f_2 : \mathbb{Q} \to \mathbb{Q}$; $f_2(x) = n\dot{x}$, and $f_3 : \mathbb{R} \to \mathbb{R}$; $f_3(x) = n\dot{x}$, where n is a fixed integer $n \neq 0, 1$ are all additive mappings, (they are all defined by the same formula) as it can be easily checked. Note that f_2 is an extension of f_1 to the set $\mathbb{Q}$. f_3 is an extension of f_2 to the set $\mathbb{R}$. If x_1, x_2 are any real numbers, then $f_3(x_1 + x_2) = n(x_1 + x_2) = nx_1 + nx_2 = f_3(x_1) + f_3(x)$; f_3 is additive and so are f_1 and f_2. However none of these mappings is multiplicative. For instance, $f_3(x_1 \cdot x_2) = n \cdot x_1 \cdot x_2$. Then for $x_1 \cdot x_2 \neq 0$, we have, $nx_1x_2 \neq (nx_1)(nx_2) = n^2x_1x_2$, since $n^2 \neq n$ by virtue of $n \neq 0, 1$. Also because n is nonzero, it is easy to see that all three mappings f_1, f_2, f_3 are injective or one-to-one. Furthermore both f_2 and f_3 are also surjective, while f_1 is not onto. Indeed given any number y in $\mathbb{Q}$ or in $\mathbb{R}$ there is a number x in $\mathbb{Q}$ or $\mathbb{R}$ such that $n \cdot x = y$. Simply take $x = n^{-1} \cdot y$ since any nonzero element in both $\mathbb{Q}$ and $\mathbb{R}$ is a unit while n^{-1} does not exist in $\mathbb{Z}$ unless n is 1 or -1 (the only two units of $\mathbb{Z}$). Also, note that the range of the mapping f_1 consists of all multiples of n. So those integers that are not divisible by n, will not be in the range of f_1. In summary we see that both mappings f_3 and f_2 are additive bijections, while f_1 is only an additive injection.

Example 29: One can prove that every additive f mapping from the ring $\mathbb{Z}$ (under the usual operations of addition and multiplication) to itself, must be of the form $f(x) = k \cdot x$, for every $x \in \mathbb{Z}$, from some fixed integer k. To see why, let f be an additive mapping: $f : \mathbb{Z} \to \mathbb{Z}$ and $f(x_1 + x_2) = f(x_1) + f(x_2)$, for all $x_1, x_2 \in \mathbb{Z}$.

$$\text{Consider } \begin{aligned} &f(2) = f(1+1) = f(1) + f(1); \\ &f(3) = f(2+1) = f(2) + f(1) = (f(1) + f(1)) + f(1) = 3 \cdot f(1). \end{aligned}$$

Continuing in this way we see that for every positive integer n we have $f(n) = n \cdot f(1)$. Next we showed that $f(0) = 0$. Indeed, $f(0) = f(0+0) \Rightarrow f(0) = f(0) + f(0) \Rightarrow f(0) = 0$. Now consider $f(-1)$. We know that $f(0) = 0 \Rightarrow f(1 + (-1)) = 0 \Rightarrow f(1) + f(-1) = 0 \Longrightarrow f(-1) = -f(1)$. If m is a negative integer, we can put $m = -n$, where n is a positive integer; $f(m) = f(-n) = f[\underbrace{(-1) + (-1) + \ldots + (-1)}_{n \text{ times}}] = f(-1) + f(-1) = \ldots + f(-1) = n \cdot f(-1) = n \cdot (-f(1)) = -n \cdot f(1) = mf(1)$

(this step is justified by applying the additive property successively $(n-1)$ times). Thus, we have shown that for any integer l, whether positive, zero, or negative, we have $f(l) = l \cdot f(1)$; since $f(1)$ is a fixed integer,, say $f(1) = k$, we see that $f(x) = x \cdot f(1) = k \cdot x, \quad \forall x \in \mathbb{Z}$. We point out here that it is also true that any additive mapping g from the ring $\mathbb{Q}$ to itself must also be a similar form: $g(x) = r \cdot x$, for every $x \in \mathbb{Q}$ where r is some fixed rational number. The argument is similar as in the case of the mapping f above. First one proves (as we did previously), that for every integer n we must have $g(n) = n \cdot g(1)$; let $g(1) = r$. Pick any rational number $x = p/q$, p and q integers, $q \neq 0$. We may assume $p > 0$ (since if $p/q < 0$, then we may take $p > 0$ and $q < 0$). By applying the additivity of the mapping g, $(p-1)$ times, we have

$$g(p/q) = g(\underbrace{\frac{1}{q} + \ldots + \frac{1}{q}}_{p \text{ times}}) = \underbrace{g(1/q) + \ldots + g(1/q)}_{p \text{ times}} = p \cdot g(1/q);$$

so we have $g(p/q) = p \cdot g(1/q)$. Now multiply the right-hand side by q and divide it by q as well:

$$g(p/q) = \frac{p \cdot q}{q} \cdot g(1/q);$$

But $$q \cdot g(1/q) = \underbrace{g(1/q) + \ldots + g(1/q)}_{q \text{ times}} = g(\underbrace{1/q + \ldots + 1/q}_{q \text{ times}}) = g(q/q) = g(1).$$

Hence $$g(p/q) = p/q \cdot (q \cdot g(1/q)) = p/q \cdot g(1) = (p/q) \cdot r, \text{ since } r = g(1).$$

We have shown that for every rational number x, $g(x) = x \cdot g(1) = x \cdot r$. We note here that when one goes from the set $\mathbb{Q}$ to the set of reals $\mathbb{R}$, it is no longer true that every additive mapping h from $\mathbb{R}$ to itself is one of the form, $h(x) = x \cdot h(1)$, $\forall x \in \mathbb{R}$.

There exist other types of additive mappings of $\mathbb{R}$ to itself, not of the above form. This topic is beyond the scope of this book.

Example 30: Recall the ring of example 3 which we will denote by $n\mathbb{Z}$; $n\mathbb{Z} = \{x \in \mathbb{Z} \text{ and } x = n \cdot y,\ y \in \mathbb{Z}\}$, where n is some fixed positive integer with $n \geq 2$. The mapping $f : \mathbb{Z} \to n\mathbb{Z}$; $f(x) = n \cdot x$, is an additive mapping, but not multiplicative as it is easily seen. It is also surjective.

Example 31: For given real constants m and b, the linear function, familiar from Calculus, $f : \mathbb{R} \to \mathbb{R}$; $f(x) = mx+b$, is neither additive nor multiplicative when $b \neq 0$. For $b = 0$ it is additive for any value of the constant m (the slope of the straight line with equation $y = mx + b$); and only for $b = 0$ and $m = 1$ is f both additive and multiplicative, that is, a homomorphism. Of course in that case, $f(x) = x$, so f is in that case the identity mapping on $\mathbb{R}$ and hence, trivially, an isomorphism. Also note that for $m \neq 0$, f is always injective and (for $m \neq 0$) surjective.

Example 32: The only homorphisms from the ring $\mathbb{Z}$ (with the usual addition and multiplication) are the identity and zero mappings. This is easily seen from Example

29. In that example we saw that if $f : \mathbb{Z} \to \mathbb{Z}$ is an additive mapping, then $f(x) = k \cdot x$, $\forall x \in \mathbb{Z}$; where $k = f(1)$ is a fixed integer. If f is also multiplicative, it must satisfy $f(x_1x_2) = f(x_1)f(x_2) \Leftrightarrow k \cdot x_1x_2 = kx_1kx_2$; for all $x_1, x_2 \in \mathbb{Z}$; $k \cdot x_1x_2 = k^2 \cdot x_1 \cdot x_2$; by taking $x_1x_2 \neq 0$ we conclude $k = k^2 \Leftrightarrow k \cdot (k-1) = 0$; $k = 0$ or 1. So that f is either the zero mapping or the identity mapping.

The same is true for $\mathbb{Q}$. Its only homomorphisms are the zero and identity mappings.

Example 33: Let R be a ring with operations $\oplus$ and $\odot$, and b a fixed element of R. The set $b \odot R = \{x \in R \mid x = b \odot y,\ y \in R\}$, is a subring of R because it is closed under multiplication $\odot$ and subtraction of elements $\ominus$; refer to Theorem 7. Indeed if $x_1, x_2 \in R$; then $x_1 \ominus x_2 = (b \odot y_1) \ominus (b \odot y_2) = b \odot (y_1 \ominus y_2) \in b \odot R$; also $x_1 \odot x_2 = (b \odot y_1) \odot (b \odot y_2) \Rightarrow x_1 \odot x_2 = b \odot [y_1 \odot (b \odot y_2)]$, by the associative property of multiplication. Thus, $x_1 \odot x_2$ is also a member of the set $b \odot R$. We conclude that the above set is a subring of R. The mapping $f : R \to b \odot R$; $f(x) = b \odot x$, $\forall x \in R$, is an additive surjective mapping, as it is easily shown: $f(x_1 \oplus x_2) = b \odot (x_1 \oplus x_2) = (b \odot x_1) \oplus (b \odot x_2) = f(x_1) \oplus f(x_2)$, by the distributive Axiom 8.

Example 34: Given two isomorphic rings, it is important to realize that most bijections from one ring to the other will not be isomorphism. For instance if we take $R_1 = \mathbb{Z} = R_2$, then obviously $\mathbb{Z}$ is isomorphic to itself. Furthermore, for each integer k, the mapping $f_k : \mathbb{Z} \to \mathbb{Z}$; $f_k(x) = x + k$, is a bijection, as it is easily shown. Only f_0 (that is, f_k, for $k = 0$) is an isomorphism; $f + 0 = I_{\mathbb{Z}} =$ identity map on $\mathbb{Z}$. In fact, interestingly, all the other mappings f_k, with $k \neq 0$, fail to be additive or multiplicative.

Example 35: The set $S = \{0, 2, 4, 6, 8\}$ is a subring of $\mathbb{Z}_{10}$, refer to Example 15. The identity mapping I_S on S; $I_S : S \to \mathbb{Z}_{10}$; $I_S(x) = x$, is an injective homomorphism. The mapping f defined as, $f : \mathbb{Z}_{10} \to S$; $f(x) = 2 \odot x =$ remainder of the division of the integer $2x$ with 10 is a surjective homomorphism (verify this!).

Example 36: If m is an odd positive integer, one can show that the set $S = \{0, 2, \ldots, m, \ldots, 2(m-1)\}$ is a subring of the ring $\mathbb{Z}_{2m}$. They have different identity elements: S has identity $1_S = m + 1$, while $\mathbb{Z}_{2m}$ has the usual identify $1_R = 1$. Verify this! Note that this example is a generalization of the previous Example 35 (in that example $m = 5$; $1_s = 6$). The mapping $I_S : S \to \mathbb{Z}_{2m} : I_S(x) = x$, is an injective homorphism. The mapping $f : \mathbb{Z}_{2m} \to S$; $f(x) = 2 \odot x =$ remainder of the division of the integer $2x$ with $2m$, is a surjective homomorphism.

Example 37: The subring $S = \{0, 2, 4, 6, 8\}$ of $\mathbb{Z}_{10}$, is isomorphic to the ring $\mathbb{Z}_5 = \{0, 1, 2, 3, 4\}$, under the operations $\oplus$ and $\odot$ as defined in Example 6. Consider the mapping $f : S \to \mathbb{Z}_5$, defined by $f(0) = 0, f(2) = 2,\ f(4) = 4,\ f(6) = 1,\ f(8) = 3$. Note that an isomorphism must carry not only the zero element of one ring to the zero element of the other, but also the identity element of one to the identity element of the other. In our case, $f(6) = 1$. Taking this, a starting points and using the multiplication table of Example 15, one can figure out the images of the rest of the elements of S; using the fact that f must be both additive and multiplicative we have,

$$f(2) = f(6 \oplus 6) = f(6) \oplus f(6) = 1 \oplus 1 = 2$$
$$f(4) = f(2 \oplus 2) = f(2) \oplus f(2) = 2 \oplus 2 = 4$$
$$f(8) = f(4 \oplus 4) = f(4) \oplus f(4) = 4 \oplus 4 = 3$$

We leave it to the reader to verify that$f$ is a homomorphism. Obviously, by definition, the mapping f is one-to-one and onto, hence a bijection. This will establish that f is an isomorphism.

Example 38: The subring $S = \{0, 2, \ldots 2(m-1)\}$, m an odd positive integer, of the ring $\mathbb{Z}_{2m}$ (see Example 36), is isomorphic to the ring $\mathbb{Z}_m$. This is a generalization of Example 37. Note that since $(m+1)$ is the identity of S, if f is an isomorphism from S to $\mathbb{Z}_{2m}$, it must satisfy $f(m+1) = 1$ (recall $\mathbb{Z}_m = \{0, 1, \ldots, m-1\}$, with the usual operations as descried in Example 6). As in Example 37, by starting with $f(m+1) = 1$ and using the fact that f must be additive, one can, in theory determine what the images of the remaining $(m-1)$ elements of S must be: $f(0) = 0$, $f(2), \ldots, f(m-1)$, $f(m+1) = 1, \ldots, f[2(m-1)]$. So, perhaps not surprisingly there is really only one isomorphism from S to $\mathbb{Z}_m$ – see Remark 9. If one knows the value of m, one can find the specific images above. For the general, m formulas for the images in terms of m could be found, but that would be complicated.

We now prove two theorems that shed light on the important properties of additive and multiplicative mappings between two rings. The first theorem lists the basic properties of additive mappings. The second gives a listing of basic properties of multiplicative mappings. As a consequence, a homomorphism will satisfy both theorems.

Theorem 10 (additive) Let R_1 be a ring with operations of addition $\oplus$ and multiplication $\ominus$ and R_2 a ring with operations of addition $*$ and multiplication $\otimes$; and with neutral elements 0_{R_1} and 0_{R_2}, respectively. Let $\ominus$ and/be the symbols of subtraction for the two rings, respectively. Suppose that f is an additive mapping from R_1 to R_2. The following properties of f hold true:

(a) $f(0_{R_1}) = 0_{R_2}$.

(b) $f(\ominus x) = \backslash f(x)$, for every $x \in R_1$.

(c) $f(x_1 \ominus x_2) = f(x_1) \backslash f(x_2)$, for all $x_1, x_2 \in R_1$.

(d) $f(nx) = nf(x)$, for every $x \in R_1$ and every positive integer n.

Proof: (a) Since $0_{R_1} = 0_{R_1} \oplus 0_{R_1}$ and f is additive, we have $f(0_{R_1}) = f(0_{R_1}) \oplus f(0_{R_1}) \Rightarrow f(0_{R_1}) = f(0_{R_1}) * f(0_{R_1})$; adding the opposite $\backslash f(0_{R_1})$ of $f(0_{R_1})$ to both sides of the last equation (remember $f(0_{R_1})$ is an element of R_2) we obtain,

$$\backslash f(0_{R_1}) * f(0_{R_1}) = \backslash f(0_{R_1}) * (f(0_{R_1}) * f(0_{R_1})) \Rightarrow$$
$$\Rightarrow \backslash f(0_{R_1}) * f(0_{R_1}) = (\backslash f(0_{R_1}) * f(0_{R_1})) * f(0_{R_1}).$$

But $\backslash f(0_{R_1}) * f(0_{R_1}) = 0_{R_2}$. Hence the last equation implies,

$$0_{R_2} = 0_{R_2} * f(0_{R_1}); \text{ but } 0_{R_2} * f(0_{R_1}) = f(0_{R_1}).$$

Since 0_{R_2} is the neutral element of the ring R_2.

We conclude that $0_{R_2} = f(0_{R_1})$. ■

(b) Start with $x \ominus x = 0_{R_1}$; $f(x \ominus x) = f(0_{R_1})$. By Part (a) we know that $f(0_{R_1}) = 0_{R_2}$ so that we obtain,

$$f(x \ominus x) = 0_{R_2}.$$

However, the element $x \ominus x$ is none other than $x \oplus (\ominus x)$. We have, $f(x \ominus x) = 0_{R_2} \Rightarrow f(x \oplus (\ominus x)) = 0_{R_2} \Rightarrow$ (since f is additive) $f(x) \oplus f(\ominus x) = 0_{R_2}$.

The last statement says that the element $f(\ominus x)$ is opposite of $f(x)$. According to Theorem 6(i), the opposite of each element is unique. Thus $f(\ominus x) = \backslash f(x)$, since $\backslash f(x)$ is the opposite of $f(x)$ in the ring R_2. ■

(c) We have $f(x_1 \ominus x_x) = f(x_1 \oplus (\ominus x_2)) = f(x_1) * f(\ominus x_2)$; however by part (b) we know that $f(\ominus x_2) = \backslash f(x_2)$, so that $f(x_1 \ominus x_2) = f(x_1) * f(\ominus x_2) = f(x_1) * (\backslash f(x_2)) = f(x_1) \backslash f(x_2)$. ■

(d) According to Definition 10(i), nx simply means $\underbrace{x \oplus \ldots \oplus x}_{n \text{ times}}$; so that $f(nx) = f(nx) = \underbrace{(x \oplus \ldots \oplus x)}_{n \text{ times}}$. By applying the additivity of the mapping $(n-1)$ times successively we get

$$f(nx) = f(\underbrace{x \oplus \ldots \oplus x}_{n \text{ times}}) = f[x \oplus (\underbrace{x \oplus \ldots \oplus x}_{(n-1) \text{ times}})] =$$

$$= f(x) * f(\underbrace{x \oplus \ldots \oplus x}_{(n-1) \text{ times}}) = \ldots = \underbrace{f(x) * \ldots * f(x)}_{n \text{ times}}$$

■

Theorem 11 (multiplicative) Let R_1 be a ring with operations of addition $\oplus$ and multiplication $\odot$; R_2 a ring with operations of addition $*$ and multiplication $\otimes$; and $0_{R_1}, 0_{R_2}$ the neutral elements of the rings R_1 and R_2, respectively. Suppose that f is a multiplicative mapping from R_1 to R_2. The following properties of f hold true:

(i) If R_1 has an identity element 1_{R_1}, then $f(1_{R_1})$ is an identity element for the image of R_1 under f (but not necessarily an identity for the ring R_2). That is, for the range R_f of the mapping f. In other words, $(\forall y \in Im(R_1) = R_f) \Rightarrow (y \otimes f(1_{R_1}) = y = f(1_{R_1}) \otimes y)$.

(ii) If f is a surjective mapping and R_1 has an identity element 1_{R_1}, then $f(1_{R_1})$ is the identity element of R_2; $f(1_{R_1}) = 1_{R_2}$.

(iii) If f is a surjective mapping and R_1 has an identity element 1_{R_1}, then $f(u^{-1}) = [f(u)]^{-1}$, for every unit u of R_1.

(iv) $f(x^n) = [f(x)]^n$, for every positive integer n and every element x of R_1.

(v) If f is an injective mapping and a is a zero divisor of the ring R_1, then $f(a)$ is a zero divisor of the ring R_2.

Proof. (i) Let y be an element in the range of the mapping f. By definition $y = f(x)$, for some $x \in R_1$. Since 1_{R_1} is the identity element of R_1, we have $1_{R_1} \odot x =$

$x = x \odot 1_{R_1} \Rightarrow f(1_{R_1} \odot x) = f(x) = f(x \odot 1_{R_1}) \Rightarrow f(1_{R_1}) \otimes f(x) = f(x) = f(x) \otimes f(1_{R_1})$, since f is multiplicative. But $y = f(x)$ so that, $f(1_{R_1}) \otimes y = y = y \otimes f(1_{R_1})$. ∎

(ii) If f is surjective, then the range of f is the entire ring $R_2 : Im(R_1) = R_f = R_2$. Then according to part (i), $f(1_{R_1})$ is the identity element (we know that it has to be unique) for R_2; that is, $f(1_{R_1}) = 1_{R_2}$. ∎

(iii) If u is a unit, then by definition $u \odot u^{-1} = 1_{R_1} = u^{-1} \odot u \Rightarrow f(u \odot u^{-1}) = f(1_{R_1}) = f(u^{-1} \odot u) \Rightarrow f(u) \otimes f(u^{-1}) = f(1_{R_1}) = f(u^{-1}) \otimes f(u)$. Since f is surjective, by part (ii), we must have $f(1_{R_1}) = 1_{R_2}$. Hence $f(u) \otimes f(u^{-1} = 1_{R_2} = f(u^{-1} \otimes f(u)$. Then by Definition 7, this proves that $f(u)$ is a unit in the ring R_2 and $f(u^{-1})$ is the multiplicative inverse of $f(u)$. So by Definition 9(ii) we have, $f(u^{-1}) = [f(u)]^{-1}$. ∎

(iv) The proof of this part rests on applying the multiplicativeness of the mapping f, $(n-1)$ times. By definition 10(ii), $x^n = \underbrace{x \odot \ldots \odot x}_{n \text{ times}}$, so that $f(x^n) = f(\underbrace{x \odot \ldots \odot x}_{n \text{ times}}) = f[x \odot (\underbrace{x \odot \ldots \odot x}_{(n-1) \text{ times}}] = f(x)f(\underbrace{x \odot \ldots \odot x}_{(n-1) \text{ times}}) = \ldots = \underbrace{f(x) \otimes \ldots \otimes f(x)}_{n \text{ times}} = [f(x)]^n$.

(v) First observe that if f is injective or one-to-one, then the only element in R_1 whose image under f is the zero element in R_2, 0_{R_2}, is 0_{R_1}. We know from Theorem 10(a), that $f(0_{R_1}) = 0_{R_2}$ thus we can write (f is injective) $\Rightarrow (f(x) = 0_{R_2} \Rightarrow x = 0_{R_1})$.

Let a be a zero divisor of the ring R_1. Then, according to Definition 6, $a \neq 0_{R_1}$; and there exists $b \in R_1$ such that $b \neq 0_{R_1}$ and $a \cdot b = 0_{R_1}$. Now apply the mapping f: $a \odot b = 0_{R_1} \Rightarrow f(a \cdot b) = f(0_{R_1}) \Rightarrow f(a) \otimes f(b) = 0_{R_2}$. But f is one-to-one and since $a \neq 0_{R_1}$, $b \neq 0_{R_1}$, and $f(0_{R_1}) = 0_{R_2}$, it readily follows that $f(a) \neq 0_{R_2}$ and $f(b) \neq 0_{R_2}$. This latter conclusion couched with the equation $f(a) \otimes f(b) = 0_{R_2}$, proves according to Definition 6 that $f(a)$ is a zero divisor of the ring R_2.

Below we give a summary of the basic properties of additive and multiplicative maps:

(a) The zero element of the first ring is mapped onto the zero element of the second ring.

(b) The opposite of any element in the first ring is mapped onto the opposite of its image in the second ring.

(c) The image of the difference of any two elements in the first ring is equal to the difference of their respective images in the second ring.

(d) The image of the sum of any element (in the first ring) n times to itself is equal to the sum of its image n times to itself, in the second ring

Multiplicative

(i) The image of the identity element in the first ring, is the identity for the image set under the mapping f; that is, it is the identity element of the range of f (but not necessarily for the ring R_2).

(ii) If f is surjective, then the identity element of R_1 is mapped onto the identity element of R_2.

(iii) If f is surjective, the image of the multiplicative inverse of a unit in the first ring, is the multiplicative inverse of the image of the unit, in the second ring.

(iv) The image of the product of any element (in the first ring) n times itself, is equal to the product of its image n times itself, in the second ring.

(v) If f is injective, the image of a zero divisor in the first ring, is a zero divisor in the second ring.

Important Statement

If the mapping f is a homomorphism, then it must satisfy all nine properties. The four additive properties (a) through (d) and the five multiplicative properties (i) through (v).

Remark 9: A ring R with identity element 1_R, is called **identity generated**, if each of each elements can be obtained from 1_R by adding 1_R to itself a (finite) number of times. In other words, for every element x of R, there exists a positive integer k such that $\underbrace{1_R \oplus 1_R \oplus \ldots \oplus 1_R}_{k \text{ times}} = x$. If two **finite** identity generated rings R_1 and R_2 with identity elements are isomorphic, then there is only one isomorphism f from R_1 to R_2. The inverse mapping f^{-1}, is the only isomorphism from R_2 to R_1. This is not hard to see because an isomorphism (since it is multiplicative) must carry the identity element of one ring onto the identity of the other; $f(1_{R_1}) = 1_{R_2}$. Furthermore, say R_1 has n elements. Obviously, R_2 must also have n elements, since f is, in particular, a bijection. Among the n elements of R_1 are the neutral element 0_{R_1} and the identity 1_{R_1}. Consider the elements 0_{R_1}, 1_{R_1}, $\underbrace{1_{R_1} \oplus 1_{R_1}}_{2 \text{ times}}$, $\underbrace{1_{R_1} \oplus 1_{R_1} \oplus 1_{R_1}}_{\text{three times}}$, $\ldots$, $\underbrace{1_{R_1} \oplus 1_{R_1} \oplus \ldots \oplus 1_{R_1}}_{(n-2) \text{ times}}$. Here we assume $n \geq 3$ (if $n = 2$, then R is a ring with elements 0_{R_1} and 1_{R_1}). Since $R_1 \simeq R_2$, then obviously there is only one isomorphism $f : 0_{R_1} \to 0_{R_2}$ and $1_{R_1} \to 1_{R_2}$. Since R_1 is a finite ring, there must be a least or smallest positive integer k such that the sum $\underbrace{1_{R_1} \oplus 1_{R_1} \oplus \ldots \oplus 1_{R_1}}_{k \text{ times}}$ is equal to 0_{R_1} (because keep adding the identity to itself must at some point, since R_1 is finite, coincide with one of the previous such sums; say $\underbrace{1_{R_1} \oplus \ldots \oplus 1_{R_1}}_{l \text{ times}} = \underbrace{1_{R_1} \oplus \ldots \oplus 1_{R_1}}_{m \text{ times}} \Rightarrow$ (by subtracting $\underbrace{1_{R_1} \oplus \ldots \oplus 1_{R_1}}_{(l-m) \text{ times}} = 0_{R_1}$; with $l > m$)).

Now if R_1 has n elements, then the smallest positive integer k for which $\underbrace{1_{R_1} \oplus \ldots \oplus 1_{R_1}}_{k \text{ times}} = 0_{R_1}$, must be n. That is $k = n$. To see why, first observe that if $k \neq n$, since R_1 has only n elements. k cannot be less than n: by the definition of the integer k. The k elements

$\otimes$	(0,0)	(0,1)	(0,2)	(1,0)	(1,1)	(1,2)
(0,0)	(0,0)	(0,0)	(0,0)	(0,0)	(0,0)	(0,0)
(0,1)	(0,0)	(0,1)	(0,2)	(0,0)	(0,1)	(0,2)
(0,2)	(0,0)	(0,2)	(0,1)	(0,0)	(0,2)	(0,1)
(1,0)	(0,0)	(0,0)	(0,0)	(1,0)	(1,0)	(1,0)
(1,0)	(0,0)	(0,0)	(0,0)	(1,0)	(1,0)	(1,0)
(1,1)	(0,0)	(0,1)	(0,2)	(1,0)	(1,1)	(1,2)
(1,2)	(0,0)	(0,2)	(0,1)	(1,0)	(1,2)	(1,1)

Figure 4

Thus mapping $f : \mathbb{Z}_6 \to \mathbb{Z}_2 \times \mathbb{Z}_3$; $(x) = ([x]_2,\ [x]_3)$, is an isomorphism. Here (as in Example 39) $[x]_2$ denotes the remainder of the division of the integer x with 2 and $[x]_3$ the remainder of the division of x with 3. A calculation shows that,

$$f(0) = ([0]_2,\ [0]_3) = (0,0);\ f(1) = ([1]_2,\ [1]_3) = (1,1);$$
$$f(2) = ([2]_2,\ [2]_3) = (0,2);\ f(3) = ([3]_2,\ [3]_3) = (1,0);$$
$$f(4) = ([4]_2,\ [4]_3) = (0,1);\ f(5) = ([5]_2,\ [5]_3) = (1,2).$$

The reader should verify that f is an isomorphism.

Example 41: Using the notation of Examples 39 and 40, the following two mappings are homomorphisms:

$$f_1 : \mathbb{Z} \to \mathbb{Z}_n;\ f(x) = [x]_n$$
$$f_2 : \mathbb{Z}_m \to \mathbb{Z}_n;\ f(x) = [x]_n$$

The mapping $f_k : \mathbb{Z}_m \to \mathbb{Z}_n$; $f(x) = [k \cdot x]_n$, where k is a fixed positive integer $k \geq 2$, is only an additive mapping, not a homomorphism.

Example 42: The following two mappings are only additive, not homomorphisms.

For given positive integers m and n, with $m \neq n$, the mapping $f : m\mathbb{Z} \to n\mathbb{Z}$; $f(x) = \frac{n}{m} \cdot x$ (he rings $m\mathbb{Z}$ and $n\mathbb{Z}$ are both subrings of $\mathbb{Z}$, refer to Example 3 and Example 30), is only additive. Let k and l be fixed positive integers with at least one k and l being greater than 1 (that is $k \geq 2$ or $l \geq 2$). Consider the mapping $g : \mathbb{Q} \to \mathbb{Z} \times \mathbb{Z}$; $f\left(\frac{x_1}{x_2}\right) = (k \cdot x_1,\ lx_2)$; where the rational number $x = \frac{x_1}{x_2}$ is in lowest terms. That is the greatest common of the integers x_1 and x_2 is equal to 1; and of course $x_2 \neq 0$. The mapping g is additive, but not multiplicative, hence not a homomorphism.

Example 43: Going back to Example 7, where we encountered the ring $M_2(\mathbb{R})$ of all 2×2 matrices with real number entries and with operations the usual matrix addition and matrix multiplication. The following six sets are subrings of $M_2(\mathbb{R})$ as it can be easily verified:

$$S_1 = \left\{ \begin{pmatrix} x & 0 \\ 0 & 0 \end{pmatrix} \middle| x \in \mathbb{R} \right\},\quad S_2 = \left\{ \begin{pmatrix} 0 & x \\ 0 & 0 \end{pmatrix} \middle| x \in \mathbb{R} \right\},$$

$$S_3 = \left\{ \begin{pmatrix} 0 & 0 \\ x & 0 \end{pmatrix} \middle| x \in \mathbb{R} \right\}, \quad S_4 = \left\{ \begin{pmatrix} 0 & 0 \\ 0 & x \end{pmatrix} \middle| x \in \mathbb{R} \right\},$$

$$S_5 = \left\{ \begin{pmatrix} x & 0 \\ 0 & x \end{pmatrix} \middle| x \in \mathbb{R} \right\}, \quad S_5 = \left\{ \begin{pmatrix} 0 & x \\ x & 0 \end{pmatrix} \middle| x \in \mathbb{R} \right\}.$$

All six of them, S_1 through S_6 are isomorphic to the ring $\mathbb{R}$. The verification of this is left to the reader.

Example 44: Recall from Example 22, the subring S of $M_2(\mathbb{R})$, defined by $S = \left\{ \begin{pmatrix} x & y \\ -y & x \end{pmatrix} \middle| x \in \mathbb{R},\ y \in \mathbb{R} \right\}$, under the usual operations of matrix addition and multiplication. Under the same operations of matrix addition and multiplication, the subset $T = \left\{ \begin{pmatrix} x & y \\ y & -x \end{pmatrix} \middle| x \in \mathbb{R},\ y \in \mathbb{R} \right\}$ of $M_2(\mathbb{R})$, is also a subring of $M_2(\mathbb{R})$. Consider the two mappings,

$$f : S \to S;\ f\left(\begin{pmatrix} x & y \\ -y & x \end{pmatrix}\right) = \begin{pmatrix} x & -y \\ y & x \end{pmatrix} \qquad \text{(this lies in } S, \text{ since } -(-y) = y)$$

$$g : S \to T;\ g\left(\begin{pmatrix} x & y \\ -y & x \end{pmatrix}\right) = \begin{pmatrix} x & y \\ y & -x \end{pmatrix}$$

Both of these mappings are additive as it can be easily verified. It is also fairly obvious that they are both injective mappings and they are both surjective as well. In other words, both f and g are additive bijections. However only f is multiplicative; g fails to be multiplicative. Let us see why. Take any two elements of

$$S;\ \begin{pmatrix} x_1 & y_1 \\ -y_1 & x_1 \end{pmatrix} \text{ and } \begin{pmatrix} x_2 & y_2 \\ -y_2 & x_2 \end{pmatrix}$$

We have $\begin{pmatrix} x_1 & y_1 \\ -y_1 & x_1 \end{pmatrix} \odot \begin{pmatrix} x_2 & y_2 \\ -y_2 & x_2 \end{pmatrix} = \begin{pmatrix} (x_1x_2 - y_1y_2) & (x_1y_2 + y_1x_2) \\ (-y_1x_2 - x_1y_2) & (-y_1y_2 + x_1x_2) \end{pmatrix}.$

So that

$$f\left(\begin{pmatrix} x_1 & y_1 \\ -y_1 & x_1 \end{pmatrix} \odot \begin{pmatrix} x_2 & y_x \\ -y_2 & x_2 \end{pmatrix}\right) =$$

$$f\begin{pmatrix} (x_1x_2 - y_1y_2) & (x_1x_2 + y_1x_2) \\ (-y_1x_2 - x_1y_2) & (-y_1y_2 + x_1x_2) \end{pmatrix} = \begin{pmatrix} (x_1x_2 - y_1y_2) & -(x_1y_2 + y_1x_1) \\ (y_1x_2 + x_1y_2) & (-y_1y_2 + x_1x_2) \end{pmatrix}.$$

Now we find the images $f\left(\begin{pmatrix} x_1 & y_1 \\ -y_1 & x_1 \end{pmatrix}\right) = \begin{pmatrix} x_1 & -y_1 \\ y_1 & x_1 \end{pmatrix}$ and f $\left(\begin{pmatrix} x_2 & y_2 \\ -y_2 & x_2 \end{pmatrix}\right) = \begin{pmatrix} x_2 & -y_2 \\ y_2 & y_2 \end{pmatrix}$ Let us multiply these two images:

$$f\left(\begin{pmatrix} x_1 & y_1 \\ -y_1 & x_1 \end{pmatrix}\right) \odot f\left(\begin{pmatrix} x_2 & y_2 \\ -y_2 & x_2 \end{pmatrix}\right) = \begin{pmatrix} x_1 & -y_1 \\ y_1 & x_1 \end{pmatrix} \odot \begin{pmatrix} x_2 & -y_2 \\ y_2 & x_2 \end{pmatrix} =$$

$$\begin{pmatrix} (x_1x_2 - y_1y_2) & -(x_1y_2 + y_1x_2) \\ (y_1x_2 + x_1y_2) & (-y_1y_2 + x_1x_2) \end{pmatrix},$$

which is the image of the product $\begin{pmatrix} x_1 & y_1 \\ -y_1 & x_1 \end{pmatrix} \odot \begin{pmatrix} x_2 & y_2 \\ -y_2 & x_2 \end{pmatrix}$ as we have already seen. Thus f is also multiplicative; and since it is additive we conclude that f is a homomorphism and since it is also a bijection; we conclude that f is an isomorphism. Now consider the mapping g. We have,

$$g\left(\begin{pmatrix} x_1 & y_1 \\ -y_1 & x_1 \end{pmatrix}\right) = \begin{pmatrix} x_1 & y_1 \\ y_1 & -x_1 \end{pmatrix} \text{ and } g\left(\begin{pmatrix} x_2 & y_2 \\ -y_2 & x_2 \end{pmatrix}\right) = \begin{pmatrix} x_2 & y_2 \\ y_2 & -x_2 \end{pmatrix}$$

and

$$g\left(\begin{pmatrix} x_1 & y_1 \\ -y_1 & x_1 \end{pmatrix} \odot g\begin{pmatrix} x_2 & y_2 \\ -y_2 & x_2 \end{pmatrix}\right) = g\begin{pmatrix} (x_1x_2 - y_1y_2) & (x_1y_2 + y_1x_2) \\ (-y_1x_2 - x_1y_2) & (-y_1y_2 + x_1x_2) \end{pmatrix}$$

$$= \begin{pmatrix} x_1x_2 - y_1y_2 & (x_1y_2 + y_1x_2) \\ (-y_1x_2 - x_1y_2) & -(-y_1y_2 + x_1x_2) \end{pmatrix} = \begin{pmatrix} x_1x_2 - y_1y_2 & (x_1y_2 + y_1x_2) \\ (-y_1x_2 - x_1y_2) & (y_1y_2 - x_1x_2) \end{pmatrix}$$

We calculate the product of the images:

$$g\left(\begin{pmatrix} x_1 & y_1 \\ -y_1 & x_1 \end{pmatrix}\right) \odot g\left(\begin{pmatrix} x_2 & y_2 \\ -y_2 & x_2 \end{pmatrix}\right) = \begin{pmatrix} x_1 & y_1 \\ y_1 & -x_1 \end{pmatrix} \odot \begin{pmatrix} x_2 & y_2 \\ y_2 & -x_2 \end{pmatrix} =$$

$$= \begin{pmatrix} (x_1x_2 + y_1y_2) & (x_1y_2 - y_1x_2) \\ (y_1x_2 - x_1y_2) & (y_1y_2 + x_1x_2) \end{pmatrix}.$$

Obviously the matrix $\begin{pmatrix} (x_1x_2 + y_1y_2) & (x_1y_2 - y_1x_2) \\ (y_1x_2 - x_1y_2) & (y_1y_2 + x_1x_2) \end{pmatrix}$ is **not** equal to the matrix $\begin{pmatrix} (x_1x_2 - y_1y_2) & (x_1y_2 + y_1x_2) \\ (-y_1x_2 - x_1y_2) & (y_1y_2 - x_1x_2) \end{pmatrix}$, for all reals x_1, x_2, y_1, y_2. In fact, an analysis shows that these two matrices are equal, that is, they have all their corresponding entries equal only when $y_1 = 0 = x_1$; or when $y_1 = 0 = x_2$; or for $y_2 = 0 = x_2$. Thus, g is not a multiplicative mapping and so it is no a homomorphism.

Example 45: Let b a fixed element of a ring R with operations $\oplus$ and $\odot$; and $\ominus$ the subtraction symbol. The mapping $f : R \to R$; $f(x) = b \odot x$, is additive. As we saw in Example 33 (in that example we consider the mapping $f : R \to b \odot R$; $f(x) = b \odot x$, but that is the exact same mapping). An interesting question arises: for which type of elements b is this mapping also multiplicative? In other word,s when is f a homomorphism? Below we examine this question only for certain types of **commutative rings** R. So assume that R is commutative. If f is also multiplicative, then $f(x_1 \odot x_2) = f(x_1) \odot f(x_2)$, for all $x_1, x_2 \in R$. This implies $b \odot (x_1 \odot x_2) = (b \odot x_1) \odot (b \odot x_2) \Rightarrow$ (since R is commutative) $b \odot (x_1 \odot x_2) = (b \odot b) \odot (x_1 \odot x_2)$ $\Rightarrow b \odot (x_1 \odot x_2) \ominus (b \odot b) \odot (x_1 \odot x_2) = 0_R$ (the neutral element) $\Rightarrow$ (by Axiom 8) $[b \ominus (b \odot b)] \odot (x_1 \odot x_2) = 0_R$; for all $x_1, x_2 \in R$. If we assume that R has an identity element 1_R, then by taking $x_1 = x_2 = 1_R$, the last equation implies,

$$b \ominus (b \odot b) = 0_R \Rightarrow b = b \odot b.$$

We would arrive at the same conclusion if we assumed that R contains at least one element a that is not a zero divisor (even though R may not have an identity element). By taking $x_1 = a = x_2$, we then have $x_1 \odot x_2 = a \odot a$; but $a \odot a$ is not a zero divisor since a is not a zero divisor (why?). Thus from

$$[b \ominus (b \odot b] \odot (x_1 \odot x_2) = 0_R; \text{ with } x_1 = x_2 = a;$$

we still conclude $b \ominus (b \odot b) = 0_R$ (since $a \odot a$ is not a zero divisor) which again gives $b = b \odot b$; or $b = b^2$. the element b is called **idempotent**; the term refers to any element of a ring, with the property that its square equals itself. We now see that in a commutative ring R, which has an identity element; or contains at least one element which is not a zero divisor; the mapping $f : R \to R$; $f(x) = b \odot x$, is a homomorphism if, and only if, b is an idempotent element. Note that 0_R is idempotent: $0_R^2 = 0_R \odot 0_R = 0_R$. So is the identity element 1_R (if R has one) : $1_R^2 = 1_R \odot 1_R = 1_R$. Note that for $b = 0_R$ or $b = 1_R$, the corresponding homomorphisms f are simply the zero and identity homomorphisms with which we are already familiar with. What about cases where b is not the zero element or the identity?

In all three rings $\mathbb{Z}, \mathbb{Q}$ and $\mathbb{R}$, the only idempotent elements are 0 and 1. The same is true for $\mathbb{Z}_n$, when $n = p$ =prime. However when n is a composite integer, there are in most cases other idempotent elements. In $\mathbb{Z}_n$, b (an integer between 0 and $(n-1)$) will be idempotent if and only if $b^2 - b$ = zero element in $\mathbb{Z}_n$ which means $b^2 - b$ must be a multiple of n (or divisible by n). By using the tools of number theory (Chapter 4), one can prove that if n has at least two distinct prime divisors, then $\mathbb{Z}_n$ will have idempotent elements other than 0 and 1. However if n is a power of a prime p, the only idempotent elements in $\mathbb{Z}_n$'s are 0 and 1. To illustrate this situation, let us suppose for simplicity that $n = p_1p_2$; the product of two different primes. We would like to find a integer b; $0 \le b \le n-1$, such that $b^2 - b$ = multiple of $n = kp_1p_2$, for some integer k; $b(b-1) = kp_1p_2$; if we can select b such that b is a multiple of p_1 and $b-1$ a multiple of p_2 (or vice-versa), our problem is solved.

As we will see in Chapter 4, the tools of number theory ensure us that this selection is always possible. For now let us give a numerical example. Take $n = 12, p_1 = 3, p_2 = 4$. The number $b = 4$ is idempotent: $4^2 - 4 = 12$ = a multiple of 12. Another such value is $b = 9$; $b^2 - b = 9(9-1) = 72$ = a multiple of 12. In fact $b = 0, 1, 4,$ and 9 are all the idempotent elements of the $\mathbb{Z}_{12}$. So the mappings $f_1 : \mathbb{Z}_{12} \to \mathbb{Z}_{12}$; $f_1(x) = 4 \odot x$, and $f_2 : \mathbb{Z}_{12} \rightarrow \mathbb{Z}_{12};\ f_2(x) = 9 \odot x$, are both homorphisms. We finish this Section by proving the Theorem below.

Theorem 12 Let R_1 be a ring with operations $\oplus$ and $\odot$. Neutral element 0_{R_1}; and R_2 a ring with operations $*$ and $\otimes$, and neutral element 0_{R_2}. Suppose that f is a homomorphism from the ring R_1 to the ring R_2. The following hold true:

(i) The image of R_1 under the mapping f is a subring of the ring R_2. In other words the range of f is a subring of R_2. (If f is surjective then obviously the range of f is the entire ring R_2.)

(ii) The subset of R_1(called the kernel of f) consisting of all elements of R_1 whose image is the zero element 0_{R_2} of R_2, form a subring of R_1. If f is injective, the kernel of f is the singleton set $\{0_{R_1}\}$.

Proof. (i) We make use of Theorem 7. Pick any two elements lying in the range of the mapping f; say y_1 and y_2. Then by definition $y_1 = f(x_1),\ y_2 = f(x_2)$, where x_1, x_2 are elements of R_1. Here we use $\backslash$ as a symbol of subtraction R_2 and $\ominus$ in R_1. We have $y_1 \backslash y_2 = f(x_1) \backslash f(x_2) = f(x_1 \ominus x_2)$, since f is a homomorphism. But $x_1 \ominus x_2 \in R_1 \Rightarrow f(x_1 \ominus x_2) \in$ Range of f. Hence $y_1 \backslash y_2$ is inside the range of f. The image set under f, is closed under subtraction $\backslash$.

Next we prove that it is also closed under multiplication $\otimes$. Indeed, $y_1 \otimes y_2 = f(x_1) \otimes f(x_2) = f(x_1 \odot x_2)$, since f is a homomorphism. Therefore, according to Theorem 7, the image set or range of f, is a subring of R_2. ■

(ii) This set is usually called the **kernel** of f, denoted by **kerf**. Obviously the set is not empty since $f(0_{R_1}) = 0_{R_2}$. Pick any two elements in kerf. Say $x_1, x_2 \in$ kerf. Then they must satisfy $f(x_1) = 0_{R_2} = f(x_2) \Rightarrow f(x_1) \backslash f(x_2) = 0_{R_2} \Rightarrow f(x_1 \ominus x_2) = 0_{R_2}$, which proves that $x_1 \ominus x_2 \in$ kerf. Thus, kerf is closed under subtraction of elements. Moreover, $f(x_1 \odot x_2) = f(x_1) \otimes f(x_2) = 0_{R_2} \otimes 0_{R_2} = 0_{R_2}$, which proves that $x_1 \odot x_2 \in$ kerf. Hence the kernel of f is also closed under multiplication of elements according to Theorem 7, kerf is a subring of R_1. Clearly if f is injective or one-to-one only the zero element 0_{R_1} can be mapped onto 0_{R_2}; kerf$= \{0_{R_1}\}$. ■

2.4 Practice Problems

P1. Let R be a ring with operations $\oplus$ and $\odot$; and $R \times R$ the Cartesian product ring (as defined in Theorem 2). Consider the subset S of $R \times R$ defined as $S = \{(x, x) | x \in R\}$.

(a) Show that S is a subring of $R \times R$.

(b) If $R = \mathbb{Z}_6$, list the elements of S.

Solution: (a) We make use of Theorem 7. Pick any two elements (x_1, x_2) and (x_2, x_2) of S; $x_1, x_2 \in R$. Let $\ominus$ be the symbol of subtraction in R; and $\backslash$ and $\overline{x}$ the symbols of subtraction and multiplication, respectively, in the ring $R \times R$. We have

$$(x_1, x_2) \backslash (x_2, x_2) = (x_1 \ominus x_2,\ x_1 \ominus x_2) \in R \times R$$
$$\text{and}\quad (x_1, x_1) \overline{x} (x_2, x_2) = (x_1 \odot x_2, x_1 \odot x_2) \in R \times R.$$

Since S is closed under both subtraction and multiplication, it follows by Theorem 7 that S is a subring of $R \times R$

(b) $S = \{(0,0), (1,1), (2,2), (3,3), (4,4), (5,5)\}$.

P2. Let $\mathbb{Z}[\sqrt{2}] = \{a + b\sqrt{2} | a, b \in \mathbb{Z}\}$. Show that the set $\mathbb{Z}[\sqrt{2}]$ is a subring of $\mathbb{R}$. Is it a field?

Solution: Let $a_1 + b_1\sqrt{2}$ and $a_2 + b_2\sqrt{2}$ be two elements of $\mathbb{Z}[\sqrt{2}]$. Their difference is $(a_1 - a_2) + (b_1 - b_2)\sqrt{2} \in \mathbb{Z}[\sqrt{2}]$, so the given set is closed under subtraction of elements. Next $(a_1 + b_1\sqrt{2}) \cdot (a_2 + b_2 \cdot \sqrt{2}) = (a_1a_2 + 2b_1b_2) + (a_1b_2 + a_2b_1) \cdot \sqrt{2} \in \mathbb{Z}[\sqrt{2}]$, since obviously both $(a_1a_2 + 2b_1b_2)$ and $(a_1b_2 + a_2b_1)$ are integers. Thus $\mathbb{Z}[\sqrt{2}]$ is also closed under multiplication, and hence it is a subring of $\mathbb{R}$, by Theorem 7. Obviously $\mathbb{Z}[\sqrt{2}]$ is a commutative ring since it is a subring of a commutative ring $\mathbb{R}$. The subring $\mathbb{Z}[\sqrt{2}]$ will be a field if, and only if each nonzero element of $\mathbb{Z}[\sqrt{2}]$ has a multiplicative inverse, i.e., if it is a unit. For given $a, b \in \mathbb{Z}$, **not both zero**, consider the equation: $(a + b\sqrt{2})(x + y\sqrt{2}) = 1$. If this equation has a solution with x and y integers, (and it will have to be unique – since the multiplicative inverse of unit must be unique), then our proof will be complete. Multiplying out the lefthand side we obtain

$$(ax + 2by) + (ay + bx)\sqrt{2} = 1 \Leftrightarrow \left\{ \begin{array}{c} ax + 2by = 1 \\ bx + ay = 0 \end{array} \right\}.$$

If we multiply the first equation by a and the second by $-2b$ and we add the two, we end up with $(a^2 - 2b^2) \cdot x = a$; but $a^2 - 2b^2$ is not equal to zero. Since if it were we would have $2 = a^2/b^2$, assuming $b \neq 0$; $\sqrt{2} = \frac{a}{b}$ which is impossible since $\sqrt{2}$ is an irrational number. On the other hand, for $b = 0$ we must have $a \neq 0$, since by assumption not both a and b are zero. Thus, in both cases, $a^2 - 2b^2 \neq 0$. Therefore from $(a^2 - 2b^2) \cdot x = a$; $x = \frac{a}{a^2 - 2b^2}$; solving for y (from the second equation) we obtain $y = -\frac{b}{a}\left(\frac{a}{a^2 - 2b^2}\right)$, or $y = \frac{-b}{a^2 - 2b^2}$, assuming $a \neq 0$. If $a = 0$, then $b \neq 0$, and the system produces $x = 0$ and $y = \frac{1}{2b}$. Note that this is also obtained from the formulas $x = \frac{a}{a^2 - 2b^2}$ and $y = \frac{-b}{a^2 - 2b^2}$ by simply putting $a = 0$ (and with $b \neq 0$) in these formulas. The question now is are the numbers $\frac{a}{a^2 - 2b^2}$ and $\frac{-b}{a^2 - 2b^2}$ integers? We know that the element $\frac{a}{a^2-2b^2} + \left(\frac{-b}{a^2-2b^2}\right) \cdot \sqrt{2}$ is the multiplicative inverse of $a + b\sqrt{2}$ in $\mathbb{R}$ (but not necessarily in $\mathbb{Z}[\sqrt{2}]$). It is easy to see that for many values of a and b neither $\frac{a}{a^2-2b^2}$ nor $\frac{-b}{a^2-2b^2}$ is an integer. For example take $a = 3$ and $b = 1$: $\frac{a}{a^2-2b^2} = \frac{3}{q-2} = \frac{3}{7}$ and $-\frac{1}{9-2} = -\frac{1}{7}$. One can find numerous such combinations of integer values of a and$b$ for which neither $\frac{a}{a^2-2b^2}$ nor $\frac{-b}{a^2-2b^2}$ is an integer. Many more such combinations for which one of $\frac{a}{a^2-2b^2}$ and $\frac{-b}{a^2-2b^2}$ is not an integer. Thus $\mathbb{Z}[\sqrt{2}]$ is not **closed** under multiplicative inverses of units. Hence, it is **not** a field.

Note: This problem can be generalized. Let k be a fixed positive integer. The set $\mathbb{Z}[k] = \{a + b\sqrt{k} | a, b \in \mathbb{Z}\}$ is a subring of $\mathbb{R}$ (but not a subfield of $\mathbb{R}$). Note that if k is a perfect square (i.e. an integer square), then $\mathbb{Z}[k] = \mathbb{Z}$. If a ring S is a subring of a ring R and if **both** R and S are fields, then S is called a **subfield** of R. In practice problem P2 we saw that $\mathbb{Z}[\sqrt{2}]$ is a subring of $\mathbb{R}$ but not a subfield of $\mathbb{R}$.

P3. Show that the set $\mathbb{Q}[\sqrt{2}] = \{a + b\sqrt{2} | a, b \in \mathbb{Q}\}$ is a subfield of $\mathbb{R}$.

Solution: Without having to repeat the work we did in P2, we see that $\mathbb{Q}[\sqrt{2}]$ is a subring of $\mathbb{R}$ (same arguments as in P2). The difference here is that each nonzero element of $\mathbb{Q}[\sqrt{2}]$ has a multiplicative inverse in $\mathbb{Q}[\sqrt{2}]$: if $a+b\sqrt{2}$ is a unit, meaning that not both rationals a and b are zero; the element $\frac{a}{a^2-2b^2} + \left(\frac{-b}{a^2-2b^2}\right)\sqrt{2}$ is the multiplicative inverse of $a + b\sqrt{2}$. Let us verify that:

$$(a + b\sqrt{2}) \cdot \left(\frac{a}{a^2-2b^2} - \frac{b\sqrt{2}}{a^2-2b^2}\right) = \frac{a^2}{a^2-2b^2} - \frac{ab\sqrt{2}}{a^2-2b^2} + \frac{ba\sqrt{2}}{a^2-2b^2} - \frac{2b^2}{a^2-2b^2} =$$
$$= \frac{a^2-2b^2}{a^2-2b^2} = 1.$$

Note that since a and b are rationals, so are the numbers $\frac{a}{a^2-2b^2}$ and $\frac{-b}{a^2-2b^2}$. Thus $\mathbb{Q}[\sqrt{2}]$ is closed under the taking of multiplicative inverses. Thus $\mathbb{Q}[\sqrt{2}]$ is a field and since $\mathbb{R}$ is a field we conclude that $\mathbb{Q}[\sqrt{2}]$ is a subfield of $\mathbb{R}$. Also note that $a^2 - 2b^2 \neq 0$. Since the equal sign would imply $\sqrt{2} = a/b =$ rational (if $b \neq 0$; if $b = 0$, then $a \neq 0$, and so $a^2 - 2b^2 \neq 0$), which cannot happen because $\sqrt{2}$ is irrational.

P4. This example is from Calculus. Let R be the ring of all continuous mappings from the set of reals $\mathbb{R}$ to itself. Consider $S = \{f \in R | f(a) = 0\}$, the set of all mappings in R, which vanish at some fixed number a. Prove that S is a subring of the ring R.

Solution: We make use of Theorem 7. If f_1 and f_2 are mappings in S, then $f_1(a) = 0 = f_2(a)$. The difference mapping lies in S since $(f_1 - f_2)(a) = f_1(a) - f_2(a) = 0 - 0 = 0$. Also $(f_1 \cdot f_2)(a) = f_1(a) \cdot f_2(a) = 0 \cdot 0 = 0$. Hence S is closed under multiplication. Since S is closed under both subtraction and multiplication of elements, it is a subring of R.

P5: The set $S = \{0, 3, 6, 9.12, 15\}$ is a subring of $\mathbb{Z}_{18}$ (verity this). Does S have an identity element?

Solution: 3 and 6 are zero divisors in S. Since $3 \odot 6 = 6 \odot 3 = 0$, so neither of them can be an identity in S (obviously any ring with more than one element must have its identity element distinct from its zero element). Likewise 12 is a zero divisor in $S : 12 \odot 12 = 0$. So if S has an identity element, it must be either 9 or 15. For instance $\underbrace{9 \odot 3 = 9 = 3 \odot 9}_{\neq 3}$. So 9 cannot be the identity element. What about 15? In order to be an identity, it would have to satisfy $15 \odot x = x = 15 \odot x$, for every $x = 0, 3, 6, 9, 12, 15$. If we take $x = 3$: $\underbrace{15 \odot 3 = 9 = 3 \odot 15}_{\neq 3}$. So 15 fails to be an identity element.

Conclusion: The subring S of $\mathbb{Z}_{18}$ has **no** identity element (even though $\mathbb{Z}_{18}$ has 1 as its identity element).

P6: Prove that in any ring R with at least two elements (and with identity 1_R), the identity element 1_R cannot be a zero divisor.

Solution: Indeed, since R has at least two elements, it follows that $1_R \neq 0_R$. The identity element 1_R is a unit since $1_R \odot 1_R = 1_R$, by Definition 7. Thus by Theorem 4(i), since 1_R is a unit, we conclude that 1_R is not a zero divisor.

P7: (NOTE: This has already been proven in Theorem 8(iii) and (iv). We do it again for practicing purposes.) Prove that if a ring R has no zero divisors then the **cancelation property** holds (familiar from high school algebra). For **any** three elements x_1, x_2, x_3 in R with $x_1 \neq 0_R$; if $x_1 \odot x_2 = x_1 \odot x_3$, then $x_2 = x_3$. Likewise, if $x_2 \odot x_1 = x_3 \odot x_1$, then $x_2 = x_3$. Conclude, that if R is an integral domain, then it must have the cancelation property.

Solution: If $x_1 \odot x_2 = x_1 \odot x_3$, then $(x_1 \odot x_2) \ominus (x_1 \odot x_3) = 0_R$; $x_1 \odot (x_2 \ominus x_3) = 0_R$. But $x_1 \neq 0_R$, which means that since R has no zero divisors, a nonzero element will produce zero only if it is multiplied by zero. We infer that $x_2 \ominus x_3 = 0_R \Rightarrow x_2 = x_3$. The implication $(x_2 \odot x_1 = x_3 \odot x_1) \Rightarrow x_2 = x_3$ is proved similarly. Finally if R is an integral domain, it is by Definition 8(i), a commutative ring with identity and no zero divisors. Since it contains no zero divisors, it must have the cancellation property as we have already shown above.

P8: Let $\mathbb{R}^+$ be the set of positive reals, in interval notation, $\mathbb{R}^+ = (0, \infty)$. Define two operations on $\mathbb{R}^+$: addition $\oplus$ defined by,

$$x_1 \oplus x_2 = x_1 \cdot x_2, \quad \text{for all } x_1, x_3 \in \mathbb{R}^+.$$

Multiplication denoted by $\odot$ and defined as follows: $x_1 \odot x_2 = x_1^{\log x_2}$, where $\log x_2$ stands for the usual common logarithm (i.e. logarithm base 10) from calculus.

(a) Is $\mathbb{R}$, equipped with the above two operations, a ring?

(b) If it is a ring, is it commutative?

(c) If it is a commutative ring, is it a field?

Solution: (a) Let us look at the eight Axioms for a ring.

(a) **Axiom 1** says that a ring must be closed under addition. Since $x_2 \oplus x_2 = x_1 x_2$ is a positive real whenever x_1 and x_2 are positive reals, yes $\mathbb{R}^+$ is closed under $\oplus$.

(2) **Axiom 2**: We have $x_1 \oplus (x_2 \oplus x_3) = x_1 \oplus (x_1 \cdot x_2) = x_1 \cdot (x_2 x_3)$ and $(x_1 \oplus x_2) \oplus x_3 = (x_1 x_2) \oplus x_3 = (x_1 x_2) \cdot x_3$, but $x_1 \cdot (x_2 x_3) = (x_1 x_2) x_3$, since ordinary multiplication among real numbers is associative. Thus $x_1 \oplus (x_2 \oplus x_3) = (x_1 \oplus x_2) \oplus x_3$. Yes, Axiom 2 is satisfied.

(3) **Axiom 3**: Obviously $x_1 \oplus x_2 = x_1 x_2$ and $x_2 \oplus x_1 = x_2 x_1$ since ordinary multiplication is commutative, $x_1 x_2 = x_2 x_1$. Thus $x_1 \oplus x_2 = x_2 \oplus x_1$; addition $\oplus$ is commutative, Axiom 3 is satisfied.

(4) **Axiom 4**: The number 1 is the neutral element of this ring since $x \oplus 1 = x \cdot 1 = x$ and $1 \oplus x = 1 \cdot x = x$. Yes, Axiom 4 is satisfied with $0_R = 1$.

(5) **Axiom 5**: If x is positive real, then $\frac{1}{x}$ is the opposite element of x under $\oplus$. Indeed, $x \oplus \frac{1}{x} = x \cdot \frac{1}{x} = 1 = 0_R$ and $\frac{1}{x} \oplus x = \frac{1}{x} \cdot x = 1 = 0_R$. Yes, Axiom 5 is satisfied.

(6) **Axiom 6**: $x_1 \odot x_2 = x_1^{\log x_2}$. Since x_1 is a positive real, the power $x_1^{\log x_2}$ is also a positive real, regardless of whether the value of $\log x_2$ is positive, zero, or negative. Hence $\mathbb{R}^+$ is closed under $\odot$. Axiom 6 is satisfied.

(7) **Axiom 7**: $x_1 \odot (x_2 \odot x_3) = x_1 \odot (x_2^{\log x_3}) = x_1^{\log(x_2 \log x_3)}$. By the properties of logarithms, $\log(x_2^{\log x_3}) = (\log x_3)(\log x_2)$. Hence$x_1 \odot (x_2 \odot x_3) = x_1^{(\log x_3) \cdot (\log x_2)}$. Now consider the product $(x_1 \odot x_2) \odot x_3 = (x_1^{\log x_2}) \odot x_3 = (x_1^{\log x_2})^{\log x_3} = x_1^{(\log x_2)(\log x_3)}$. We conclude that $x_1 \odot (x_2 \odot x_3) = (x_1 \odot x_2) \odot x_3$. Yes, Axiom 7 is satisfied.

(8) **Axiom 8**: $x_1 \odot (x_2 \oplus x_3) = x_1 \odot (x_1 x_3) = x_1^{\log(x_2 x_3)}$. According to the property of logarithms, $\log(x_2 x_3) = \log x_2 + \log x_3$. Hence $x_1^{\log(x_2 x_3)} = x_1^{(\log x_2 + \log x_3)} = x_1^{\log x_2} \cdot x_1^{\log x_3} =$ By definition of $\oplus$, $x_1^{\log x_2} \cdot x_1^{\log x_3} = x_1^{\log x_2} \oplus x_1^{\log x_3}$. By definition of $\odot$, $x_1^{\log x_2} = x_1 \odot x_2$ and $x_1^{\log x_3} = x_1 \odot x_3$. Therefore $x_2^{\log(x_2 x_3)} = x_1^{\log x_2} \cdot x_1^{\log x_3} = x_1^{\log x_2} \oplus x_1^{\log x_3} = (x_1 \odot x_2) \oplus (x_1 \odot x_3)$. We conclude that, $x_1 \odot (x_2 \oplus x_3) = (x_1 \odot x_2) \oplus (x_1 \oplus x_3)$. The other part of the Axiom, namely that $(x_1 \oplus x_2) \odot x_3 = (x_1 \odot x_3) \oplus (x_2 \odot x_3)$ is proved similarly. Yes, Axiom 8 is satisfied.

Since all eight Axioms are satisfied we conclude that $\mathbb{R}^+$ is a ring under the operations $\oplus$ and $\odot$.

(b) Let us check to see whether $\mathbb{R}^+$ is commutative. We have $x_1 \odot x_2 = x_1^{\log x_2}$; and $x_2 \odot x_1 = x_2^{\log x_1}$. Even though these two powers do not appear to be equal, in fact they are; this is tricky: $x_1^{\log x_2} = 10^{\log(x_1^{\log x_2})}$, by the definition of the logarithm base 10: but $\log\left(x_1^{\log x_2}\right) = (\log x_2)(\log x_1)$ by one of the properties of logarithms. Thus, $x_1^{\log x_2} = 10^{\log[x_1{}^{\log x_2}]} = 10^{(\log x_2)(\log x_1)} = (10^{\log x_2})^{\log x_1} = x_2^{\log x_1}$. This proves that $x_1 \odot x_2 = x_1^{\log x_2} = x_2^{\log x_1} = x_2 \odot x_1$. Yes, multiplications is commutative; $\mathbb{R}^+$ is a commutative ring.

(c) Given $x \in \mathbb{R}^+$, $x \neq 1 = 0_R$ (so x is a nonzero element). Does x have a multiplicative inverse? First, does $\mathbb{R}^+$ have an identity element? Yes, it does; 10 is the identity element (the logarithmic base number). Let us verify this. $x \odot 10 = x^{\log 10} = x^1 = x$ and $10 \odot x = x$, since $x \odot 10 = 10 \odot x$ (by part (b), multiplication is commutative). thus 10 is the identity element of $\mathbb{R}^+$. For x' to be the multiplicative inverse of x, we must have $x \odot x' = 10 = x' \odot x$. Since multiplication commutes, let us just take $x \odot x' = 10 \Leftrightarrow x^{\log x'} = 10$ (take logs of both sides): $\log(x^{\log x'}) = \log 10 \Leftrightarrow (\log x')(\log x) = 1; \log x' = \frac{1}{\log x}$ (note that $\log x \neq 0$, since $x \neq 1$). Consequently we obtain, $10^{\log x'} = 10^{(1/\log x)}$; $x' = 10^{(1/\log x)}$. It is now clear that for each $x \in \mathbb{R}^+$, $x \neq 1$, the number $10^{(1/\log x)}$ is the multiplicative inverse of x. We can verify this,

$x \odot x' = x \odot 10^{(1/\log x)} = x^{\log 10/\log x} = x^{\frac{1}{\log x}} = 10^{\log x/\log x} = 10^1 = 10 =$ identity element of $\mathbb{R}^+$. Yes, $\mathbb{R}^+$ is a field with 10 being its identity element.

P8. Prove or disprove: (i) If R_1 and R_2 are integral domains, so is $R_1 \times R_2$.

(ii) If R_1 and R_2 are fields, so is $R_1 \times R_2$.

Solution: Both of these statements are false. Take (i). If a_1 is a nonzero element of R_1 (since it is an integral domain, it has at least two elements so that $0_{R_1} \neq 1_{R_1}$) and a_2 a nonzero element $R_2(0_{R_2} \neq 1_{R_2})$ then take the product, $(a_1, 0_{R_2})x(0_{R_1}, a_2) =$ (refer to the notation of Theorem 2) $= (a_1 \odot 0_{R_1}, 0_{R_2} \otimes a_2) = (0_{R_1}, 0_{R_2}) =$ zero element of the ring $R_1 \times R_2$. Since neither $(a_1, 0_{R_2})$ nor $(0_{R_1}, a_2)$ is the zero element in $R_1 \times R_2$, it follows that both of these elements are zero divisors of the ring $R_1 \times R_2$. Hence $R_1 \times R_2$ is not an integral domain.

(ii) $R_1 \times R_2$ cannot be a field since it is not an integral domain to begin with (Refer to Theorem 5).

P9. Let R be a division ring.

(i) Prove that if a nonzero element b of R is idempotent, that is, it satisfies $b \odot b = b$, then b must be the identity element; $b = 1_R$.

(ii) Prove that if an element u of R satisfies $u \odot u' = 1_R$, for some $u' \in R$, then u is actually a unit. That is, it also satisfies $u' \odot u = 1_R$.

Solution:

(i) Since R is a division ring, every nonzero element of R is a unit. Thus since b is a unit, there exists $b' \in R$ such that (by Definition 7) $b \odot b' = 1_R = b' \odot b$. Apply this to the given condition: $b \odot b = b \Rightarrow (b \odot b) \odot b' = b \odot b' \Rightarrow b \odot (b \odot b') = b \odot b' \Rightarrow b \odot 1_R = 1_R \Rightarrow b = 1_R$.

(ii) The assumption is that $u \odot u' = 1_R$. Take $b = u' \odot u$. We compute $b \odot b = (u' \odot u) \odot (u' \odot u) = [(u' \odot u) \odot u'] \odot u = [u' \odot (u \odot u')] \odot u = [u' \odot 1_R] \odot u = u' \odot u = b$.Note that $b = u' \odot u$ cannot be zero because $u' \odot u = 0_R$ would simply imply that either u' or u (or both) is zero (because a division ring has no zero divisors – every nonzero element is a unit and thus a nonzero divisor by Theorem 4(i)). If either u' or u is zero then $u \odot u' = 0_R$ and since we also have $u \odot u' = 1_R$, we would have $1_R = 0_R$ which is impossible, since a division ring has at leat two elements (only a trivial ring with one element satisfies $0_R = 1_R$). We have shown that $b \odot b = b$ and that $b = u' \odot u$ is not zero. By Part (i), the element b must be the identity element: $b = 1_R$; $u' \odot u = 1_R$. Since u satisfies $u \odot u' = 1_R = u' \odot u$ (and since obviously $u \neq 0_R$, because $0_R \neq 1$), u is, by Definition 7, a unit.

P10. Show that $\mathbb{Z}_n$, under the operations of $\oplus$ and $\odot$ as defined in Example 6, is a commutative ring with identity.

Solution: Let n be a positive integer, $n \geq 2$. (The case $n = 1$ is trivial: $\mathbb{Z}_1 = \{0\}$ is a trivial ring.) We will prove that the set $\mathbb{Z}_n = \{0, 1, \ldots, (n-1)\}$, equipped with the operations $\oplus$ and $\odot$ defined below, is a commutative ring with identity; $x_1 \oplus x_2 =$ the remainder of the division of the integer $(x_1 + x_2)$ with n and $x_1 \odot x_2 =$ the remainder of the division of the integer $x_1 \cdot x_2$ with n.

Axiom 1: Clearly $\mathbb{Z}_n$ is closed under addition $\oplus$ since by definition, the remainder of the division (of any integer) with n, is an integer between 0 and $(n-1)$. Hence

Axiom 1 is satisfied.

Axiom 2: Let x_1, x_2, x_3 be any three members of $\mathbb{Z}_n$. That is, x_1, x_2, x_3 are integers between 0 and $(n-1)$. Let $r = x_2 \oplus x_3$ and $r' = x_1 \oplus x_2$. Since r and r' are the remainders of the divisions of the integers $(x_2 + x_3)$ and $(x_1 + x_2)$, respectively, with n, we must have $x_2 + x_3 = q \cdot n + r$ and $x_1 + x_2 = q' \cdot n + r'$, where q and q' are the quotients of those two divisions. Keep in mind that all five integers x_1, x_2, x_3, r, and r' are integers between 0 and $(n-1)$. We have $x_1 \oplus (x_2 \oplus x_3) = x_1 \oplus r$; and $(x_1 \oplus x_2) \oplus x_3 = r' \oplus x_3$. Now, $x_1 \oplus r$ is, by definition, the remainder of the division of the integer $(x_1 + r)$ with n; and $r' \oplus x_3$ is the remainder of the division of the integer $(r' + x_3)$ with n. From $x_2 + x_3 = q \cdot n + r \Rightarrow x_1 + x_2 + x_3 = x_1 + qn + r = qn + (x_1 + r)$. From $x_1 + x_2 = q'n + r' \Rightarrow x_1 + x_2 + x_3 = q'n + (r' + x_3)$. If we perform the divisions of $(x_1 + r)$ and $(r' + x_3)$ with n we obtain, $x_1 + r = q_1 \cdot n + v_1$ and $r' + x_3 = q_2 \cdot n + v_2$; v_1 and v_2 being the remainders. Going back we see that

$$\begin{aligned} x_1 + x_2 + x_3 = qn + (x_1 + r) \Rightarrow \quad & x_1 + x_2 + x_3 = qn + q_1 \cdot n + v_1; \\ & x_1 + x_2 + x_3 = (q + q_1)n + v_1 \end{aligned} \tag{1}$$

Likewise,

$$\begin{aligned} x_1 + x_2 + x_3 = q'n + (r' + x_3) \Rightarrow \quad & x_1 + x_2 + x_3 = q'n + q_2 n + v_2; \\ & x_1 + x_2 + x_3 = (q' + q_2)n + v_2 \end{aligned} \tag{2}$$

Since v_1, v_2 are integers between 0 and $(n-1)$ both (1) and (2) say that v_1, v_2 are remainders of the division of $(x_1 + x_2 + x_3)$ with n. As we know both the quotient and remainder of the division of an integer with n, are unique. This proves that $v_1 = v_2$. But $v_1 = x_1 \oplus r$, $v_2 = r' \oplus x_3$, by definition. Hence $x_1 \oplus r = r' \oplus x_3$ and by going back we see that $x_1 \oplus (x_2 \oplus x_3) = (x_1 \oplus x_2) \oplus x_3$. Yes, Axiom 2 is satisfied.

Axiom 3: This is immediate, since $x_1 \oplus x_2 =$ remainder of the division of the integer $(x_1 + x_2)$ with n; but $x_1 + x_2 = x_2 + x_1$. So that,$x_2 \oplus x_1 =$ remainder of the division of the integer $(x_1 + x_2)$ with n. Hence $x_1 \oplus x_2 = x_2 \oplus x_1$. Axiom 3 is satisfied.

Axiom 4: The number 0 is the neutral element since $x \oplus 0 =$ remainder of the division of the integer $(x + 0)$ with n, but $x + 0 = x$. Since x is an integer between 0 and$(n-1)$, clearly x is the remainder of the division of x with n: $x = 0 \cdot q + x$. We conclude that $x \oplus 0 = x$. Since the addition is already proven to be commutative, we see that Axiom 4 is satisfied.

Axiom 5: If x is an integer between 0 and $(n-1)$, then $(n-x)$ s also an integer between 0 and $(n-1)$ and thus also a member of $\mathbb{Z}_n$; and $x \oplus (n-x) =$ remainder of the integer $x + (n-x)$ with n, but $x + (n-x) = n$ so that the remainder of n with itself is zero. Hence $x \oplus (n-x) = 0$ and since the addition is commutative we also have $(n-x) \oplus x = 0$. Yes, Axiom 5 is satisfied.

Axiom 6: $\mathbb{Z}_n$ is closed under multiplication $\odot$, since if $x_1, x_2 \in \mathbb{Z}_n$ then $x_1 \odot x_2 =$ remainder of the division of the integer $x_1 x_2$ with n. The remainder of (any integer) with n is an integer between 0 and $(n-1)$, and hence a member of $\mathbb{Z}_n$. Yes, Axiom 6 is satisfied.

Axiom 7: The proof that the multiplication operation $\odot$ is associative is similar to the proof that the addition operation $\oplus$ is associative (see Axiom 3), and it is left to the reader.

Axiom 8: Let x_1, x_2, x_3 be members of $\mathbb{Z}_n$. That is, integers between 0 and $(n-1)$. We want to show that, $x_1 \odot (x_2 \oplus x_3) = (x_1 \odot x_2) \oplus (x_1 \odot x_3)$.

Let r_2 be the remainder of the division of the integer x_1x_2 with n. Then $x_1x_2 = q_2 \cdot n + r_2$, where q_2 is the quotient of the division; and by definition,$x_1 \odot x_2 = r_2$. Likewise if r_3 is the remainder of the division of the integer x_1x_3 with n and q_3 the quotient, we have

$$x_1x_3 = q_3 \cdot n + r_3 \quad \text{and} \quad x_1 \odot x_3 = r_3.$$

If we add the two equations,

$$\left.\begin{array}{r} x_1x_2 = q_2 \cdot n + r_2 \\ x_1x_3 = q_3 \cdot n + r_3 \end{array}\right\} \Rightarrow x_1x_2 + x_1x_3 = (q_2n + r_2) + (q_3 \cdot n + r_3) \Rightarrow x_1 \cdot (x_2 + x_3) =$$

$$= (q_2 + q_3)n + (r_2 + r_3);$$

this clearly shows that the remainders of the divisions of the integers $x_1 \cdot (x_2 + x_3)$ and $(r_2 + r_3)$ with n, are equal (verify: if two integers differ by a multiple of n, their remainders, when they are divided by n, must be equal), in other words the remainder of the division of $x_1 \cdot (x_2 + x_3)$ with n is simply the integer $r_2 \oplus r_3$. Furthermore the integers $x_1 \cdot (x_2 + x_3)$ and $x_1 \cdot (x_2 \oplus x_3)$ have the same remainder; consequently the remainder of the division of $x_1 \cdot (x_2 \oplus x_3)$ with n, is precisely $r_2 \oplus r_3$. Then, by definition, $x_1 \odot (x_2 \oplus x_3) = r_2 \oplus r_3 \Rightarrow$ (see above) $\Rightarrow x_1 \odot (x_2 \oplus x_3) = (x_1 \odot x_2) \oplus (x_1 \oplus x_3)$ and we are done.

The other part of Axiom 8, namely the statement $(x_1 \oplus x_2) \odot x_3 = (x_1 \odot x_3) \oplus (x_2 \odot x_3)$ is proved similarly.

To conclude this practice problem, note that $\mathbb{Z}_n$ is commutative: $x_1 \odot x_2 =$ remainder of the division of x_1x_2 with n; but $x_2x_2 = x_2x_1$; hence $x_1 \odot x_2 =$ remainder of the division of x_2x_1 with n; and so $x_1 \odot x_2 = x_2 \odot x_1$. The identity element of $\mathbb{Z}_n$ is the integer 1: indeed, $x \odot 1 =$ remainder of the division of $x \cdot 1$ with n; but $x \cdot 1 = x = 1 \cdot x$; but x is an integer between 0 and $(n-1)$, which means that the remainder of the division of x with n, is equal to x itself; we conclude $x \odot 1 = 1 = 1 \odot x$.

P12. Let n be a positive integer, $n \geq 2$, and d a division of n. Prove that the set,

$$S = \left\{ x \,\middle|\, x \in \mathbb{Z}_n, \; x = d \cdot y, \; y \in \mathbb{Z}, \; \text{and } 0 \leq y \leq \frac{n}{d} - 1 \right\}$$

is a subring of $\mathbb{Z}_n$ (refer to Example 16).

Solution: We employ Theorem 7. All we have to show is that S is closed under multiplication $\odot$ and subtraction $\ominus$ of elements. If x_1, x_2 are elements of S, then by definition, $x_1 = d \cdot y_1$ and $x_2 = d \cdot y_2$, with $y_1, y_2 \in \mathbb{Z}$ and $0 \leq y_1, y_2 \leq \frac{n}{d} - 1$. We have, $x_1 \ominus x_2 =$ remainder of the division of the integer $(x_1 - x_2)$ with n. If we denote by r

that remainder, then we only need to show that r is divisible by d, since we already know (by definition) that r is a member of $\mathbb{Z}_n$. But $x_1 = d \cdot y_1$ and $x_2 = dy_2$ imply $x_1 - x_2 = d \cdot (y_1 - y_2)$.

$$\left.\begin{array}{l}\text{But } 0 \le x_1 \le n-1 \\ \text{and } 0 \le x_2 \le n-1\end{array}\right\} \Rightarrow \left.\begin{array}{l} 0 \le x_1 \le n-1 \\ -(n-1) \le -x_2 \le 0 \end{array}\right\} \Rightarrow$$
$$-(n-1) \le x_1 - x_2 \le n-1;$$

which proves that the remainder r of the division of the integer $(x_1 - x_2)$ with n is either equal to itself or equal to minus itself, depending on whether $(x_1 - x_2)$ is positive or negative, respectively. thus in either case, it follows from $x_1 - x_2 = d \cdot (y_1 - y_2)$ that the remainder r of the division of $(x_1 - x_2)$ with n is divisible by d. Since r is a member of $\mathbb{Z}_n$ it follows that r is also a member of S (observe from the definition that S consists precisely of those members of $\mathbb{Z}_n$ that are multiples of d).

To prove that the product $x_1 \odot x_2$ is also a member of S is even easier, since from $x_2 = dy_1$ and $x_2 = dy_2$ we infer $x_1x_2 = ddy_1y_2$; d must divide the remainder of the division of x_1x_2 with n. Indeed, if v is that remainder, we have $x_1x_2 = q \cdot n + v$; $0 \le v \le n-1$; q the quotient of the division. We obtain, $d(dy_1y_2) = qn + v$; but d is a divisor of n; put $n = dk$, $k \in \mathbb{Z}$; we have $d(dy_1y_2) - qn = v \Rightarrow d(dy_1y_2) - qdk = v \Rightarrow d(dy_1y_2 - qk) = v$, which proves that d is a divisor of v. thus $x_1 \odot x_2 = v =$ a multiple of $d \Rightarrow x_1 \odot x_2 \in S$. The proof is complete.

P13. As we have seen earlier, an element b in a ring R is called idempotent if it satisfies $b^2 = b$; that is, $b \odot b = b$. In any ring with identity 1_R (and with $1_R \ne 0_R$; that is a ring with at least two elements), both 1_R and the zero element 0_R are idempotent since $1_R \odot 1_R = 1_R$ and $0_R \odot 0_R = 0_R$.

(a) In the ring $M_2(\mathbb{R})$ find all the idempotent elements. In other words find all the solutions to the equation $x^2 = x$, with $x \in M_2(\mathbb{Z})$.

(b) What are the idempotent elements of $\mathbb{Z}_{12}$?

(c) Can an integral domain have any idempotent elements other than 0_R and 1_R?

(d) In a commutative ring is the product of two idempotent elements, idempotent?

Solution: (a) Suppose that $\begin{pmatrix} a & b \\ c & d \end{pmatrix}$ in $M_2(\mathbb{R})$ is idempotent. It satisfies $\begin{pmatrix} a & b \\ c & d \end{pmatrix}\cdot$ $\begin{pmatrix} a & b \\ c & d \end{pmatrix} = \begin{pmatrix} a & b \\ c & d \end{pmatrix} \Leftrightarrow \begin{pmatrix} (a^2+bc) & (ab+bd) \\ (ca+dc) & (cb+d^2) \end{pmatrix} = \begin{pmatrix} a & b \\ c & d \end{pmatrix}$; this yields four simultaneous equations in a, b, c and d:
$$\left.\begin{array}{l} a^2 + bc = a \\ (a+d)c = c \\ (a+d)b = b \\ cb + d^2 = d \end{array}\right\}$$

The two middle equations provide a starting point. If the quantity $a + d$ is equal to 1; $a + d = 1$, then the two middle equations become the identities $c = c$ and $b = b$, they are always true. So first assume $a + d = 1$. So if $a + d = 1$, then $d = 1 - a$ the first and fourth equations imply $bc = a - a^2$ and $cb = (1-a) - (1-a)^2$, respectively. Hence we obtain $a - a^2 = (1-a) - (1-a)^2 \Leftrightarrow a - a^2 = 1 - a - (1 - 2a + a^2)$

$a - a^2 = 1 - a - 1 + 2a - a^2 \Leftrightarrow 0 = 0$ which is true. Thus for $d = 1 - a$ and the values of b and c such that $bc = a - a^2$, the matrices $\begin{pmatrix} a & b \\ c & d \end{pmatrix}$ are idempotent. Going back to the system of four equations we see that if $a + d \neq 1$, then the second and third equations imply $c = b = 0$. Indeed $(a+d)c = c \Leftrightarrow (a+d)c - c = 0 \Leftrightarrow (a+d-1)c = 0$, which for $a + d - 1 \neq 0$, implies $c = 0$. Likewise $(a + d)b = b$ implies $b = c = 0$ with $b = c = 0$, the first and fourth equations yield $a^2 = a$ and $d^2 = d$ which produced $a = 0$ or $a = 1$ and $d = 0$ or $d = 1$. We summarize as follows:

1) The matrices $\begin{pmatrix} 1 & 0 \\ 0 & 1 \end{pmatrix}$, $\begin{pmatrix} 1 & 0 \\ 0 & 0 \end{pmatrix}$, $\begin{pmatrix} 0 & 0 \\ 0 & 1 \end{pmatrix}$, $\begin{pmatrix} 0 & 0 \\ 0 & 0 \end{pmatrix}$ are all idempotent.

2) Any matrix of the form $\begin{pmatrix} a & b \\ c & 1-a \end{pmatrix}$ and such that $bc = a^2 - a$; $bc = a(a-1)$, is also idempotent. Note that for $a = 0, 1$, at least one of b and c must be zero. So we can classify these matrices further as follows: $\begin{pmatrix} 1 & 0 \\ c & 0 \end{pmatrix}$, $\begin{pmatrix} 1 & b \\ 0 & 0 \end{pmatrix}$, $\begin{pmatrix} 0 & 0 \\ c & 1 \end{pmatrix}$, $\begin{pmatrix} 0 & b \\ 0 & 1 \end{pmatrix}$; and $\begin{pmatrix} a & b \\ \frac{a(1-a)}{b} & 1-a \end{pmatrix}$, with $a \neq 0, 1$

(b) If a member x of $\mathbb{Z}_{12}$ is idempotent, then x is an integer between 0 and 11; and $x \odot x = x = x \odot 1 \Leftrightarrow (x \odot x) = (x \odot 1) = 0_R = 0$; $x \odot (x \ominus 1) = 0$; this last equation says that the integer $x(x-1)$ must be a multiple of 12. Since the only divisor in common between the integers x and $(x-1)$ is 1, it follows that there are exactly four possibilities:

1) x is a multiple of 12. This simply means that $x = 0$.

2) $x - 1$ is a multiple of 12 which means that $x = 1$.

3) x is a multiple of 3 and $x - 1$ is a multiple of 4.

Among the integers 0 through 11 the multiples of 3 are the numbers 0,3,6, and 9; but only the value $x = 9$ is such that $x - 1$ is a multiple of 4; thus $x = 9$.

4) x is a multiple of 4 and $x - 1$ is a multiple of 3. The multiples of 4, between 0 and 11, are the numbers 0,4, and 8; but only $x = 4$ is such that $x - 1$ is a multiple of 3. Hence $x = 3$.

Summarizing, we see that the idempotent elements of $\mathbb{Z}_{12}$ are the numbers 0,1,4, and 9.

(c) No, an integral domain can only have 0_R and 1_R as idempotent elements. Indeed, suppose $x \odot x = x$ is an integral domain R, then $(x \odot x) \ominus x = 0_R$; but $x = x \odot 1_R$. Thus $(x \odot x) \ominus (x \odot 1_R) = 0_R \Leftrightarrow x \odot (x \ominus 1_R) = 0_R$. Since R is an integral domain, the last equation implies that (at least) one of the two factors must be zero: $x = 0_R$ or $x \ominus 1_r = 0_R$ which implies $x = 1_R$.

(d) The answer to this question is in the affirmative. Indeed, suppose $x_1^2 = x_1 \odot x_1 = x_1$ and $x_2^2 = x_2 \odot x_2 = x_2$. Because multiplication is assumed commutative, we can switch around product factors as we please. We have

$$\begin{aligned}(x_1 \odot x_2)^2 &= (x_1 \odot x_2) \odot (x_1 \odot x_2) = [(x_1 \odot x_2) \odot x_1] \odot x_2 \\ &= [x_1 \odot (x_2 \odot x_1)] \odot x_2 = [x_1 \odot (x_1 \odot x_2)] \odot x_2 = \\ &= [(x_1 \odot x_1) \odot x_2] \odot x_2 \ (x_1 \odot x_1) \odot (x_2 \odot x_2);\end{aligned}$$

and since $x_1 \odot x_1 = x_1$ and $x_2 \odot x_2 = x_2$, we conclude that, $(x_1 \odot x_2)^2 = (x_1 \odot x_1) \odot (x_2 \odot x_2) = x_1 \odot x_2$, which proves that the element $x_1 \odot x_2$ is also idempotent.

P14. If R_1 and R_2 are subrings of a ring R.

(a) Is $S_1 \cap S_2$ a subring of R?

(b) Is $S_1 \cup S_2$ a subring of R?

Solution: (a) According to Theorem 1 a subset of R is a subring if, and only if, it satisfies four conditions: it is closed under addition $\oplus$, under multiplication $\odot$, it contains the neutral element 0_R, and it is closed under the taking of opposites. The two subsets S_1 and S_2 do satisfy these two conditions, since each of them is assumed to be a subring; let s_1, s_2 be any two elements in the intersection set $S_1 \cap S_2$ which means $s_1, s_2 \in S_1$; and $s_1, s_2 \in S_2$. The sum $(s_1 \oplus s_2)$ is in both S_1 and S_2, since both S_1 and S_2 are subrings; the same is true for the product $(s_1 \odot s_2)$; so we have $(s_1 \oplus s_2) \in S_1 \cap S_2$ and $(s_1 \odot s_2) \in S_1 \cap S_2$ since 0_R belongs to both S_1 and S_2, it follows that $0_R \in S_1 \cap S_2$. Finally if $s \in S_1 \cap S_2$, then the opposite element $(\ominus s)$ lies on both S_1 and S_2 since both subrings are closed under opposites. Thus $(\ominus s) \in S_1 \cap S_2$. Yes, $S_1 \cap S_2$ is a subring of R.

(b) The answer is in the negative. One can find numerous examples in which the union $S_1 \cup S_2$ fails to satisfy all four conditions of Theorem 1. Note that $S_1 \cup S_2$ always satisfies one condition, namely it contains 0_R. For example, take $R = \mathbb{Z}$, $S_1 = 2\mathbb{Z} = \{x|x = 2y, y \in \mathbb{Z}\}$ and $S_2 = 3\mathbb{Z} = \{x|x = 3y, y \in \mathbb{Z}\}$. In other words S_1 is the set of even integers and S_2 is the set of all integers which are multiples of 3. They are both subrings of $\mathbb{Z}$, as we have already seen. But $S_1 \cup S_2$ is not closed under addition of elements. Generally, the sum of an even integer with an integer which is a multiple of 3, may produce an integer that is neither. For instance $2 + 3 = 5$ which does not belong to $S_1 \cup S_2$; likewise $4 + 3 = 7 \notin S_1 \cup S_2$. Note though that $S_1 \cup S_2$is closed under multiplication.

In conclusion, we see that $S_1 \cup S_2$ is not a subring of $\mathbb{Z}$.

P15. Prove or disprove: the set of units in a ring R is a subring of R.

Solution: For a ring R to have at least one unit, there are two necessary conditions. First the ring must contain at least two elements. Secondly it must contain an identity element 1_R. Since the zero element cannot be a unit (because by Theorem 3 $x \odot 0_R = 0_r = 0_R \odot x$, for every $x \in R$ – so zero multiplied by an element can never produce the identity element), it follows that the set of units in a ring R cannot be a subring R. It fails requirement or condition (a) of Theorem 1. Interestingly the set of units is closed under multiplication (verify this), but not under addition. (Generally speaking – pick any known ring, typically the sum of two units is not a unit.)

P16. Prove that if u is a unit in $\mathbb{Z}_n$ then the greatest common divisor between u and n is 1.

Solution: Since u is a unit in $\mathbb{Z}_n$, u must be nonzero and between 0 and $(n-1)$. Thus u must be some integer between 1 and $(n-1)$. Let u' be the multiplicative inverse of u. By Definition 7, $u \odot u' = 1 = u' \odot u$. since $u \odot u'$ is simply the remainder of the division of the integer $u \cdot u'$ (u' is also some integer between 1 and $(n-1)$) with n, we have $u \cdot u' = q \cdot n + r$; $r = u' \odot u$. But $r = 1 \Rightarrow u \cdot u' = q \cdot n + 1$; q the quotient of the division. Thus, $u \cdot u' - qn = 1$. If d is the greatest common divisor of u and n, we can set $u = d \cdot u_1$ and $n = d \cdot n_1$, for some positive integers (whose greatest common factor must be 1) u_1 and n_1.

Substituting for u and n in $uu' - qn = 1$ we obtain, $d \cdot u_1 \cdot u' - q \cdot d \cdot n_1 = 1 \Rightarrow d \cdot (u_1 \cdot u' - q \cdot n_1) = 1$; but d is a positive integer, the last equation shows that d divides 1. Hence $d = 1$, we are done. As we will see in Chapter 4, the Converse Statement of this problem is also true.

P17. Let R be a ring. Show the following:

(a) If $a \odot b$ is a zero divisor of R, then at least one of a or b must be a zero divisor of R.

(b) If R is a commutative ring and at least one of a or b is a zero divisor and $a \odot b \neq 0_R$, then $a \odot b$ must also be a zero divisor.

Solution: a) Since $a \odot b$ is a zero divisor, then according to Definition 6, there must exist an element c of R, with $c \neq 0_R$, and such that either $(a \odot b) \odot c = 0_R$; or $c \odot (a \odot b) = 0_R$; or both Suppose that $(a \odot b) \odot c = 0_R$. by associativity of multiplication we conclude $a \odot (b \odot c) = 0_R$. Now neither a nor b can be zero since $a \odot b$ is a zero divisor by hypothesis (if a or b were zero then $a \odot b$ would also be zero, which cannot happen since $a \odot b$ is assumed to be a zero divisor).

Thus all three a, b, and c are nonzero: $a, b, c \neq 0_R$. Now consider $a \odot (b \odot c) = 0_R$. If $b \odot c \neq 0_R$, then by virtue of $a \neq 0_R$, the last equation or statement proves that the element a is a zero divisor, according to Definition 6. The second possibility is $b \odot c = 0_R$; then $b, c \neq 0_R$, according to Definition 6, b must be a zero divisor. The case $c \odot (a \odot b) = 0_R$ is treated similarly.

(b) Suppose that a is a zero divisor. Then $a \neq 0_R$ and there must exist $k \in R$ such that $k \neq 0_R$ and either $a \odot k = 0_R$ or $k \odot a = 0_R$ (or both). R is assumed commutative so that actually,

$$a \odot k = 0_R = k \odot a.$$

Form $a \odot k = 0_R \Rightarrow b \odot (a \odot k) = b \odot 0_R \rightarrow (b \odot a) \odot k = 0_R$ (By Theorem 3). Since multiplication in R is commutative, we have $a \odot b = b \odot a$ and so from the last statement above, $(b \odot a) \odot k = 0_R \Rightarrow (a \odot b) \odot k = 0_R$. Since $a \odot b \neq 0_R$ by hypothesis and on account of $k \neq 0_R$, we conclude that the element $a \odot b$ is a zero divisor.

The proof is exactly the same for the case b is a zero divisor.

P18. Let S be a subring of a ring R with identity.

(a) If both S and R are fields, and 1_S and 1_R are the identity elements of S and$R$, show that $1_S = 1_R$.

(b) If S has an identity 1_S (and S has at least two elements) show that either

$1_S = 1_R$ or otherwise 1_S is a zero divisor of R.

Solution: (a) Let $1'_S$ be the multiplicative inverse of 1_S in R. We have $1_S \odot 1'_S = 1_R = 1;'_S \odot 1_S$. Also, since 1_R is the identity in R, we have $1_S \odot 1_R = 1_S = 1_R \odot 1_S$. Remember, multiplication is commutative in both S and R, since they are fields. Now consider, $1_S \odot 1_R = 1_S$; we can replace 1_S by $1_S \odot 1_S$ since obviously $1_S = 1_s \odot 1_S$. We have

$$1_S \odot 1_R = 1_S \odot 1_S \Rightarrow$$

$$\Rightarrow (1_S \odot 1_R) \ominus (1_S \odot 1_S) = 0_R \Rightarrow 1_s \odot (1_R \ominus 1_S) = 0_R \Rightarrow$$

$$\Rightarrow 1'_S \odot [1_S \odot (1_R \ominus 1_S)] = 0_R \Rightarrow$$

$$\Rightarrow (1'_S \odot 1_s) \odot (1_R \ominus 1_s) = 0_R; \text{ but } 1'_S \odot 1_S = 1_R.$$

$$\text{Hence } 1_R \odot (1_R \ominus 1_S) = 0_R \Rightarrow 1_R \ominus 1_S = 0_R \Rightarrow 1_R = 1_S.$$

(b) Since S has an identity 1_S and 1_R is the identity of R we must have $1_S \odot 1_R = 1_S = 1_R \odot 1_S$ and $1_S \odot 1_S = 1_S$. From $1_S \odot 1_R = 1_S \odot 1_S \Rightarrow (1_S \odot 1_R) \ominus (1_S \odot 1_s) = 0_R \Rightarrow 1_S \odot (1_R \ominus 1_S) = 0_R$ (In the same manner, it is also true that $(1_R \ominus 1_S) \odot 1_S = 0_R$.) Now look at the last equation or statement. Clearly $1_S \neq 0_R$ (because 0_R is also the zero element of the subring S and S has at leat two elements). So if $1_R \ominus 1_S \neq 0_R$, then the last equation above proves that 1_S is a zero divisor of the ring R. The other possibility is $1_R \ominus 1_R = 0_R$ which implies $1_R = 1_S$. We are done.

P19. Let R be a commutative ring with identity 1_R and b a fixed element of R. The subset, $S = \{x \in R \mid x = b \odot y,\ y \in R\}$ is a subring of R. (You may refer to example 33.) Show that if a unit u of R is also an element of the subring S, then $S = R$.

Solution: If u, a unit of R, lies in S, then $u = b \odot y$, for some element y of R. Let u' be the multiplicative inverse of u: $u \odot u' = 1_R = u' \odot u$. We have,

$$u = b \odot y \Rightarrow u' \odot u = u' \odot (b \odot y) \Rightarrow u' \odot u = (u' \odot b) \odot y;$$

and since $u' \odot u = 1_R$ we now have, $1_R = (u' \odot b) \odot y$; R is commutative which proves that 1_R lies in S (since $u' \odot y$ is simply an element of R.) Then $1_R \in S$ and let us set $1_R = b \odot t$ (where $t = u' \odot y$). If x is any element of R, then $x = 1_R \odot x = x \odot 1_R$, since 1_R is the identity of R. Now substitute for $1_R = b \odot t$ in the equation $x = 1_R \odot x$. We obtain $x = (b \odot t) \odot x \Rightarrow x = b \odot (t \odot x)$, which proves $x \in S$. Since every element x of R must lie in S, we conclude that $R$is a subset of S; however S is a subset of R, since it is a subring. Hence the two are equal, $S = R$.

P20. Any ring R in which every element is idempotent, is called a **Boolean ring**. In other words, a Boolean ring R is a ring that satisfies the condition $x^2 = x$; $x \odot x = x$, for every $x \in R$.

(a) Prove that if R is Boolean, then $x \oplus x = 0_R$, for every $x \in R$.

(b) Prove that if R is Boolean, R must be commutative.

Solution: (a) Let $x \in R$. We expand the square

$$(x \oplus x)^2;\ (x \oplus x)^2 = (x \oplus x) \odot (x \oplus x) = [x \odot (x \oplus x)] \oplus [x \odot (x \oplus x)] =$$

$$= [(x \odot x) \oplus (x \odot x)] \oplus [(x \odot x) \oplus (x \odot x)]$$

$$(x \odot x) \oplus (x \odot x) \oplus (x \odot x) \oplus (x \odot x)$$

and since by hypothesis we know that $x \odot x = x$, we now have, $(x \oplus x)^2 = (x \oplus x) \oplus (x \oplus x)$; but $(x \oplus x)^2 = x \oplus x$ since every element of$R$ is idempotent. Hence, we must have $(x \oplus x) = (x \oplus x) \oplus (x \oplus x)$. Adding the opposite element of $(x \oplus x)$ to both sides of the last equation and using the associative property of addition (the reader should fill in the details) we arrive at $x \oplus x = 0_R$.

(b) We want to show that $a \odot b = b \odot a,\ \forall a, b \in R$. Expand $(a \oplus b)^2 = (a \oplus b) \odot (a \oplus b) = [a \odot (a \oplus b)] \oplus [b \odot (a \oplus b)] \Rightarrow (a \oplus b)^2 = (a \odot a) \oplus (a \odot b) \oplus (b \odot a) \oplus (b \odot b)$ but $(a \oplus b)^2 = a \oplus b$, $a \odot a = a$, and $b \odot b = b$, by the fact that every element in R is idempotent. Consequently, the last equation above yields, $a \oplus b = a \oplus (a \odot b) \oplus (b \odot a) \oplus b$; and since addition $\oplus$ is commutative we now have, $(a \oplus b) = (a \oplus b) \oplus (a \odot b) \oplus (b \odot a)$. Adding the opposite of $(a \oplus b)$ to both sides gives,

$$0_R = (a \odot b) \oplus (b \odot a).$$

By part (a) of this problem, each element is opposite itself. So if we add $a \odot b$ to both sides of the last equation we get $(a \odot b) \oplus 0_R = [(a \odot b) \oplus (a \odot b)] \oplus (b \odot a) \Rightarrow a \odot b = 0_R \oplus (b \odot a)$; $a \odot b = b \odot a$. We are done.

P21. Let R_1 be a ring with operations $\oplus$ and $\odot$. R_2 a ring with operations $+$ and $\otimes$ and $R_1 \times R_2$ with operations (as defined in Theorem 2) $*$ and $\overline{x}$. Let S be the subring of $R_1 \times R_2$ consisting of all elements of the form $(x,\ 0_{R_2})$, $x \in R_1$. Show that the mapping defined by $f : R_1 \to S$; $f(x) = (x,\ 0_{R_2})$ is an isomorphism.

Solution: By inspection, this mapping is one-to-one and onto. Thus, in order to prove that f is an isomorphism, we need only show that it is a homorphism. Let $x_1, x_2 \in R_1$. We have $f(x_1 \oplus x_2) = (x_1 \oplus x_2,\ 0_{R_2})$, by definition of the mapping R. But $(x_1 \oplus x_2, 0_{R_2}) = (x_1, 0_{R_2}) * (x_2, 0_{R_2}) = f(x_1) * f(x_2)$; which proves that F is additive. and $f(x_1 \odot x_2) = (x_1 \odot x_2, 0_{R_2}) = (x_1 \odot x_2, 0_{R_2} \otimes 0_{R_2}) = (x_1, 0_{R_2})\overline{x}(x_2, 0_{R_2}) = f(x_1)\overline{x}f(x_2)$, which proves f to be a multiplicative mapping; f is an homomorphism and the proof is complete.

P22. (i) Give an example of a ring R and an additive mapping $f : R \to R$, which is not multiplicative.

(ii) Give an example of a ring R and a mapping $G : R \to R$ which is multiplicative but not additive.

Solution: Several examples for both (i) and (ii) have already been given in this Chapter.

(i) Take $R = \mathbb{Z}$. The mapping $f : \mathbb{Z} \to \mathbb{Z}$; $f(x) = 2x$ is additive but not multiplicative as it is easily seen.

(ii)Take $R = \mathbb{Z}$. The mapping $g : \mathbb{R} \to \mathbb{R}$; $g(x) = x^2$ (more generally $g(x) = x^n$, n a fixed integer ≥ 2) is multiplicative but not additive since $g(x_1x_2) = (x_1x_2)^2 = x_1^2x_2^2 = g(x_1)g(x_2)$, but $g(x_1 + x_2) = (x_1 + x_2)^2 = x_1^2 + 2x_1x_2 + x_2^2 \neq x_1^2 + x_2^2 = g(x_1) + g(x_2)$ unless $x_1x_2 = 0$ (so $g(x_1 + x_2) = g(x_1) + g(x_2)$ holds only when one of the x_1 or x_2 is zero. So it does not hold for all $x_1, x_2 \in \mathbb{R}$)).

P23. Give an example of a ring R and a mapping $f : R \to R$ such that $f(x_1 \oplus x_2) = f(x_1) \odot f(x_2)$, for all $x_1, x_2 \in R$.

Solution: Take $R = \mathbb{R}$, the set of all reals, and $f : \mathbb{R} \to \mathbb{R}$; $f(x) = a^x$ where a is some fixed positive real, $a \neq 1$. This is the usual exponential function that the reader is probably familiar with from calculus. We have,

$$f(x_1 + x_2) = a^{x_1+x_2} = a^{x_1} \cdot a^{x_2} = f(x_1) \cdot f(x_2)$$

P24. If $f : R_1 \to R_2$ is a homorphism from a ring R_1 to a ring R_2, is it true that every zero divisor b of R_1 is mapped onto a zero divisor of R_2?

Solution: Let $\odot$ be the multiplication symbol in R_1, $\otimes$ the multiplication symbol in R_2. If b is a zero divisor of R_1, then $b \neq 0_{R_1}$ and there exists an element c in R_1, with $c \neq 0_{R_1}$, and such that $b \odot c = 0_{R_1}$. Applying the fact that a homomorphism is in particular multiplicative we have, $f(b \odot c) = f(0_{R_1}) \Rightarrow f(b) \otimes f(c) = f(0_{R_1})$. By Theorem 10 we know that $f(0_{R_1}) = 0_{R_2}$. Hence $f(b) \otimes f(c) = 0_{R_2}$. We cannot conclude that the image $f(b)$ is a zero divisor in R_2 since $f(b)$ maybe the zero element 0_{R_2}. So the answer to the problems question is in the negative. Even if $f(b) \neq 0_{R_2}$, the image $f(c)$ may equal 0_{R_2}. Only if we know that $f(b) \neq 0_{R_2}$ and $f(c) \neq 0_{R_2}$ (for all the elements c in R_1 such that $b \odot c = 0_{R_1}$) can we conclude that $f(b)$ is a zero divisor in R_2.

P25. Let R_1, R_2 and R_3 be rings. For simplicity we will use the same symbols for the two operations in all three rings. $\oplus$ for the addition operation and $\odot$ for the multiplication operation. Prove that if the mappings $f : R_1 \to R_2$ and $g : R_2 \to R_3$ are both homomorphisms, then so is the composite mapping $g \circ f$. Conclude that if both f and g are isomorphisms, so is $g \circ f$.

Solution: (1) Additive: let $x_1, x_2 \in R_1$. Since f is an additive we have $f(x_1 \oplus x_2) = f(x_1) \oplus f(x_2)$. Since g is an additive we have $g(y_1 \oplus y_2) = g(y_1) \oplus g(y_2)$, for all $y_1, y_2 \in R_2$; taking $y_1 = f(x_1)$ and $f(x_2) = y_2$ we have $(g \circ f)(x_1 \oplus x_2) = g[f(x_1 \oplus x_2)] = g(f(x_1) \oplus f(x_2)) = g(f(x_1)) \oplus g(f(x_2)) = (g \circ f)(x_1) \oplus (g \circ f)(x_2)$, which proves $g \circ f$ to be additive.

(2) Multiplicative: $(g \circ f)(x_2 \odot x_2) = g[f(x_1 \odot x_2)] = g(f(x_1) \odot f(x_2)) = g(f(x_1)) \odot g(f(x_2)) = (g \circ f)(x_1) \odot (g \circ f)(x_2)$, which proves $g \circ f$ to be multiplicative. Thus $g \circ f$ is a homomorphism. Obviously, if both f and g are isomorphisms, they are in particular bijections (one-to-one and onto mappings). Thus $g \circ f$ is also a bijection as it is easily shown, and since $g \circ f$ is also a homomorphism, it follows that $g \circ f$ must be an isomorphism.

P26. Let $f : R_1 \to R_2$ be a homomorphism from a ring R_1 to a ring R_2. Suppose that S_2 is a subring of R_2. Show that the subset $S_1 = \{x \in R_1 \mid f(x) \in S_2\}$ is a subring of R_1.

Solution: If we can show that S_1 is closed under subtraction and multiplication of elements, then, according to Theorem 7, S_1 must be a subring of R_1. Let $\oplus$ and $\odot$ stand for addition and multiplication in R_1, and $*$ and $\otimes$ stand for addition and multiplication in R_2; and $*$ and $\otimes$ stand for addition and multiplication in R_2; and $\ominus$, $/$ the subtraction symbols.

If x_1 and x_2 are elements of S_1 such that $f(x_1)$ and $f(x_2)$ are in S_2, then, since S_2 is a subring of R_2; and f is a homomorphism, so Theorem 10(c) says that we must have $f(x_1)/f(x_2) = f(x_1 \ominus x_2) \in S_2 \Rightarrow$ (by the definition of the set S_1) $x_1 \ominus x_2 \in S_1$, which proves that S_1 is closed under subtraction of elements. Next consider the product $f(x_1) \otimes f(x_2)$. This product must be in S_2 because S_2 is closed under multiplication and due to the fact that f is a homomorphism we have $f(x_1) \otimes f(x_2) = f(x_1 \odot x_2)$, which proves that $f(x_1 \odot x_2) \in S_2 \Rightarrow$ (by the definition of the set S_1) $x_1 \odot x_2 \in S_1$. S_1 is closed under multiplication. Hence since it is also closed under subtraction of elements, we conclude that S_1 is a subring of R_1.

P27. Suppose that m and n are distinct positive integers. Recall the sets $m\mathbb{Z} = \{x \in \mathbb{Z} \mid x = my,\ y \in \mathbb{Z}\}$ and $n\mathbb{Z} = \{x \in \mathbb{Z} \mid x = ny,\ y \in \mathbb{Z}\}$, both being subrings of $\mathbb{Z}$ (as we have already seen) under the usual operations of addition and multiplication. Prove that $m\mathbb{Z}$ and $n\mathbb{Z}$ are not isomorphic rings.

Solution: Observe that there are at least two bijections between $m\mathbb{Z}$ and $n\mathbb{Z}$. These two bijections are $f_1 : m\mathbb{Z} \to n\mathbb{Z};\ f_1(x) = \frac{n}{m} \cdot x$, and $f_2 : n\mathbb{Z} \to m\mathbb{Z};\ f_2(x) = \frac{m}{n} \cdot x$. Observe that f_1 and f_2 are inverse mappings.

Nevertheless, as we see below, these two rings are not isomorphic, they have different structures. When one tries to prove that two rings are not isomorphic, one may employ either of two techniques:

(1) Show that one ring has a property that the other does not. Or (2) assume the existence of a mapping (from one ring onto the other) that is an isomorphism and prove that this assumption leads to a contradiction.

Since there is no easily discernible property in either ring to use in our proof, we will use the second method or technique. Assume f to be an isomorphism from $m\mathbb{Z}$ onto $n\mathbb{Z}$. That means f is a bijection and a homomorphism. Since f is a homomorphism, according to Theorem 10(d) we must have, $f(kx) = kf(x)$, for every $x \in m\mathbb{Z}$ and every positive integer k. According to Theorem 10(d), $f(-x) = -f(x)$, for every $x \in m\mathbb{Z}$. Thus if l is a negative integer in $m\mathbb{Z}$ by putting $l = -k$, where k is a positive integer, we have $f(lx) = f(-kx) = -f(kx) = -kf(x)$, for every $x \in m\mathbb{Z}$. In short, $f(tx) = tf(x)$, for every $t \in \mathbb{Z}$ and every $x \in m\mathbb{Z}$. Now $x = m$ is an element of $m\mathbb{Z}$. So if we set $x = m$ in the last equation we see that $f(tm) = tf(m)$, which shows the mapping f is completely determined by the image $f(m)$. Since $f(m) \in n\mathbb{Z}$, we can put $f(m) = m_1 \cdot n$, for some integer m_1. Thus if $tm, t \in \mathbb{Z}$, is any element of $\mathbb{Z}_m$ we must have $f(tm) = tf(m) = tm_1 \cdot n$. We claim that m_1 must be either 1 or -1. This is because f (as an isomorphism) is a surjective mapping: indeed if n_1 is an integer other than 1 or -1, then the image set or range of f will consist of all the integer multiples of $m_1 \cdot n$ and will not include those multiples of n which are not divisible by m_1. In particular the range of f will not include the integers n and $-n$. We know that f is surjective which means range of f must be the entire set

$n\mathbb{Z}$. Thus, $m_1 = 1$ or -1. So $f(tm) = tn$ or $t \cdot (-n)$; but t can be any integer, so altogether we have $f(tm) = tn$, for every $t \in \mathbb{Z}$.

Now we use the fact that f is multiplicative:

$$f(x_1x_2) = f(x_1)f(x_2), \text{ for all } x_1, x_2 \in m\mathbb{Z}.$$

Setting $x_1 = my_1, x_2 = my_2$ we obtain, $f(m\cdot(my_1y_2)) = f(my_1)f(my_2)$. Applying the fact that $f(tm) = t\cdot n$, for every $t \in \mathbb{Z}$, we conclude $my_1y_2n = y_1\cdot n\cdot y_2\cdot n \Rightarrow$ (since n is positive) $my_1y_2 = ny_1y_2$; taking $y_1y_2 \neq 0$, we arrive at $m = n$ which contradicts the assumption $m \neq n$. We conclude that there is no isomorphism from $m\mathbb{Z}$ onto $n\mathbb{Z}$.

2.5 Exercise Set

1. The following subsets of $\mathbb{Z}$, with ordinary addition and multiplication, satisfy all but one of the axioms for a ring. In each case below, which of the eight axioms fails?

   (i) The set of all odd integers together with zero.

   (ii) The set of nonnegative integers.

2. Which of the following sets below are subrings of $M_2(\mathbb{R})$? Identify those that have an identity element.

   (i) The set of all matrices of the form $\begin{pmatrix} 0 & x \\ 0 & 0 \end{pmatrix}$, with $x \in \mathbb{Q}$.

   (ii) The set of all matrices of the form $\begin{pmatrix} x_1 & x_2 \\ 0 & x_3 \end{pmatrix}$, with $x_1, x_2, x_3 \in \mathbb{Z}$.

   (iii) The set of all matrices of the form $\begin{pmatrix} x & 0 \\ x & 0 \end{pmatrix}$, with $x \in \mathbb{R}$.

   (iv) The set of all matrices of the form $\begin{pmatrix} x & x \\ 0 & 0 \end{pmatrix}$, $x \in \mathbb{R}$.

   (v) The set of all matrices of the form $\begin{pmatrix} x & x \\ y & y \end{pmatrix}$, with $x, y \in \mathbb{R}$.

   (vi) The set of all matrices of the form $\begin{pmatrix} x & y \\ x & y \end{pmatrix}$, with $x, y \in \mathbb{R}$.

   (vii) The set of all matrices of the form $\begin{pmatrix} x & 0 \\ 0 & 2x \end{pmatrix}$, with $x \in \mathbb{R}$.

   (viii) The set of all matrices of the form $\begin{pmatrix} x & 1 \\ 0 & 0 \end{pmatrix}$, with $x \in \mathbb{Q}$.

3. Let $R = \{0, i, a, b\}$ be a set equipped with addition and multiplication defined by the tables illustrated below. One can prove that R is indeed a ring. All eight axioms are satisfied.

   (i) Which is the neutral or zero element?

   (ii) Does R have an identity element? If yes, identify it.

   (iii) Is R a commutative ring?

   (iv) Is R a ring with no zero divisors or does it have zero divisors? If it does, identify at least one such pair of zero divisors.

   (v) Is R an integral domain?

   (vi) Is R a division ring?

(vii) Is R a field?

(viii) Identify all idempotent elements. (x is idempotent if $x^2 = x$).

$\oplus$	0	i	a	b
0	0	i	a	b
i	i	0	b	a
a	a	b	0	i
b	b	a	i	0

$\odot$	0	i	a	b
0	0	0	0	0
i	0	i	a	b
a	0	a	a	0
b	0	b	0	b

4. The set $R = \{0, i, a, b\}$ with operations defined in the two tables below, is a ring. Show that R is actually a field

$\oplus$	0	i	a	b
0	0	i	a	b
i	i	0	b	a
a	a	b	0	i
b	b	a	i	0

$\odot$	0	i	a	b
0	0	0	0	0
i	0	i	a	b
a	0	a	b	i
b	0	b	i	a

Figure 5

5. Let R_1 and R_2 be rings and consider the subsets S_1 and S_2 of the ring $R_1 \times R_2$:

$$s_1 = \{x, 0_{R_2}) \mid x \in R_1\}, \quad S_2 = \{(0_{R_1}, x) \mid x \in R_2\},$$

where 0_{R_1} and 0_{R_2} are the zero elements of R_1 and R_2, respectively.

(i) Show that S_1 and S_2 are subrings of $R_1 \times R_2$

(ii) If $R_1 = \mathbb{Z}_3$ and $R_2 = \mathbb{Z}_5$, describe the sets S_1 and S_2. Are the subrings S_1 and S_2 isomorphic to $\mathbb{Z}_3$ and $\mathbb{Z}_5$, respectively?

6. Let S be a fixed set and $P(S)$ the set of all subsets of S. Define addition $\oplus$ and multiplication $\odot$ as follows: $x_1 \oplus x_2 = (x_1 - x_2) \cup (x_2 - x_1)$, for any two subsets x_1 and x_2 of S. Here the difference set $(x - y)$ is defined to be the set of all elements of x that do not lie in y (i.e. $x - y$ consists of those elements of x that do not belong to the intersection $x \cap y$) and $x_1 \odot x_2 = x_1 \cap x_2$, for all $x_1, x_2 \in P(S)$.

(i) Prove that $P(S)$ equipped with these two operations is a commutative ring with neutral element $0_{P(S)} = \emptyset =$ empty set and identity element $1_{P(S)} = S$.

(ii) Show that every element of the ring $P(S)$ is idempotent. That is $x^2 = x \odot x = x$, for every subset x of S.

(iii) Show that $x \oplus x = 0_{P(S)}$, $\forall x \in P(S)$.

(iv) If $S = \{a, b, c\}$, set up an addition and a multiplication table for the ring $P(S)$.

Note: In part (a) you may give, for some Axioms (especially for the Axiom of additive associativity and distributivity) informal proofs using Venn diagrams.

7. Define a new multiplication in $\mathbb{Z}$ by the rule $x_1 \odot x_2 = 1$, for all $x_1, x_2\ in \mathbb{Z}$. Under ordinary addition and this new multiplication, is $\mathbb{Z}$ a ring? If not, which of the eight Axioms fail? If it is a ring, is it commutative?

8. Repeat exercise 7, by replacing the new multiplication in that exercise, with the multiplication $x_1 \odot x_2 = 0$, for all $x_1, x_2 \in \mathbb{Z}$.

9. Let $2\mathbb{Z}$ be the set of even integers. Define a new multiplication as follows: $x_1 \odot x_2 = \frac{x_1 x_2}{2}$, for all $x_1, x_2 \in 2\mathbb{Z}$. Under this new multiplication and ordinary addition, prove that $2\mathbb{Z}$ is a commutative ring with identity.

10. Define a new addition and a new multiplication in $\mathbb{Z}$ by the rules,

$$x_1 \oplus x_2 = x_1 + x_2 - 1 \text{ and } x_1 \odot x_2 = x_1 + x_2 - x_1 x_2, \text{ for all } x_1, x_2 \in \mathbb{Z}.$$

Prove that $\mathbb{Z}$ equipped with these two operations is an integral domain.

11. In $\mathbb{Q}$ define two new operations:

$$x_1 \oplus x_2 = x_1 + x_2 + 1 \text{ and } x_1 \odot x_2 = x_1 + x_2 + x_1 x_2, \text{ for all } x_1, x_2 \in \mathbb{Q}.$$

Prove that $\mathbb{Q}$ equipped with these two operations is a commutative ring with identity. Is it an integral domain? Is it a field? Identify all the units in this ring.

12. The addition tables and parts of the multiplication tables for two rings (one with three elements and the other with four elements) are illustrated below. In each case, use the distributive laws to complete the two

(i) $R = \{a, b, c\}$.

$\oplus$	a	b	c
a	a	b	c
b	b	c	a
c	c	a	b

$\odot$	a	b	c
a	a	a	a
b	a	c	
c	a		

(ii) $R = \{x, y, z, w\}$

$\oplus$	x	y	z	w
x	y	z	w	x
y	z	w	x	y
z	w	x	y	z
w	x	y	z	w

$\odot$	x	y	z	w
x	y		w	
y		w		w
z		w	y	w
w	w	w	w	w

Figure 6

13. Let $M(\mathbb{Z}_2)$ be the set of all 2×2 matrices with entries in $\mathbb{Z}_2$. Show that $M(\mathbb{Z}_2)$ is a noncommutative ring with identity, that contains 16 elements.

14. Let $S_1 = \{0, 2\}$, a subset of $\mathbb{Z}_4$; and $S_2 = \{0, 3, 6, 9\}$, a subset of $\mathbb{Z}_{12}$. Prove that $S_1 \times S_2$ is a subring of $\mathbb{Z}_4 \times \mathbb{Z}_{12}$.

15. (From calculus) Let c be a fixed real number, a constant. Consider the functions f and g from $\mathbb{R}$ to $\mathbb{R}$ defined as follows:

$$f : \mathbb{R} \to \mathbb{R};\ f(x) = \begin{cases} 0, & \text{if } x < c \\ x - c, & \text{if } x \geq c \end{cases};\ \text{and,}$$

$$g : \mathbb{R} \to \mathbb{R};\ g(x) = \left\{ \begin{array}{ll} c - x, & \text{if } x < c \\ 0, & \text{if } x \geq c \end{array} \right)$$

Verify, by using well-known methods from Calculus, that these are continuous functions. Therefore they are members of the ring of continuous functions (from $\mathbb{R}$ to $\mathbb{R}$)F. Show that the product function $f \cdot g$ is the zero function from $\mathbb{R} \to \mathbb{R}$; $f \cdot g = 0_F$. This establishes that the ring F is not an integral domain. Why?

16. Let d be a positive integer that is not a perfect square. That is, d is not the square of an integer. Consider the set $\mathbb{Q}(\sqrt{d}) = \{a + b\sqrt{d} \mid a, b \in \mathbb{Q})$, a subset of the real numbers. One can prove that $\mathbb{Q}(\sqrt{d})$is a subring of $\mathbb{R}$ and with some identity element 1 as $\mathbb{R}$. Show that every nonzero element in $\mathbb{Q}(\sqrt{d})$ is a unit. This establishes that actually $\mathbb{Q}(\sqrt{d})$ is a subfield of $\mathbb{R}$.

Hint: If $a + b\sqrt{d}$ is an element of $\mathbb{Q}(\sqrt{d})$, with at least one of a and b being nonzero, then the reciprocal of $a + b\sqrt{d}$; $\frac{1}{a+b\sqrt{d}}$ is an element of $\mathbb{R}$. Multiply this reciprocal element by $1 = \frac{a-b\sqrt{d}}{a-b\sqrt{d}}$ to show that $\frac{1}{a+b\sqrt{d}}$ actually lies in $\mathbb{Q}(\sqrt{d})$. (Note that $a - b\sqrt{d}$ is nonzero. Why?)

17. (For students familiar with the ring $\mathbb{C}$ of complex numbers.) In the ring $M_2(\mathbb{C})$ (all 2×2 matrices whose entries are comples numbers) let,

$$\mathbf{1} = \begin{pmatrix} 1 & 0 \\ 0 & 1 \end{pmatrix},\ \boldsymbol{i} = \begin{pmatrix} i & 0 \\ 0 & -i \end{pmatrix},\ \boldsymbol{j} = \begin{pmatrix} 0 & 1 \\ -1 & 0 \end{pmatrix},\ \boldsymbol{k} = \begin{pmatrix} 0 & i \\ i & 0 \end{pmatrix}.$$

The product of a real number and a matrix, is defined to be the matrix that comes about by the rule described as follows: $r \cdot \begin{pmatrix} x_1 & x_2 \\ x_3 & x_4 \end{pmatrix} = \begin{pmatrix} rx_1 & rx_2 \\ rx_3 & rx_4 \end{pmatrix}$.
The set H of **real quaternions** consists of all matrices of the form

$$\begin{aligned} a\mathbf{1} + b\mathbf{i} + c\mathbf{j} + d\mathbf{k} &= \begin{pmatrix} 1 & 0 \\ 0 & 1 \end{pmatrix} + b\begin{pmatrix} i & 0 \\ 0 & -i \end{pmatrix} + c\begin{pmatrix} 0 & 1 \\ -1 & 0 \end{pmatrix} + d\begin{pmatrix} 0 & i \\ i & 0 \end{pmatrix} \\ &= \begin{pmatrix} a & 0 \\ 0 & a \end{pmatrix} + \begin{pmatrix} bi & 0 \\ 0 & -bi \end{pmatrix} + \begin{pmatrix} 0 & c \\ -c & 0 \end{pmatrix} + \begin{pmatrix} 0 & di \\ di & 0 \end{pmatrix} \\ &= \begin{pmatrix} a+bi & c+di \\ -c+di & a-bi \end{pmatrix}, \text{ where } a,b,c,d \in \mathbb{R} \end{aligned}$$

(i) Show that $\mathbf{i}^2 = \mathbf{j}^2 = \mathbf{k}^2 = -\mathbf{1}$, $\mathbf{i} = -\mathbf{j}\mathbf{k}$, $\mathbf{j}\mathbf{k} = -\mathbf{k}\mathbf{j} = \mathbf{i}$ and $\mathbf{k}\mathbf{i} = -\mathbf{i}\mathbf{k} = \mathbf{j}$.

(ii) Show that H is a subring of $M_2(\mathbb{Z})$. More specifically show that H is a noncommutative ring with identity $\mathbf{1}$.

(iii) Show that H is actually a division ring. **Hint**: Show that if $e = a^2 + b^2 + c^2 + d^2$, then the matrix $\frac{1}{e}\begin{pmatrix} a-bi & -c-di \\ c-di & a+bi \end{pmatrix}$ is the multiplicative inverse of the element $\begin{pmatrix} a+bi & c+di \\ -c+di & a-bi \end{pmatrix}$. Here we assume that at least one of the real numbers a, b, c, or d is nonzero.

(iv) Show that the equation $x^2 = -\mathbf{1}$ has infinitely many solutions in H.

Hint: Consider quaternions of the form $\begin{pmatrix} bi & c+di \\ -c+di & -bi \end{pmatrix}$, where $b^2 + c^2 + d^2 = 1$.

18. Consider $\mathbb{R} \times \mathbb{R}$ with the usual coordinatewise addition and a new multiplicated given by the rule $(x_1, y_1) \odot (x_2, y_2) = (x_1x_2 - y_1y_2, x_1y_2 + y_1x_2)$. Prove that under these operations $\mathbb{R} \times \mathbb{R}$ is a field.

19. **(i)** Show that if a matrix K in $M_2(\mathbb{R})$, has the property that $K \cdot A = A \cdot K$, for every matrix A in $M_2(\mathbb{R})$, then the matrix K must be of the form $K = \begin{pmatrix} k & 0 \\ 0 & k \end{pmatrix}$, where k is some real number. Such matrices are called **scalar** matrices.

(ii) Conversely, prove that every scalar matric K has that property. That is, $K \cdot A = A \cdot K$, $\forall A \in M_2(\mathbb{R})$.

(iii) Show that the set of scalar matrices is a subring of $M_2(\mathbb{R})$.

20. Let R be a ring and $a, b, \in R$. In each case below expand the given product.

(i) $(a \oplus b) \odot (a \ominus b) = ?$

(ii) $(a \oplus b)^3 = ?$

(iii) What are the answers in parts (a) and (b) if R is commutative?

21. In the ring $M_2(\mathbb{R})$, let

$$A = \begin{pmatrix} 1 & 2 \\ 0 & 1 \end{pmatrix}, \; B = \begin{pmatrix} 4 & 1 \\ 1 & -1 \end{pmatrix}, \; C = \begin{pmatrix} 2 & 3 \\ 1 & -1 \end{pmatrix}, \; D = \begin{pmatrix} 4 & -7 \\ 1 & -3 \end{pmatrix}.$$

Show that the matric C is a solution to the equation $A \cdot X = B$ and the matrix D is a solution to the equation $X \cdot A = B$. Does either of these two equations have any other solutions?

22. (i) Let R be any ring and b a fixed element of R. Let $S_1 = \{x \in R \mid x \odot b = 0_R\}$ and $S_2 = \{x \in R \mid b \odot x = 0_R\}$ (when R is commutative, $S_1 = S_2$). Prove that both S_1 and S_2 are subrings of R.

(ii) Let $B = \begin{pmatrix} 0 & 0 \\ 1 & 0 \end{pmatrix}$ and let S_1, S_2 be the sets defined as $S_1 = \{X \in M_2(\mathbb{R}) \mid X \cdot B = \begin{pmatrix} 0 & 0 \\ 0 & 0 \end{pmatrix}\}$, $S_2 = \{X \in M_2(\mathbb{R}) \mid B \cdot X = \begin{pmatrix} 0 & 0 \\ 0 & 0 \end{pmatrix}\}$. Prove that both S_1 and S_2 are subrings of $M_2(\mathbb{R})$.

(iii) Let $S_1 = \{x \in \mathbb{Z}_{10} \mid x \odot 5 = 0\}$ and $S_2 = \{x \in \mathbb{Z}_{10} \mid 5 \odot x = 0\}$. Prove that both S_1 and S_2 are subrings of $\mathbb{Z}_{10}$.

23. Let R be a ring and let $C = \{x \in R \mid x \odot t = t \odot x$, for every $t \in R\}$. Note that C is nonempty since it contains the zero element 0_R. The set S is called the **center** of the ring R.

(i) Prove that C is a subring of R.

(ii) If $R = M_2(\mathbb{R})$, find two elements in the center C of R and three elements of $M_2(\mathbb{R})$ that do not lie in C. Use exercise 19.

24. Let R be a ring with identity 1_R and let $S = \{n \cdot 1_R \mid n \in \mathbb{Z}\}$. (Recall that if n is a positive integer; $n \cdot 1_R = \underbrace{1_R \oplus \ldots \oplus 1_R}_{n \text{ times}}$; if $n = 0$, $0 \cdot 1_R = 0_R$, and if n is a negative integer then $n \cdot 1_R = -\underbrace{(1_R \oplus \ldots \oplus 1_R)}_{(-n) \text{ times}}$.) Show that the set S is a subring of R.

25. Let R be a ring and $a, b, \in R$; m and n positive integers.

(i) Show that $a^m a^n = a^{m+n}$ and $(a^m)^n = a^{mn}$.

v(ii) Show that if R is commutative, $(ab)^n = a^n b^n$.

26. Let R_1 and R_2 be rings with identity elements 1_{R_1} and 1_{R_2}. Describe the units in the ring $R_1 \times R_2$.

(i) Let R be a ring and $a, b, \in R$; m and n nonnegative integers. Show the following:

(a) $(m+n)a = ma + na$.

(b) $m(a \oplus b) = ma \oplus mb$.

(c) $m(a \odot b) = (ma) \odot b = a \odot (mb)$.

(d) $(ma) \odot (nb) = mn(a \odot b)$.

(ii) Do part (i) when m and n are any integers.

27. Let R be a ring with identity, consisting exactly of four elements; $r = \{0_R, 1_R, a, b\}$. Suppose that a and b are units. Write out the multiplication table for R.

28. Let R be a commutative ring with identity 1_R and no zero divisors. Prove that if $a \odot b$ is a unit in R, then both a and b must be units as well.

In exercises 30-32, the following definition is needed. Let R be a ring with identity. If there is a smallest positive integer n such that $n \cdot 1_R = 0_R$, then R is said to have **characteristic n**. If no such n exists, R is said to have **characteristic zero**.

29. **(i)** Show that $\mathbb{Z}$ has characteristic zero and $\mathbb{Z}_n$ has characteristic n.

(ii) Find the characteristic of the ring $\mathbb{Z}_4 \times \mathbb{Z}_6$. Can you generalize: what must the characteristic of $\mathbb{Z}_m \times \mathbb{Z}_n$ be?

30. Prove that every finite ring with identity has characteristic n, for some positive integer n.

31. Let R be a ring with characteristic n, where n is a positive integer.

(i) Show that $nx = 0_R$, for every $x \in R$.

(ii) If R is an integral domain, show that n must be a prime number.

(iii) Does the result of part (ii) still hold if R is replaced with a ring identity and o zero divisors (for instance a noncommutative ring)?

32. Let R be a ring without identity and S the set $R \times \mathbb{Z}$. Define a new addition $*$ and multiplication $\otimes$ on S as follows:

$$(x_1, y_1) * (x_2, y_2) = (x_1 \oplus x_2, y_1 + y_2), \text{ for all } x_1, x_2 \in R \text{ and all } y_1, y_2 \in \mathbb{Z}$$

and

$(x_1, y_1) \otimes (x_2, y_2) = ((x_1 \odot x_2) \oplus (y_1 x_2) \oplus (y_2 x_1), y_1 y_2)$ for all $x_z, x_2 \in R$ and all $y_1, y_2 \in \mathbb{Z}$.

(Here of course, $\oplus$ and $\odot$ stand for the operations of addition and multiplication in the ring R.)

(i) Prove that S is a ring with identity the element $(0_R, 1)$ and zero element the pair $(0_R, 0)$.

(ii) Let S_1 consist of all elements of the form $(x, 0)$ in S. Show that S_1 is a subring of S.

(iii) Let $S_2 = \{(0_R, y) \mid y \in \mathbb{Z}\}$. Is S_2 a subring of S?

33. Use tables to show that $\mathbb{Z}_2 \times \mathbb{Z}_2$ is isomorphic to the ring of exercise 4. Describe an isomorphism between these two rings.

34. Let S be the subring $\{0, 2, 4, 6, 8\}$ of $\mathbb{Z}_{10}$. Use tables to show that the following bijection f from $\mathbb{Z}_5$ onto S is not an isomorphism:

$$f : \mathbb{Z}_5 \to S;\ f(0) = 0, f(1) = 2,\ f(2) = 4,\ f(3) = 6,\ f(4) = 8.$$

35. Prove that the field $\mathbb{R}$ of real numbers is isomorphic to the ring of all 2×2 matrices of the form $\begin{pmatrix} 0 & 0 \\ 0 & x \end{pmatrix}$, $x \in \mathbb{R}$.

36. Let $\mathbb{Q}(\sqrt{2}) = \{a + b\sqrt{2} \mid a, b \in \mathbb{Q}\}$. Show that the function $f : \mathbb{Q}(\sqrt{2}) \to \mathbb{Q}(\sqrt{2})$; $f(a + b\sqrt{2}) = a - b\sqrt{2}$, an isomorphism from $\mathbb{Q})\sqrt{2})$ onto itself. Recall that $\mathbb{Q}(\sqrt{2})$ is a field, a subfield of $\mathbb{R}$ (see exercise 16).

37. Prove that the ring of all matrices of the form $\begin{pmatrix} x & 0 \\ 0 & x \end{pmatrix}$ (a subring of $M_2(\mathbb{R})$) is isomorphic to $\mathbb{R}$.

38. If $f : \mathbb{Z} \to \mathbb{Z}$ is an isomorphism, show that f must be the identity map.

39. Below you are given twelve mappings, in each case from a ring to a ring. In each case determine whether the given mapping is additive, multiplicative, or both. That is, a homorphism or none of the above.

(i) $f : \mathbb{Z} \to \mathbb{Z};\ f(x) = 6x$.

(ii) $f : \mathbb{Z}_2 \to \mathbb{Z}_2;\ f(x) = -x$.

(iii) $f : \mathbb{Z} \to \mathbb{Z};\ f(x) = -x$.

(iv) $f : \mathbb{Q} \to \mathbb{Q};\ f(x) = \frac{1}{x^2+1}$

(v) $f : \mathbb{Q} \to \mathbb{Q};\ f(x) = kx$, where k is a fixed rational number.

(vi) $f : \mathbb{R} \to M_2(\mathbb{R})$; $f(x) = \begin{pmatrix} -x & 0 \\ x & 0 \end{pmatrix}$.

(vii) $f : \mathbb{R} \to M_2(\mathbb{R})$; $f(x) = \begin{pmatrix} 2x & 0 \\ 0 & 3x \end{pmatrix}$.

(viii) $f : \mathbb{Z}_{12} \to \mathbb{Z}_4$; $f(x) = [x]_4$, where $[x]_4$ denotes the remainder of the division of the integer x with 4.

(ix) $f : \mathbb{R} \to \mathbb{R}$; $f(x) = x^2$.

(x) $f : \mathbb{Q} \to \mathbb{Q}$; $f(x) = x^3$.

(xi) $f : \mathbb{Z} \to \mathbb{Z}$; $f(x) = x^4$.

(xii) $f : \mathbb{R} \to M_2(\mathbb{R})$; $f(x) = \begin{pmatrix} 0 & x \\ 0 & -x \end{pmatrix}$.

40. (For students familiar with complex numbers.) Show that the following function, from the ring of complex numbers $\mathbb{C}$ to itself, is a bijection $f : \mathbb{C} \to \mathbb{C}$; $f(a + bi) = a - bi$. This mapping is called the complex conjugation function. Prove that this mapping is actually a homomorphism and since it is also a bijection, it is an isomorphism.

41. Let $S = \{0, 4, 8, 12, 16, 20, 24\}$.

(i) Show that S is a subring of $\mathbb{Z}_{28}$.

(ii) Prove that the mapping $f : \mathbb{Z}_7 \to S$; $f(x) = [8x]_{28}$ is an isomorphism. Here $[8x]_{28}$ denotes the remainder of the division of the integer $8x$ with 28.

42. Let E be the set of even integers equipped with the ordinary addition of integers, and a new multiplication $\otimes$ defined by $x_1 \otimes x_2 = \frac{x_1 \cdot x_2}{2}$, for all $x_1, x_2 \in E$.

(i) Show that the set E equipped with these two operations is a ring.

(ii) Show that the mapping $f : E \to \mathbb{Z}$; $f(x) = \frac{x}{2}$, is an isomorphism.

43. Let $\mathbb{Z}^*$ denote the ring of integers with the $\oplus$ and $\odot$ operations defined in exercise 10. Show that $\mathbb{Z}^*$ is isomorphic to $\mathbb{Z}$.

44. Let $\hat{\mathbb{Z}}$ denote the ring of integers equipped with two operations $\oplus$ and $\odot$ defined as follows:

$x_1 \oplus x_2 = x_1 + x_2 - 1$, for all $x_1, x_2 \in \mathbb{Z}$; $x_1 \odot x_2 = x_1x_2 - (x_1 + x_2) + 2$, for all $x_1, x_2 \in \mathbb{Z}$.

(i) Prove that $\hat{\mathbb{Z}}$ is a commutative ring with the zero element being the integer 1 and the identity element being the integer 2.

(ii) Prove that $\hat{\mathbb{Z}}$ is isomorphic to $\mathbb{Z}$.

Hint: Consider the mapping $\mathbb{Z} \to \hat{\mathbb{Z}}$; $f(x) - x + 1$. Note that by inspection, this mapping maps the zero element of $\mathbb{Z}$ onto the zero element of $\hat{\mathbb{Z}}$; and the identity element of $\mathbb{Z}$ onto the identity element of $\hat{\mathbb{Z}}$. It is also, by inspection, a bijection. Next, you must show that it is both an additive and multiplicative mapping, hence a homomorphism.

45. (For students familiar with the ring of complex numbers $\mathbb{C}$.) Show that the ring $\mathbb{R} \times \mathbb{R}$ is isomorphic to the ring $\mathbb{C}$.

46. Define a new multiplication $\otimes$ on $\mathbb{R} \times \mathbb{R}$ as follows: $(a, b) \otimes (c, d) = (ac, bc)$, for all $a, b, c, d \in \mathbb{R}$.

(i) Prove that $\mathbb{R} \times \mathbb{R}$ equipped with this new multiplication and the usual coordinatewise addition, is a ring with zero element the pair $(0, 0)$. We denote this ring by $(\mathbb{R} \times \mathbb{R})^*$.

(ii) Show that this ring has no identity element.

(iii) Show that this ring is not a commutative ring.

(iv) Prove that this ring is isomorphic to the ring S of all matrices of the form $\begin{pmatrix} a & 0 \\ b & 0 \end{pmatrix}$ (a subring of $M_2(\mathbb{Z})$) **Hint**: Let $f : (\mathbb{R} \times \mathbb{R})^* \to S$; $f((a, b)) = \begin{pmatrix} a & 0 \\ b & 0 \end{pmatrix}$. Show that f is an isomorphism.

47. Let $M_2(\mathbb{Z})$ be the ring of all 2×2 matrices with integer entries; and I be the subring of $M_2(\mathbb{Z})$, consisting of all matrices of the form $\begin{pmatrix} a & 0 \\ b & c \end{pmatrix}$; $a, b, c \in \mathbb{Z}$. Show that the mapping $F : I \to \mathbb{Z}$; $f\left(\begin{pmatrix} a & 0 \\ b & c \end{pmatrix}\right) = a$ is a surjective homomorphism but it fails to be injective and so it is not an isomorphism.

48. Consider the mapping $f :\to M_2(\mathbb{R})$; $f(x) = \begin{pmatrix} 0 & 0 \\ -x & x \end{pmatrix}$. Prove that f is an injective homomorphism, but not a surjective mapping. Thus it is not an isomorphism.

49. **(i)** Give an example of a homomorphism $f : R \to S$, from a ring R to a ring S, such that R has an identity element, but S does not.

(ii) Give an example of a homomorphism $f : R \to S$ such that R has no identity element, but S does. In both (i) and (ii) try to give examples other than the zero homomorphism.

50. Let $f : R \to S$ be an isomorphism of rings and let $g : S \to R$ be the inverse function of f. Prove that g is also an isomorphism. **Hint**: To show $g(x_1 * x_2) = g(x_1) \oplus g(x_2)$ (here we denote addition in S by $*$ and addition in R by $\oplus$).

Consider the images of the left- and right-hand side under f and use the facts that f is a homomorphism and the composite function $f \circ g$ is the identity mapping on S. Repeat this approach in order to show $g(x_1 \otimes x_2) = g(x_1) \odot g(x_2)$ (here we denote multiplication in S by $\otimes$ and multiplication in R by $\odot$).

51. Let $f : R \to S$ be a homomorphism of rings and let $T = \{x \in R \mid f(x) = 0_S\}$. Prove that T is a subring of R.

52. Let F be the ring of all continuous functions form $\mathbb{R}$ to itself and c a fixed real number. Consider the mapping $\varphi : F \to \mathbb{R}$; $\varphi(f) = f(c)$. Show that φ is a surjective homomorphism. Is the mapping φ an isomorphism? Why or why not?

53. If $f : R \to S$ is an isomorphism of rings, which of the following properties are preserved by this isomorphism? Justify your answers.

(i) $b \in R$ is a zero divisor.

(ii) $a \in R$ is idempotent (i.e. it satisfies $a^2 = a$).

(iii) R is an integral domain.

54. In each case below show that the first ring is not isomorphic to the second. To do so you may use either of two methods:

(i) Show that one ring has a property that the other does not have.

(ii) Assume the existence of an isomorphism between the two rings and show that this assumption leads to a contradiction or an impossibility.

(a) E (the set of even integers) and $\mathbb{Z}$.

(b) $\mathbb{Z}_4 \times \mathbb{Z}_6$ and $\mathbb{Z}_{24}$.

Hint: List all the zero divisors in both rings and conclude that these two rings have a different number of zero divisors.

(c) $\mathbb{Z} \times \mathbb{Z}_2$ and $\mathbb{Z}$.

(d) $\mathbb{Z}$ and $\mathbb{R}$.

Hint: Any isomorphism from $\mathbb{Q}$ to $\mathbb{R}$ would have to map the identity element of $\mathbb{Q}$ (i.e. the number 1) onto the identity element of $\mathbb{R}$ (the number 1 again). First prove this statement or use a relevant Theorem to justify it. Proceed further by using the additivity and multiplicativity that such a mapping must have. Conclude that such an isomorphism would in fact be the identity mapping on $\mathbb{Q}$, effectively contradicting the assumption that is surjective (an isomorphism is in particular onto); the image of such a mapping would be just $\mathbb{Q}$ itself failing to cover all of $\mathbb{R}$.

(e) $\mathbb{Z}_4 \times \mathbb{Z}_4$ and $\mathbb{Z}_{16}$.

55. **(i)** If $f : R \to S$ is a homomorphism of rings show that for any $x \in R$ and any $n \in \mathbb{Z}$, $f(nx) = nf(x)$.

(ii) Show that isomorphic rings with identity must have the same characteristic (refer to exercises 30-32).

(iii) If $f : R \to S$ is a homomorphism of rings with identity, is it true that R and S must have the same characteristic?

Chapter 3

Real Numbers & their Fundamental Properties

3.1 The Axioms of Real Numbers.

Over years of schooling, students become familiar with some of the properties of the real numbers and they learn how to perform operations involving real numbers. Seldom, however, are students presented with all the Axioms that the set of real numbers satisfy.

The real numbers that we are accustomed with, form a non-empty set $\mathbb{R}$ equipped with two binary operations, addition "+" and multiplication "·". The set $\mathbb{R}$ satisfies the following fifteen axioms.

Axiom 1: $\mathbb{R}$ is closed under the addition operation: for every a and b in $\mathbb{R}$, the sum $a+b$ is also in $\mathbb{R}$.

Axiom 2: The addition operation is Associative: that is for every a, b, and c in $\mathbb{R}$, we have $(a+b)+c = a+(b+c)$. Therefore, the sum $a+b+c$ is unambiguously defined.

Axiom 3: The addition operation is commutative: for every a, b in $\mathbb{R}$, $a+b = b+a$.

Axiom 4: There exists a real number, called a zero number and denoted by 0, such that $a+0 = a = a+0$, for every real number a. (Refer to Remark 1 below, the zero number or element is unique).

Axiom 5: For every real number a, there exists another real number a', called an opposite of a such that $a + a' = 0$. (Refer to Remark 1 below. The opposite of a number is unique.)

Axiom 6: $\mathbb{R}$ is closed under multiplication: that if a, b are in $\mathbb{R}$, so is $a \cdot b$, called the product of a and b.

Axiom 7: The multiplication operation is Associative: for every a, b, c in $\mathbb{R}$, $a \cdot (b \cdot c) = (a \cdot b) \cdot c$. Thus the product $a \cdot b \cdot c$ is unambiguously defined.

Axiom 8: The multiplication operation is commutative: for every a, b in $\mathbb{R}$, $a \cdot b = b \cdot a$.

Axiom 9: There exists a real number, called one or an identity denoted by 1, such that $a \cdot 1 = a = 1 \cdot a$ for every a in $\mathbb{R}$. (Refer to Remark 1 below. The identity element is unique.)

Axiom 10: For every real number a, which is not a zero number (a zero number is defined in Axiom 4), there is another real number a', such that $a \cdot a' = 1$, where 1 is an identity (as defined in Axiom 9). We call a' an inverse of a (a' is unique. See Remark 3 below. So a' is the inverse of a).

Axiom 11: The multiplication operation is distributive with respect to the addition operation: for every a, b, c in $\mathbb{R}$, $a \cdot (b + c) = a \cdot b + a \cdot c$

Axiom 12: The set $\mathbb{R}$ is totally ordered (Refer to Chapter 1 for these two concepts.) under the usual ordering of "$>$": (which reads "larger than") for every a, b in $\mathbb{R}$ exactly one of the following three statements must be true: either $a > b$ (it reads "a is larger than b") or $a = b$ (it reads "a is equal to b") or $b > a$ (it reads "b is larger than a".) Here we insert the following definitions: "$<$" reads: "less than"

Definition 1: $a < b$ if and only if $b > a$.

Definition 2: $a \leq b$ if and only if (it is not $a > b$).

Axiom 13: If a, b, c, d are in $\mathbb{R}$ and ($a > b$ and $c \geq d$), then $a + c > b + d$.

Axiom 14: If $a > b$ and $c > 0$, then $a \cdot c > b \cdot c$.

Axiom 15: (This Axiom holds for any bounded above nonempty subset S of $\mathbb{R}$.) Let $S \subset \mathbb{R}$, i.e. S a subset of $\mathbb{R}$. If the set $E = \{x | x \in \mathbb{R}$ and $s \leq x$, for every $s \in S\}$ is nonempty, then it has a minimum element. That is, there exists a member m of E such that, $m \leq \epsilon$, for every ϵ in E.

IMPORTANT REMARKS on the Fifteen Axioms

Remark 1: Note that since $\mathbb{R}$ satisfies Axioms 1 through 7, Axiom 9, and Axiom 11, it follows that $\mathbb{R}$ is a ring with an identity. In Chapter 2, we already showed that in a ring with an identity, the zero element is unique, and the identity element is also unique. Moreover, the opposite of an element (Axiom 5) is also unique. (This was also proven in Chapter 2.) Thus the zero number 0 and the number 1 we are accustomed with, are the unique zero and identity elements of $\mathbb{R}$, respectively.

Remark 2: Since $\mathbb{R}$ satisfies Axiom 8, it follows that $\mathbb{R}$ is a commutative ring as well.

Remark 3: Since $\mathbb{R}$ satisfies Axiom 10 (in addition to Axioms 1-9 and Axiom 11), it follows that $\mathbb{R}$ is a field, and therefore an integral domain as well (See Chapter 2. Every field is an integral domain, though the converse is false.) Moreover a' in Axiom 10 is unique, called the inverse of a. The uniqueness of a' was proved in Chapter 2 for any field. The inverse a' of a is denoted by a^{-1}, and also by $1/a$.

Remark 4: Since $\mathbb{R}$ also satisfies Axiom 12, it follows that $\mathbb{R}$ is a totally ordered field. In addition, since $\mathbb{R}$ satisfies Axioms 13-15, $\mathbb{R}$ is also what is called a complete field. (Refer to Chapter 1, for the concept of a totally ordered set.) However, $\mathbb{R}$ is not a well-ordered set, as we have seen in Chapter 1. (It does not satisfy the well-ordering property.) For instance, the set $S\{x | x \in \mathbb{R}]$ and $0 < x \leq 1\}$ has no least element, even though it is a set bounded below.

Remark 5: It can be proven that every other ring-field structure F that satisfies Axioms 1 through 15, under two operations of addition "+" and multiplication "·", must isomorphic to $\mathbb{R}$. (Refer to Chapter 1 for the definition of isomorphism.) It has essentially the same structure as $\mathbb{R}$ does. It must have all the properties that $\mathbb{R}$ has and no other properties that $\mathbb{R}$ does not have. By relabeling the elements of F, as well as the two operations of addition and multiplication in F, we will simply obtain the set $\mathbb{R}$ with its usual operations of addition and multiplication. Hence, $\mathbb{R}$ is essentially a unique structure satisfying Axioms 1-15.

3.2 Properties of Real Numbers

We now list a number of algebraic properties of the real numbers. These are an immediate consequence of Axioms 1-7, Axioms 9 and 11, which make the reals into a ring. All the properties listed below have already been proven in Chapter 2.

Property 1: If $a + c = b + c$, then $a = b$.

Definition 3: The real number $a + (-b)$, we call the difference of a, b and we denote it by $a - b$; $a - b := a + (-b)$, where $(-b)$ is the (unique) opposite of b. Note (see Chapter 2) that $a - b$ is the unique solution to the equation $x + b = a$.

Property 2: $-(-a) = a$ and $(-1) \cdot a = -a$.

Property 3: $a - (-b) = b$.

Property 4: $-(a + b) = (-a) + (-b)$.

Property 5: $-(a - b) = b - a$.

Property 6: $(a + b) - c = a + (b - c)$.

Property 7: $(a - b) + c = a - (b - c)$.

Property 8: $a \cdot 0 = 0$.

Property 9: $a \cdot (-b) = -(ab)$.

Property 10: $(-a) \cdot (-b) = ab$.

Property 11: $a(b - c) = ab - ac$.

The following property is due to the fact that $\mathbb{R}$ is integral domain which in turn follows from the fact that $\mathbb{R}$ is a field. This property is also proven in Chapter 2.

Property 12: If $ab = 0$, then either $a = 0$ or $b = 0$ or both. A consequence of Property 12 is the next property.

Property 13: If $ac = b \cdot c$ and $b \cdot c \neq 0$, then $a = b$. This is proven in Chapter 2.

3.3 Fractions or Ratios

Definition 4: Let a and b be real numbers, $b \neq 0$. The unique real number $a \cdot b^{-1}$, where b^{-1} is the inverse of b, is called the **ratio of a over b**. It is also denoted by $\frac{a}{b}$ so we write $a \cdot b^{-1} = \frac{a}{b}$. The ratio $\frac{a}{b}$ is also called **the fraction with numerator a and denominator b.**

Remark 6: An alternative notation to a/b would be $a : b$, where ":" stands for the usual division symbol we are accustomed with so we can write:

$$(a : b) \underset{\to \text{Definition}}{:=} \frac{a}{b} = a \cdot b^{-1} = a \cdot \left(\frac{1}{b}\right)$$

Properties of Fractions

Property 14: $\left(\frac{a}{b}\right) \cdot b = a$.

Proof: $\left(\frac{a}{b}\right) \cdot b = (a \cdot b^{-1}) \cdot b \underset{\text{by Ax. 7}}{=} a \cdot (b^{-1} \cdot b) \underset{\text{by Ax. 8 and 10}}{=} a \cdot (1) \underset{\text{by Ax. 9}}{=} a.$

Property 15: If $b \cdot d \neq 0$ (which means $b \neq 0$ and $d \neq 0$ by Property 8), then $\left(\frac{a}{b}\right) \cdot \left(\frac{c}{d}\right) = \left(\frac{a \cdot d}{b \cdot d}\right)$.

Proof: We have $\left(\frac{a}{b}\right) \cdot \left(\frac{c}{d}\right) \cdot (b \cdot d) \underset{\text{by Ax. 8}}{=} \left(\frac{a}{b}\right) \cdot \left(\frac{c}{d}\right) \cdot (d \cdot b) \underset{\text{by Ax. 7}}{=} \left(\frac{a}{b}\right) \cdot \left(\frac{c}{d} \cdot d\right) \cdot b \underset{\text{by Prop. 14}}{=}$ $\left(\frac{a}{b}\right) \cdot c \cdot b \underset{\text{by Ax. 8}}{=} \left(\frac{a}{b}\right) \cdot b \cdot c \underset{\text{by Prop. 14}}{=} \left(\frac{a}{b} \cdot b\right) \cdot c = a \cdot c.$

Hence, we have shown that $\left(\frac{a}{b}\right) \cdot \left(\frac{c}{d}\right) \cdot (b \cdot d) = a \cdot c$; but $b \cdot d \neq 0$ and so if we multiply both sides by inverse of $b \cdot d$, which is equal to $\frac{1}{b \cdot d}$ we obtain

$$\left[\left(\frac{a}{b}\right) \cdot \left(\frac{c}{d}\right) \cdot (b \cdot d)\right] \cdot \left(\frac{1}{b \cdot d}\right) = (a \cdot c) \cdot \left(\frac{1}{b \cdot d}\right); \left(\frac{a}{b}\right) \cdot \left(\frac{c}{d}\right) \cdot \underbrace{\left[(b \cdot d) \cdot \left(\frac{1}{b \cdot d}\right)\right]}_{\text{equal to 1}} = \frac{a \cdot c}{b \cdot d},$$

so that

$$\left(\frac{a}{b}\right) \cdot \left(\frac{c}{d}\right) = \frac{a \cdot c}{b \cdot d}.$$

Property 16: If $b \cdot c \neq 0$, then $\frac{(a \cdot c)}{(b \cdot c)} = \frac{a}{b}$.

Proof: This is an immediate consequence of property 15, just take $c = d$.

Property 17: If $a \cdot b \neq 0$, then $\left(\frac{a}{b}\right)^{-1} = \frac{b}{a}$.

Proof: Apply Property 15 to the product $\left(\frac{a}{b}\right) \cdot \left(\frac{b}{a}\right)$; we have,

$$\left(\frac{a}{b}\right) \cdot \left(\frac{b}{a}\right) = \frac{a \cdot b}{b \cdot a} \underset{\text{by Ax. 8}}{=} \frac{a \cdot b}{a \cdot b} = (a \cdot b) \cdot \left(\frac{1}{a \cdot b}\right) \underset{\text{by Ax.10}}{=} 1.$$

Property 18: If b, c and d are all nonzero real numbers, then

$$\left(\frac{a}{b}\right) : \left(\frac{c}{d}\right) = \left(\frac{a}{b}\right) \cdot \left(\frac{c}{d}\right)^{-1} = \frac{(a \cdot d)}{(b \cdot c}).$$

Proof: By definition (See Remark 6 under Definition 4.)

$$\left(\frac{a}{b}\right) : \left(\frac{c}{d}\right) = \left(\frac{a}{b}\right) \cdot \left(\frac{c}{d}\right)^{-1}; \text{ but by Property 17, } \left(\frac{c}{d}\right)^{-1} = \frac{d}{c}.$$

Thus $\left(\frac{a}{b}\right) : \left(\frac{c}{d}\right) = \left(\frac{a}{b}\right) \cdot \left(\frac{d}{c}\right) \underset{\text{by Prop. 15}}{=} \frac{a \cdot d}{b \cdot c}$.

Property 19: If $b \cdot d \neq 0$, then $\left(\frac{a}{b}\right) + \left(\frac{c}{d}\right) = \frac{(ad+bc)}{(b \cdot d)}$. Likewise $\left(\frac{a}{b}\right) - \left(\frac{c}{d}\right) = \frac{(ad-bc)}{(bd)}$

Proof: $\left(\frac{a}{b} + \frac{c}{d}\right) \cdot (bd) \underset{\text{by Ax. 8}}{=} \left(\frac{a}{b}\right) \cdot (b \cdot d) + \left(\frac{d}{d}\right) \cdot (b \cdot d) \underset{\text{by Ax. 11}}{=} \left(\frac{a}{b}\right) \cdot (b \cdot d) + \left(\frac{c}{d}\right) \cdot (d \cdot b) =$ $\left[\left(\frac{a}{b}\right) \cdot b\right] \cdot d + \left[\left(\frac{c}{d}\right) \cdot d\right] \cdot b \underset{\text{by Property 14}}{=} a \cdot d + c \cdot b$.

Hence we have shown that, $\left(\frac{a}{b} + \frac{c}{d}\right) \cdot (b \cdot d) = a \cdot d + c \cdot b$. Multiplying both sides by $(b \cdot d)^{-1}$, $\left[\left(\frac{a}{b} + \frac{c}{d}\right) \cdot (bd)\right] \cdot \left[(bd)^{-1}\right] = (a \cdot d + c \cdot b) \cdot (bd)^{-1}$; by Axiom 7 we have

$$\left(\frac{a}{b} + \frac{c}{d}\right) \underbrace{\left[(bd) \cdot (bd)^{-1}\right]}_{\text{equals 1, by Ax. 10}} = (ad + c \cdot b)(bd)^{-1}; \ \left(\frac{a}{b} + \frac{c}{d}\right) \cdot 1 = (ad + c \cdot b) \cdot \left(\frac{1}{b \cdot d}\right),$$

and by Axiom 9 and Remark 6 we arrive at, $\frac{a}{b} + \frac{c}{d} = \frac{ad+cb}{bd}$.

Remark 7: Now that we have used some of the Axioms to prove some of the fundamental algebraic properties of real numbers, **we do not have to formally refer to these Axioms or to these fundamental properties, Properties 1-19,** every time we want to prove an algebraic statement. We can just apply them without referring to them.

Analogies

Definition 5: We say that the four real numbers a, b, c, and d form an analogy if and only if,

(i) $a \cdot b \cdot c \cdot d \neq 0$, i.e., they are all nonzero, and

(ii) $\frac{a}{b} = \frac{c}{d}$.

The real number a and d are called the **extreme terms** of the analogy, whereas b and c are called the **middle terms** of the analogy.

The number $k = \frac{a}{b}$ is called the **constant of the analogy**

Example 1: The reals $a = 6,\ b = 4,\ c = 3$, and $d = 2$ form an analogy because $\frac{6}{4} = \frac{3}{2}$.

Properties of Analogies

Suppose the real numbers a, b, c, d form an analogy.

Property 20: $ad = bc$.

Property 21: The real numbers d, b, c, a also form an analogy.

Property 22: The real numbers a, c, b, d also form an analogy.

Property 23: The real numbers b, a, d, c also form an analogy.

Property 24: $\frac{a+b}{b} = \frac{c+d}{d}$ (also $\frac{a-b}{b} = \frac{c-d}{d}$).

Property 25: $\frac{a}{a+b} = \frac{c}{c+d}$ (also $\frac{a}{a-b} = \frac{c}{c-d}$, assuming $a \neq b$ and $c \neq d$).

Property 26: $\frac{a+b}{a-b} = \frac{c+d}{c-d}$, assuming $a \neq b$ and $c \neq d$.

Remark 8: Property 24 can also be presented as: "$a+b, b, c+d, d$ also form an analogy". Similar adaptations can be applied to Properties 25 and 26.

Below we offer proofs to Properties 20, 21, 24, and 26. The other three are left to the student as an exercise.

Proof of Property 20: From $\frac{a}{b} = \frac{c}{d}$ it follows that,

$$\frac{ad}{bd} = \frac{cb}{db} \Rightarrow (ad)\cdot\left(\frac{1}{bd}\right) = (cb)\cdot\left(\frac{1}{db}\right) \Rightarrow \text{(by the cancelation Property 13)} \Rightarrow ad = cb.$$

Proof of Property 21: By Property 20, we have

$$a \cdot d = c \cdot b \Rightarrow (ad)\cdot\left(\frac{1}{a\cdot b}\right) = (b\cdot c)\cdot\left(\frac{1}{a\cdot b}\right) \Rightarrow \frac{ad}{ab} = \frac{b\cdot c}{ab}; \frac{ad}{ab} = \frac{bc}{ba} \Rightarrow \frac{d}{b} = \frac{c}{a},$$

which proves that the real numbers d, b, c, a form an analogy.

Proof of Property 24: From $\frac{a}{b} = \frac{c}{d}$ it follows that,

$$\frac{a}{b} + 1 = \frac{c}{d} + 1 \Rightarrow \frac{a}{b} + \frac{b}{b} = \frac{c}{d} + \frac{d}{d} \Rightarrow \frac{a+b}{b} = \frac{c+d}{d}.$$

Proof of Property 25: By applying Property 24 we have $\frac{a+b}{b} = \frac{c+d}{d}$ and $\frac{a-b}{b} = \frac{c-d}{d}$. By applying Property 18

$$\left(\frac{a+b}{b}\right) : \left(\frac{a-b}{b}\right) = \left(\frac{c+d}{d}\right) : \left(\frac{c-d}{d}\right) \Rightarrow \frac{(a+b)\cdot b}{b\cdot(a-b)} = \frac{(c+d)\cdot d}{d\cdot(c-d)} \Rightarrow \frac{a+b}{a-b} = \frac{c+d}{c-d}.$$

We finish this section by establishing Theorem 1.

Theorem 1

If a, b, c, d, e and f are real numbers that satisfy $d \cdot e \cdot f \neq 0$, and $\frac{a}{b} = \frac{c}{d} = \frac{e}{f} = k$, then $k = \frac{k_1a+k_2c+k_3e}{k_1b+k_2d+k_3f}$, where k_1, k_2, k_3 are any real numbers such that $k_1b+k_2d+k_3f \neq 0$. More generally, if n is a positive integer $n \geq 3$, and $a_1, a_2, a_3, \ldots, a_n, b_1, b_2, b_3 \ldots, b_n$ are real numbers that satisfy $\underbrace{b_1 \cdot b_2 \cdot b_3 \ldots b_n}_{\text{product of the } b's} \neq 0$, then $\frac{a_1}{b_1} = \frac{a_2}{b_2} = \frac{a_3}{b_3} = \frac{a_1}{b_1} = \frac{a_2}{b_2} = \frac{a_2}{b_3} = \ldots = \frac{a_n}{b_n} = l = \frac{a_n}{b_n} - = k$, then we must also have $k = \frac{k_1a_1+k_2a_2+\ldots+k_na_n}{k_1b_1+k_2b_2+\ldots+k_nb_n}$, for any real numbers $k_1, k_2, \ldots, k_n$ such that $k_1b_1 + k_2b_2 + \ldots + k_nb_n \neq 0$.

Proof: Suppose $\frac{a}{b} = \frac{c}{d} = \frac{e}{f} = k$. We have, $a = b \cdot k$, $c = d \cdot k$, $e = f \cdot k$. From $a = b \cdot k \Rightarrow k_1 \cdot a = k_1 \cdot b \cdot k$; likewise $k_2 \cdot c = k_2 \cdot d \cdot k$ and $k_3 \cdot e = k_3 \cdot f \cdot k$. Hence, $k_1a + k_2b + k_3c = k_1bk + k_2dk + k_3fk$; $k_1a + k_2b + k_3c = (k_1b + k_2d + k_3f) \cdot k$. Since

$k_1b + k_2d + k_3f \neq 0$ by hypothesis, we can divide both sides of the last equation by $k_1b + k_2d + k_3f$ to obtain

$$k = \frac{k_1a + k_2b + k_3c}{k_1b + k_2d + k_3f}$$

and the result is proven. The generalized case for $n \geq 3$ (e.g. $n = 4: \ a_1, a_2, a_3, a_4, b_1, b_2, b_3, b_4$) is proven in a similar manner. There is no essential difference in the proof, except that there will be more than three equal fractions for instance for $n = 4: \ \frac{a_1}{b_1} = \frac{a_2}{b_2} = \frac{a_3}{b_3} = \frac{a_4}{b_4} = k$.

Remark 9: If we apply Theorem 1 with $k_1 = k_2 = k_3 = 1$, then it follows that $\frac{a}{b} = \frac{c}{d} = \frac{e}{f} = k$ implies $\frac{a+c+e}{b+d+f} = k$. The same is true for any $n \geq 3: \ \frac{a_1}{b_1} = \frac{a_2}{b_2} = \frac{a_3}{b_3} = \ldots = \frac{a_n}{b_n} = k$ implies $k = \frac{a_1+a_2+a_3+\ldots+a_n}{b_1+b_2+b_3+\ldots+b_n}$

3.4 Properties of Ordering and of Inequalities

Definition 6: We say that,

(a) The real number a is positive, if and only if $0 < a$.

(b) The real number a is negative, if and only if $a < 0$.

(c) The real number a is nonpositive, if and only if $a \leq 0$.

(d) The real number a is nonnegative, if and only if $0 \leq a$.

(e) The real numbers a and b have the same sign if and only if they are both positive or they are both negative.

(f) The real numbers a and b have opposite signs if and only if a is positive and b negative or vice-versa.

Property 27: If $a < b$, then $a \leq b$, but not conversely.

Proof: Because if it is not $a \leq b$, then by Axiom 12, we must have $a > b$ contrary to hypothesis. Also, the converse is false since $a = b$ does not imply $a < b$.

Property 28: If $a > b$, then $a \geq b$, but not conversely.

Proof: Similar with the previous proof.

Property 29: If $a = b$, then ($a \leq b$ and $a \geq b$), and conversely.

Proof: Similar with the previous one.

Property 30: If $a > b$, then $a - b > 0$, and conversely.

Proof: If $a > b$, then $a + (-b) > b + (-b)$, by Axiom 13. Thus, since $b + (-b) = 0$ (by Axiom 5), we have $a + (-b) > 0$; but $a + (-b) = a - b$, by Definition 3. Thus $a - b > 0$. Conversely, if $a - b > 0$, then $[a + (-b)] + b > 0 + b \Rightarrow$ by (Axiom 2) $a + [(-b) + b)] > 0 + b \Rightarrow$ by Axiom 5); $a + 0 > 0 + b$ (by Axiom 4), $a > b$.

Property 31 (Transitive property): If $a > b$ and $b > c$, then $a > c$.

Proof: Applying Property 30 we have $a - b > 0$, and $b - c > 0$. Applying Axiom 14 gives $(a - b) + (b - c) > 0$ and by the Associative Axiom (Axiom 2) of addition we obtain $a + (-b + b) - c > 0 \Rightarrow a + 0 - c > 0 \Rightarrow$ (Axiom 4) $\Rightarrow a - c > 0 \Rightarrow$ (Property 30) $a > c$.

Property 32: If $a < 0$, then $-a > 0$.

Proof: If $a < 0$, then $a + (-a) < 0 + (-a) \Rightarrow$ (by Axiom 5) $0 < 0 + (-a) \Rightarrow$ (by Axiom 4) $0 < (-a); 0 < -a; -a > 0$ (by Definition 1).

Definition 7: We define $\underline{a \cdot a = a^2}$ and $\underline{a^2 \cdot a = a^3}$; $a \cdot a \cdot a = a^3$. More generally if n is a positive integer we define $a^n = \underbrace{a \cdot a \cdot a \ldots a}_{n \text{ times}}$.

Property 33: (i) If $a \neq 0$, then $a^2 > 0$.

(ii) If $a = 0$, then $a^n = 0$, where n is any positive integer.

(iii) If $a > 0$, then $a^n > 0$, where n is any positive integer.

(iv) If $a < 0$, then $a^n > 0$, if n is an even positive integer.

(v) If $a < 0$, then $a^n < 0$, if n is an odd positive integer.

(vi) $(ab)^n = a^n b^n = b^n a^n$ n any positive integer.

Proof: (i) If $a > 0$, then by Axiom 14, $a \cdot a > a \cdot 0$; but $a \cdot 0 = 0$ by Property 8. Hence $a^2 = a \cdot a > 0$. If $a < 0$, then $-a > 0$ by Property 32. Apply Axiom 14 again we obtain $(-a) \cdot (-a) > (-a) \cdot 0$; but $(-a) \cdot 0 = 0 \cdot (-a) = 0$. Thus, $(-a) \cdot (-a) > 0$ and by Property 10 we have $(-a)(-a) = a^2$; so that $a^2 > 0$.

(ii) If $a = 0$, then $a^n = a = 0$, if $n = 1$. If $n \geq 2$, then $n - 1$ is a positive integer; $a^n = \underbrace{a \cdot a \ldots a}_{n \text{ times}} = \underbrace{a \cdot a \ldots a}_{(n-1) \text{ times}} \cdot a$; but the product $\underbrace{a \ldots a}_{(n-1) \text{ times}} \cdot a$ is zero, because a=0 and by Property 8. Hence $a^n = 0 \cdot a = a \cdot 0 = 0$, by Property 8 again.

(iii) If $n = 1$, then $a^n = a^1 = a > 0$, by hypothesis. If $n \geq 2$, then $n - 1$ is a positive integer: $a^n = \underbrace{a \ldots a}_{n \text{ times}}$. Since the last produce has at least two factors we can apply Axiom 14 $(n-1)$ times as follows: $a \cdot a > 0$. Then $(a \cdot a) \cdot a > 0$; $((a \cdot a) \cdot a) \cdot a > 0$ and so on) to show that $a^n = \underbrace{a \ldots a}_{\text{times}}$ is indeed positive.

(iv) If $a < 0$ and n is an even positive integer, then $n = 2k$, where k is a positive integer. We write $a^n = \underbrace{a \ldots a}_{n \text{ times}} = \underbrace{(a \cdot a) \cdot (a \cdot a) \ldots (a \cdot a)}_{k \text{ factors each being equal to } a \cdot a}$. It is clear that by Definition 7, the product $\underbrace{(a \cdot a) \cdot (a \cdot a) \ldots (a \cdot a)}_{k \text{ times}}$ is equal to $(a \cdot a)^k$, since k is a positive integer. Thus $a^n = (a \cdot a)^k$; but $a \cdot a > 0$ by (i); and by (iii) it follows that $(a \cdot a)^k > 0$. Thus $a^n > 0$.

(v) If $a < 0$, and n is an odd positive integer, then if $n = 1$, it is clear that $a^n = a^1 = a < 0$. If $n \geq 3$, then we can set $n = 2k + 1$, where k is a positive integer. Applying a grouping similar to the one we did in (iv) we have,

$$a^n = \underbrace{(a \cdot a) \cdot (a \cdot a) \ldots (a \cdot a) \cdot a}_{k \text{ factors each equal to } (a \cdot a)} = \underbrace{(a \ldots a)}_{2k \text{ factors}} \cdot a = a^{2k} \cdot a$$

By (iv) we know that since $n - 1 = 2k$ is an even positive integer, the real number a^{2k} must be positive. Also $a < 0$ implies $-a > 0$, by Property 32. Hence, $a^{2k} \cdot (-a) > 0$; but by Property 9 we have $a^{2k} \cdot (-a) = -(a^{2k} \cdot a)$, which shows that $-(a^{2k} \cdot a) > 0 \Rightarrow -\left(a^{2k+1}\right) > 0$; $-(a^n) > 0$. According to Axiom 12 and Definition 6 the real number a^n must either be positive, zero, or negative. If $a^n > 0$ then $-(a^n) < 0$ by Property 32, contradicting the already deduced statement that $-(a^n) > 0$. The second possibility $a^n = 0$ cannot happen either, because if $a^n = 0$ then $(-1) \cdot a^n = (-1) \cdot 0$, but $(-1) \cdot 0 = 0$ by Property 8 and $(-1)a^n = -(a^n)$ by Property 2. Consequently, $-(a^n) = 0$ contradicting $-(a^n) > 0$ again. Hence we conclude the third possibility, namely that

$$a^n < 0.$$

Property 34: $1 > 0$.

Proof: by Axiom 9, $1 \cdot 1 = 1$. Thus $1^2 = 1$ and so by Property 33(i), it follows that $1 > 0$.

Property 35: If $a > b$ and $c < 0$, then $a \cdot c < b \cdot c$.

Proof: Since $c < 0 \Rightarrow$ (by Property 32) $-c > 0$ and by Axiom 14, $a \cdot (-c) > b \cdot (-c)$. By Property 9 $\Rightarrow a \cdot (-c) = -(ac)$ and $b \cdot (-c) = -(bc)$. The last inequality implies $-(a \cdot c) > -(b \cdot c) \Rightarrow$ (by Property 30) $-(a \cdot c) - [-(b \cdot c)] > 0$. But $-[-(b \cdot c)] = bc$, by Property 2. Hence, $-(a \cdot c) + (b \cdot c) > 0$, by Axiom 3; $(b \cdot c) - (a \cdot c) > 0 \Rightarrow$ (by Property 30) $b \cdot c > a \cdot c$ which is equivalent to $a \cdot c < b \cdot c$ (by Definition 1).

Remark 10: It readily follows from Property 35 and Definition 6, parts (e) and (f) that the real numbers a and b will have the same sign if and only if $ab > 0$. They will have opposite signs if and only if $ab < 0$ (to see this, apply Property 35 by replacing b with 0 and replacing c with b; or replacing a with b and replacing c with a; or replacing a with 0, b with a and c with b. If both a and b are positive, then $a > 0$ and $b > 0$, and by Axiom 14 $\Rightarrow a \cdot b > 0 \cdot 0 = 0$).

Property 36: If $a > b$, then there a unique real number $x > 0$ such that $a = b + x$, and conversely.

Proof: If $a > b$, then $a - b > 0$ (by Property 30), let $x = a - b$. We have

$$b + x = b + (a - b) \underset{\text{by Ax. 3}}{=} b + (-b + a) \underset{\text{by Ax. 2}}{=} (b + (-b)) + a \underset{\text{by Ax. 5}}{=} 0 + a \underset{\text{by Ax. 4}}{=} a.$$

Conversely, if $a = b + x$, with $x > 0$, then $a + (-b) = (-b) + (b + x)$; $a - b = (-b + b) + x \Rightarrow a - b = 0 + x \Rightarrow a - b = x$; since $x > 0$, we have $a - b > 0 \Rightarrow a > b$.

Property 37: Let a be any nonzero real number. Then, a and $\frac{1}{a}$ have the same sign.

Proof: $a \cdot \left(\frac{1}{a}\right) = 1 > 0$ (by Property 34). Since $a \cdot \left(\frac{1}{a}\right) > 0$, it follows by Remark 10 that a and $\frac{1}{a}$ must have the same sign.

Property 38: If $a + c > b + c$, then $a > b$

Proof: If $a + c > b + c$, then $(a + c) + (-c) > (b + c) + (-c) \Rightarrow a + [c + (-c)] > b + [c + (-c)] \Rightarrow a + 0 > b + 0 \Rightarrow a > b$.

Property 39: (i) If $ac > bc$ and $c > 0$, then $a > b$.

(ii) If $ac > bc$ and $c < 0$, then $a < b$.

Proof: (i) Since $c > 0$, it follows from Property 37 that $1/c > 0$. From $\left(ac > bc \text{ and} \frac{1}{c} > 0\right) \rightarrow$ (by Axiom 14), $(ac) \cdot \frac{1}{c} > (b \cdot c) \cdot \frac{1}{c} \Rightarrow$ (by Axiom 7) $a \cdot \left(c \cdot \frac{1}{c}\right) > b \cdot \left(c \cdot \frac{1}{c}\right) \Rightarrow a \cdot 1 > b \cdot 1 \Rightarrow a > b$ (by Axiom 9).

Property 40: If $a < b$, $c < d$, and both b and c are positive, then $a \cdot c < b \cdot d$.

Proof: By Axiom 14, $a < b$ and $c > 0$ imply $a \cdot c < b \cdot c(1)$. By the same axiom, Axiom 14 the conditions $c < d$ and $b > 0$ imply $b \cdot c < b \cdot d(2)$. Combining inequalities (1) and (2) and applying Property 31 we deduce that $a \cdot c < b \cdot d$.

Property 41: (i) If $a < b$, then $-a > -b$.

(ii) If $0 < a < b$, then $\frac{1}{a} > \frac{1}{b}$.

Proof: (i) Since $1 > 0$ by Property 34, it follows by Property 35 that $(-1) \cdot 1 < (-1) \cdot 0 = 0$; $(-1) \cdot 1 < 0$; but $(-1) \cdot 1 = -1$, by Property 2. Hence $-1 < 0$. Now we have assumed $a < b$. by applying Property 35 again we conclude $(-1) \cdot a > (-1) \cdot b \Rightarrow$ (by Property 2) $-a > -b$.

(ii) If $0 < a < b$, then $0 < a$ and $0 < b$ in particular; and so $a > 0$ and $b > 0$ by Definition1. Applying Axiom 14 we obtain $a \cdot b > 0 \cdot 0 = 0$; $a \cdot b > 0$. Next, by Property 37 we infer that $\frac{1}{a \cdot b} > 0$. Since we have $a < b$ and $\frac{1}{a \cdot b} > 0$ or $b > a$ and $\frac{1}{a \cdot b} > 0$, we can apply Axiom 14 again to obtain $b \cdot \left(\frac{1}{a \cdot b}\right) > a \cdot \left(\frac{1}{a \cdot b}\right) \Rightarrow \frac{b}{a \cdot b} > \frac{a}{a \cdot b} \Rightarrow$ (by Property 16) $\frac{1}{a} > \frac{1}{b}$.

Example 2: $0 < 6 < 26 \Rightarrow \frac{1}{6} > \frac{1}{26}$.

Remark 11: It is important to realize that Property 41 (ii) is false if one drops the assumption $0 < a$. In other words, the inequality $a < b$ **does not imply** $\frac{1}{a} > \frac{1}{b}$. For example, if a is negative and b positive it **is true** that $a < b$, but it is **not true** that $\frac{1}{a} > \frac{1}{b}$. Now, if both a and b are negative and $a < b$, then it is **true** that $\frac{1}{a} > \frac{1}{b}$. Why? Because in view $a < 0$ and $b < 0$, we have $-a > 0$ and $-b > 0$. Also from $a < b \Rightarrow -a > -b$. So, all together, $0 < -b < -a$. Applying Property 41 (ii) we arrive at $-\frac{1}{b} > \frac{1}{-a}$.

By Property 41(i), $-\left(\frac{1}{-b}\right) < -\left(\frac{1}{-a}\right)$; $\frac{(-1)}{(-1) \cdot a} \Rightarrow \frac{1}{b} < \frac{1}{a}$ or simply $\frac{1}{a} > \frac{1}{b}$.

Property 42: (i) If $a > 1$, then $\frac{1}{a} < 1$.

(ii) If $0 < a < 1$, then $\frac{1}{a} > 1$.

Proof: (i) Since $a > 1 > 0$, we apply Property 41 (ii) to obtain $\frac{1}{a} < \frac{1}{1}$; but $\frac{1}{1} = 1$ (since $1 \cdot 1 = 1$). Thus, $\frac{1}{a} < 1$.

(ii) If $0 < a < 1$, again by Property 41 (ii) $\Rightarrow \frac{1}{a} > \frac{1}{1} = 1$; $\frac{1}{a} > 1$.

Property 43: If $a < b$, $b > 0$, and $0 < c < d$, then $\frac{a}{d} < \frac{b}{c}$.

Proof: From $0 < c < d$ it follows by Property 41(ii) that $\frac{1}{c} > \frac{1}{d}$ or $\frac{1}{d} < \frac{1}{c}$. It is also clear that $\frac{1}{d} > 0$ because we know that $d > 0$ (from $0 < c$ and $c < d$ and Property 31, it immediately follows $d > 0$). Property 37 tells us that d and $\frac{1}{d}$ must have the same sign, thus $d > 0$ implies $\frac{1}{d} > 0$. In summary we have $0 < \frac{1}{d}$, $\frac{1}{d} < \frac{1}{c}$, $b > 0$, and $a < b$. Applying Property 40, (with c replaced by $\frac{1}{d}$ and d replaced by $\frac{1}{c}$), we conclude that $a \cdot \frac{1}{d} < b \cdot \frac{1}{c}$ or simply $\frac{a}{d} < \frac{b}{c}$.

3.5 Some Important Theorems

Below we list theorems that are widely used in the algebra of real numbers.

Theorem 2: (i) If a, b are real numbers, $(a+b)^2 = a^2 + 2ab + b^2$.

(ii) More generally, if n is a positive integer greater than or equal to 2, and $a_1, a_2, \ldots, a_n$ are reals,,

$$(a+1+a_2+\ldots+a_n)^2 = a_1^2+a_2^2+\ldots+a_n^2+2\cdot(\text{sum of all products of the form } a_i \cdot a_j),$$

where the indices i and j satisfy $i \neq j$, $i = 1, 2, \ldots, n$ and $j = 1, 2, \ldots, n$. The above described sum can be written as

$$\sum_{\substack{1 \leq i,j \leq n \\ i \neq j}} a_i a_j$$

This sum contains $\frac{(n^2-n)}{2}$ product terms. (See Remark 13 below.)

Proof: (i) $(a+b)^2 = (a+b)\cdot(a+b) = (a+b)\cdot a + (a+b)\cdot b = a\cdot(a+b) + b\cdot(a+b) = a\cdot a + a\cdot b + b\cdot a + b\cdot b = a^2 + ab + ba + b^2 = a^2 + ab + ab + b^2 = a^2 + 2ab + b^2$. Note that we have used Axioms 8 and 11 to prove this.

Remark 13: Given n numbers, $a_1, \ldots, a_n$, the number of choices we can make by choosing two (distinct ones) among them is $\frac{n(n-1)}{2} = \frac{n^2-n}{2}$. This is the binomial coefficient $\binom{n}{2}$ ("n choose two"). More of this in Chapter 5.

(ii) $(a+b+c)^2 = [(a+b)+c]^2$. We can now apply part (i):

$$[(a+b)+c]^2 = (a+b)^2 + 2\cdot(a+b)\cdot c + c^2 \underset{\text{by part (i) again}}{=} a^2 + 2ab + b^2 +$$
$$2\cdot[(a+b)\cdot c] + c^2 =$$
$$= a^2 + 2ab + b^2 + 2\cdot(ac+bc) + c^2 = a^2 + 2ab + b^2 + 2ac + 2bc + c^2 =$$
$$a^2 + b^2 + c^2 + 2(ab + ac + bc).$$

(iii) It is easier to prove this part by using the method of mathematical induction, which will be introduced in the next Chapter, than trying to prove it by using the head on approach.

Theorem 3: (i) If a and b are reals,

(ii) $a^3 - b^3 = (a-b)(a^2 + ab + b^2)$.

(iii) $a^3 + b^3 = (a+b)(a^2 - ab + b^2)$.

(iv) If n is any positive integer greater than or equal to 2 and a, b any reals, then $a^n - b^n = (a-b)\cdot\left(\underbrace{a^{n-1} + a^{n-2}\cdot b^1 + \ldots + a^1\cdot b^{n-2} + b^{n-1}}_{n \text{ terms}}\right)$. In particular when $b = 1$, $a^n - 1 = (a-1)(a^{n-1} + a^{n-2} + \ldots + a + 1)$.

(v) If n is any **odd** integer greater than or equal to 3, and a, b real numbers,

$$a^n + b^n = (a+b)\left(\underbrace{a^{n-1} - a^{n-2} \cdot b^1 + \ldots a^1 \cdot b^{n-2} + b^{n-1}}_{n \text{ terms alternating sign}}\right).$$

Remark 12: Each term in the sum $a^{n-1} + a^{n-2} \cdot b^1 + \ldots + a^1 \cdot b^{n-2} + b^{n-1}$, is of the form $a^i \cdot b^{(n-1)-i}$, where i runs from $n-1$ to 0(start with $i = n-1$, then $i = n-2, \ldots, i = 1$, $i = 0$). For technical purposes we define $a^0 = 1$ (as long as a is not zero).

Example 2: $$a^4 - b^4 = (a-b)(a^3 + a^2 \cdot b^1 + a^1 \cdot b^2 + b^3)$$ $$a^4 - b^4 = (a-b)(a^3 + a^2 \cdot b + a \cdot b^2 + b^3)$$

It is also true that $a^4 - b^4 = a^2 \cdot a^2 - b^2 \cdot b^2 \underset{\text{by (i)}}{=} (a^2)^2 - (b^2)^2 = (a^2 - b^2)(a^2 + b^2) = (a-b)(a+b)(a^2+b^2)$. Note that if you multiply out, you will find that $(a+b)(a^2+b^2) = a^3 + a^2b + ab^2 + b^3$!

Example 3: $$a^5 + b^5 = (a+b)(a^4 - a^3 \cdot b^1 + a^2b^2 - a^1b^3 + b^4)$$ $$a^5 + b^5 = (a+b)(a^4 - a^3b + a^2b^2 - ab^3 + b^4)$$ $$a^7 + b^7 = (a+b)(a^6 - a^5b + a^4b^2 - a^3b^3 + a^2b^4 - ab^5 + b^6)$$

Proof: We leave the proofs of (i), (ii), and (iii) to the students. One can verify these by starting from the right-hand side and using Axioms 8 and 11.

(iv) By Axiom 8, $(a-b) \cdot (a^{n-1} + a^{n-2} \cdot b^1 + \ldots + a^1 \cdot b^{n-2} + b^{n-1}) = (a^{n-1} + a^{n-2} \cdot b^1 + \ldots \frac{1}{a} \cdot b^{n-2} + b^{n-1}) \cdot [a + (-b)] \underset{\text{by Axiom 11}}{=} (a^{n-1} + a^{n-2} \cdot b^1 + \ldots + a^1 \cdot b^{n-2} + b^{n-1}) \cdot a + (a^{n-1} \cdot b^1 + \ldots + a^1 \cdot b^{n-2} + b^{n-1}) \cdot (-b) \underset{\text{by Ax. 8}}{=} a \cdot (a^{n-1} + a^{n-2} \cdot b^1 + \ldots + a^1 \cdot b^{n-2} + b^{n-1}) + (-b) \cdot (a^{n-1} + a^{n-2} \cdot b^1 + \ldots + a^1 \cdot b^{n-2} + b^{n-1}) \underset{\text{by applying Ax. 11 (n-2) times}}{=}$ $a \cdot a^{n-1} + a \cdot a^{n-2} \cdot b^1 + \ldots + a \cdot a^i \cdot b^{(n-1)-i} + \ldots + a \cdot a^1 b^{n-2} + a \cdot b^{n-1} + (-b)a^{n-1} + (-b)a^{n-2}b^1 + \ldots + (-b)a^i \cdot b^{(n-1)-i} + \ldots + (-b)a^1 \cdot b^{n-2} + (-b) \cdot b^{n-1} \underset{\text{by Ax. 8 and Properties 4 and 9}}{=}$ $a^n + (a^{n-1} \cdot b^1 + \ldots + a^{(i+1)} \cdot b^{n-(1+i)} + \ldots + a^2 \cdot b^{n=2} + ab^{n-1}) - (a^{n-1} \cdot b^1 + a^{n-2} \cdot b^2 + \ldots + a^i \cdot b^{n-i} + \ldots + a^1 \cdot b^{n-1}) - b^n$

The two sums in parentheses, each containing $(n-1)$ product terms have identical terms, each product term being of the form $a^j \cdot b^{n-j}$, where j runs from $(n-1)$ to 1. Therefore the two sums in parentheses cancel each other out and we end up with $a^n - b^n$.

(v) This can be easily proven with the aid of part (iv). Since n is odd,$(-1)^n = -1$ and $(-b)^n = -b^n)$ because by definition,

$$(-b)^n = \underbrace{(-b)\ldots(-b)}_{n \text{ times}} \underset{\text{by Property (2)}}{=} \underbrace{[(-1) \cdot b] \cdot [(-1) \cdot b] \ldots [(-1) \cdot b]}_{n \text{ times}}$$

$$\underset{\text{by Ax. 8}}{=} \underbrace{[(-1)\ldots(-1)]}_{n \text{ times}} \cdot \underbrace{[(-1)\ldots(-1)]}_{n \text{ times}} \cdot \underbrace{[b \ldots b]}_{n \text{ times}} \underset{\text{since n is odd}}{=} (-1) \cdot b^n = -(b^n).$$

Consequently, since $(-b)^n = -(b^n)$ for any real b and n odd positive integer, it is also true that $[-(-b)]^n = -(-b)^n$. Now apply part (iv) with b replaced by $(-b)$:

$$a^n + b^n = a^n + [-(-b)]^n = a^n - (-b)^n \underset{\text{by part iv}}{=} [a - (-b)] \cdot [a^{n-1} + a^{n-2} \cdot (-b)^1 + \ldots + a^1 \cdot (-b)^{n-2} + (-b)^{n-1}]$$

Note that since n is odd, the numbers $n-2, n-4, \ldots, \underbrace{n-(n-1)} = 1$, are all odd; so that,

$$(-b)^{n-2} = -(b^{n-2}),\ (-b)^{n-4} = -(b^{n-4}), \ldots, (-b)^3 = -(b^3), (-b)^1 = -(b).$$

This shows that the above factorization takes the form

$$a^n + b^n = (a+b)\left(a^{n-1} - a^{n-2} \cdot b^1 + a^{n-3} \cdot b^2 - a^{n-4} \cdot b^3 + \ldots - a^3 \cdot b^{n-4} + a^2 \cdot b^{n-3} - a^1 \cdot b^{n-2} + b^{n-1}\right),$$

because the integers $n-1, n-3, \ldots, \underbrace{n-(n-2)} = 2$, are all **even** which means that

$$(-b)^{n-1} = b^{n-1}, (-b)^{n-3} = b^{n-3}, \ldots, (-b)^2 = (-b)(-b) = (-1)(-1)(b \cdot b) = b^2$$

Theorem 4 (Language) (i): If a_1, a_2, b_1, b_2 are real numbers then $(a_1^2 + a_2^2)(b_1^2 + b_2^2) - (a_1 b_1 + a_2 b_2)^2 = (a_1 b_2 - a_2 b_1)^2$.

(ii) If n is a positive integer, $n \geq 2$, and $a_1, a_2, \ldots, a_n, b_1, b_2, \ldots, b_n$, are real numbers. Then,

$$(a_1^2 a_2^2 + \ldots a_n^2) \cdot (b_1^2 + b_2^2 + \ldots b_n^2) - (a_1 b_1 + a_2 b_2 + \ldots + a_n b_n)^2 = \sum_{\substack{i \neq j \\ 1 \leq i,j \leq n}} (a_i \cdot b_j - a_j \cdot b_k)^2$$

Note: The sum on the right-hand side of the above equality can be written in expanded form as follows:

$$\sum_{\substack{i \neq j \\ 1 \leq i,j \leq n}} (a_i \cdot b_j - a_j \cdot b_i)^2 = (a_1 b_2 - a_2 b_1)^2 + (a_1 b_3 - a_3 b_1)^2 + \ldots + (a_1 b_n - a_n b_1)^2$$

$$+ (a_2 b_3 - a_3 b_2)^2 + (a_2 b_4 - a_4 b_2)^2 + \ldots + (a_2 b_n - a_n b_2)^2$$

$$+ \ldots + (a_i b_{i+1} - a_{i+1} \cdot b_i)^2 + \ldots + (a_i \cdot b_n - a_n \cdot b_i)^2$$

$$+ \ldots + (a_{n-1} \cdot b_n - a_n \cdot b_{n-1})^2.$$

Example 4: Write out the equality statement in Lagrange's Theorem for $n = 3$ and $n = 4$; write out the statement in expanded form.

For $n = 3$:

$$(a_1^2 + a_2^2 + a_3^2) \cdot (b_1^2 + b_2^2 + b_3^2) - (a_1b_1 + a_2b_2 + a_3b_3)^2 = (a_1b_2 - a_2b_1)^2 + (a_1b_3 - a_3b_1)^2 + (a_2b_3 - a_3b_2)^2$$

For $n = 4$:

$$(a_1^2 + a_2^2 + a_3^2 + a_4^2) \cdot (b_1^2 + b_2^2 + b_3^2 + b_4^2) - (a_1b_1 + a_2b_2 + a_3b_3 + a_4b_4)^2 = (a_1b_2 - a_2b_1)^2 + (a_1b_3 - a_3b_1)^2 + (a_1b_4 - a_4b_1)^2 + (a_2b_3 - a_3b_2)^2 + (a_2b_4 - a_4b_2)^2 + (a_3b_4 - a_4b_3)^2$$

Proof (i): We have

$$(a_1^2 + a_2^2)(b_1^2) + b_2^2) - (a_1b_1 + a_2b_2)^2 =$$

$$= a_1^2 \cdot b_1^2 + a_1^2 \cdot b_2^2 + a_2^2 \cdot b_1^2 + a_2^2 \cdot b_2^2 - (a_1b_1 + a_2b_2)^2 =$$

$$\underset{\text{byTh.2}}{=} a_1^2b_1^2 + a_1^2 \cdot b_2^2 + a_2^2 \cdot b_1^2 + a_2^2 \cdot b_2^2 - [(a_1b)^2 + 2(a_1b_1)(a_2b_2) + (a_2b_2)^2]$$

$$\underset{\text{byAx.8}}{=} a_1^2 \cdot b_1^2 + a_1^2 \cdot b_2^2 + a_2^2 \cdot b_1^2 + a_2^2 \cdot b_2^2 - (a_1^2 \cdot b_1^2 + 2a_1a_2b_1b_2 + a_2^2 \cdot b_2^2]$$

(Note that by Ax. 8, $(a \cdot b)^2 = (a \cdot b)(a \cdot b) = a \cdot a \cdot b \cdot b = a^2 \cdot b^2$)

$$= a_1^2 \cdot b_1^2 + a_1^2 \cdot b_2^2 + a_2^2 \cdot b_1^2 + a_2^2 \cdot b_2^2 - a_1^2 \cdot b_1^2 - 2a_1 \cdot a_2b_1b_2 - a_2^2 \cdot b_2^2$$
$$= a_1^2 \cdot b_2^2 + a_2^2b_1^2 - 2a_1 \cdot a_2 \cdot b_1 \cdot b_2$$

$$\underset{\text{by Ax.8}}{=} (a_1b_2)^2 + (a_2 \cdot b_1)^2 - 2(a_1b_2) \cdot (a_2 \cdot b_1) \underset{\text{by Th.2 (i)}}{=} (a_1b_2 - a_2b_1)^2$$

(ii) We omit the proof here. We do point out thought that the proof is essentially the same. The calculation and grouping of terms we did in part (i) to arrive at $(a_1b_2 - a_2b_1)^2$ must be done over and over again. The typical expression will be

$$a_i^2 \cdot b_j^2 + a_j^2 \cdot b_i^2 - 2a_ia_jb_ib_j = (a_ib_j - a_jb_i)^2 \text{ with } i \neq j.$$

Try it for $n = 3$ and $n = 4$!

Theorem 5: (Schwarz's Inequality) If n is a positive integer, $n \geq 2$and $a_1, \ldots,$ $a_n, b_1, \ldots, b_n$ are real numbers, then

$$(a_1^2 + b_2^2 + \ldots + a_n^2)(b_1^2 + b_2^2 + \ldots b_n^2) \geq (a_1b_1 + a_2b_2 + \ldots + a_nb_n)^2$$

Proof: This is an immediate consequence of Theorem 4(ii), since by that theorem we must have

$$(a_1^2+a_2^2+\ldots+a_n)^2\cdot(b_1^2+b_2^2+\ldots+b_n)^2 = (a_1b_1+a_2b_2+\ldots+a_nb_n)^2 + \sum_{\substack{i\neq j\\ 1\leq i,j\leq n}} (a_i\cdot b_j - a_jb_i)^2$$

but the sum of the squares $(a_ib_j - a_jb_i)^2$ cannot be negative, since each square is nonnegative (by Property 33) and hence the sum of nonnegative reals must be nonnegative (by Axiom 13). The result follows immediately.

Remark 14: It should be pointed out that Theorems 2, 3, and 4 hold true for any commutative ring (refer to Chapter 2, a ring with commutative multiplication operation).

Theorem 6: If n is a positive integer, $n \geq 2$, and $a_1, a_2, \ldots, a_1$ are real numbers, then

$$(n-1)\cdot[a_1^2 + a_2^2 + \ldots + a_n^2] \geq 2\cdot\underbrace{[\pm a_1a_2 \pm \ldots \pm a_ia_j \pm \ldots \pm a_{n-1}\cdot a_n]}_{\text{sum contains } \frac{\text{n(n}-1)}{2} \text{ terms}},$$

where the sum on the right-hand side contains all possible product terms of the form $a_i \cdot a_j$, with $i \neq j$ and $1 \leq i, j \leq n$, and the choices of plus or minus signs are completely arbitrary. In particular, for $n = 2$, $a_1^2 + a_2^2 \geq \pm 2a_1a_2$ and for $n = 3$, $a_1^2 + a_2^2 + a_3^2 \geq a_1a_2 \pm a_1a_33 \pm a_2a_3$.

Proof: The key case is $n = 2$. We are asked to show that $a_1^2 + a_2^2 \geq \pm 2a_1a_2$ which is equivalent to $a_1^2 + a_2^2 \mp 2a_1a_2 \geq 0$ which in turn is equivalent to (by Theorem 2(i), if minus holds, $a_1^2 + (-a_2)^2 + 2a_1 \cdot (-a_2) \geq 0$, then write $(a_1 \mp a_2)^2 \geq 0$, which is always true since the square of any real number must be nonnegative. Next suppose that $n \geq 3$. Of the n real numbers, if we pick any two of them, say a_i and a_j, with $i \neq j$, then according to what we have already shown (case $n = 2$), we must have $a_i^2 + a_j^2 \geq \pm 2a_i \cdot a_j$. Listing **all possible** such inequalities, there $\frac{n(n-1)}{2}$ of them, it would look like:

$$\left.\begin{array}{l} a_1^2+a_2^2 \geq \pm 2a_1a_2 \\ a_1^2+a_3^2 \geq \pm 2a_1a_3 \\ \vdots \\ a_1^2+a_n^2 \geq \pm 2a_1a_n \end{array}\right\} (n-1) \text{ of them}$$

$$\left.\begin{array}{l} a_2^2+a_3^2 \geq +2a_2a_3 \\ \vdots \\ a_2^2+a_n^2 \geq \pm 2a_2a_n \end{array}\right\} (n-2) \text{ of them}$$

$\vdots$

Last inequality : $a_{n-1}^2 + a_n^2 \geq \pm a_{n-1}a_n$

Total number of inequalities: $(n-1)+(n-2)$ $+\ldots+2+1 = \frac{n(n-1)}{2}$

This can be proven by a counting argument or by the method of mathematical induction which we study in Chapter 4.

If add up memberwise all $\frac{n(n-1)}{2}$ inequalities we obtain $(n-1)\cdot[a_1^2+a_2^2+\ldots+a_n^2] \geq 2\cdot[\pm a_1a_2+\ldots\pm a_1a_n\pm\ldots\pm a_2a_1\pm\ldots\pm a_{n-1}\cdot a_n]$ because each square term on the left-hand side of the entire list of the $\frac{n(n-1)}{2}$ inequalities occurs exactly $(n-1)$ times. Clearly for $n=3$, we obtain

$$2\cdot(a_1^2+a_2^2-a_3^2) \geq 2\cdot(\pm a_1a_2 \pm a_1a_3 \pm a_2a_3);\ \text{ or } a_1^2+a_2^2+a_3^2 \geq \pm a_1a_2 \pm a_1a_3 \pm a_2a_3.$$

The proof is complete.

Remark 15: Note for $n=2$, the equal sign in $a_1^2+a_2^2 \geq 2a_1a_2$ will hold exactly when $a_1=a_2$ (just look at the proof, $(a_1-a_2)^2 \geq 0$). And in $a_1^2+a_2^2 \geq -2a_1a_2$, the equal sign will hold exactly when$a_1=-a_2$. For $n \geq 3$, the equal sign (in Theorem 6)will hold exactly when $a_1^2=a_2^2=a_3^2=\ldots=a_n^2=a^2$ some real number and the choice of sign in each term $\pm a_i\cdot a_j$ is such that $\pm a_i\cdot a_j = a^2$.

Theorem 7: Let n, a positive integer, be greater than or equal to 2. If $a_1, a_2, \ldots, a_n$ are all nonzero real numbers, then

$$(a_1^2+a_2^2+\ldots+a_n^2)\cdot\left(\frac{1}{a_1^2}+\frac{1}{a_2^2}+\ldots+\frac{1}{a_n^2}\right) \geq n^2$$

Proof: This theorem is an immediate application of Schwarz's Inequality, Theorem 5. We can apply Theorem 5, with $b_1=\frac{1}{a_1}, b_2=\frac{1}{a_2}, \ldots, b_n=\frac{1}{a_n}$ so that $b_1^2=\frac{1}{a_1^2}$, $b_2^2=\frac{1}{a_2^2}, \ldots, b_n^2=\frac{1}{a_n^2}$.. We have

$$(a_a^2 + a_2^2 + \ldots + a_n^2) \cdot \left(\frac{1}{a_1^2} + \frac{1}{a_2^2} + \ldots + \frac{1}{a_n^2} \right) \geq \left(a_1 \cdot \frac{1}{a_1} + a_2 \cdot \frac{1}{a_2} + \ldots + a_n \cdot \frac{1}{a_n} \right)$$

$$(a_1^2 + a_2^2 + \ldots + a_n^2) \cdot \left(\frac{1}{a_1^2} + \frac{1}{a_2^2} + \ldots + \frac{1}{a_n^2} \right) \geq \underbrace{(1 + 1 + \ldots + 1)^2}_{n \text{ terms}};$$

$$(a_1^2 + a_2^2 + \ldots + a_n^2) \cdot \left(\frac{1}{a_1^2} + \frac{1}{a_2^2} + \ldots + \frac{1}{a_n^2} \right) \geq n^2$$

Theorem 8 (Cauchy) If a, b, c are real numbers then

$$a^3 + b^3 + c^3 - 3abc = \frac{1}{2} \cdot (a + b + c) \cdot \left[(a - b)^2 + (b - c)^2 + (c - a)^2\right]$$

Remark 16: (i) An interesting Collorary of this theorem is that if the sum of three real numbers is equal to zero, then the sum of their cubes is equal to three times their product.

(ii) If a, b, c are integers and $a + b + c$ is even, then Cauchy Theorem provides a factorization formula. For example, say $a = 14, b = 11, c = 7$, then, we have,

$$(14)^3 + (11)^3 + (7)^3 - 3(14)(11)(7) = \frac{1}{2} \cdot (14 + 11 + 7) \cdot [(14 - 11)^2 + (11 - 7)^2 +$$

$$(7 - 14)^2 = \frac{1}{2} \cdot (32) \cdot [3^2 + 4^2 + 7^2] = 16 \cdot [9 + 16 + 49] = (16) \cdot (74) =$$
$$2^4 \cdot 2 \cdot 37 = 2^5 \cdot (37)$$

Proof: Even though the expansion of $(a + b)^3$ is a special case of the binomial theorem which will be discussed in Chapter 5, we need this expansion in our proof. We have $(a + b)^3 = (a + b)(a + b)^2 = (a + b)(a^2 + 2ab + b^2)$, by Theorem 2. Applying Axioms 8 and 11 we can multiply out, collect alike terms and obtain $(a + b)^3 = a^3 + 3a^2b + 3ab^2 + b^3$; $(a + b)^3 = a^3 + b^3 + 3ab(a + b) \Rightarrow a^3 + b^3 = (a + b)^3 - 3ab(a + b)$. We now have

$$\begin{aligned} a^3 + b^3 + c^3 - 3abc &= (a + b)^3 - 3ab(a + b) + c^3 - 3abc \\ a^3 + b^3 + c^3 - 3abc &= (a + b)^3 + c^3 - 3ab \cdot (a + b + c) \end{aligned} \qquad (1)$$

But$(a + b + c)^3 = [(a + b) + c]^3 = (a + b)^3 + 3(a + b)^2 \cdot c + 3(a + b)c^2 + c^3$

$$\begin{aligned} &(a + b + c)^3 = (a + b)^3 + c^3 + 3(a + b) \cdot c \cdot [(a + b) + c] \Rightarrow (a + b)^3 + c^3 = \\ &(a + b + c)^3 - 3(a + b) \cdot c \cdot [(a + b) + c] \end{aligned} \qquad (2)$$

Combining (2) and (1) (substituting for $(a + b)^3 + c^3$ in (1)), we obtain

$$a^2 + b^3 + c^3 - 3abc = (a + b + c)^3 - 3(a + b) \cdot c \cdot (a + b + c) - 3ab(a + b + c) \qquad (3)$$

Factoring out the common factor $(a+b+c)$ from each product term on the right-hand side of (3) gives

$$a^3 + b^3 + c^3 - 3abc = (a + b + c) \cdot [(a + b + c)^2 - 3(a + b)c - 3ab] \qquad (4)$$

But (by Theorem 2, for $n = 3$) $(a+b+c)^2 = a^2+b^2+c^2+2(ab+ac+bc)$. Substituting for $(a + b + c)^2$ in (4) we now have

$$\begin{aligned}
a^3 + b^3 + c^3 - 3abc &= (a + b + c) \cdot [a^2 + b^2 + c^2 + 2(ab + ac + bc) \\
&\quad -3ac - 3bc - 3ab]. \\
a^3 + b^3 + c^3 - 3abc &= (a + b + c) \cdot [a^2 + b^2 + c^2 - ab - ac - bc]; \\
a^3 + b^3 + c^3 - 3abc &= \tfrac{1}{2} \cdot (a + b + c) \cdot [2a^2 + 2b^2 + 2c^2 - 2ab - 2ac - 2bc]; \\
a^3 + b^3 + c^3 - 3abc &= \tfrac{1}{2} \cdot (a + b + c) \cdot [(a^2 - 2ab + b^2) + (b^2 - 2bc + c^2) \\
&\quad +(c^2 - 2ac + a^2)] \\
a^3 + b^3 + c^3 - 3abc &= \tfrac{1}{2} \cdot (a + b + c) \cdot [(a - b)^2 + (b - c)^2 + (c - a)^2],
\end{aligned}$$

and the proof is complete.

3.6 Absolute values

Definition 8: Let a be a real number. The absolute value of a, denoted by $|a|$ is the nonnegative real number given by

$$|a| = \begin{cases} -a, \text{ if } a < 0 \\ 0, \text{ if } a = 0 \\ a, \text{ if } a > 0 \end{cases}$$

We see that if a is nonnegative, $a \geq 0$, then $|a|$ is simply a itself, $|a| = a$. If a is a negative real number, then $|a| = -a$, the opposite of a, which is positive; e.g. $|-3| = -(-3) = 3$; and $|3| = 3$.

Theorem 9: Let $a, b, a_1, \dots, a_n, b_1, \dots, b_n$ be real numbers, n a positive integer greater than or equal to 2.

(i) $|a| = |-a|$

(ii) $-|a| \leq a \leq |a|$. Also, $|x| \leq |a|$ if and only if $-|a| \leq x \leq |a|$

(iii) $|a|^2 = a^2$

(iv) Suppose $a \cdot b \geq 0$. Then $|a| = |b|$ if and only if $a = b$

(v) Suppose $a \cdot b < 0$. Then $|a| = |b|$if and only if $a = -b$

(vi) (Triangle Inequalities) $||a| - |b|| \leq |a \pm b| \leq |a| + |b|$

(vii) (Triangle Inequality) $|a_1 + a_2 + \dots + a_n| \leq |a_1| + |a_2| + \dots + |a_n|$

(viii) $|a \cdot b| = |a| \cdot |b|$

(ix) $|a_1 \cdot a_2 \dots a_n| = |a_1| \cdot |a_2| \dots |a_n|$

(x) $|\frac{a}{b}| = \frac{|a|}{|b|}$, provided $b \neq 0$.

Proof: (i) This is an immediate consequence of Definition 8. For if $a \geq 0$ then $|a| = a$ but $-a \leq 0$ so $|-a| = -(-a) = a$. Thus $|a| = |-a|$ and if $a < 0$ then $|a| = -a$. But $-a > 0$ so that $|-a| = -a$, thus again $|a| = |-a|$.

(ii) If $a \geq 0$, then $|a| = a$. But always $-|a| \leq 0$, so $-|a| \leq 0 \leq a = |a|$ which implies $-|a| \leq a \leq |a|$.

If $a < 0$, then $|a| = -a$; $-|a| = a$ which implies $-|a| = a < 0 < |a|$ and therefore the statement $-|a| \leq a \leq |a|$ is once again true.

Next suppose that $|x| \leq |a|$. If $x \geq 0$, then $|x| = x$ and therefore we obtain $0 \leq x \leq |a|$; but $-|a| \leq 0 \Rightarrow -|a| \leq 0 \leq x \leq |a| \Rightarrow -|a| \leq x \leq |a|$. If $x < 0$, then $|x| = -x$ and so $|x| \leq |a|$ implies $-x \leq |a| \Rightarrow$ (multiply both sides by (-1), apply Properties 35 and 2 to obtain $-(-x) \geq -|a|$; $x \geq -|a|$. Thus, $-|a| \leq x < 0 \leq |a|$ and so again we have arrived at $-|a| \leq x \leq |a|$.

(iii) If $a \geq 0$, then $|a| = a \Rightarrow |a|^2 = a^2$. If $a < 0$, we have $|a| = -a \Rightarrow |a|^2 = (-a)^2 = a^2$, by Property 10.

(iv) and (v) If $|a| = |b|$, then $|a|^2 = |b|^2$ and by (iii) we have $a^2 = b^2 \Leftrightarrow a^2 - b^2 \Leftrightarrow (a - b) \cdot (a + b) = 0 \Leftrightarrow (a - b = 0 \text{ or } a + b = 0) \Leftrightarrow (a = b \text{ or } a = -b)$. If $a = b$ and $a \neq 0$, then b is also nonzero and we have $a \cdot b = a \cdot a = a^2 > 0$ since $a \neq 0$. If $a = 0$, then $a = b$ gives $b = 0$ as well and of course $a \cdot b = 0$. The converse is obviously true: namely, if $a \cdot b \geq 0$ and $a = b$, then $|a| = |b|$. If the second possibility in ($a = b$ or $a = -b$) holds, namely $a = -b$ with $b \neq 0$ (because for $b = 0$ we are back to the case $a = b = 0$), then $a \cdot b = (-b) \cdot (b) = -(b \cdot b) = -(b^2) < 0$, since $b \neq 0$. The converse is obviously true. If $ab < 0$ and $a = -b$, then $|a| = |-b| = |b|$, by (i).

(vi) First we prove that $|a+b| \leq |a|+|b|$. Indeed, from part (ii) we have $-|a| \leq a \leq |a|$ and $-|b| \leq b \leq |b|$. Adding these two inequalities memberwise yields $-(|a|+|b|) \leq a+b \leq |a|+|b|$ which is equivalent, according to part (ii), with $|a+b| \leq ||a|+|b||$. But $|a|+|b| \geq 0$ since obviously $|a| \geq 0$ and $|b| \geq 0$ which means that $||a|+|b|| = |a|+|b|$ so that the latter inequality becomes $|a+b| \leq |a|+|b|$ which is what we wanted to prove. To show that $|a-b| \leq |a|+|b|$, we can simply replace b by $-b$: $|a-b| = |a+(-b)| \leq |a|+|-b| = |a|+|b|$ since $|b| = |-b|$ by part (i). We have so far proven $|a \pm b| \leq |a|+|b|$.

Next we want to show that $||a|-|b|| \leq |a \pm b|$. To do this, let us prove first $||a|-|b|| \leq |a-b|$, by writing $a = (a-b)+b$. What this accomplishes is $|a| = |(a-b)+b| \leq |a-b|+|b|$ by what we have already proved (namely that $|A+B| \leq |A|+|B|$. Apply it with $A = a-b$ and $B = b$). Clearly

$$|a| \leq |a-b|+|b| \text{ implies } |a|-|b| \leq |a-b|. \tag{5}$$

By using the same trick as before, we can write $b = (b-a)+a \Rightarrow |b| = |(b-a)+a| \leq |b-a|+|a|$. Again $|b| \leq |b-a|+|a|$ implies $|b|-|a| \leq |b-a|$ but $|b-a| = |-(a-b)| = |a-b|$ by part (i). Hence $|b|-|a| \leq |a-b|$. By multiplying both sides of the last inequality with -1 we obtain

$$|a|-|b| \geq -|a-b| \text{ or simply } -|a-b| \leq |a|-|b| \tag{6}$$

Combining inequalities (5) and (6)

$$-|a-b| \leq |a|-|b| \leq |a-b| \tag{7}$$

Part (ii) of the theorem and (7) now give

$$||a|-|b|| \leq |a-b|. \tag{8}$$

which is part of what we wanted to show. To also show $||a|-|b|| \leq |a+b|$ simply apply (8): $||a|-|b|| = ||a|-|-b|| \leq |a-(-b)||$. Thus

$$||a|-|b|| \leq |a+b|. \tag{9}$$

Inequalities (8) and (9) simply say that $||a|-|b|| \leq |a \pm b|$, and the proof is complete.

(vii) This is a generalization of the proof given for $|a+b| \leq |a|+|b|$. For $n = 2$, we already know from part (vi) that

$$|a_1+a_2| \leq |a_1|+|a_2|$$

For $n = 3$, we can apply part (vi) twice as follows:

$$|a_1 + a_2 + a_3| = |(a_1 + a_2) + a_3| \leq |a_1 + a_2| + |a_3|.$$

but $|a_1 + a_2| \leq |a_1| + |a_2|$, hence $|a_1 + a_2| + |a_3| \leq (|a_1| + |a_2|) + |a_3|$, by Axiom 13. We now have

$$\left.\begin{array}{rl} & |a_1 + a_2 + a_3| \leq |a_1 + a_2| + |a_3| \\ \textit{and} & |a_1 + a_2| + |a_3| \leq |a_1| + |a_2| + |a_3| \end{array}\right\} \Rightarrow \text{(by Property 31)}$$

$$\Rightarrow \qquad |a_1 + a_2 + a_3| \leq |a_1| + |a_2| + |a_3|.$$

Given any positive integer $n \geq 3$, we can establish the inequality

$$|a_1 + a_2 + a_3 + \ldots + a_n| \leq |a_1| + |a_2| + |a_3| + \ldots + |a_n|,$$

by simply applying part (vi) (the case $n = 2$), $(n - 1)$ times - for this we would need to list $(n - 1)$ inequalities (as we did for $n = 3$) and then apply the transitive property (Property 31) to obtain the result. It is easier though to use the method of mathematical induction which will be studied in Chapter 4.

(viii) We have four cases.

Case 1: $a \geq 0$, $b \geq 0$, then $ab \geq 0$ as well. $|a| = a, |b| = b, |ab| = ab$. We have $|ab| = ab = |a||b|$.

Case 2: $a < 0, b < 0$, then $ab > 0$. $|a| = -a, |b = -b, |ab| = ab$. We have $|ab| = ab = (-1)^2 \cdot ab = (-1)a \cdot (-1)b = (-a) \cdot (-b) = |a||b|$.

Case 3: $a < 0, b \geq 0$ then $ab \leq 0$. $|a| = -a, |b| = b, |ab| = -(ab)|ab| = -(ab) = (-a)(b) = |a| \cdot |b|$.

Case 4: $a \geq 0, b < 0$ then $ab \leq 0$. Proof similar with the one offered in Case 3.

(ix) this is an easy generalization of part (viii). Just apply part (viii), $(n - 1)$ times.

(x) Since $b \neq 0$ we have $\frac{a}{b} \cdot b = a \Rightarrow \left|\frac{a}{b} \cdot b\right| = |a| \Rightarrow$ (by Part (viii)) $\left|\frac{a}{b}\right| \cdot |b| = |a| \Rightarrow \left(\left|\frac{a}{b}\right| \cdot |b|\right) \cdot \frac{1}{|b|} = |a| \cdot \frac{a}{|b|} \Rightarrow \left|\frac{a}{b}\right| \cdot \left(|b| \cdot \frac{1}{|b|}\right) = \frac{|a|}{|b|} \Rightarrow \left|\frac{a}{b}\right| \cdot 1 = \frac{|a|}{|b|} \Rightarrow \left|\frac{a}{b}\right| = \frac{|a|}{|b|}$. Note that since $b \neq 0$ by the definition of the absolute value, we must have $|b| \neq 0$ and thus $|b|$ possesses a unique inverse, namely $\frac{1}{|b|}$ (because clearly $|b| \cdot \frac{1}{|b|} = \frac{|b|}{|b|} = 1$).

3.7 Practice Problems

P1. Show that if the real numbers a, b, c, satisfy the conditions $a^2+b^2+c^2-2(ab+ac+bc)=0$ and $a+b+c=0$, then $a=b=c=0$.

Solution: The first condition can be written $(a-b)^2+c^2-2c\cdot(a+b)=0$. Adding $(a+b)^2$ to both sides,

$$(a-b)^2+\underbrace{c^2-2c\cdot(a+b)+(a+b)^2}_{\text{a perfect square}}=(a+b)^2 \Rightarrow (a-b)^2+[c-(a+b)]^2=(a+b)^2 \quad (1)$$

Now apply the second condition $a+b+c=0$; $c=-(a+b)$. Substituting c in equation (1) we have

$$(a-b)^2+3\cdot(a+b)^2=0 \quad (2)$$

But $(a-b)^2 \geq 0$ and $3(a+b)^2 \geq 0$. Therefore (2) implies $a-b=0$ and $a+b=0$ since if either of $a-b$ or $a+b$ were nonzero then at least one of the two terms in the left-hand side of (2) would be positive and hence the sum would be positive. Clearly, $a-b=0=a+b$ imply $2a=0$ and $2b=0$ and from $a+b+c=0 \Rightarrow c=0$; $a=b=c=0$.

P2. Use practice problem P1 to show that if the real numbers x, y and z satisfy $(y-z)^2+(z-x)^2+(x-y)^2=(y+z-2x)^2+(z+x-2y)^2+(x+y-2z)$ then $x=y=z$.

Solution If we set $a=y-z$, $b=z-x$, and $c=x-y$, then the sum $a+b+c$ is zero; $a+b+c=(y-z)+(z-x)+(x-y)=0$ and the exercise's condition becomes

$$a^2+b^2+c^2=[(y-x)+(z-x)]^2+[(z-y)+(x-y)]^2+[(x-z)+(y-z)]^2 \Rightarrow$$

$$a^2+b^2+c^2=(-c+b)^2+(-a+c)^2+(-b+a)^2 \Rightarrow$$

$$a^2+b^2+c^2=c^2-2cb+b^2+a^2-2ac+c^2+b^2-2ba+a^2$$

$$\Rightarrow a^2+b^2+c^2-2(ab+bc+ac)=0.$$

Thus, since the hypothesis of practice problem P1 is satisfied, we must have $a=b=c=0$ which means $y-z=z-x=x-y=0 \Rightarrow x=y=z$.

P3. Prove that if $abcd \neq 0$ and $(a^2+b^2)\cdot(c^2+d^2)=4abcd$, then $|a|=|b|$ and $|c|=|d|$ (which means $a=\pm b$ and $c=\pm d$).

Solution:

$$\begin{aligned}&(a^2+b^2)(c^2+d^2)=4abcd \Rightarrow a^2c^2+a^2d^2+b^2c^2+b^2d^2=4abcd \Rightarrow\\ &\Rightarrow ((ac)^2+(bd)^2-2(ac)(d))+((ad)^2+(bc)^2-2(ad)(bc))=0 \Rightarrow\\ &\Rightarrow (ac-bd)^2+(ad-bc)^2=0.\end{aligned} \quad (1)$$

Since the left-hand side of (1) is the sum of two squares and the right-hand side is zero, each of the squares must be zero which means $ac - bd = 0$ and $ad - bc = 0$. Thus

$$\left.\begin{array}{l} ac = bd \text{ and} \\ ad = bc \end{array}\right\} \Rightarrow (ac)(ad) = (bd)(bc) \Rightarrow a^2 \cdot cd = b^2 \cdot cd \Rightarrow (a^2 - b^2) \cdot c \cdot d = 0.$$

Since $abcd \neq 0$ we have, $cd \neq 0$, then $a^2 - b^2 = 0 \Rightarrow (a-b)(a+b) = 0$, which implies $a = b$ or $a = -b$; $a = \pm b$ which is equivalent to $|a| = |b|$. Going back to $ac = bd$ (or to $ad = bc$) we have $|ac| = |bd| \Rightarrow |a||c| = |b||d|$; $|a||c| = |a||d|$ we can cancel $|a|$ from both sides, since $|a| \neq 0$. Thus $|c| = |d| \Leftrightarrow c = \pm d$. The proof is complete.

P4. Prove that if $2(a^2 + b^2 - ab) \cdot (c^2 + d^2 - cd) = a^2c^2 + b^2d^2$, then ($a = b$ and $c = d$) or ($a = b = 0, c, d$ arbitrary) or (a, b arbitrary $c = d = 0$).

Solution Multiplying out the left-hand side of the given condition

$$2(a^2 + b^2)(c^2 + d^2) - 2cd(a^2 + b^2) - 2ab(c^2 + d^2) + 2abcd = a^2c^2 + b^2d^2$$
$$\Rightarrow 2a^2c^2 + 2a^2d^2 + 2b^2c^2 + 2b^2d^2 - 2cda^2 - 2cdb^2 - 2abc^2 - 2abd^2 + 2abcd = a^2c^2 + b^2d^2$$

$$\Rightarrow (2a^2c^2 - a^2c^2) + (2b^2d^2 - b^2d^2) + 2a^2d^2 + 2b^2c^2 - 2(ac)(ad) - 2(cb)(db) - 2(ac)(bc) - 2(ad)(bd) + 2abcd = 0$$

$$[(ac)^2 - 2(ac)(ad) + (ad)^2] + [(bd)^2 - 2(bd)(cb) + (cb)^2] +$$

$$+[(ac)^2 - 2(ac)(bc) + (bc)^2] + [(ad)^2 - 2(ad)(bd) + (bd)^2] - [(ac)^2 + (bd)^2 - 2(ac)(bd)] = 0 \quad (1)$$

$$[ac - ad]^2 + [bd - cb]^2 + [ac - bc]^2 + [ad - bd]^2 - [ac - bd]^2 = 0$$

$$[a(c-d)]^2 + [b(d-c)]^2 + [c(a-b)]^2 + [d(a-b)]^2 - [ac - bd]^2 = 0$$

$$a^2 \cdot (c-d)^2 + b^2 \cdot (c-d)^2 + c^2 \cdot (a-b)^2 + d^2 \cdot (a-b)^2 - (ac - bd)^2 = 0.$$

We have come to a point in the proof where some trick or artifice maybe needed to avoid going around in circles. If we look at (1), we see that both differences $c - d$ and $a - b$ appear prominently in that equation. Let us put

$$\left.\begin{array}{ll} a - b = k \ ; & a = b + k \\ \text{and } c - d = \lambda & c = d + \lambda \ ; \end{array}\right\} \quad (2)$$

Substituting for a and c in (1) we obtain

$$(b+k)^2\cdot\lambda^2+b^2\cdot\lambda^2+(d+\lambda)^2\cdot k^2+d^2\cdot k^2-[bd+b\lambda+k\cdot d+k\cdot\lambda-bd]^2=0$$

Expanding the squares in the last equation

$$(b^2+2bk+k^2)\cdot\lambda^2+b^2\lambda^2+(d^2+2d\lambda+\lambda^2)\cdot k^2+d^2k^2-(b\lambda+kd+k\lambda)^2=0$$

$$b^2\lambda^2+2bk\lambda^2+k^2\lambda^2+b^2\lambda^2+d^2k^2+2d\lambda k^2+\lambda^2k^2+d^2k^2$$

$$-b^2\lambda^2-k^2d^2-k^2\lambda^2-2bdk\lambda-2bk\cdot\lambda^2-2d\lambda\cdot k^2=0 \tag{3}$$

Looking at equation (3) we see that only four terms do not cancel out:

$$d^2k^2+b^2\cdot\lambda^2+k^2\lambda^2-2bk\lambda^2=0$$

$$(dk)^2+\underbrace{(b\lambda)^2+(k\lambda)^2-2(b\lambda)(k\lambda)}_{\text{a perfect square}}=0 \tag{4}$$

$$(dk)^2+(b\lambda-k\lambda)^2=0$$

$$(dk)^2+[(b-k)\lambda]^2=0$$

Clearly (4) implies $dk=0=(b-k)\lambda$. There are four possibilities:

Possibility 1: $d=0$ and $b-k=0\Rightarrow d=0$ and $b=k=a-b$

Possibility 2: $d=0$ and $\lambda=0\Rightarrow d=0$and $c-d=0$; $d=0=c$

Possibility 3: $k=0$ and $b-k=0\Rightarrow a-b=0$ and $b-(a-b)=0$

Possibility 4: $k=0$and $\lambda=0\Rightarrow a-b=0=c-d$; $a=b$ and $c=d$.

Possibility 1: implies $d=0$ and $b=a-b\Rightarrow a=2b$. If we go back to the original condition of the problem we see that

$2[(2b)^2+b^2-(2b)(b)]\cdot c^2=(2b)^2\cdot c^2\Rightarrow 2\cdot[4b^2+b^2-2b^2]\cdot c^2=4b^2c^2\Rightarrow 6b^2c^2=4b^2c^2\Rightarrow b^2c^2=0\Rightarrow(b=0\text{ or }c=0)$.

If $c=0$, then $d=c=0$, and a and b are arbitrary. If $b=0$ then $a=2b=0$ so that in this case we have $a=b=d=0$ and c arbitrary.

Possibility 2: implies $d=0$ and $c-d=0$; $0=d=c$. So the original condition is satisfied, a and b are arbitrary.

Possibility 3: implies $a-b=0$ and $b-(a-b)=0$ so that $a-b=0$ and $b-0=0$; $a-b=0$ and $b=0$; $a=b=0$ and c,d are arbitrary.

Possibility 4: implies $a-b=0=c-d\Rightarrow a=b$ and $c=d$.

The proof is completed.

This problem is an example of an algebraically long-winded and involved exercise, which sheds some light on the algebraic tricks sometimes used to tackle such problems.

P5. If the real numbers x, y, z satisfy the conditions $xyz = x + y + z = a$ and $xy + yz + zx = 2a$, then they must also satisfy $x^2y^2 + y^2z^2 + z^2x^2 = 2a^2$.

Solution We can employ the identify of Theorem 2(ii):

$$(xy + yz + zx)^2 = (xy)^2 + (yz)^2 + (zx)^2 + 2\cdot$$

$$[(xy)(yz) + (yz)(zx)+)xy)(zx)] \tag{1}$$

$$(xy) + yz + zx)^2 = x^2y^2 + y^2z^2 + z^2x^2 + 2xyz(x + y + z).$$

Now apply the given conditions to (1):

$$(2a)^2 = x^2y^2 + y^2z^2 + z^2x^2 + 2 \cdot a \cdot a$$

$$\Rightarrow 4a^2 - 2a^2 = x^2y^2 + y^2x^2 \qquad x^2y^2 + y^2z^2 + z^2 + z^2x^2 = 2a^2.$$

P6. Prove that if $\frac{x}{\alpha} + \frac{y}{\beta} + \frac{z}{\gamma} = 1$ and $\frac{\alpha}{x} + \frac{\beta}{y} + \frac{\gamma}{z} = 0$, then $\frac{x^2}{\alpha^2} + \frac{y^2}{\beta^2} + \frac{z^2}{\gamma^2} = 1$.

Solution Square both sides of the first condition to obtain

$$\frac{x^2}{\alpha^2} + \frac{y^2}{\beta^2} + \frac{z^2}{\gamma^2} + 2\left(\frac{xy}{\alpha\beta} + \frac{xz}{\alpha\gamma} + \frac{yz}{\beta\gamma}\right) =$$

$$\frac{x^2}{\alpha^2} + \frac{y^2}{\beta^2} + \frac{z^2}{\gamma^2} + \frac{2\cdot(xy\gamma + xz\beta + yz\alpha)}{\alpha\beta\gamma} = 1 \tag{1}$$

Now apply the second condition of the problem

$$\frac{\alpha}{x} + \frac{\beta}{y} + \frac{\gamma}{z} = 0 \Rightarrow \frac{\alpha yz + \beta xz + \gamma xy}{xyz} = 0 \Rightarrow \alpha yz + \beta xz + \gamma xy = 0 \tag{2}$$

Combining (1) and (2) we immediately arrive at

$$\frac{x^2}{\alpha^2} + \frac{y^2}{\beta^2} + \frac{z^2}{\gamma^2} = 1.$$

P7. If x, y, z are all distinct and they satisfy

$y^2 + z^2 + kyz = z^2 + x^2 + kzx = x^2 + y^2 + kxy = \lambda$, for some k and λ, prove that $\lambda = \frac{1}{2}(x^2 + y^2 + z^2)$.

Solution. The first condition $y^2 + 2^2 + kyz = z^2 + x^2 + kzx$ implies

$$y^2 - x^2 + k \cdot z(y - x) = 0; \quad (y - x)(y + x) + kz(y - x) = 0 \Rightarrow (y - x) \cdot [y + x + kz] = 0.$$

Since y and x are distinct, $y - x \neq 0$ and therefore the last equation implies that

$$y + x + kz = 0 \tag{1}$$

Working similarly, we see that $y^2 + z^2 + kyz = x^2 + y^2 + kxy$ implies

$$z + x + ky = 0 \tag{2}$$

Likewise $z^2 + x^2 + kzx = x^2 + y^2 + kxy$ implies

$$z + y + kx = 0 \tag{3}$$

By subtracting (2) from (1) we obtain

$$y - z + kz - ky = 0 \Rightarrow y - z - k(y - z) = 0 \Rightarrow (y - z) \cdot (1 - k) = 0$$

and since $y \neq z$, it follows $1 - k = 0$; $k = 1$. Consequently any of (1), (2) or (3) yields

$$x + y + z = 0 \tag{4}$$

Squaring both sides of (4) gives

$$x^2 + y^2 + z^2 + 2(xy + xz + yz) = 0 \Rightarrow xy + xz + yz = -\frac{1}{2}(x^2 + y^2 + z^2) \tag{5}$$

Going back to the original conditions of the exercise and substituting for $k = 1$,

$$\left.\begin{aligned} y^2 + z^2 + yz = \lambda \\ z^2 + x^2 + zx = \lambda \\ x^2 + y^2 + xy = \lambda \end{aligned}\right\} \tag{6}$$

Adding (memberwise) the three equations in (6) we arrive at

$$2(x^2 + y^2 + z^2) + (yz + zx + xy) = 3\lambda \tag{7}$$

Applying (5) to (7) produces

$$2(x^2+y^2+z^2) - \frac{1}{2}(x^2+y^2+z^2) = 3\lambda \Rightarrow \frac{3}{2}(x^2+y^2+z^2) = 3\lambda \Rightarrow \lambda = \frac{1}{2}(x^2+y^2+z^2),$$

the proof is complete.

P8. Show that if three real numbers α, β, γ satisfy the conditions $\alpha + \beta + \gamma = 1$, $\alpha^2 + \beta^2 + \gamma^2 = 1$, and $\alpha^3 + \beta^3 + \gamma^3 = 1$, then two of these numbers must be zero and the third equal to one.

Solution We apply Cauchy's Theorem, Theorem 8.

$$\alpha^3+\beta^3+\gamma^3-3\alpha\beta = \frac{(\alpha+\beta+\gamma)}{2}[(\alpha-\beta)^2+(\beta-\gamma)^2+(\gamma-\alpha^2];$$

$$\alpha^3+\beta^3+\gamma^3-3\alpha\beta\gamma = \frac{(\alpha+\beta+\gamma}{2}[2a^2+2\beta^2+2\gamma^2-2(\alpha\beta+\beta\gamma+\gamma\alpha)]; \quad (1)$$

$$\alpha^3+\beta^3+\gamma 3-3\alpha\beta\gamma = (\alpha+\beta+\gamma)=[\alpha^2+\beta^2+\gamma^2-\alpha\beta-\beta\gamma-\gamma\gamma\alpha].$$

According to hypothesis, $\alpha+\beta+\gamma=1=\alpha^2+\beta^2+\gamma^2=\alpha^3+\beta^3+\gamma^3$. Hence (1) implies

$$1-3\alpha\beta\gamma = 1\cdot[1-(\alpha\beta+\beta\gamma+\gamma\alpha)];$$

$$1-3\alpha\beta\gamma = 1-(\alpha\beta+\beta\gamma+\gamma\alpha); \quad (2)$$

$$3\alpha\beta\gamma = \alpha\beta+\beta\gamma+\gamma\alpha$$

But $(\alpha+\beta+\gamma)^2=\alpha^2+\beta^2+\gamma^2+2(\alpha\beta+\beta\gamma+\gamma\alpha) \Rightarrow 1^2=1+2(\alpha\beta+\beta\gamma+\gamma+\alpha)) \Rightarrow 0=\alpha\beta+\beta\gamma+\gamma\alpha$; and so (2) implies $3\alpha\beta\gamma=0$; $\alpha=0$ or $\beta=0$ or $\gamma=0$. If $\alpha=0$, the problem's original conditions imply $\beta+\gamma=1$, $\beta^2+\gamma^2=1$, $\beta^3+\gamma^3=1$. But $\beta^2+\gamma^2=(\beta+\gamma)^2-2\beta\gamma \Rightarrow 1=1^2-2\beta\gamma \Rightarrow \beta\gamma=0$; $\beta=0$ or $\gamma=0$. If $\beta=0$, then $\gamma=1$. So we see that $3\alpha\beta\gamma=0$ implies one of three possibilities $(\alpha=\beta=0,\ \gamma=1)$ or $(\alpha=\gamma=0,\ \beta=1)$ or $(\beta=\gamma=0,\ \alpha=1)$. We are done.

P9. Let a, b, c be real numbers. Prove the inequalities $a^2+b^2+c^2 \geq \pm(ab+ac+bc)$. When do the plus sign and equal sign hold simultaneously? When does the minus sign and equal sign hold simultaneously?

Solution According to Theorem 6

$$\begin{aligned} a^2+b^2 &\geq 2ab \\ a^2+c^2 &\geq 2ac \\ b^2+c^2 &\geq 2bc \end{aligned} \quad (1)$$

Adding the three inequalities in (1) we obtain

$$\begin{aligned} 2(a^2+b^2+c^2) &\geq 2(ab+ac+bc) \\ a^2+b^2+c^2 &\geq ab+ac+bc \end{aligned} \quad (2)$$

Likewise by Theorem 6 we must also have

$$\left.\begin{aligned} a^2+b^2 &\geq -2ab \\ a^2+c^2 &\geq -2ac \\ b^2+c^2 &\geq -2bc \end{aligned}\right\} \quad (3)$$

Adding the three inequalities in (3),

$$\begin{aligned} 2(a^2+b^2+c^2) &\geq -2(ab+ac+bc); \\ a^2+b^2+c^2 &\geq -(ab+ac+bc) \end{aligned} \quad (4)$$

combining (2) and (4) we get

$$a^2 + b^2 + c^2 \geq \pm(ab + ac + bc) \tag{5}$$

Note that (5) is equivalent to $a^2 + b^2 + c^2 \geq |ab + ac + bc|$.

If the plus sign holds in (5), then (5) simply becomes (2). In order that the equal sign hold in (2), **all three** equal signs must hold in (1), refer to **Remark 15**. Which means that the equal sign will hold in (2) if and only if $\boldsymbol{a = b = c}$ (because in such a case we must have $(a-b)^2 = (b-c)^2 = (c-a)^2 = 0$).

On the other hand the equal sign will hold in (4) if and only if all **three** equal signs hold in (3) which means $(a+b)^2 = (a+c)^2 = (b+c)^2 = 0$, which implies $a + b = a + c = b + c = 0 \Rightarrow \boldsymbol{a = b = c = 0}$.

P10. If the real numbers a, b, c, d form an analogy, show that

(a) $\frac{ka^2+\lambda b^2}{ka^2-\lambda b^2} = \frac{kc^2+\lambda d^2}{kc^2-\lambda d^2}$, for any nonzero real numbers k, λ such that λ/k is not equal to the square of the constant of the analogy (that a, b, c, d form)

(b) $\frac{c^2d+cd^2}{a^2b+ab^2} = \frac{(c+d)^2}{(a+b)^2}$, provided that the constant of the analogy is not equal to -1.

Solution. By Definition 5, since a, b, c, d form an analogy, we must have $abcd \neq 0$ and $ad = bc$

$$\frac{a}{b} = \frac{c}{d} \tag{1}$$

(a) Since k, λ are nonzero and $abcd \neq 0$, it follows $(ka)(kb) \cdot (\lambda c)(\lambda d) \neq 0$. Moreover, (1) shows that

$$\frac{ka^2}{kb^2} = \frac{\lambda c^2}{\lambda d^2} \Rightarrow (ka^2)(\lambda d^2) = (kb^2)(\lambda c^2) \tag{2}$$

Conditions (2) and $(ka^2)(kb^2)(\lambda c^2)(\lambda d^2) \neq 0$ imply that the real numbers $ka^2, kb^2, \lambda c^2, \lambda d^2$ form an analogy. Furthermore we must have $ka^2 - \lambda b^2 \neq 0$ and $kc^2 - \lambda d^2 \neq 0$ (because if it were $ka^2 - \lambda b^2 = 0$ then $\lambda/k = a^2/b^2 = (a/b)^2 =$ square of the constant of the analogy that a, b, c, d form, contrary to the hypothesis of the exercise). Therefore by applying **Property 26** we obtain

$$\frac{ka^2 + \lambda b^2}{ka^2 - \lambda b^2} = \frac{kc^2 + \lambda d^2}{kc^2 - \lambda d^2}$$

(b) Let t be the constant of the analogy, $\frac{a}{b} = \frac{c}{d} = t$. Since by assumption $t \neq -1$, it immediately follows that $a + b \neq 0$ and $c + d \neq 0$. We have $a = tb$ and $c = td$. We compute,

$$\frac{c^2 \cdot d + c \cdot d^2}{a^2 \cdot b + a \cdot b^2} = \frac{cd(c+d)}{ab(a+b)} = \frac{(td)d(td+d)}{(tb)b(tb+b)} =$$

$$\frac{t \cdot d^2 \cdot (t+1)d}{t \cdot b^2 \cdot (t+1)b} = \frac{d^3}{b^3} = \left(\frac{d}{b}\right)^3.$$

Next we have

$$\frac{(c+d)^3}{(a+b)^2} = \frac{(td+d)^3}{(tb+b)^3} = \frac{((t+1)\cdot d)^3}{((t+1)\cdot b)^3} = \frac{(t+1)^3\cdot d^3}{(t+1)^3 b^3} = \frac{d^3}{b^3}$$

We have proved that $\frac{c^2\cdot d+c\cdot d^2}{a^2\cdot b+a\cdot b^2} = \frac{(c+d)^3}{(a+b)^3} = \left(\frac{d}{b}\right)^3$.

P11. Let n be a positive integer, $n \geq 2$ and $a_1, a_2, \ldots, a_n, b_1, b_2, \ldots, b_n$, all nonzero real numbers. Find the necessary and sufficient condition that these real numbers must satisfy in order that

$$(a_1^2 + a_2^2 + \ldots + a_n^2)\cdot(b_1^2 + b_2^2 + \ldots + b_n^2) = (a_1b_1 + a_2b_2 + \ldots + a_nb_n)^2$$

Solution According to Schwarz's inequality (theorem 5) we have

$$(a_1^2 + a_2^2 + \ldots + a_n^2)\cdot(b_1^2 + b_2^2 + \ldots + b_n^2) \geq (a_1b_1 + a_2b_2 + \ldots + a_nb_n)^2 \qquad (1)$$

Of we look at the proof of (1), which involves Lagrange's Theorem (Theorem 4), we see that the left-hand side of (1) is equal to the right-hand side of (1) plus all the terms of the form $(a_ib_j - a_jb_i)^2$, with $i \neq j$ and $1 \leq i, j \leq n$. Thus, the equal sign will hold in (1) if and only if $a_ib_j - a_jb_i = 0 \Leftrightarrow a_ib_j = a_jb_i$, for i, j with $i \neq j$ and $1 \leq i, j \leq n$ which means since $a_ib_j \cdot a_jb_i \neq 0$, that the reals a_i, b_i, a_j, b_j must form an analogy: $\frac{a_i}{b_i} = \frac{a_j}{b_j}$, for all $i, j, 1 \leq i, j \leq n$ and $i \neq j$. So the answer to this exercise is

$$\frac{a_1}{b_1} = \frac{a_2}{b_2} = \ldots = \frac{a_n}{b_n}.$$

P12. Show that if the nonzero reals $\alpha, \beta, \gamma, x, y, z$ satisfy

$$(\alpha^2 + \beta^2 + \gamma^2)\cdot(x^2 + y^2 + z^2) = (\alpha x + \beta y + \gamma z)^2, \text{ then } \frac{x}{\alpha} = \frac{y}{\beta} = \frac{z}{\gamma}.$$

Solution This result is an immediate consequence of the previous practice P11, with $n = 3, a_1 = \alpha, a_2 = \beta, a_3 = \gamma, \beta_1 = x, \beta_2 = y$, and $\beta_3 = z$.

P13. If the real numbers $\alpha, \beta, \gamma, x, y$, satisfy the conditions $\frac{x}{y+z} = \alpha$, $\frac{y}{z+x} = \beta$, $\frac{z}{x+y} = \gamma$, and $x + y + z \neq 0$, prove that $\frac{\alpha}{1+\alpha} + \frac{\beta}{1+\beta} + \frac{\gamma}{1+\gamma} = 1$.

Solution Let $k = x + y + z$, with $k \neq 0$.

$$\frac{x}{y+z} = \alpha \Rightarrow \frac{x}{y+z} + 1 = \alpha + 1; \ \frac{x}{y+z} + \frac{y+z}{y+z} = \alpha + 1 \Rightarrow$$

$$\Rightarrow \frac{x+y+z}{y+z} = \alpha + 1; \text{ we have } \frac{k}{y+z} = \alpha + 1$$

which in turn implies

$$\frac{y+z}{k} = \frac{1}{\alpha+1} \tag{1}$$

We also have the original condition

$$\frac{x}{y+z} = \alpha \tag{2}$$

Multiplying (1) and (2) gives

$$\left(\frac{y+z}{k}\right) \cdot \left(\frac{x}{y+z}\right) = \frac{\alpha}{1+\alpha}; \quad \frac{x}{k} = \frac{\alpha}{1+\alpha} \tag{3}$$

Similarly we derive

$$\frac{y}{k} = \frac{\beta}{1+\beta} \tag{4}$$

$$\frac{z}{k} = \frac{\gamma}{1+\gamma} \tag{5}$$

Adding (3), (4), and (5) gives,

$$\frac{\alpha}{1+\alpha} + \frac{\beta}{1+\beta} + \frac{\gamma}{1+\gamma} = \frac{x+y+z}{k} = \frac{k}{k} = 1; \text{ the proof is complete}$$

P14. Let $\alpha_1, \alpha_2, \alpha_3, \alpha_4, \dots, \alpha_{2n-1}, \alpha_{2n}$ be a positive integers, $n \geq 2$, and suppose the real numbers, $\alpha_1, \alpha_2, \alpha_3, \alpha_4, \dots, \alpha_{2n-1}, \alpha_{2n}$ satisfy the condition $\frac{\alpha_1}{\alpha_2} = \frac{\alpha_3}{\alpha_4} = \dots = \frac{\alpha_{2n-1}}{\alpha_{2n}}$. Prove that they must also satisfy the condition

$$\frac{\alpha_1^2+\alpha_2^2}{\alpha_1+\alpha_2} + \frac{\alpha_3^2+\alpha_4^2}{\alpha_3+\alpha_4} + \dots + \frac{\alpha_{2n-1}^2+\alpha_{2n}^2}{\alpha_{2n-1}+\alpha_{2n}} =$$

$$\frac{(\alpha_1+\alpha_3+\dots+\alpha_{2n-1})^2 + (\alpha_2+\alpha_4+\dots+\alpha_{2n})}{\alpha_1+\alpha_2+\alpha_3+\alpha_4+\dots+\alpha_{2n-1}+\alpha_{2n}}$$

Solution. Put

$$k = \frac{\alpha_1}{\alpha_2} = \frac{\alpha_3}{\alpha_4} = \dots = \frac{\alpha_{2n-1}}{\alpha_{2n}} \Rightarrow (\alpha_1 = k\alpha_2, \alpha_3 = k\alpha_4, \dots, \alpha_{2n-1} = k \cdot \alpha_{2n}).$$

We have

$$\frac{\alpha_1^2+\alpha_2^2}{\alpha_1+\alpha_2} = \frac{k^2\alpha_2^2+\alpha_2^2}{k\alpha_2+\alpha_2} = \frac{\alpha_2^2\cdot(k^2+1)}{\alpha_2\cdot(k+1)} = \alpha_2\cdot\frac{(k^2+1)}{k+1}$$

$$\frac{\alpha_3^2+\alpha_4^2}{\alpha_3+\alpha_4} = \ldots = \alpha_4\frac{(k^2+1)}{k+1}$$

$$\vdots$$

$$\frac{\alpha_{2n-1}^2}{+}\alpha_{2n}^2\alpha_{2n-1}+\alpha_{2n} = \ldots = \alpha_{2n}\cdot\frac{(k^2+1)}{k+1}$$

Therefore the left-hand side of the equality we are asked to prove, is equal to

$$(\alpha_2+\alpha_4+\ldots+\alpha_{2n})\cdot\frac{(k^2+1)}{k+1}.$$

Now let us calculate the right-hand side of this exercise's condition we are trying to prove:

$$\frac{(\alpha_1+\alpha_3+\ldots+\alpha_{2n-1})^2+(\alpha_2+\alpha_4+\ldots+\alpha_{2n})^2}{\alpha_1+\alpha_2+\alpha_3+\alpha_4+\ldots+\alpha_{2n-1}+\alpha_{2n}}$$

$$= \frac{(k\alpha_2+k\alpha_4+\ldots+k\alpha_{2n})^2+(\alpha_2+\dot{\alpha}_4+\ldots+\alpha_{2n})^2}{ka_2+\alpha_2+k\alpha_4+\alpha_4+\ldots+k\alpha_{2n}+\alpha_{2n}}$$

$$= \frac{k^2\cdot(\alpha_2+\alpha_4+\ldots+\alpha_{2n})^2+(\alpha_2+\alpha_4+\ldots+\alpha_{2n})^2}{k(\alpha_2+\alpha_4+\ldots+\alpha_{2n})+(\alpha_2+\alpha_4+\ldots+\alpha_{2n})}$$

$$= \frac{(1+k^2)\cdot(\alpha_2+\alpha_4+\ldots+\alpha_{2n})^2}{(1+k)\cdot(\alpha_2+\alpha_4+\ldots+\alpha_{2n})} = \frac{(k^2+1)}{k+1}\cdot(\alpha_2\alpha_4+\ldots+\alpha_{2n}),$$

which is the same quantity we have above. The two sides of the exercise's condition we are trying to prove, are indeed equal.

P15. Prove that if $\alpha < \beta+\gamma$, $\beta < \gamma+\alpha$, and $\gamma < \alpha+\beta$ then the real numbers $\alpha, \beta, , \gamma$ must be all positive.

Solution. $\beta < \gamma+\alpha$ implies $\beta-\gamma < \alpha$ and $\gamma < \alpha+\beta \Rightarrow \gamma-\beta < \alpha$; since $\beta-\gamma < \alpha$ and $-(\beta-\gamma) < \alpha$, it follows that $|\beta-\gamma| < \alpha \Rightarrow 0 < \alpha$. Similarly we deduce $0 < \beta$, and $0 < \gamma$.

P16. Suppose the reals α, β, γ, satisfy the conditions

$$\frac{\alpha}{\beta} < 1,\ \beta > 0,\ \gamma > 0.$$

Show that

$$\frac{\alpha+\gamma}{\beta+\gamma} > \frac{\alpha}{\beta}$$

Solution. Since $\beta > 0$ and $\gamma > 0 \Rightarrow \beta + \gamma > 0$. Thus $\beta \cdot (\beta + \gamma) > 0$. Because $\beta \cdot (\beta + \gamma) > 0$ the inequality we are asked to prove is equivalent to

$$\beta \cdot (\beta + \gamma) \cdot \left(\tfrac{\alpha+\gamma}{\beta+\gamma}\right) > \beta \cdot (\beta + \gamma) \cdot \tfrac{\alpha}{\beta} \Leftrightarrow \beta \cdot (\alpha + \gamma) > (\beta + \gamma) \cdot \alpha \Leftrightarrow$$
$$\Leftrightarrow \beta \cdot \alpha + \beta \cdot \gamma > \beta \cdot \alpha + \gamma\alpha \Leftrightarrow \beta\gamma > \gamma\alpha$$

since $\beta\gamma > 0$, the last inequality is equivalent to

$$\frac{1}{\beta\gamma} \cdot (\beta\gamma) > \frac{1}{\beta\gamma} \cdot (\gamma\alpha) \Leftrightarrow 1 > \frac{\alpha}{\beta}$$

which is true by hypothesis. The proof is complete.

Note: One can start with $\frac{\alpha}{\beta} < 1$. Multiply both sides by

$$\beta\gamma : (\beta\gamma)\tfrac{\alpha}{\beta} < \beta\gamma \Rightarrow \gamma\alpha < \beta\gamma \Rightarrow \beta\alpha + \gamma\alpha < \beta\alpha + \beta\gamma \Rightarrow \alpha(\beta + \gamma) < \beta \cdot (\alpha + \gamma)$$

$$\Rightarrow \text{ (since } \beta \cdot (\beta + \gamma) > 0)) \tfrac{\alpha(\beta+\gamma)}{\beta(\beta+\gamma)} < \tfrac{\beta \cdot (\alpha+\gamma)}{\beta(\beta+\gamma)} \Rightarrow \tfrac{\alpha}{\beta} < \tfrac{\alpha+\gamma}{\beta+\gamma}$$

P17. Prove $(\alpha\beta + \beta\gamma + \gamma\alpha)^2 \geq 3\alpha\beta\gamma(\alpha + \beta + \gamma)$

Solution. By P9 we know that

$$(\alpha\beta)^2 + (\alpha\gamma)^2 + (\beta\gamma)^2 \geq (\alpha\beta)(\alpha\gamma) + (\alpha\beta)(\beta\gamma) + (\alpha\gamma)(\beta\gamma)$$

$$\Rightarrow \quad (\alpha\beta)^2 + (\alpha\gamma)^2 + (\beta\gamma)^2 \geq \alpha\beta\gamma(\alpha + \beta + \gamma)$$

$$\Rightarrow \quad (\alpha\beta)^2 + (\alpha\gamma)^2 + (\beta\gamma)^2 \geq 3\alpha\beta\gamma(\alpha + \beta + \gamma) - 2\alpha\beta\gamma(\alpha + \beta + \gamma) \Rightarrow$$

$$\Rightarrow \quad (\alpha\beta)^2 + (\alpha\gamma)^2 + (\beta\gamma)^2 + 2\alpha\beta\gamma(\alpha + \beta + \gamma) \geq 3\alpha\beta\gamma(\alpha + \beta + \gamma) \Rightarrow$$

$$\Rightarrow \quad (\alpha\beta)^2 + (\alpha\gamma)^2 + (\beta\gamma)^2 + 2 \cdot [(\alpha\beta)(\alpha\gamma) + (\alpha\beta)(\beta\gamma) + (\alpha\gamma)(\beta\gamma)] \geq 3\alpha\beta\gamma(\alpha + \beta + \gamma)$$

$$\Rightarrow \quad \text{(by Th. 2(ii))}(\alpha\beta + \alpha\gamma + \beta\gamma)^2 \geq 3\alpha\beta\gamma(\alpha + \beta + \gamma).$$

P18. Show that $a^2b^2 + b^2c^2 + c^2a^2 \geq abc(a + b + c)$.
When does the equal sigh hold?

Solution. This is an immediate consequence of problem P9. We have

$$(ab)^2 + (bc)^2 + (ca)^2 \geq (ab)(bc) + (ab)(ca) + (bc)(ca) \Rightarrow a^2b^2 + b^2c^2 + c^2a^2 \geq abc \cdot (a + b + c)$$

According to problem P9 the equal sign holds exactly when $ab = bc = ca$. If $abc \neq 0$, then $ab = bc = ca$ clearly implies $a = b = c$. If $abc = 0$ then $a = 0$ or $b = 0$

or $c = 0$. Suppose $a = 0$. From $ab = bc = ca$, it follows that $bc = 0$. So at least one of b, c must be zero.

Conclusion: The equal sign will hold in the above inequality if and only if either ($abc \neq 0$ and $a = b = c$) **or** ($abc = 0$, with two of the reals among a, b, c, being zero, the third arbitrary).

P19. Show that if a, b, c are the side lengths of a triangle, they must satisfy $a^2 + b^2 + c^2 < 2(ab + ac + bc)$.

Solution. Since a, b, c are the side lengths of a triangle they must satisfy $a < b+c$, $b < a + c$, and $c < a + b$. (These three inequalities do imply $a > 0$, $b > 0$, and $c > 0$, by problem P15.) Since $a - b < c$ and $b - a < c$, we have $a - b < c$ and $-(a - b) < c$. Thus $|a - b| < c$.

$$\text{Similarly,} \quad \left.\begin{array}{l} |a-b| < c \\ |b-c| < a \\ |c-a| < b \end{array}\right\} \qquad (1)$$

By Theorem 9(iii), the square of any real number is equal to the square of its absolute value. On the other hand, since $0 \leq |a - b| < c$, it is clear by Property 31 (Transitive property) that $0 < c$.

Consider

$$\left.\begin{array}{r} |a-b| < c \\ \text{and} \quad |a-b| < c \end{array}\right\} \qquad (2)$$

same inequality ↑

If $a - b \neq 0$ then $0 < |a - b|$ and since also $0 < c$, we can conclude from Property 40 that inequalities (2) imply,

$$|a-b| \cdot |a-b| < c \cdot c$$

$$|a-b|^2 < c^2$$

If $a - b = 0$, then since $c > 0$, we still have $|a - b|^2 < c^2$. In a similar manner we can obtain from (1)

$$\left.\begin{array}{l} |a-b|^2 < c^2 \\ \\ |b-c|^2 < a^2 \\ \\ |c-a|^2 < b^2 \end{array}\right\} \qquad (3)$$

By Theorem 9(iii), $|a - b|^2 = (a - b)^2$, $|b - c|^2 = (b - c)^2$, $|c - a|^2 = (c - a)^2$. Thus by (3),

$$\Rightarrow \left.\begin{array}{l} (a-b)^2 < c^2 \\ (b-c)^2 < a^2 \\ (c-a)^2 < b^2 \end{array}\right\} \Rightarrow \left.\begin{array}{l} a^2 - 2ab + b^2 < c^2 \\ b^2 - 2c + c^2 < a^2 \\ c^2 - 2ac + a^2 < b^2 \end{array}\right\} \Rightarrow$$

$\Rightarrow$ (adding the three inequalities memberwise) $2(a^2+b^2+c^2) - 2(ab+bc+ac) < a^2+b^2+c^2$

$\Rightarrow 2(a^2+b^2+c^2) - (a^2+b^2+c^2) < 2(ab+bc+ac)$;

$\Rightarrow a^2+b^2+c^2 < 2(ab+bc+ac)$; the result is proven.

P20. Prove that $a+b+c>0$ implies $a^3+b^3+c^3 \geq 3abc$.

Solution. This is an immediate consequence of Cauchy's Theorem, Theorem 8 since

$$a^3+b^3+c^3-3abc = \underbrace{\frac{(a+b+c)}{2}}_{>0} \cdot \underbrace{\left[(a-b)^2+(b-c)^2+(c-a)^2\right]}_{\geq 0},$$

and so $a^3+b^3+c^3-3abc \geq 0$.

P21. Suppose that the real numbers a,b,c are positive. Prove that $ab(a+b)+bc(b+c)+ac(a+c) \geq 6abc$. When does the equal sign hold?

Solution. We know from Theorem 6 that

$$\left.\begin{array}{r} a^2+b^2 \geq 2ab \\ b^2+c^2 \geq 2bc \\ c^2+a^2 \geq 2ac \end{array}\right\} \qquad (1)$$

Since a,b,c are all positive, the three products ab, bc, and ca are also positive. If we divide each of the inequalities in (1) by the products ab, bc, ca, respectively, then Property 39(i) and (1) tell us that

$$\left.\begin{array}{r} \dfrac{a^2+b^2}{ab} \geq \dfrac{2ab}{ab} \\[2ex] \dfrac{b^2+c^2}{2bc} \geq \dfrac{2bc}{bc} \\[2ex] \dfrac{c^2+a^2}{2ca} \geq \dfrac{2ca}{ca} \end{array}\right\} \Rightarrow \left.\begin{array}{r} \dfrac{a}{b}+\dfrac{b}{a} \geq 2 \\[2ex] \dfrac{b}{c}+\dfrac{c}{b} \geq 2 \\[2ex] \dfrac{c}{a}+\dfrac{a}{c} \geq 2 \end{array}\right\} \qquad (2)$$

Adding the three inequalities in (2) yields

$$\frac{a}{b}+\frac{b}{a}+\frac{b}{c}+\frac{c}{a}+\frac{c}{a}+\frac{a}{c} \geq 6 \qquad (3)$$

Since $abc>0$, multiplication of both sides of (3) by abc leads to

$$a^2c+b^2c+b^2a+c^2a+c^2b+a^2b \geq 6 \Rightarrow ab(a+b)+bc(b+c)+ac(a+c) \geq 6abc,$$

the result is proven.

Note: The key to the method of proof implemented above, lies with the product abc. If one divides the exercise's inequality by abc, then the method of proof becomes more apparent. Finally, in order that the equal sign holds in the problem's inequality, it is necessary (and sufficient) that all three equal signs hold in (1) which according to **Remark 15**, happens exactly when $a = b$, $b = c$, and $c = a$. That is, when $a = b = c$.

P22. Let $\alpha, \beta, \gamma, x, y, z$ be real numbers. Suppose that α, β, γ have the same sign and x, y, z also have the same sign. Prove that

$$(\alpha x + \beta y + \gamma z)\left(\frac{\alpha}{x} + \frac{\beta}{y} + \frac{\gamma}{z}\right) \geq (\alpha + \beta + \gamma)^2$$

When does the equal sign hold?

Solution. According to hypothesis there are four possibilities:

Possibility 1: $(\alpha > 0,\ \beta > 0,\ \gamma > 0)$ and $(x > 0,\ y > 0,\ z > 0$

Possibility 2: $(\alpha > 0,\ \beta > 0,\ \gamma > 0)$ and $(x < 0,\ y < 0,\ z < 0)$

Possibility 3: $(\alpha < 0,\ \beta < 0,\ \gamma < 0)$ and $(x > 0,\ y > 0, z > 0)$

Possibility 4: $(\alpha < 0,\ \beta < 0,\ \gamma < 0)$ and $(x < 0,\ y < 0,\ z < 0)$

Observe that in all four possibilities

$$(\alpha x + \beta y + \gamma z)\left(\frac{\alpha}{x} + \frac{\beta}{y} + \frac{\gamma}{z}\right) = (|\alpha x| + |\beta y| + |\gamma z|)\left(\left|\frac{\alpha}{x}\right| + \left|\frac{\beta}{y}\right| + \left|\frac{\gamma}{z}\right|\right).$$

This should be clear because if **Possibility 1** occurs then

$$|\alpha x| = \alpha x,\ \ |\beta x| = \beta x,\ \ |\gamma z| = \gamma z,\ \ \left|\frac{\alpha}{x}\right| = \frac{\alpha}{x},\ \ \left|\frac{\beta}{y}\right| = \beta/y,\ \ \left|\frac{\gamma}{z}\right| = \frac{\gamma}{z}$$

If **Possibility 2** occurs, then

$$|\alpha x| = -\alpha x, |\beta x| = -\beta x,\ \ |\gamma z| = -\gamma z,\ \ \left|\frac{\alpha}{x}\right| = -\alpha/x,\ \ |\beta/y| = -\beta/y,\ \ \left|\frac{\gamma}{z}\right| = -\gamma/z.$$

Same thing if **Possibility 3** occurs. Finally, for **Possibility 4** the situation is exactly the same as in **Possibility 1**, namely

$$|\alpha x| = \alpha x,\ \ |\beta x| = \beta x, |\gamma z| = \gamma z,\ \ |\alpha/x| = \alpha/x,\ \ |\beta/y| = \beta/y,\ \ |\gamma/z| = \gamma/z.$$

Consequently, in **all** cases we have

$$P = (\alpha x + \beta x + \gamma z) \cdot \left(\frac{\alpha}{x} + \frac{\beta}{y} + \frac{\gamma}{z}\right) = (|\alpha x| + |\beta y| + |\gamma z|) \cdot \left(\left|\frac{\alpha}{x}\right| + \left|\frac{\beta}{y}\right| + \left|\frac{\gamma}{z}\right|\right) \quad (1)$$

If we multiply out the left-hand side of (1) we have

$$P = |\alpha|^2+|\beta|^2+|\gamma|^2+|\alpha\beta|\cdot\left(\frac{|x|}{|y|}+\frac{|y|}{|x|}\right)+|\alpha\gamma|\cdot\left(\frac{|x|}{|z|}+\frac{|z|}{|x|}\right)+|\beta\gamma|\cdot\left(\frac{|y|}{|z|}+\frac{|z|}{|y|}\right) \quad (2)$$

But

$$\frac{|x|}{|y|}+\frac{|y|}{|x|} \geq 2, \quad \frac{|x|}{|z|}+\frac{|z|}{|x|} \geq 2, \quad \frac{|y|}{|z|}+\frac{|z|}{|y|}, \quad (3)$$

since

$$|x|^2+|y|^2 \geq 2|y||x|, \ |x|^2+|z|^2 \geq 2|z||x|, \ |y|^2+|z|^2 \geq 2|y||z|$$

by Theorem 6 combined with $xyz \neq 0$ (so that we can divide the latter three inequalities by $|y||x|$, $|z||x|$, and $|y||z|$ to obtain (3)). Combining (2) and (3) produces,

$$P \geq |\alpha|^2+|\beta|^2+|\gamma|^2+2|\alpha||\beta|+2|\alpha||\gamma|+2|\beta||\gamma|;$$
$$P \geq (|\alpha|+|\beta|+|\gamma|)^2 = (\alpha+\beta+\gamma)^2,$$

the last equality being true because α, β, γ have the same sign and therefore,

$$|\alpha|+|\beta|+|\gamma| = \pm(\alpha+\beta+\gamma) \Rightarrow (|\alpha|+|\beta|+|\gamma|)^2 = [\pm(\alpha+\beta+\gamma)]^2 = (\alpha+\beta+\gamma)^2$$

The result is proven. Now let us answer the second question, when does the equal sign hold? For the equal sign to hold in the inequality of the problem, it is necessary and sufficient that **all three** equal signs hold in (2) which means $|x| = |y| = |z|$ (because in such a case $(|x|-|y|)^2 = (|x|-|z|)^2 = (|y|-|z|)^2 = 0$). But x, y, z have the same sign so $|x| = |y| = |z|$ would simply imply either $x = y = z$ or $-x = -y = -z$. In both cases $\boxed{x = y = z}$ is the necessary and sufficient condition for the equal sign to hold.

P23. Show that if $\alpha, \beta\gamma$ are all positive or all negative $(\alpha+\beta+\gamma)\left(\frac{1}{\alpha}+\frac{1}{\beta}+\frac{1}{\gamma}\right) \geq 9$. When does the equal sign hold?

Solution. This is an immediate consequence of the previous problem, P22. If the α, β, γ in Problem 22 are all replaced by 1 (i.e. take $\alpha = \beta = \gamma = 1$) and if the x, y, z are replaced by α, β, γ, respectively, then problem 22 says that

$$(1\cdot\alpha+1\cdot\beta+1\cdot\gamma)\cdot\left(\frac{1}{\alpha}+\frac{1}{\beta}+\frac{1}{\gamma}\right) \geq (1+1+1)^2; \ (\alpha+\beta+\gamma)\cdot\left(\frac{1}{\alpha}+\frac{1}{\beta}+\frac{1}{\gamma}\right) \geq 9.$$

Again, according to P22, equality will occur exactly when $\alpha = \beta = \gamma$.

P24. In this problem assume that for every **nonnegative** real number a, $a \geq 0$, there exists a unique nonnegative real x such that $x^2 = a$; x is called the square-root of a and it is denoted by $\sqrt{a}$; $x = \sqrt{a}$. Also, assume that for **every** real number a, there exists a unique real number y such that $y^3 = a$. y is called the cubic root of a and it is denoted by $\sqrt[3]{a}$. We write $y = \sqrt[3]{a}$. Assume that the product of the square roots of any number of real numbers is equal to the square root of their product. Likewise for the cube root. Using the above assumptions, Theorem 6, and Problem P20 show that

(a) If a and b are positive reals

$$\frac{a+b}{2} \geq \sqrt{ab} \geq \frac{2}{1/a + 1/b}$$

the equal signs holding exactly when $a = b$.

and (b) If a, b, c are positive reals

$$\frac{a+b+c}{3} \geq \sqrt[3]{abc} \geq \frac{3}{\frac{1}{a}+\frac{1}{b}+\frac{1}{c}}$$

the equal signs occuring exactly when $a = b = c$.

Note: We will see in Chapter 6 that given any positive integer $n \geq 2$, and positive reals $a_1, a_2, \ldots, a_n$,

$$\frac{a_1 + a_2 + \ldots + a_n}{n} \geq \sqrt[n]{a_1 a_2 \ldots a_n} \geq \frac{n}{\frac{1}{a_1}+\frac{1}{a_2}+\ldots+\frac{1}{a_n}}$$

This is called the $\underbrace{\text{arithmetic mean}}_{\text{A.M.}}$ - $\underbrace{\text{geometric mean}}_{\text{G.M.}}$ - $\underbrace{\text{harmonic mean}}_{\text{H.M.}}$ inequality.

Solution. (a) We apply Theorem 6. First observe that by the definition of the square root, we have $a = (\sqrt{a})^2$, $b = (\sqrt{b})^2$. Thus, $a + b = (\sqrt{a})^2 + (\sqrt{b})^2$ and by Theorem 6

$$\begin{aligned}
&(\sqrt{a})^2 + (\sqrt{b})^2 \geq 2(\sqrt{a})(\sqrt{b});\\
&a + b \geq 2\sqrt{a}\sqrt{b};\\
&a + b \geq 2\sqrt{ab};\\
&\tfrac{a+b}{2} \geq \sqrt{ab}.
\end{aligned} \tag{1}$$

Next,

$$\frac{1}{a} + \frac{1}{b} = \frac{1}{\sqrt{a})^2} + \frac{1}{(\sqrt{b})^2} = \left(\frac{1}{\sqrt{a}}\right)^2 + \left(\frac{1}{\sqrt{b}}\right)^2.$$

Applying Theorem 6 again,

$$\left(\frac{1}{\sqrt{a}}\right)^2 + \left(\frac{1}{\sqrt{b}}\right)^2 \geq 2 \cdot \left(\frac{1}{\sqrt{a}}\right) \cdot \left(\frac{1}{\sqrt{b}}\right) \Rightarrow \frac{1}{a} + \frac{1}{b} > 2 \cdot \frac{1}{\sqrt{ab}} \geq 0$$

apply **Property 41(ii)** to get

$$\frac{1}{\frac{1}{a}+\frac{1}{b}} \leq \frac{1}{\left(2/\sqrt{ab}\right)};$$

$$\frac{1}{\frac{1}{a}+\frac{1}{b}} \leq \left(\frac{1}{2/\sqrt{ab}}\right) \cdot \frac{\sqrt{ab}}{\sqrt{ab}}; \text{ (Property16)} \tag{2}$$

$$\frac{1}{\frac{1}{a}+\frac{1}{b}} \leq \frac{\sqrt{ab}}{2} \Rightarrow \frac{2}{\frac{1}{a}+\frac{1}{b}} \leq \sqrt{ab}$$

Combining (1) and (2) we arrive at

$$\frac{a+b}{2} \geq \sqrt{ab} \geq \frac{2}{\frac{1}{a}+\frac{1}{b}}. \tag{3}$$

(b) We have $a+b+c = (\sqrt[3]{a})^3 + \left(\sqrt[3]{b}\right)^3 + (\sqrt[3]{c})^3$. Since $\sqrt[3]{a} > 0$, $\sqrt[3]{b} > 0$, $\sqrt[3]{c} > 0$ (because $a > 0$ and $(\sqrt[3]{a})^3 = a$; likewise for b and c), the sum $\sqrt[3]{a}+\sqrt[3]{b}+\sqrt[3]{c}$ is positive. By Problem P20 we have,

$$\begin{aligned} &a+b+c = (\sqrt[3]{a})^3 + \left(\sqrt[3]{b}\right)^3 + (\sqrt[3]{c})^3 \geq 3(\sqrt[3]{a})\left(\sqrt[3]{b}\right)(\sqrt[3]{c}) \\ &a+b+c \geq 3\sqrt[3]{abc} \Rightarrow \tfrac{a+b+c}{3} \geq \sqrt[3]{abc} \end{aligned} \tag{4}$$

Likewise,

$$\begin{aligned}
&\frac{1}{a}+\frac{1}{b}+\frac{1}{c} = \frac{1}{(\sqrt[3]{a})^3} + \frac{1}{\left(\sqrt[3]{b}\right)^3} + \frac{1}{(\sqrt[3]{c})^3} \\
&\frac{1}{a}+\frac{1}{b}+\frac{1}{c} = \left(\frac{1}{\sqrt[3]{a}}\right)^3 + \left(\frac{1}{\sqrt[3]{b}}\right)^3 + \left(\frac{1}{\sqrt[3]{c}}\right)^3 \geq 3 \cdot \left(\frac{1}{\sqrt[3]{a}}\right)\left(\frac{1}{\sqrt[3]{b}}\right)\left(\frac{1}{\sqrt[3]{c}}\right) \Rightarrow \\
&\Rightarrow \frac{1}{a}+\frac{1}{b}+\frac{1}{c} \geq \frac{3}{\sqrt[3]{abc}} > 0; \text{ using Property 41(ii),} \\
&\frac{1}{\frac{1}{a}}+\frac{1}{b}+\frac{1}{c} \leq \frac{1}{\left(\frac{3}{\sqrt[3]{abc}}\right)} \\
&\frac{1}{\frac{1}{a}}+\frac{1}{b}+\frac{1}{c} \leq \frac{1}{\left(\frac{3}{\sqrt[3]{abc}}\right)} \cdot \frac{\sqrt[3]{abc}}{\sqrt[3]{abc}}; \text{ (Property 16)} \\
&\frac{1}{\frac{1}{a}}+\frac{1}{b}+\frac{1}{c} \leq \frac{\sqrt[3]{abc}}{3} \Rightarrow \frac{3}{\frac{1}{a}}+\frac{1}{b}+\frac{1}{c} \leq \sqrt[3]{abc}
\end{aligned} \tag{5}$$

Combining (4) and (5) we arrive at

$$\frac{a+b+c}{3} \geq \sqrt[3]{abc} \geq \frac{3}{\frac{1}{a}+\frac{1}{b}+\frac{1}{c}} \tag{6}$$

To conclude, let us investigate when the equal signs hold. First consider (6). If we look at the very short solution of problem P20, we see that in $a^3+b^3+c^3 \geq 3abc$ the equal sign holds exactly when $a=b=c$. Therefore in $(\sqrt[3]{a})^2+(\sqrt[3]{b})^3+(\sqrt[3]{c}) \geq 3\cdot\sqrt[3]{a}\sqrt[3]{b}\sqrt[3]{c}$, the equal sign holds exactly when $\sqrt[3]{a}=\sqrt[3]{b}=\sqrt[3]{c} \overset{\text{by cubing}}{\Rightarrow} a=b=c$. Hence $\frac{a+b+c}{3} \geq \sqrt[3]{abc}$ holds exactly when $a=b=c$. A similar argument shows that in $\sqrt[3]{abc} \geq \frac{3}{\frac{1}{a}+\frac{1}{b}+\frac{1}{c}}$, the equal sign will occur exactly when $\frac{1}{\sqrt[3]{a}} = \frac{1}{\sqrt[3]{b}}] = \frac{1}{\sqrt[3]{c}} \Rightarrow a=b=c$, once more.

Next consider (3),

$$\frac{a+b}{2} \geq \sqrt{ab} \geq \frac{2}{\frac{1}{a}+\frac{1}{b}}.$$

A look at Theorem 6 reveals that the first equal sign occurs exactly when $\sqrt{a} = \sqrt{b} \Rightarrow (\sqrt{a})^2 = (\sqrt{b})^2$; $a=b$. And the second equal sign occurs when $\frac{1}{\sqrt{a}} = \frac{1}{\sqrt{b}}$, again implying $a=b$. We are done.

P25. Prove that if the real numbers α, β, γ satisfy the conditions $\alpha < \beta+\gamma$, $\beta < \alpha+\gamma$, $\gamma < \alpha+\beta$, then $\frac{2}{3}(\alpha+\beta+\gamma)\cdot(\alpha^2+\beta^2+\gamma^2) > \alpha^3+\beta^3+\gamma^3+\alpha\beta\gamma$

Solution. First observe that the conditions $\alpha < \beta+\gamma$, $\beta < \alpha+\gamma$, $\gamma < \alpha+\beta$ imply that α, β, γmust be positive. This is easy to show and we have already done this in problem P15. The next item we need to do is to explore the inequality we are asked to prove: to find an **equivalent** inequality that relates to the theorems of this Chapter in a more recognizable way so that we can prove it. We have,

$$\begin{aligned}
&\tfrac{2}{3}(\alpha+\beta+\gamma)\cdot(\alpha^2+\beta^2+\gamma^2) > (\alpha^3+\beta^3+\gamma^3+\alpha\beta\gamma) \\
\Leftrightarrow\ & 2(\alpha+\beta+\gamma)\cdot(\alpha^2+\beta^2+\gamma^2) > 3(\alpha^3+\beta^3+\gamma^3+\alpha\beta\gamma) \\
\Leftrightarrow\ & 2(\alpha^3+\beta^2+\gamma^3)+2\alpha\beta(\alpha+\beta)+2\alpha\gamma(\alpha+\gamma)+2\beta\gamma(\beta+\gamma) > \\
& 3(\alpha^3+\beta^3+\gamma^3)+3\alpha\beta\gamma \\
\Leftrightarrow\ & 2\alpha\beta(\alpha+\beta)+2\alpha\gamma(\alpha+\gamma)+2\beta\gamma(\beta+\gamma)(\beta+\gamma) > \alpha^3+\beta^3+\gamma^3+3\alpha\beta\gamma
\end{aligned} \tag{1}$$

So the problem's inequality is equivalent to (1). If we prove (1) we are done. Since $a > 0$, $\beta < 0$, $\gamma > 0$, the product $\alpha\beta\gamma$ is also positive. Thus (1) is equivalent to (divide both sides by $\alpha\beta\gamma$),

$$\begin{aligned}
&2\left(\frac{\alpha}{\gamma}+\frac{\beta}{\gamma}\right)+2\left(\frac{\alpha}{\beta}+\frac{\gamma}{\beta}\right)+2\left(\frac{\beta}{\alpha}+\frac{\gamma}{\alpha}\right) > \left(\frac{\alpha}{\beta}\right)\cdot\left(\frac{\alpha}{\gamma}\right)+ \\
&\left(\frac{\beta}{\alpha}\right)\left(\frac{\beta}{\gamma}\right)+\left(\frac{\gamma}{\alpha}\right)\left(\frac{\gamma}{\beta}\right)+3
\end{aligned} \tag{2}$$

Let $\frac{\alpha}{\beta} = x$, $\frac{\beta}{\gamma} = y$, $\frac{\gamma}{\alpha} = z$. Clearly,$x, y, z$ are all positive since the reals α, β, γ are. Inequality (2) becomes

$$2(x+y+z)+2\left(\frac{1}{x}+\frac{1}{y}+\frac{1}{z}\right) > \frac{x}{z}+\frac{y}{x}+\frac{z}{y}+3 \tag{3}$$

Let us go back to the inequalities $\alpha < \beta+\gamma$, $\beta < \gamma+\alpha$, $\gamma < \alpha+\beta$. Take for instance $\gamma < \alpha+\beta$; since $\alpha, \beta, \gamma > 0$, we can multiply both sides with $\frac{\gamma}{\alpha\beta}$ and obtain

$$\left(\frac{\gamma}{\alpha}\right)\cdot\left(\frac{\gamma}{\beta}\right) < \frac{\gamma}{\alpha\beta}(\alpha+\beta);$$

$$\left(\frac{\gamma}{\alpha}\right)\cdot\left(\frac{\gamma}{\beta}\right) < \frac{\gamma}{\beta}+\frac{\gamma}{\alpha}; \tag{4}$$

$$\text{But } \frac{\gamma}{\alpha} = z,\ \frac{\beta}{\gamma} = y \text{ so that } \frac{z}{y} < \frac{1}{y} + z$$

Likewise the inequality $\beta < \gamma + \alpha$ implies (by multiplying both sides by $\frac{\beta}{\alpha\gamma}$),

$$\frac{y}{x} < \frac{1}{x} + y \tag{5}$$

And the inequality $\gamma < \alpha + \beta$ implies,

$$\frac{x}{z} < \frac{1}{z} + x \tag{6}$$

Remember, our objective is to prove (3). However, inequalities (4), (5), and (6) already hold **true**. If we add (4), (5), and (6) we obtain

$$(x+y+z)+\left(\frac{1}{x}+\frac{1}{y}+\frac{1}{z}\right) > \frac{z}{y}+\frac{y}{x}+\frac{x}{z}, \tag{7}$$

and thus,

$$2(x+y+z)+2\left(\frac{1}{x}+\frac{1}{y}+\frac{1}{z}\right) > 2\left(\frac{z}{y}+\frac{y}{x}+\frac{x}{z}\right) \tag{8}$$

Now, problem P24, part (b) comes into play. Since $\frac{a_1+a_2+a_3}{3} \geq \sqrt[3]{a_1a_2a_3}$ for any positive reals, by setting $a_1 = \frac{z}{y}$, $a_2 = \frac{y}{x}$, $a_3 = \frac{x}{z}$, we see that $a_1a_2a_3 = \frac{z}{y}\cdot\frac{y}{x}\cdot\frac{x}{z} = 1$! And so we obtain

$$\begin{aligned} &\frac{\frac{z}{y}+\frac{y}{x}+\frac{x}{z}}{3} \geq \sqrt[3]{1} = 1; \\ &\frac{z}{y}+\frac{y}{x}+\frac{x}{z} \geq 3 \end{aligned} \tag{9}$$

Adding $\left(\frac{z}{y}+\frac{y}{x}+\frac{x}{z}\right)$ to both sides of (9) we arrive at

$$2\left(\frac{z}{y}+\frac{y}{x}+\frac{x}{z}\right) \geq \frac{z}{y}+\frac{y}{x}+\frac{x}{z}+3 \tag{10}$$

Inequality (3) which is our aim of proving, is an immediate consequence of inequalities (8) and (10). The proof is complete!

This definitely is a challenging problem requiring the skill of exploration couched with applying theorems most effective to solving the problem. It is our hope that the student will develop these skills during this course!

P26. Show that if $|b_i| \leq 1$ for $i = 1, 2, \ldots, n$, then

$$|a_1b_1 + a_2b_2 + \ldots + a_nb_n| \leq |a_1| + |a_2| + \ldots + |a_n|$$

Solution. Applying the triangle inequality (Theorem (9vii)),

$$|a_1b_1a_2b_2 + \ldots + a_nb_n| \leq |a_1b_1| + |a_2b_2| + \ldots| a_nb_n| \tag{1}$$

But $|a_ib_i| = |a_i||b_i| \leq |a_i|$, since $0 \leq |b_i| \leq 1$, for all $i = 1, 2, \ldots, n$. Therefore

$$\left.\begin{array}{l} |a_1, b_1| \leq |a_1| \\ \vdots \\ |a_2b_2| \leq |a_2| \\ \vdots \\ |a_nb_n| \leq |a_n| \end{array}\right\}$$

$$\Rightarrow \text{(by adding these inequalities)} |a_1b_1| + |a_2b_2| + \ldots + |a_nb_n| \leq |a_1| + |a_2| + \ldots + |a_n|. \tag{2}$$

By combining (1) and (2) we arrive at $|a_1b_1+a_2b_2+\ldots+a_nb_n| \leq |a_1|+|a_2|+\ldots+|a_n|$.

P27. If $|a_1| + |a_2| + \ldots + |a_n| = 1$ and $|b_1| \leq 1$, for $i = 1, 2, \ldots, n$, show that $|a_1b_1 + a_2b_2 + \ldots + a_nb_n| \leq 11$.

Solution. This is an immediate consequence of the previous practice problem P26.

P28. Show that $||x + \alpha| - |x + \beta|| \leq |\alpha - \beta| \leq y - \alpha| + |y - \beta|$, for any real numbers x, y, α, β.

Solution. Observe that $(x+\alpha) - (x+\beta) = \alpha - \beta$, hence by the triangle inequalities (Theorem 9(vi)) we have

$$||x+\alpha| - |x+\beta|| \leq |(x+\alpha) - (x+\beta)| = |\alpha - \beta| \tag{1}$$

Moreover, $\alpha - \beta = (\alpha - y) + (y - \beta)$. Applying the triangle inequalities again,

$$|\alpha - \beta| = |(\alpha - y) + (y - \beta)| \leq |\alpha - y| + |y - \beta| = |(y - \alpha)| + |y - \beta|$$

We have shown

$$|\alpha - \beta| \le |y - \alpha| + |y - \beta| \tag{2}$$

Combine (1) and (2) and the result readily follows.

P29. For every real number a, $||a| + a| - ||a| - a| = 2a$.

Solution. If $a < 0$, then $|a| = -a$; $||a| + a| - ||a| - a| = |-a + a| - |-a - a|| = |0| - |-2a|| = -2a = -(-2a) = 2a$. If $a \ge 0$, then $|a| = a \Rightarrow ||a| + a| - ||a| - a| = |a + a| - |a - a| = |2a| - |0| = 2a$.

P30. Suppose that x, y and z are non zero real numbers. Let $\alpha = \frac{x}{|y|+|z|}$, $\beta = \frac{y}{|x|+|z|}$, $\gamma = \frac{z}{|x|+|y|}$. Prove the following

(a) $|\alpha\beta\gamma| \le 1/8$.

(b) $\frac{1}{|\alpha|} + \frac{1}{|\beta|} + \frac{1}{|\gamma|} \ge 6$.

Solution (a).

$$\frac{1}{|\alpha\beta\gamma|} = \frac{1}{|\alpha|} \cdot \frac{1}{|\beta|} \cdot \frac{1}{|\gamma|} = \left(\frac{|y|}{|x|} + \frac{|z|}{||x|}\right) \cdot \left(\frac{|x|}{|y|} + \frac{|z|}{|y|}\right) \cdot \left(\frac{|x|}{|z|} + \frac{|y|}{|x|}\right) =$$

$$= 2 + \left(\frac{|y|}{|x|} + \frac{|x|}{|y|}\right) + \left(\frac{|y|}{|z|} + \frac{|z|}{|y|}\right) + \left(\frac{|z|}{|x|} + \frac{|x|}{|z|}\right) \tag{1}$$

Theorem 6 tells us that $|x|^2 + |y|^2 \ge 2|x||y|$ and since $xy \ne 0$ (because $xyz \ne 0$ by hypothesis), we infer $\frac{|x|^2}{|x||y|} + \frac{|y|^2}{|x||y|} \ge 2$.

In a similar manner we deduce

$$\left.\begin{aligned} \frac{|x|}{|y|} + \frac{|y|}{|x|} \ge 2 \\ \frac{|y|}{|z|} + \frac{|z|}{|y|} \ge 2 \\ \frac{|z|}{|x|} + \frac{|x|}{|z|} \ge 2 \end{aligned}\right\} \tag{2}$$

Adding the three inequalities in (2) memberwise we obtain

$$\frac{|x|}{|y|} + \frac{|y|}{|x|} + \frac{|y|}{|z|} + \frac{|z|}{|y|} + \frac{|z|}{|x|} + \frac{|x|}{|z|} \ge 6 \tag{3}$$

Combining (1) and (3) yields

$\frac{1}{|\alpha\beta\gamma|} \ge 2+2+2+2 = 8 \rightarrow |\alpha\beta\gamma| \le \frac{1}{8}$. Note that the equal sign will hold precisely when $|x| = |y| = |z|$ (which means $|\alpha| = |\beta| = |\gamma| = \frac{1}{2}$.)

(b) We have $\frac{1}{|\alpha|} = \frac{|y|}{|x|} + \frac{|z|}{|x|}$, $\frac{1}{|\beta|} = \frac{|z|}{|y|} + \frac{|x|}{|y|}$, $\frac{1}{|\gamma|} = \frac{|x|}{|z|} + \frac{|y|}{|z|}$, so that $\frac{1}{|\alpha|} + \frac{1}{|\beta|} + \frac{1}{|\gamma|} = \underbrace{\left(\frac{|y|}{|x|} + \frac{|x|}{|y|}\right)}_{\geq 2} + \underbrace{\left(\frac{|z|}{|y|} + \frac{|y|}{|z|}\right)}_{\geq 2} + \underbrace{\left(\frac{|x|}{|z|} + \frac{|z|}{|x|}\right)}_{\geq 2}$; $\frac{1}{|\alpha|} + \frac{1}{|\beta|} + \frac{1}{|\gamma|} \geq 2 + 2 + 2 = 6$

P31. If the nonzero reals α, β, γ satisfy the condition

$$\frac{|\alpha|\beta\gamma| + \beta|\gamma\alpha| + \gamma|\alpha\beta||}{|\alpha\beta\gamma|} = 3,$$

prove that the three reals must have the same sign.

Solution. One method of proof is to show that if we assume that α, β, γ do not have the same sign, then under the given condition, this leads to a contradiction. So, let us assume that α, β, γ do **not** have the same sign.

Case 1: Two among α, β, γ are positive, the third negative. This means three subcases are possible: ($\alpha > 0, \beta > 0,\ \gamma < 0$), or $\alpha > 0,\ \gamma > 0,\ \beta < 0$), or $\alpha < 0,\ \beta > 0,\ \gamma > 0$). Because of the symmetry in the given condition (i.e. if switch the position of any two letters among α, β, γ in the given condition, the condition remains the same), it suffices to only treat one of these three subcases (the other two are treated similarly).

Assume $\alpha > 0,\ \beta > 0,\ \gamma < 0$; thus $|\beta\gamma| = -(\beta\gamma),\ |\gamma\alpha| = -(\gamma\alpha),\ |\alpha\beta| = \alpha\beta$. The given condition implies

$$\frac{|\alpha|\beta\gamma| + \beta|\gamma\alpha| + \gamma|\alpha\beta||}{|\alpha\beta\gamma|} = 3$$

$$\frac{|\alpha(-\beta\gamma) + \beta\cdot(-\gamma\alpha) + \gamma(\alpha\beta)|}{|\alpha\beta\gamma|} = 3 \Rightarrow \frac{|-\alpha\beta\gamma - \alpha\beta\gamma + \alpha\beta\gamma|}{|\alpha\beta\gamma|} = 3 \Rightarrow \frac{|\alpha\beta\gamma|}{|\alpha\beta\gamma|} = 3;\ 1 = 3$$

an obvious contradiction.

Case 2. Two among α, β, γ are negative, the third positive. There are three subcases: ($\alpha < 0,\ \beta < 0,\ \gamma > 0$) or ($\alpha < 0,\ \gamma < 0,\ \beta > 0$) or ($\beta < 0, \gamma < 0, \alpha > 0$). Again, it suffices to treat one of these three subcases. Assume that $\alpha < 0,\ \beta < 0,\ \gamma > 0$. We then have $|\beta\gamma| = -\beta\gamma,\ |\gamma\alpha| = -\gamma\alpha,\ |\alpha\beta| = \alpha\beta$. The given condition implies,

$$\frac{|\alpha(-\beta\gamma) + \beta(-\gamma\alpha) + \gamma(\alpha\beta)|}{|\alpha\beta\gamma|} = 3 \Rightarrow \frac{|-\alpha\beta\gamma - \alpha\beta\gamma + \alpha\beta\gamma|}{|\alpha\beta\gamma|} = 3 \Rightarrow 1 = 3,$$

same contradiction as before.

P32(a). Prove that the equal sign in the triangle inequality

$$|a + b| \leq |a| + |b|$$

occurs precisely when $ab \geq 0$. In other words it occurs if and only if a and b have the same sign or when one of the zero.

(b) Prove that the equal sign in the triangle inequality,

$$||a| - |b|| \leq |a - b|$$

also occurs when a and b have the same sign or when one of them is zero.

Solution. (a) Suppose that $ab > 0$. Ifa and b are both positive, then $|a| = a$, $|b| = b$, and $a + b > 0 \Rightarrow |a + b| = a + b$. Therefore $|a + b| = a + b = |a| + |b|$. Next, if a and b are both negative and so $|a| = -a$, $|b| = -b$; but also $|a + b| = -(a + b)$, since $a + b < 0$. Therefore $|a + b| = -(a + b) = |a| + |b|$. To continue, if either of a, b is zero then it is obvious that $|a + b| = |a| + |b|$. Finally, if $ab < 0$, the equal sign cannot hold. For if it were

$$\begin{aligned} & |a+b| &=& \ |a| + |b| \\ \Rightarrow\ & |a+b|^2 &=& \ (|a| + |b|)^2; \\ & (a+b)^2 &=& \ (|a| + |b|)^2; \\ & a^2 + 2ab + b^2 &=& \ |a|^2 + |b|^2 + 2|a||b|; \\ & a^2 + 2ab + b^2 &=& \ a^2 + b^2 + 2|ab|; \\ \Rightarrow\ & ab &=& \ |ab| \Rightarrow ab \geq 0 \end{aligned}$$

which shows that $ab \geq 0$ is a necessary condition for the equal sign to hold. It is also sufficient as we have already seen.

(b) Similar proof, left to the student.

P33. Prove that if $a^4 + b^4 + c^4 + d^4 = 4|abcd|$, then $a = b = c = d$.

Solution. We know that $a^4 + b^4 = (a^2) + (b^2)^2 = (a^2 - b^2)^2 + 2a^2b^2$. Likewise $c^4 + d^4 = (c^2 - d^2)^2 + 2c^2d^2$. We have

$$a^4 + b^4 + c^4 + d^4 = 4|abcd|$$

$$\Rightarrow (a^2 - b^2)^2 + (c^2 - d^2)^2 + 2(a^2b^2 + c^2d^2) = 4|abcd| \Rightarrow$$

$$\Rightarrow (a^2 - b^2)^2 + (c^2 - d^2)^2 + 2(a^2b^2 + c^2d^2 - 2|abcd|) = 0; \qquad (1)$$

$$(a^2 - b^2)^2 + (c^2 - d^2)^2 + 2 \cdot [|ab^2 + |cd|^2 - 2|ab||cd|] = 0;$$

$$(a^2 - b^2)^2 + (c^2 - d^2)^2 + 2 \cdot (|ab| - |cd|)^2 = 0$$

The left-hand side of (1) is a sum of three **non-negative** terms. Since the right-hand side of (1) is equal to zero, each of the three terms on the left must be zero:

$$(a^2-b^2)^2 = (c^2-d^2)^2 = 2(|ab|-|cd|)^2 = 0$$

$$\Rightarrow \; a^2-b^2=0,\; c^2-d^2=0,\; |ab|-|cd|=0; \qquad (2)$$

$$\Rightarrow \; (a-b)(a+b)=0,\; (c-d)(c+d)=0,\; |ab|-|cd|=0$$

The first two conditions in (2) say that $a=b$ or $a=-b$ and $c=d$ or $c=-d$. In effect they say that $|a|=|b|$ and $|c|=|d|$; therefore, the third condition in (2) implies

$$|a||b|-|c||d|=0;$$

$$|a||a|-|c||c|=0 \Rightarrow |a|^2-|c|^2=0 \Rightarrow$$

$$\Rightarrow (|a|-|c|)(|a|+|c|)=0;$$

unless both a and c are zero, $|a|+|c|>0$. Thus we conclude $|a=|c|$ (which is also true when both a and c are zero). We have prove that

$$|a|=|b|, \quad |c|=|d|, \text{ and } |a|=|c|.$$

Altogether, $|a|=|b|=|c|=|d|$. The proof is complete.

Note: This problem could be more easily solved by applying the arithmetic mean-geometric mean inequality for $n=4$ which will be established in Chapter 6.

P34. If $\frac{\alpha}{\beta}=\frac{\beta}{\gamma}=\frac{\gamma}{\delta}>0$, prove that $|\alpha-\delta|\geq 3\cdot|\beta-\gamma|$.

Solution. The condition $\frac{\alpha}{\beta}=\frac{\beta}{\gamma}=\frac{\gamma}{\delta}>0$ clearly implies that the real numbers $\alpha,\beta,\gamma,\delta$ must have the same sigh: either all positive or all negative.

From $\frac{\alpha}{\beta}=\frac{\beta}{\gamma}\Rightarrow\alpha=\frac{\beta^2}{\gamma}$ and from $\frac{\beta}{\gamma}=\frac{\gamma}{\delta}\Rightarrow\delta=\frac{\gamma^2}{\beta}$. Consequently,

$$|\alpha-\delta| = \left|\frac{\beta^2}{\gamma}-\frac{\gamma^2}{\beta}\right| = \left|\frac{\beta^2-\gamma^3}{\gamma\beta}\right| \qquad (1)$$

We can apply the identity (Theorem 3(ii)), $\beta^3-\gamma^3=(\beta-\gamma)(\beta^2+\beta\gamma+\gamma^2)$; (1) gives

$$|\alpha-\delta| = \left|\frac{(\beta-\gamma)(\beta^2+\beta\gamma+\gamma^2)}{\beta\gamma}\right|;$$

$$|\alpha-\delta| = |\beta-\gamma|\cdot\left|\frac{\beta^2}{\beta\gamma}+\frac{\beta\gamma}{\beta\gamma}+\frac{\gamma^2}{\beta\gamma}\right|; \qquad (2)$$

$$|\alpha-\delta| = |\beta-\gamma|\cdot\left|\left(\frac{\beta}{\gamma}+\frac{\gamma}{\beta}\right)+1\right|.$$

Since β and γ have the same sign, the product $\beta\gamma$ must be positive, so that the inequality

$\beta^2 + \gamma^2 \geq 2\beta\gamma$ implies $\frac{\beta^2+\gamma^2}{\beta\gamma} \geq 2 \Rightarrow$

$$\Rightarrow \frac{\beta^2}{\beta\gamma} + \frac{\gamma^2}{\beta\gamma} \geq 2 \Rightarrow \frac{\beta}{\gamma} + \frac{\gamma}{\beta} \geq 2 \Rightarrow \frac{\beta}{\gamma} + \frac{\gamma}{\beta} + 1 \geq 3 \Rightarrow \quad (3)$$

$$\Rightarrow |\beta/\gamma + \gamma/\beta + 1| \geq 3.$$

Applying (2) and (3) yields

$$|\alpha - \delta| \geq |\beta - \gamma| \cdot 3;$$

$$|\alpha - \delta| \geq 3 \cdot |\beta - \gamma|$$

the proof is complete.

Note: As we will see in Chapter 22, the condition $\frac{\alpha}{\beta} = \frac{\beta}{\gamma} = \frac{\gamma}{\delta} > 0$ in effect says that the real numbers $\alpha, \beta, \gamma, \delta$ have the same sign and that they are four **consecutive** terms of a geometric progression.

P35. Find all the integers a and b that satisfy the three conditions below:
(i) $b \neq 0$, (ii) $\frac{a}{b}$ is an integer, and
(iii) $4(a^2 + b^2) < 17|ab|$

Solution Since we require $\frac{a}{b}$ to be an integer, let us put $\frac{a}{b} = k$; $a = b \cdot k$, where k is some integer. Substituting for $a = bk$ in condition (iii) gives,

$$4b^2 \cdot (k^2 + 1) < 17 \cdot b^2|k|; \text{ since } b \neq 0, \text{ we have } b^2 > 0.$$

Dividing both sides by b^2 we arrive at

$$4(k^2 + 1) < 17 \cdot |k| \quad (1)$$

Condition (1) shows that any integer k that satisfies (1) must be nonzero. Thus $|k| \geq 1$. Since $|k| > 0$, dividing both sides of (1) by $4|k|$ yields

$$\frac{4(k^2+1)}{4|k|} < \frac{17|k|}{4|k|};$$

$$|k| + \frac{1}{|k|} < \frac{17}{4} \quad (2)$$

A quick calculation shows the only positive integer values of $|k|$ that satisfy (2) are $|k| = 1, 2, 3$. Hence,

$$k = -3, -2, -1, 1, 2, 3. \quad (3)$$

and since $a = bk$ we arrive at the following conclusion: The integers a and b satisfy the problem's requirements if and only if they satisfy precisely one of the following six conditions: $(a=-b), b\neq 0)$ or $(a=-2b,\ b\neq 0)$, or $(a=-3b, b\neq 0)$ or $(a=b, b\neq 0)$ or $(a=2b,\ b\neq 0)$, or $(a=3b,\ b\neq 0)$.

P36. Prove that $|a|<|b|$ if and only if $a^2<b^2$.

Solution. First suppose that $|a|<|b|$. If $a=0$, then $0<|b|$, so $b\neq 0$ which in turn implies $b^2>0$ and we are done since $a^2=0$ in this case. Now suppose that, $0<|a|<|b|\Rightarrow$ (by Property 40) $|a|\cdot|a|<|b|\cdot|b|$. Rewriting $0<|a|<|b|\Rightarrow|a|^2<|b|^2$ by (Theorem 9(iii)); $a^2<b^2$ and we are done. Next, let us prove the converse statement: if $a^2<b^2$, then $|a|<|b|$. The assumption

$$a^2<b^2 \text{ implies } 0<b^2-a^2\Rightarrow 0<|b|^2-|a|^2\Rightarrow 0<(|b|-|a|)\cdot(|b|+|a|) \quad (1)$$

The factor $|b|+|a|$ must be positive since it is always $|b|+|a|\geq 0$; in virtue of $|a|\geq 0$ and $|b|\geq 0$. But $|b|+|a|$ cannot be zero, for this would imply $a=b=0$ violating the assumption $a^2<b^2$. Thus $|b|+|a|>0$. In turn (1) now implies $|b|-|a|>0\Rightarrow|b|>|a|$. The proof is complete.

P37. In each of the following four cases, prove that each of the three given inequalities follows from the other two

(i) $|\alpha+\lambda\beta|<|\lambda\alpha+\beta|,\ |\alpha|<|\beta|,\ |\lambda|<1$

(ii) $|\alpha+\lambda\beta|<|\lambda\alpha+\beta|,\ |\alpha|>|\beta|,\ |\lambda|>1$

(iii) $|\alpha+\lambda\beta|>|\lambda\alpha+\beta|,\ |\alpha|<|\beta|,\ |\lambda|>1$

(iv) $|\alpha+\lambda\beta|>|\lambda\alpha+\beta|,\ |\alpha|>|\beta|,\ |\lambda|<1$

Solution. We only prove (i). The other three can be proved by using the same method described below. Note that

$$\begin{aligned}&|\alpha+\lambda\beta|^2-|\lambda\alpha+\beta|^2=(\alpha+\lambda\beta)^2-(\lambda\alpha+\beta)^2=\alpha^2+2\lambda\alpha\beta+\lambda^2\beta^2-\lambda^2\alpha^2\\&-2\lambda\alpha\beta-\beta^2;\\&|\alpha+\lambda\beta|^2-|\lambda\alpha+\beta|^2=\alpha^2\cdot(1-\lambda^2)-\beta^2\cdot(1-\lambda^2);\\&|\alpha+\lambda\beta|^2-|\lambda\alpha+\beta|^2=(\alpha^2-\beta^2)\cdot(1-\lambda^2)\end{aligned}\quad (1)$$

According to problem P36, the left-hand side of (1) is negative if and only if $|\alpha+\lambda\beta|<|\lambda\alpha+\beta|$. Likewise, $\alpha^2-\beta^2<0\Leftrightarrow|\alpha|<|\beta|$ and $1-\lambda^2>0$ if and only if $|\lambda|<1$. Now consider the three inequalities in (i). If $|\alpha|<|\beta|$ and $|\lambda|<1$ hold true, then the right-hand side of (1) will be negative. Thus so will be the left-hand side of (1) which means that $|\alpha+\lambda\beta|<|\lambda\alpha+\beta|$. Proceeding similarly we see that if $|\alpha+\lambda\beta|<|\lambda\alpha+\beta|$ and $|\alpha|>|\beta|$, then (1) implies $1-\lambda^2>0\Rightarrow|\lambda|<1$. And if $|\alpha+\lambda\beta|<|\lambda\alpha+\beta|$ and $|\lambda|<1$ hold true, then (1) implies $\alpha^2-\beta^2<0\Rightarrow\alpha^2<\beta^2\Rightarrow$ $|\alpha|<|\beta|$.

3.8 Exercise Set

1. If $2(\alpha^2 + \beta^2) = (\alpha + \beta)^2$, then $\alpha = \beta$.

2. If $a + b + c = 2t$, then $(t - a)^2 + (t - b)^2 + (t - c)^2 + t^2 = a^2 + b^2 + c^2$.

3. If the real numbers x satisfies the condition $(a^2+b^2)x^2 - 2b(a+c)x + (b^2+c^2) = 0$, where a, b, c are nonzero real numbers, then $x = \frac{c}{b} = \frac{b}{a}$.

4. Prove that if the real numbers a, b, c satisfy $3(a^2 + b^2 + c^2) = (a + b + c)^2$, then $a = b = c$.

5. If $(\alpha + \beta + \gamma)^2 = 3(\alpha\beta + \beta\gamma + \gamma\alpha)$, then $\alpha = \beta = \gamma$.

6. If $a + b + c = 0$, then $(a^2 + b^2 + c^2)^2 = 2(a^4 + b^4 + c^4) = 4(a^2b^2 + b^2c^2 + c^2a^2)$. **Hint**: First show that $(a + b + c)(a - b + c)(a + b - c)(a - b - c) = (a^2 + b^2 + c^2)^2 - 4(a^2b^2 + a^2c^2 + b^2c^2)$.

7. If $\alpha+\beta+\gamma = 2\tau$, prove that $\alpha(\beta-\gamma)(\tau-\alpha)^2+\beta(\gamma-\alpha)(\tau-\beta)^2+\gamma(\alpha-\beta)(\tau-\gamma)^2 = 0$.

8. Show that $(x^2 + xy + y^2)^2 - 4xy(x^2 + y^2) = (x^2 - xy + y^2)^2$.

9. Prove $(x^2 + y^2 + z^2 + w^2)^2 = (x^2 + y^2 - z^2 - w^2)^2 + (xz + yw)^2 + 4(xw - yz)^2$.

10. Suppose that the nonzero real numbers a, b, c, x, y, z, satisfy the condition, $(a^2 + b^2 + c^2) \cdot (x^2 + y^2 + z^2) = (ax + by + cz)^2$. Show that x, a, y, b must form an analogy and y, b, z, c must also form an analogy. In other words $\frac{x}{a} = \frac{y}{b} = \frac{z}{c}$.

11. Suppose that two real numbers have sum a and product b. Find, (a) The sum of their squares.
(b) The sum of their cubes.
(c) The square of their difference.

12. If $a + b + c = 0$, then $a^3 + b^3 + c^3 = 3abc$.

13. Show that if the sum of the cubes of three real numbers is equal to three times their product, then their sum must be equal to zero or the three numbers must be equal.

14. Prove that $(y - z)^3 + (z - x)^3 + (x - y)^3 = 3(y - z)(z - x)(x - y)$.

15. If $\alpha + \beta + \gamma = 3x$, then $(x - \alpha)^3 + (x - \beta)^3 + (x - \gamma)^3 = 3(x - \alpha)(x - \beta)(x - \gamma)$
Hint: Consider the sum of $x - \alpha$, $x - \beta$, and $x - \gamma$.

16. If the reals a, b, c satisfy the condition $a + b + c = 0$, show each of the following:

    (i) $(2a - b)^3 + (2b - c)^3 + (2c - a)^3 = 3(2a - b)(2b - c)(2c - a)$
    (ii) $(a + b - c)^3 + (b + c - a)^3 + (c + a - b)^3 = 3(a + b - c)(b + c - a)(c + a - b)$

(iii) $(ay+bx)^3+(ax+cy)^3+(cx+by)^3=3(ax+cy)(ay+bx)(cx+by)$

17. If $x+y+z=2a$, $xy+yz+zx=a^2$, and $xyz=a^3$, then $x^3+y^3+z^3=5a^3$.
Hint: Use Cauchy's Theorem.

18. If $x+y+z=\alpha, x^2+y^2+z^2=\beta^2$, $x^3+y^3+z^3-3xyz=\gamma^3$, then $3\gamma^3=2\alpha\beta^2-\alpha^3$.

19. If $x=\beta+\gamma$, $y=\gamma+\alpha$, $z=\alpha+\beta$, then $x^2+y^2+z^2-yz-zx-xy=\alpha^2+\beta^2+\gamma^2-\beta\gamma-\gamma\alpha-\alpha\beta$.

20. If $\frac{a}{b}=\frac{2}{3}$, then show

(i) $\frac{12a-2b}{3a+2b}=\frac{3}{2}$

(ii) $\frac{9ab-2b^2}{9a^2+5b^2}=\frac{4}{9}$

21. Suppose the nonzero reals a,b,c,d form an analogy. Prove each of the following

(i) $\frac{a^2+2ab+3b^2}{c^2+2cd+3d^2}=\frac{b(a-3b)}{d(c-3d)}$

(ii) $\frac{(a+c)(b+d)}{a+b+c+d}=\frac{ab}{a+b}+\frac{cd}{c+d}$

22. Suppose that $b^2=ac$, with a,b,c nonzero real numbers. Show that

(i) $a^2-b^2+c^2=b^4\left(\frac{1}{a^2}-\frac{1}{b^2}+\frac{1}{c^2}\right)$.

(ii) $\frac{a}{c}=\frac{(a+b)^2}{(b+c)^2}$

23. If $(\alpha+\beta+\gamma+\delta)(\alpha-\beta-\gamma+\delta)=(\alpha-\beta+\gamma-\delta)(\alpha+\beta-\gamma-\delta)$, then $\alpha\delta=\beta\gamma$.

24. If the nonzero reals a,b,x,y satisfy the condition $\frac{a^2+ax+x^2}{a^2-ax+x^2}=\frac{b^2+by+y^2}{b^2-by+y^2}$, show that they must satisfy the condition $\frac{x}{y}=\frac{a}{b}$; or the condition $\frac{b}{x}=\frac{y}{a}$.

25. If $\frac{a}{x}=\frac{b}{y}=\frac{c}{z}$, then $\frac{a^2+b^2+c^2}{x^2+y^2+z^2}=\frac{(a+b+c)^2}{(x+y+z)^2}$.

26. If $\frac{a}{b}=\frac{b}{c}=\frac{c}{d}$, then $(a^2+b^2+c^2)(b^2+c^2+d^2)=(ab+bc+cd)^2$.

27. If $\frac{a}{b}=\frac{b}{c}=\frac{c}{d}$, prove the following:
(a) $\frac{a^3+b^3+c^3}{b^3+c^3+d^3}=\frac{a}{d}$, and (b) $(b+d)\cdot c^3=ad(c^2+d^2)$.

28. If $\frac{a}{b}=\frac{b}{c}=\frac{c}{d}$, then $\left(\frac{a+bc}{d+1}\right)^2=\frac{abc}{d}$.

29. If $\frac{\mu}{x}=\frac{\nu}{y}=\frac{\rho}{z}$ and $\frac{x^2}{a^2}+\frac{y^2}{b^2}+\frac{z^2}{c^2}=1$, then $\frac{\mu^2}{a^2}+\frac{\nu^2}{b^2}+\frac{\rho^2}{c^2}=\frac{\mu^2+\nu^2+\rho^2}{x^2+y^2+z^2}$.

30. Suppose that the reals a,b,c,d form an analogy and that the reals x,y,z,w also form an analogy. Prove that in order for the condition $\frac{a+x}{b+y}=\frac{c+z}{d+w}$ to hold true, it is necessary and sufficient (i.e. if and only if) that either of the conditions hold true: $\frac{a}{c}=\frac{z}{x}$ or $\frac{a}{b}=\frac{z}{w}$.

31. If $\frac{bz-cy}{a} = \frac{cx-az}{b} = \frac{ay-bx}{c}$, then $\frac{x}{a} = \frac{y}{b} = \frac{c}{z}$.

32. If $\alpha^2 = x_1^2 + x_2^2 + x_3^2$, $\beta^2 = y_1^2 + y_2^2 + y_3^2$, $\alpha\beta = x_1y_1 + x_2y_2 + x_3y_3$, and $y_1y_2y_3 \neq 0$, then $\frac{\alpha}{\beta} = \frac{x1}{y_1} = \frac{x_2}{y_2} = \frac{x_3}{y_3}$.

Hint: Use Lagrange's Theorem.

33. If $\frac{y-z}{y+z} = \alpha$, $\frac{z-x}{z+x} = \beta$, $\frac{x-y}{x+y} = \gamma$, then $\alpha + \beta + \gamma + \alpha\beta\gamma = 0$.

34. If $\alpha < \beta$, then $\alpha < \frac{\alpha+\beta}{2} < \beta$.

35. If $\alpha < \beta$ and λ, μ are positive reals whose sum is 1, $\lambda + \mu = 1$, then $\alpha < \lambda\alpha + \mu \cdot \beta < \beta$. Moreover, show that this is false if one of λ, μ is negative (with sum 1), by giving a counterexample.

36. If $0 < a < c$ and $0 < b < d$, then $a - \frac{1}{b} < c - \frac{1}{d}$.

37. If $a > 0,\ b > 0,\ a^2 - 4b > 0,\ 2c > a$, show that $ac > 2b$ and $c > \frac{a}{4} + \frac{b}{a}$.

38. If $a < 0,\ b < 0$, and $a^2 > b^2$, then $a < b$.

39. If $\frac{a}{b} < 1$ and $0 < c < b$, then $\frac{a-c}{b-c} < \frac{a}{b}$.

40. Prove the following: (a) $\left(\frac{a+b}{2}\right)^2 \geq ab$, (b) $a^2 \pm a + 1 > 0$.

41. If $a^2 + b^2 = 1$ and $x^2 + y^2 = 1$, then $ax + by \leq 1$.

42. Show that if $ab > 0$, then $\frac{a}{b} + \frac{b}{a} \geq 2$ and if $ab < 0$, then $\frac{a}{b} + \frac{b}{a} \leq -2$.

43. Show that $3(1 + a^2 + a^4) \geq (1 + a + a^2)^2$.

44. If $a > b > 0$, then (a) $a^3 - b^3 \geq (a-b)^3$ and (b) $a^2 + 2b^3 \geq 3ab^2$.

45. Show that if $a \geq b > 0$, then $a^3 + b^3 \geq ab(a+b)$.

46. Assume a, b to be positive reals. Prove that if a, b satisfy the inequality $\frac{a-b}{a+b} < \frac{a^2-b^2}{a^2+b^2}$, then $b < a$. Conversely, show that if $0 < b < a$, then $\frac{a-b}{a+b} < \frac{a^2-b^2}{a^2+b^2}$.

47. Prove that if a, b, c, d are positive reals, then they must satisfy the inequality $\frac{(a+b)cd}{ad+bc} \leq \frac{ac+bd}{a+b}$.

48. Prove each of the following inequalities:

(i) $(\alpha - \beta)(\alpha - \gamma) + (\beta - \gamma)(\beta - \alpha) + (\gamma - \alpha)(\gamma - \beta) \geq 0$

(ii) $(\alpha + \beta)(\alpha - \gamma) + (\beta + \gamma)(\beta - \alpha) + (\gamma + \alpha)(\gamma - \beta) \geq 0$

49. If the real numbers a, b, c, x, y, z satisfy the conditions $y^2 + zy + z^2 = a^2$, $z^2 + zx + x^2 = b^2$, $x^2 + xy + y^2 = c^2$, and $xy + yz + zx = 0$ prove that they must also satisfy (at least) one of the following four conditions: $a + b + c = 0$ or $a - b + c = 0$ or $a + b - c = 0$ or $a - b - c = 0$.

Hint: Show that the product $(a+b+c)(a-b+c)(a+b-c)(a-b-c)$ is equal to zero. To accomplish this, use the difference of two squares identity to show that this product is equal to

$$a^4 + b^4 + c^4 - 2(a^2b^2 + a^2c^2 + b^2c^2) = (a^2 - b^2 - c^2)^2 - 4b^2c^2$$

Then apply the given conditions in the hypothesis of the problem.

Chapter 4

Integers and their Basic Properties

4.1 Mathematical Induction

As we begin this Chapter, we will need to make a reference to the previous Chapter, Chapter 3. Recall from that Chapter, that the set of real numbers is essentially unique, in the sense that alls sets equipped with an addition operation, a multiplication operation, and a total ordering, and which satisfy all sixteen axioms (that the real numbers do), must be isomorphic: that is, by relabeling any one of them, one simply obtains the set of real numbers. In that sense, the set of natural numbers (the subset of the reals, which is additively generated by the identity element (or that is, the multiplicatively neutral element)) is unique. Of course one can write an entire diatribe about the true meaning of number one, as existing in a strict platonic sense, in the ideal world, independently of a sentient's being senses. That would be a philosophical issue for thought, debate, and discussion; not a mathematical one.

We now proceed with the set of natural numbers, usually denoted by $\mathbb{N}$. The set of natural numbers $\mathbb{N}$ (as a subset of the set of reals $\mathbb{R}$, equipped with the usual operations of addition and multiplication) satisfies the following three conditions:

Condition 1: $1 \in \mathbb{N}$ (that is, 1 is a natural number).

Condition 2: If $n \in \mathbb{N}$, then $n+1 \in \mathbb{N}$ (that is, if n is a natural number, so is $n+1$).

Condition 3: (The inductive axiom): Every subset S of $\mathbb{N}$, which contains 1 and satisfies condition 2, must equal $\mathbb{N}$.

Definition 1: If to every natural number n, there corresponds one and only one statement (denoted by) $P(n)$ (sometimes also symbolized by P_n), then and only then we say that the statements $P(1)$, $P(2)$, ..., $P(n), \ldots$, constitute a **sequence of statements**, also denoted by $\{P(n)\}$; the statement $P(n)$ is called the **nth statement**.

THE PRINCIPLE OF MATHEMATICAL INDUCTION:

Let $P(1), P(2), \ldots,\ P(n), \ldots$, be a sequence of statements.

If (i) $P(1)$ is a true statement and (ii) whenever $P(k)$ is assumed to hold true, then this implies that $P(k+1)$ is also a true statement, $k \in \mathbb{N}$. Then $P(n)$ is a true *statement for every $n \in \mathbb{N}$.*

Observe that the Principle of Mathematical Induction directly follows from **Condition 3** above. Indeed if a sequence of statements $P(1), P(2), \ldots, P(n), \ldots$, satisfies (i) and (ii), then let S be the subset of $\mathbb{N}$, consisting of those natural numbers n for which $P(n)$ is a true statement.

Since S contains 1, it follows that $P(1)$ is a true statement, hence (i) is satisfied. Since S satisfies **Condition 2**, it follows that whenever $P(k)$ is true, then so is $P(k+1)$; thus (ii) is also satisfied.

Remark 1: In mathematical induction, the assumption that $P(k)$ holds true (which is then used to show that $P(k+1)$ also holds true), is often called the **inductive hypothesis or assumption**.

Remark 2: It is important to realize that statement (ii) in the Principle of Mathematical Induction, by itself is NOT sufficient to establish that the statement $P(n)$ is true for every $n \in \mathbb{N}$. The first condition (i) is needed as well. For example consider the statement

$$P(n): \quad 2+4+6+\ldots+2n = n(n+1)-1, \quad \forall n \in \mathbb{N}$$

(Note that the left-hand side of the statement is simply the sum of the first n even numbers.) Note that $P(1)$ clearly fails since $2 \neq 2-1 = 1$. In fact $P(n)$ is false for every natural number n. This can be seen at a glance by the mere observation that the number $n(n+1)$ is even for every $n \in \mathbb{N}$ (because n and $n+1$ are consecutive integers, one has to be even, hence their product must be even); and thus $n(n+1)-1$ must always be odd; but $2+4+6+\ldots+2n$ is always even since it is a sum of even numbers. On the other hand, the assumption that $P(k)$ is true, $k \in \mathbb{N}$, always implies that $P(k+1)$ is true as well. Indeed, assume that $P(k)$ holds.

$$\begin{aligned} P(k): \quad & 2+\ldots+2k = k(k+1)-1 \\ & \Rightarrow 2+\ldots+2k+2(k+1) = (k+1)(k+2)-1; \text{ which is } P(k+1). \end{aligned}$$

So we see that $P(k)$ implies $P(k+1)$, even though, as explained above, $P(n)$ is false for all $n \in \mathbb{N}$. We will see in Theorem 1, in fact,

$$2+4+6+\ldots+2n = n(n+1).$$

Below, we offer an example of mathematical induction.

Example 1: Suppose that $a_1, a_2, \ldots, a_n, \ldots$ is a sequence of real numbers such that $a_1 > 1$ and $a_{n+1} = \frac{1}{2}\left(a_n + \frac{1}{a_n}\right)$, for every natural number n. Show that for every natural number the following holds:

$$1 < a_{n+1} < a_n \quad \text{and} \quad \frac{a_n - 1}{a_n + 1} = \left(\frac{a_1 - 1}{a_1 + 1}\right)^{2^{n-1}}.$$

(a) First we show that $1 < a_n$, for every natural n. $P(n): 1 < a_n$ for every natural number n. $P(1)$ holds true since $1 < a_1$, by hypothesis. Now assume that $P(k)$ holds true. We will show that $P(k)$ implies $P(k+1)$. So assume $1 < a_k$ and apply the formula

$$a_{k+1} = \frac{1}{2}\left(a_k + \frac{1}{a_k}\right).$$

In view of $0 < 1$ and $1 < a_k$, it follows that $0 < a_k$. Also, since $a_k \neq 1$ we have $(a_k - 1)^2 > 0 \Rightarrow a_k^2 - 2a_k + 1 > 0 \Rightarrow a_k^2 + 1 > 2a_k$; but $0 < \frac{1}{a_k}$, so that $a_k^2 + 1 > 2a_k \Rightarrow \frac{1}{2a_k} \cdot (a_k^2 + 1) > \frac{2a_k}{2a_k} \Rightarrow \frac{1}{2} \cdot \left(a_k + \frac{1}{a_k}\right) > 1 \Rightarrow a_{k+1} > 1$, which is the statement $P(k+1)$.

(b) Next we show that $a_{n+1} < a_n$, for every natural number n. We have, observed that

$$a_{n+1} - a_n = \frac{1}{2}\left(a_n + \frac{1}{a_n}\right) - a_n = \frac{1}{2a_n}\left(a_n^2 + 1 - 2a_n^2\right) = \frac{1 - a_n^2}{2a_n} < 0,$$

since by part (a) we have shown that $1 < a_n$ and consequently $1 - a_n^2 < 0$. Thus $a_{n+1} - a_n < 0 \Rightarrow a_{n+1} < a_n$ and we are done.

(c) $P(n) : \frac{a_n - 1}{a_n + 1} = \left(\frac{a_1 - 1}{a_1 + 1}\right)^{2^{n-1}}$, for every natural number n. $P(1) : \frac{a_1 - 1}{a_1 + 1} = \left(\frac{a_1 - 1}{a_1 + 1}\right)^{2^0}$, which is true since $2^0 = 1$. Assume that $P(k)$ holds: $\frac{a_k - 1}{a_k + 1} = \left(\frac{a_1 - 1}{a_1 + 1}\right)^{2^{k-1}}$. Now consider $\frac{a_{k+1} - 1}{a_{k+1} + 1}$ and apply $a_{k+1} = \frac{1}{2}\left(a_k + \frac{1}{a_k}\right)$:

$$\frac{a_{k+1} - 1}{a_{k+1} - 1} = \frac{\frac{1}{2}\left(a_k + \frac{1}{a_k}\right) - 1}{\frac{1}{2}\left(a_k + \frac{1}{a_k}\right) + 1} = \frac{a_k^2 + 1 - 2a_k}{a_k^2 + 1 + 2a_k} =$$

$$= \frac{(a_k - 1)^2}{(a_k + 1)^2} = \left(\frac{a_k - 1}{a_k + 1}\right)^2 = \text{by the inductive hypothesis}$$

$$= \left[\left(\frac{a_1 - 1}{a_1 + 1}\right)^{2^{k-1}}\right]^2 = \left(\frac{a_1 - 1}{a_1 + 1}\right)^{2^k}$$

and the induction process is complete. ■

We now prove the first Theorem in this Section. Undoubtedly the student will be familiar with parts of this theorem. Historically, it was Frederick Gauss who first proved the first part, at the age of ten. Instead of induction, he did this by simply observing that if you look at the first n numbers; $1, 2, \ldots, n$; the sum of the fist and last is $n + 1$; so is the sum of the second number and the number before the last: $2 + (n - 1) = n + 1$. Continuing this way, we see that if n is odd, then there is a middle number $\frac{n+1}{2}$; the sum of $\frac{n+1}{2}$ with itself is again $(n+1)$; if n is even, then there are two middle numbers, $\frac{n}{2}$ and $\frac{n}{2} + 1$; their sum is again $(n+1)$. Thus in all cases by grouping the n numbers $1, 2, \ldots, n$ into $\frac{n}{2}$ pairs each having sum $n + 1$, if n is even; or otherwise $\frac{n+1}{2}$ pairs each having sum $n + 1$, (but in this case we must subtract the middle number $\frac{n+1}{2}$ since it is repeated, $\frac{(n+1)}{2}(n+1) - \frac{(n+1)}{2} = \frac{(n+1)}{2}n)$ we see that in all cases the sum $1+2+\ldots+n$ is equal to $\frac{n(n+1)}{2}$. There is a simplier way of doing this:

Simply consider the two rows,

$$\begin{array}{llll} 1 & 2 & \ldots & (n-1) & n \\ n & (n-1) & \ldots & 2 & 1 \end{array}$$

If we add the numbers on each column we always get $n+1$; the total sum of the columns is therefore $n \cdot (n+1)$, since there are n of them. Both rows obviously have the same sum S, one is a rearrangement of the other. Thus $2S = n(n+1) \Rightarrow S = \frac{n(n+1)}{2}$.

Theorem 1 For every natural number n the following holds true

(i) $1+2+\ldots+n = \frac{n(n+1)}{2}$

(ii) $1^2+2^2+\ldots+n^2 = \frac{n(n+1)(2n+1)}{2}$

(iii) $1^3+2^3+\ldots+n^3 = \left[\frac{n(n+1)}{2}\right]^2$

(iv) $1\cdot 2+2\cdot 3+3\cdot 4+\ldots+n\cdot(n+1) = \frac{n(n+1)(n+2)}{3}$

(v) $\frac{1}{1\cdot 2}+\frac{1}{2\cdot 3}+\frac{1}{3\cdot 4}+\ldots+\frac{1}{n(n+1)} = \frac{n.}{n+1}$

(vi) $1\cdot 2\cdot 3+2\cdot 3\cdot 4+\ldots+n(n+1)(n+2) = \frac{n(n+1)(n+2)(n+3)}{4}$

(vii) $\frac{1}{1\cdot 2\cdot 3}+\frac{1}{2\cdot 3\cdot 4}+\ldots+\frac{1}{n(n+1)(n+2)} = \frac{n(n+3)}{4(n+1)(n+2)}$

Proof. (i) By induction; the statement is true for $n=1$, since $1 = \frac{1\cdot(1+1)}{2}$. Assume the statement to hold for $n=k$:

$$1+2+\ldots+k = \frac{k(k+1)}{2} \Rightarrow 1+2+\ldots+k+(k+1) = \frac{k(k+1)}{2}+(k+1) \Rightarrow$$

$$\Rightarrow 1+2\ldots+k+(k+1) = \frac{k(k+1)+2(k+1)}{2} = \frac{(k+1)(k+2)}{2};$$

$$1+2+\ldots+k+(k+1) = \frac{(k+1)[(k+1)+1]}{2};$$

the inductive step is complete.

(ii) Exercise 1

(iii) For $n=1:\ \ 1^3 = \left[\frac{1(1+1)}{2}\right]^3$, which is true.

Assume $1^3+2^3+\ldots+k^3 = \left[\frac{k(k+1)}{2}\right]^2$; add $(k+1)^3$ to both sides:

$$1^3 + 2^3 + \ldots + k^3 + (k+1)^3 = \left[\frac{k(k+1)}{2}\right]^2 + (k+1)^3;$$

$$1^3 + 2^3 + \ldots + k^3 + (k+1)^3 = \frac{k^2 \cdot (k+1)^2 + 4 \cdot (k+1)^3}{4};$$

$$1^3 + 2^3 + \ldots + k^3 + (k+1)^3 = \frac{(k+1)^2 \cdot [k^2 + 4(k+1)]}{4} = \frac{(k+1)^2 \cdot (k+2)^2}{4}$$

$$1^3 + 2^3 + \ldots + k^3 + (k+1)^3 = \left[\frac{(k+1)(k+2)}{2}\right]^2 = \left[\frac{(k+1)((k+1)+1)}{2}\right]^2,$$

and the inductive step is complete.

(iv) Exercise 2.

(v) Exercise 3.

(vi) For $n = 1$: $1 \cdot 2 \cdot 3 = \frac{1(1+1)(1+2)(1+3)}{4}$, which is true. Assume

$$1 \cdot 2 \cdot 3 + \ldots + k \cdot (k+1)(k+2) = \frac{k(k+1)(k+2)(k+3)}{4}$$

$$\Rightarrow 1 \cdot 2 \cdot 3 + \ldots + k(k+1)(k+2) + (k+1)(k+2)(k+3) =$$

$$\frac{k(k+1)(k+2)(k+3) + 4(k+1)(k+2)(k+3)}{4};$$

$$1 \cdot 2 \cdot 3 + \ldots + k(k+1)(k+2) + (k+1)(k+2)(k+3) = \frac{(k+1)(k+2)(k+3)[k+4]}{4};$$

$$1 \cdot 2 \cdot 3 + \ldots + k \cdot (k+1)(k+2) + (k+1)(k+2)(k+3) =$$

$$\frac{(k+1)[(k+1)+1][(k+1)+2][(k+1)+3]}{4};$$

the inductive step is complete.

(vii) Exercise 4.

Remark 3. The sum in part (ii) of Theorem 1, is an example of what is called a telescopic sum. Note that $\frac{1}{n)n+1)} = \frac{1}{n} - \frac{1}{n+1}$. A telescopic sum $(a_1 + \ldots + a_n)$ occurs when there exists a fixed natural number t and a sequence $b_1, \ldots, b_n, b_{n+1}$ such that $a_i = b_{i+1} - b_i$, for all $i = 1, 2, \ldots, n$. We have

$$a_1 + a_2 + \ldots + a_n = (b_2 - b_1) + (b_3 - b_2) + \ldots + (b_n - b_{n-1}) + (b_{n+1} - b_n) = b_{n+1} - b_1.$$

In the case of Theorem 1 (ii), $b_t = \frac{-1}{t}$; for $t = 1, 2, \ldots, n, n+1$. So that

$$\frac{1}{1\cdot 2}+\frac{1}{2\cdot 3}+\ldots+\frac{1}{n(n+1)}=b_{n+1}-b_1=-\frac{1}{n+1}-\left(-\frac{1}{1}\right)=1-\frac{1}{n+1}.$$

Back in Chapter 1, we saw that S is said to have n elements (n a natural number) if, and only if, there is bijection from the set $\{1,2,\ldots,n\}$ onto S. We have the following theorem.

Theorem 2 (i) For every natural number n, $n \geq 1$.

(ii) For every natural number n either $(n-1) \in \mathbb{N}$ or $n-1=0$.

(iii) For every natural number n, every set containing (exactly) n real numbers must have a maximum and a minimum element under the total order $\leq$ of the real numbers $\mathbb{R}$.

(iv) For ever natural number n, $n < n+1$.

(v) If $n \in \mathbb{N}$, then there does not exist a natural number m such that $n < m < n+1$.

(vi) The set $\mathbb{N}$ has no maximum element.

(vii) The set $\mathbb{N}$ is infinite; that is, it is not finite.

(viii) The number 1 is the minimum element of $\mathbb{N}$.

(ix) For every real number a, there exists a natural number N such that $N > a$.

(x) (Eudoxus-Archimedes) If $a, b \in \mathbb{R}$; $a > 0$, $b > 0$, then there exists a natural number N such that $N \cdot a > b$.

(xi) If S is a set of n elements, then every subset X of S is finite and $m \leq n$; where m stands for the number of elements in X.

Proof: (i) We proceed by induction. For $n=1$, $1 \geq 1$ which is true since $1=1$. Assume that the statements holds for $n=k:\ k \geq 1$; then $k+1 \geq 1+1$; but $1+1 > 1$, since $1 > 0$ (and so $1+1>1$); this was proved in Chapter 3. Thus ($k+1 \geq 1+1$ and $1+1>1$) $\Rightarrow k+1>1$ which implies $k+1 \geq 1$. Thus the statement also holds for $n=k+1$. The induction process is complete.

(ii) Since $1-1=0$, the statement holds for $n=1$; assume that the statement holds for $n=k$: which means that we assume the number $(k-1)$ is a natural number. Then by **Conditions 2 and 1** (two of the three conditions that the set $\mathbb{N}$ satisfies – see the beginning of this section), we must have that $(k-1)+1 \in \mathbb{N}$; $k \in \mathbb{N}$. But $k=(k+1)-1$, so the statement also holds for $n=k+1$.

(iii) Let $A = \{a_1, a_2, \ldots, a_n\}$ be a set containing n real numbers. We will demonstrate that A possesses a maximum element M_n and a minimum element m_n. The statement obviously holds true for $n=1$, since in that case $A=\{a_1\}$ and $M_1 = a_1 = m_1$ is both the maximum and minimum element under $\leq$. Assume that the statement holds for $n=k$: that is, any set containing exactly k reals must possess a maximum element M_k and a minimum element m_k. Now let $A_{k+1} = \{a_1, a_2, \ldots, a_k, a_{k+1}\}$ be any set containing $(k+1)$ reals.

Now according to the inductive hypothesis, the set $A_k = \{a_1, a_2, \ldots, a_k\}$ possess a maximum element M_k and a minimum element m_k under $\leq$; which means that $m_k \leq a_i \leq M_k$, for $i=1,2,\ldots,k$.

Now consider the real number a_{k+1}; there are three distinct possibilities: either $a_{k+1} < m_k$ or $m_k \le a_{k+1} \le M_k$, or $M_k < a_{k+1}$. In view of $m_k \le a_i \le M_k$ for $i = 1, 2, \ldots, k$ we have the following: if the first possibility holds, then $m_{k+1} = a_{k+1}$ is the minimum of the set A_{k+1} and $M_{k+1} = M_k$ is its maximum. If $m_k \le a_{k+1} \le M_k$, then $M_k = M_{k+1}$ is the maximum of A_{k+1}; $m_k = m_{k+1}$ is its minimum, because in this case we have $m_k \le a_i \le M_k$, for $i = 1, 2, \ldots, k, k+1$. Finally, if the third possibility holds, namely $M_k < a_{k+1}$, it is clear that $M_{k+1} = a_{k+1}$ is the maximum of the set A_{k+1}; and $m_{k+1} = m_k$ is its minimum. The inductive step is complete.

(iv) Since $0 < 1$, (by Axiom 14 in Chapter 3) we must have $0 + n < 1 + n$, for every natural number n. Thus, $n < n + 1$.

(v) We will argue by contradiction. We will assume the negation (of what we are trying to prove) to hold true and will show that this assumption leads to a contradiction. Indeed suppose that for some $n \in \mathbb{N}$, there exists $m \in \mathbb{N}$, such that $n < m < n + 1$; the opposite $(-n)$ of n exists as a real number, since we know from Chapters 2 and 3, $\mathbb{R}$ is a ring. Thus $n < m \Rightarrow n - n < m - n$; $0 < m - n$. And $m < n + 1 \Rightarrow m - n < 1$. But this constitutes a contradiction since by part (ii) (of this Theorem), we already know that for every $t \in \mathbb{N}$ we must have $t \ge 1$. In our case $t = m - n$. To complete the proof we must show that $(m - n)$ is a natural number. Recall that $n < m$ and $n, m \in \mathbb{N}$. What we need to prove is that

$$(\forall n, m \text{ with } n < m) \Rightarrow (m - n \in \mathbb{N})$$

Later in this Section we will see that this can simply be proven by using what is called double induction. First we show that, $(\forall m \in \mathbb{N}$ such that $1 < m) \Rightarrow (m - 1 \in \mathbb{N})$. This is simply the case $n = 1$. By part (ii) of this Theorem we already know that for every natural number m either $m - 1 \in \mathbb{N}$ or $m - 1 = 0$. It cannot be $m - 1 = 0$; since this would imply $m = 1$ contrary to the assumption $1 < m$. Next we go to the inductive step. Suppose that the statement we want to prove holds for $n = k$:

$$(\text{For } n = k \text{ and for every } m \in \mathbb{N} \text{ with } k < m) \Rightarrow (m - k \in \mathbb{N}).$$

By part (i) of the Theorem, we know that $m - k \ge 1$; $m \ge k + 1$. In particular if $m > k + 1$, then clearly $m \ge k + 1$. Which means that if $m > k + 1$, then $m - k > 1$; but we know that $m - k \in \mathbb{N}$, by the inductive hypothesis; hence by part (ii) of this Theorem, either the number $(m - k) - 1$ is a natural number or otherwise $(m - k) - 1 = 0$; the latter possibility cannot occur though in view of $m - k > 1$. We conclude that $((m - k) - 1) \in \mathbb{N}$. We have thus proven that (for $n = k + 1$ and for every $m \in \mathbb{N}$ such that $k + 1 < m) \Rightarrow (((m - k) - 1) \in \mathbb{N})$; or simply $m - (k + 1) \in \mathbb{N}$. The statement also holds for $n = k + 1$. The proof is s complete.

(vi) If the set $\mathbb{N}$ had a maximum element, say M, then we would have $n \le M$, for every $n \in \mathbb{N}$. In particular, for $n = M + 1$ we would have $M + 1 \le M$, which is impossible in view of part (iv) of this Theorem.

(vii) If $\mathbb{N}$ were a finite set, then according to part (iii), it would have a maximum element, which cannot happen, by part (vi).

(viii) This is an immediate consequence of part (i).

(ix) We will argue by contradiction. Suppose to the contrary that there exists some real number b such that the statement "$n \leq b$, for every $n \in \mathbb{N}$", holds true. Let $S = \{x \in \mathbb{R} \mid n \leq x, \text{ for every } n \in \mathbb{N}\}$. The set S must be nonempty since it contains at least one element, namely the real number b. According to Axiom 16 of the set of real numbers $\mathbb{R}$ (refer to Chapter 3), the set S has a minimum element, let us call it ϵ. Thus, since $\epsilon \in S$, we must also have $n \leq \epsilon$, for every natural number n. Now, according to part (ii), for every $n \in \mathbb{N}$ with $n > 1$, the number $n - 1 = m$ is a natural number; given $n > 1$, clearly $n - 1 = m$ is uniquely determined; and conversely given $m \in \mathbb{N}$, $n = m + 1$ is uniquely determined. Thus by putting $n = m + 1 \leq \epsilon$, we infer that $m \leq \epsilon - 1$, where m can be any natural number. Hence the real number $(\epsilon - 1)$ is also a member of the set S; but $\epsilon - 1 < \epsilon$, in view of $0 < 1$. This contradicts the fact that ϵ is the minimum element of S. We have arrived at a contradiction: therefore the initial assumption must be false. We have shown that given $a \in \mathbb{R}$, there exists $N \in \mathbb{N}$ such that $N > a$.

(x) This is an immediate consequence of part (ix): Apply part (ix) with $\frac{b}{a}$ in place of a: there exists $N \in \mathbb{N}$ such that $N > \frac{b}{a} \Rightarrow$ (since $a > 0$)$N \cdot a > a \cdot \frac{b}{a} \Rightarrow N \cdot a > b$.

(xi) Trivially, the statement holds for $n = 1$, that is for any singleton set $A = \{a_1\}$. suppose that the statement holds true for every set A that contains k elements. We will show inductively that it must also hold for any set containing $(k + 1)$ elements. Let $A = \{a_1, a_2, \ldots, a_k, a_{k+1}\}$ be any set containing $(k + 1)$ elements. According to the inductive hypothesis the statement holds true for the set $A' = \{a_1, a_2, \ldots, a_k\}$ which contains k elements. Let B be a subset of A.

Case 1: If a_{k+1} is not a member of B, then every element of B is among $a_1, a_2, \ldots, a_k$; in other words B is a subset of A' and by the inductive hypothesis B contains m elements, where m is a natural number such that $m \leq k$; but $k < k+1$ by part (iv). Thus $m < k + 1$.

Case 2: If a_{k+1} is a member of B, then $B = \{a_{k+1}\} \cup B'$, where B' is a subset of A'. Again by the inductive hypothesis, B' contains m' elements, where m' is a natural number such that $m' \leq k$; but then the set B contains $(m' + 1)$ elements; and $m' \leq k \Rightarrow m' + 1 \leq k + 1$. We see that in both **Case 1** and **Case 2**, the number of elements that the set B contains, does not exceed the natural number $(k + 1)$. The inductive step is complete. ■

The set of natural numbers $\mathbb{N}$ satisfies the **Well Ordering Property**; we have already come across this concept back in Chapter 1.

The Well Ordering Property: Every nonempty subset of the set of natural numbers $\mathbb{N}$, has a smallest or least element.

In what follows, we will examine the connection between the Well Ordering Property and the Principle of Mathematical Induction. First we introduce the concept of extended or complete mathematical induction.

1. **The Principle of Complete Mathematical Induction**

Let $P(1), P(2), \ldots, P(n), \ldots$ be a sequence of statements.

(a) Suppose that $P(1)$ holds true.

(b) Suppose that whenever the statements $P(1), \ldots, P(k)$ are assumed to hold true, then this implies that the statement $P(k+1)$ is also a true statement.

Then (under (a) and (b)), the statement $P(n)$ holds true for every natural number n.

2. **Its generalization.**

Let r be a fixed natural number and $P(r), P(r+1), \dots, P(n), \dots,$ a sequence of statements.

(a) Suppose that $P(r)$ holds true.

(b) Suppose that whenever the statements $P(r), \dots, P(k)$ (with $k \geq r$) are assumed to hold true, then this implies that the statement $P(k+1)$ is also a true statement.

Then (under (a) and (b)), the statement $P(n)$ holds true for every natural number $n \geq r$.

Theorem 3 The Principle of Mathematical Induction implies the Principle of Complete Mathematical Induction. The converse is also true.

Proof: Exercise 5.

We now proceed to show the connection between the Well Ordering Property and the Principle of Complete Mathematical Induction. This connection is highlighted by the next Theorem. In fact the two are logically equivalent.

Remark 4: As is the case with the Principle of Complete Mathematical Induction, the Principle of mathematical Induction can also be easily generalized by simply replacing the starting value $n = 1$ with $n = r$, where r is some fixed natural number.

Theorem 4 (i) The Principle of Mathematical Induction implies the Well Ordering Property.

(ii) The Well Ordering Property implies the Principle of Complete Mathematical Induction.

Proof (i) Let S be a nonempty subset of the set of natural numbers $\mathbb{N}$. We will prove that S must have a least element. Consider the set,

$$T = \{x \mid x \in \mathbb{N} \text{ such that } x \leq y \text{ for every } y \in S\}$$

The set T consists of those natural numbers, each of which does not exceed any of the members of S. To put it equivalently, no member of T is greater than any of the members of S. The set T is nonempty, since it contains the number 1; this follows immediately from Theorem 2 (i). Now fix a member of S, say $y_0 \in S$. By definition, ever member of T is less than or equal to y_0 : $\forall x \in T,\ x \leq y_0$. Thus T is a subset of the set of natural numbers, $\{1, \dots, y_0\}$, the set containing all the natural numbers from 1 to y_0. By Theorem 2 (xi), it is now clear that the set T has at most y_0 elements. thus T is a finite set and so by Theorem 2 (iii), the set T must have a maximum (and a minimum) element. Let M be the maximum element of T. Since $M \in T$ we must have,

$$M \leq y, \text{ for every } y \in S.$$

We claim that M must be a member of the set S. We argue by contradiction: if M

is not a member of$S$, then $M \neq y$, $\forall y \in S$. But in view of $M \leq y$, for $y \in S$, we would conclude that $M < y$, for every $y \in S$. Clearly y cannot equal 1 (in view of Theorem 2 (i) and the fact that M is a natural number); and since there is no natural number between $(y-1)$ and y (by theorem 2 (v)), the inequality $M < y$ implies that $M \leq y-1$, for every $y \in S$. Thus, $M+1 \leq y$, $\forall y \in S$; which shows by the definition of the set T that the natural number $(M+1)$ is a member of the set T; however $M < M+1$, which violates the assumption that M is the maximum element of T. Therefore, the assumption that M is not a member of S must be false: M is a member of the set S and since $M \leq y$ for every $y \in S$ (because M is also a member of the set T), we conclude that M is the minimum or least element of the set S.

(ii) We will argue by contradiction. If the Principle of Complete Mathematical Induction fails, then there must exist a sequence of statements $P(1), P(2), \ldots, P(n), \ldots$, and a natural number k_0 such that:

(a) $P(1)$ is true.

(b) whenever the statements $P(1), \ldots, P(k)$ are assumed to be true, then this implies that the statement $P(k+1)$ is also true and

(c) The statement $P(k_0)$ is not ture.

Consider the set $T = \{t \in \mathbb{N}$ such that $P(t)$ is not true$\}$. Clearly T is nonempty since it contains the natural number k_0. Therefore by the Well Ordering Property, the set T contains a smallest natural number m. Clearly $m \neq 1$ because $P(1)$ is true according to condition (a). Thus by $m > 1$ (by Theorem 2 (i)) and by Theorem 2 (ii)), $(m-1)$ is a natural number; but $(m-1)$ is not a member of T in view of $m-1 < m$ and the fact that m is the smallest element of T; consequently all the statements $P(1), \ldots, P(m-1)$ must be true; however, then, condition (b) implies that the statement $P[(m-1)+1] = P(m)$ must also be true, contrary to the fact that $P(m)$ is not true in virtue of $m \in T$. We have arrived at a contradiction. The proof is complete. ∎

Example 2: Use the Principle of complete Mathematical Induction to prove that if the sequence of real numbers $a_1, a_2, a_3, \ldots, a_n, \ldots$ satisfies the conditions $a_1 = 1 = a_2$, and $a_{n-1} + a_n = a_{n+1}$, for every natural number $n > 1$, then

$$a_n = \frac{1}{\sqrt{5}}\left[\left(\frac{1+\sqrt{5}}{2}\right)^2 - \left(\frac{1-\sqrt{5}}{2}\right)^n\right],$$

for every natural number n.

Note that by the definition of the given sequence, all of its members are natural numbers: $a_1 = 1$, $a_2 = 1$, $a_3 = 2$, $a_4 = 3$, $a_5 = 5$, $a_6 = 8$, $a_7 = 13$, $a_8 = 21$, $a_9 = 34$, $a_{10} = 55$, $a_{11} = 89$, $a_{12} = 144, \ldots$ The members of this sequence are called the **Fibonacci Numbers**; they have very special and important properties and they have been studied extensively over the last two centuries; books have been published devoted to these numbers and a major mathematical journal, **The Fibonacci Quarterly**, was founded in the early sixties, with the expressed purpose of discoveries of new properties of the Fibonacci and related numbers.

First we show that the given statement is true for $n = 1$:

$$P(1): \qquad a_1 = 1 = \frac{1}{\sqrt{5}}\left[\left(\frac{1+\sqrt{5}}{2}\right) - \left(\frac{1-\sqrt{5}}{2}\right)\right] = \frac{2\sqrt{5}}{\sqrt{5}\cdot 2} = 1,$$

which is true. Because of the way this sequence is defined (with the first two terms being independently defined, and each term, starting with the third term, being the sum of the previous two terms), we need to also show that P(2) is true – the truth of P(2) cannot be established from that of P(1), because the first two terms are independently defined.

$$a_2 = 1 = \frac{1}{\sqrt{5}}\cdot\left[\left(\frac{1+\sqrt{5}}{2}\right)^2 - \left(\frac{1-\sqrt{5}}{2}\right)^2\right];$$

$$a_2 = 1 = \frac{1}{\sqrt{5}}\cdot\left[\frac{(1+2\sqrt{5}+5) - (1-2\sqrt{5}+5)}{4}\right];$$

$$a_2 = 1 = \frac{1}{\sqrt{5}}\left[\frac{4\sqrt{5}}{4}\right]$$

which is true.

We now go to the inductive step. Assume that the statements, $P(1), P(2), \ldots, P(k-1), P(k)$, are all true. We will now show that this assumption implies that the statement $P(k+1)$ is also true. Indeed, according to the inductive hypothesis the statements $P(k-1)$ and $P(k)$ are in particular true:

$$a_{k-1} = \frac{1}{\sqrt{5}}\left[\left(\frac{1+\sqrt{5}}{2}\right)^{k-1} - \left(\frac{1-\sqrt{5}}{2}\right)^{k-1}\right]; \text{ and}$$

$$a_k = \frac{1}{\sqrt{5}}\left[\left(\frac{1+\sqrt{5}}{2}\right)^{k} - \left(\frac{1-\sqrt{5}}{2}\right)^{k}\right].$$

Next, apply the condition that the given sequence satisfies:

$$a_{k+1} = a_{k-1} + a_k \Rightarrow 2^{k+1} \cdot \sqrt{5} \cdot a_{k+1} = 2^{k+1} \cdot \sqrt{5} \cdot (a_{k-1} + a_k) \Rightarrow$$

$$\Rightarrow 2^{k+1} \cdot \sqrt{5} \cdot a_{k+1} = 2^2 \cdot \left(2^{k-1} \cdot \sqrt{5} \cdot a_{k-1}\right) + 2 \cdot (2^k \cdot \sqrt{5} \cdot a_k) \Rightarrow$$

$$\Rightarrow 2^{k+1} \cdot \sqrt{5} \cdot a_{k+1} = 2^2 \cdot \left[(1+\sqrt{5})^{k-1} - (1-\sqrt{5})^{k-1}\right] + 2 \cdot \left[(1+\sqrt{5})^k - (1-\sqrt{5})^k\right]$$

$$\rightarrow 2^{k+1} \cdot \sqrt{5} \cdot a_{k+1} = (1+\sqrt{5})^{k-1} \cdot \left[4 + 2(1+\sqrt{5})\right] - (1-\sqrt{5})^{k-1} \cdot \left[4 + 2(1-\sqrt{5})\right]$$

$$\Rightarrow 2^{k+1} \cdot \sqrt{5} \cdot a_{k+1} = (1+\sqrt{5})^{k-1} \cdot [1 + 2\sqrt{5} + 5] - (1-\sqrt{5})^{k-1} \cdot [1 - 2\sqrt{5} + 5]$$

$$\Rightarrow 2^{k+1} \cdot \sqrt{5} \cdot a_{k+1} = (1+\sqrt{5})^{k-1} \cdot (1+\sqrt{5})^2 - (1-\sqrt{5})^{k-1} \cdot (1-\sqrt{5})^2$$

$$\Rightarrow 2^{k+1} \cdot \sqrt{5} \cdot a_{k+1} = (1+\sqrt{5})^{k+1} - (1-\sqrt{5})^{k+1}$$

$$\Rightarrow a_{k+1} = \frac{1}{\sqrt{5}} \cdot \left[\left(\frac{1+\sqrt{5}}{2}\right)^{k+1} - \left(\frac{1-\sqrt{5}}{2}\right)^{k+1}\right];$$

the inductive step is complete.

Example 3: A double sequence of statements $\{P(m,n)\}$ is a sequence of statements such that for every ordered pair (m,n) of natural numbers m, and n, there corresponds one and only one statement $P(m,n)$. Such a sequence can be pictorially represented by the way of an (infinite) matrix:

$$\begin{array}{lllll} P(1,1) & P(1,2) & P(1,3)\cdots & P(1,n)\cdots \\ P(2,1) & P(2,2) & P(2,3)\cdots & P(2,n)\cdots \\ \vdots\cdots & \cdots & \cdots & \cdots \\ P(m,1) & P(m,2) & P(m,3)\cdots & P(m,n)\cdots \end{array}$$

Not surprisingly, the Principle of Mathematical Induction implies the Principle of Double Mathematical Induction stated below:

Let $\{P(m,n)\}$ be a double sequence of statements. If

(a) The statement $P(1,1)$ is ture

(b) Whenever $P(1,\lambda)$ is assumed to be ture; then this assumption implies $P(1, \lambda + 1)$ also to be true and

(c) Whenever $P(k,\lambda)$ is assumed to be true, then this assumption implies $P(k + 1, \lambda)$ also to be true.

Then (under (a),(b), and (c)), the statement $P(m,n)$ is true for every natural number m and every natural number n. In the proof of Theorem 2 (v), essentially double induction was used in the last part of the proof. In this example, we use double induction to show,

(i) For every $m \in \mathbb{N}$ and every $n \in \mathbb{N}$, such that $m < n$, we must have $m \leq n - 1$.

(ii) If $m \in \mathbb{N}$, $n \in \mathbb{N}$, with $n \geq m$, then either $n - m \in \mathbb{N}$ or $n - m = 0$.

(iii) If $m \in \mathbb{N}$, $n \in \mathbb{N}$, then both numbers $m + n$ and $m \cdot n$ are natural numbers.

Proof: We prove part (ii) first.

(ii) Since $1 - 1 = 0$, the statement $P(1, 1)$ holds true. Suppose that $P(1, \lambda)$ holds true: that is, $\lambda - 1 \in \mathbb{N}$ or $\lambda - 1 = 0$. But then $(\lambda - 1) + 1 \in \mathbb{N}$ as well, by virtue of the fact that the set satisfies **Conditions 1 and 2** (see beginning of this Section); thus, $(\lambda - 1) + 1 = (\lambda + 1) - 1$ is a natural number; thus the statement $P(1, \lambda + 1)$ also holds true. To complete the double induction, assume $P(k, \lambda)$ to hold true: we assume that $\lambda - k$ is a natural number. By Theorem 2 (ii), the number $(\lambda - k) - 1$ is either a natural number or it is zero; if $(\lambda - k) - 1$ is a natural number then $(\lambda - k) - 1 = \lambda - (k + 1)$ is a natural number which shows that the statement $P(k + 1, \lambda)$ holds true. If $(\lambda - k) - 1 = 0$; then $\lambda = k - 1$, which cannot happen under the inductive hypothesis that $\lambda - k \in \mathbb{N}$; which implies, in particular, that $\lambda - k \geq 1$ (by Theorem 2 (i)); $\lambda \geq k + 1 > k > k - 1$). Thus we conclude that $P(k + 1, \lambda)$ holds true. The Double Induction is complete.

Next we prove part (i).

(i) First we show that $P(1, n)$ holds true for every $n \in \mathbb{N}$. $P(1, n)$: "If $1 < n$, for every $n \in \mathbb{N}$, then $1 \leq n - 1$ for every $n \in \mathbb{N}$". Indeed: From $n > 1$ it follows by Theorem 1 (ii) that the number $(n - 1)$ is a natural number and hence by Theorem 2 (i), we must have $n - 1 \geq 1$.

Next, we assume that $P(m, n)$ to hold true, and we show that this assumption implies $P(m + 1, n)$ also to be true. $P(m, n)$: "If $m < n$, then $m \leq n - 1$". We will show that "If $m + 1 < n$, then $m + 1 \leq n - 1$". Indeed if $m + 1 < n$, then $m < n - 1$; but then under the assumption of $P(m, n)$ (replace n with $n - 1$) it follows that $m \leq (n - 1) - 1$; which in turn implies $m + 1 \leq n - 1$, which establishes $P(m + 1, n)$. The induction process is complete.

(iii) The proof of this part is left to the student, it is in fact listed as Exercise 6 in the exercise set of this Chapter. Back in Chapter 2, we dealt with the ring of integers $\mathbb{Z}$ and the field of rationals $\mathbb{Q}$. We mention here that $\mathbb{Z}$ is the natural ring extension of the natural numbers $\mathbb{N}$; under the usual operations of real number addition and multiplication, the ring $\mathbb{Z}$ is the "smallest" ring containing $\mathbb{N}$; in the sense that any other ring (under the same operations) containing $\mathbb{N}$, must also contain the integers $\mathbb{Z}$. Likewise, in the same sense, the field of rational numbers $\mathbb{Q}$, is the "smallest" field containing the ring $\mathbb{Z}$. We finish this Section with the Theorem stated below.

Theorem 5 (i) For every real number a, there exists a unique integer n such that $n \leq a < n + 1$.

(ii) If $a, b \in \mathbb{Q}$, $a < b$, then there exists $r \in \mathbb{Q}$, such that $a < r < b$.

(iii) If a, b are real numbers, $a < b$, then there exists a rational number r such that $a < r < b$.

Proof: (i) Let a be a real number. By theorem 2 (ix), there exists a natural number N such that $N > -a$; $N + a > 0$. By the Well Ordering Property, if S is the set consisting of those natural numbers each of which is greater than $N + a$, then S has a smallest element m; that is, m is the smallest natural number greater than $N + a$.

Moreover, by Theorem 2 (ii), the number $(m-1)$ is either a natural number or it is equal to zero. If $(m-1) \in \mathbb{N}$, then in virtue of the fact that m is the smallest natural number exceeding $(N+a)$, it follows that $m-1 \leq N+a$. Thus, if $m-1 \in \mathbb{N}$ we must have $m-1 \leq N+a < m \Rightarrow m-1-N \leq a < m-N$; $(m-N)-1 \leq a < m-N$; therefore the integer $n = m-N-1$ satisfies $n \leq a < n+1$, since $n+1 = m-N$. To show the uniqueness of the integer n, suppose that n' is some integer that also satisfies $n' \leq a < n'+1$; but then from $n \leq a < n+1$ we conclude that $n < n'+1$ and $n' < n+1$. We point out here that theorem 2 (v) can easily be extended to the set of integers $\mathbb{Z}$; there is no integer t strictly between k and $k+1$ (with $k \in \mathbb{Z}$); that is, there is no $t \in \mathbb{Z}$ such that $k < t < k+1$ (the student may want to fill in the details); thus the inequality $n < n'+1$ implies $n \leq n'$ and the inequality $n' < n+1$ likewise implies $n' \leq n$. So that ($n \leq n'$ and $n' \leq n$) imply $n' = n$.

Finally, if we go back and assume that $m-1=0$; $m=1$, then the number 1 would be the smallest natural number exceeding $N+a$; $1 > N+a$; but we also know that $0 < N+a$; hence $0 < N+a < 1$, which implies $-N < a < 1-N$; the integer $n = -N$ satisfies $n \leq a < n+1$ (because it satisfies $n < a < n+1$ in particular); the uniqueness of n is proved in a similar way as in the previous case.

(ii) This is done very easily, simply by taking $r = \frac{a+b}{2}$; The inequalities $a < \frac{a+b}{2} < b$ hold true, as it is easily verified, in view of $a < b$.

(iii) Since $a < b$, we have $0 < b-a$; by Theorem 2 (x), there exist a natural number N such that,

$$N \cdot (b-a) > 1$$

(Here we apply Theorem 2 (x) by taking $(b-a)$ in place of a and 1 in place of b.) Thus,

$$Nb > 1 + Na$$

By part (i), there exists a (unique) integer m such that $m-1 \leq Na < m \Rightarrow Na < m \leq Na+1$; and since $Na+1 < Nb$, we conclude that $Na < m < Nb \Rightarrow$ (dividing across by $N > 0$) $a < \frac{m}{N} < b$, which proves that the rational number $r = \frac{m}{N}$ lies strictly between a and b. The proof is complete. ■

We close this Section by pointing out that by using Theorem 2 (parts (i) - (ii) and (iv)-(viii)) one can prove the following:

(a) The set of integers $\mathbb{Z}$ is infinite.

(b) $\mathbb{Z}$ has neither a maximum nor a minimum element.

(c) Every subset of $\mathbb{Z}$ that is bounded above, must have a maximum element.

(d) Every subset of $\mathbb{Z}$ that is bounded below, must have a minimum element.

(e) For every $n \in \mathbb{Z}$, there is no integer k such that $n < k < n+1$.

4.2 Divisors and their Properties

Definition 2: Let a, b be integers with $b \neq 0$. We say that the nonzero integer b **divides** the integer a, if there exists another **integer c such that $a = b \cdot c$**. If that is the case we can say that the integer a is **divisible** by the integer b; we can also say that the integer a is a **multiple** of the integer b. The integer b is said to be a **divisor** of a.

The concent of divisibility is one that the student is undoubtedly familiar from his/her earlier mathematics education. The concept appears to be simple, but it has profound implications in the field of number theory, one of the branches of mathematics. As an example, both integers 12 and 14 are divisors of the integer 168 since $(12) \cdot (14) = 168$.

Notation: If the nonzero integer b divides a, we sometimes use the notation $\boldsymbol{b|a}$ and if b does not divide a, we can use the notation $\boldsymbol{b|a}$.

Theorem 6 (i) If $b|a$, then there exists a unique integer c such that $a = b \cdot c$.

(ii) If $b|a$, then $b|(-a), (-b|a, (-b)|(-a)$, and $|b|||a|$

(iii) $\pm 1|a$ and $a|a$.

(iv) (Transitivity of Divisibility) If $a|b$ and $b|c$, then $a|c$.

(v) if $a|b$ and $b|a$, then $|a| = |b|$; and conversely.

(vi) If $b \cdot c|a \cdot c$, then $b|a$; and conversely: if $b|a$, then $b \cdot c|a \cdot c$.

(vii) If $a|(b + c)$ and $a|b$, then $a|c$.

(viii) If $a|b$, then $a|k \cdot b$, for any integer k.

(ix) If $a|b_1, a|b_2, \dots, a|b_n$, then $a|(k_1b_1+k_2b_2+\dots+k_nb_n)$, for any integers $k_1, k_2, \dots, k_n$.

(x) The set of divisors of an integer nonzero a is finite. (Note: Obviously, every integer is a divisor of the zero integer 0.)

Proof: (i) Suppose $c_1, c_2 \in \mathbb{Z}$ such that $a = b \cdot c_1 = b \cdot c_2$. Since $b \neq 0$, $b \cdot c_1 = b \cdot c_2$ obviously implies $c_1 = c_2$.

(ii) If $b|a$, then $a = b \cdot c$, for a unique integer c. From this it follows $(-a) = (-b) \cdot c = b \cdot (-c)$ and $|a| = |b| \cdot |c|$. According to **Definition 2**, it is now clear that $b|(-a), (-b)|a, (-b)|(-a)$, and $|b|||a|$.

(iii) $a = 1 \cdot a$ and $(-a) = (-1) \cdot a$, which proves that $a|a$, and $\pm 1|a$.

(iv) From $a|b$, we have $b = a \cdot k_1$, for $k_1 \in \mathbb{Z}$. From $b|c$ we have $c = b \cdot k_2$, for $k_2 \in \mathbb{Z}$. Hence $c = (a \cdot k_1) \cdot k_2 = a \cdot (k_1k_2)$, which (since $k_1, k_2 \in \mathbb{Z}$) proves that $a|c$.

(v) From $a|b$ we have $b = ac_1$, for $c_1 \in \mathbb{Z}$. From $b|a$ we have $A = b \cdot c_2$, for $c_2 \in \mathbb{Z}$. Thus, $a = (a \cdot c_1) \cdot c_2 \Rightarrow a = a \cdot (c_1 \cdot c_2)$. But both a and b must be nonzero since $a|b$ and $b|a$ (according to **Definition 2**). Hence from $a = a \cdot (c_1 \cdot c_2)$ we conclude $1 = c_1 \cdot c_2 \Rightarrow | = |c_1| \cdot |c_2|$; since $|c_1|, |c_2|$ are positive integers or natural numbers, we must have, $|c_1| \geq 1$ and $|c_2| \geq 1$; this in conjunction with $|c_1| \cdot |c_2| = 1$ implies $|c_1| = |c_2| = 1$. But $c_1 \cdot c_2 = 1$, which means that c_1, c_2 are either both positive or both negative. We conclude that either $c_1 = c_2 = 1$ or $c_1 = c_2 = -1$. Hence, either $a = b$ or $a = -b$; all together we have $|a| = |b|$.

Conversely if $|a| = |b|$, we must have either $a = b = 1 \cdot b$ or $a = -b = (-1) \cdot b$; in either case, $b|a$.

(vi) Suppose that $b \cdot c | a \cdot c$; then we have $a \cdot c = (b \cdot c) \cdot k$, for some $k \in \mathbb{Z}$. But $b \cdot c \neq 0$, by **Definition 2**; thus $c \neq 0$ in particular. From $a \cdot c = (b \cdot c) \cdot k \Rightarrow a \cdot c = (b \cdot k) \cdot c$ (by the commutativity and associativity of multiplication) which, by virtue of $c \neq 0$, implies $a = b \cdot k$; $b|a$. The converse is proved as easily, it is left to the student.

(vii) From $a|b + c$ we have $b + c = k_1 \cdot a$, for some $k_1 \in \mathbb{Z}$; and from $a|b$ we have $b = a \cdot k_2$, for $k_2 \in \mathbb{Z}$. Thus, $a \cdot k_2 + c = k_1 \cdot a \Rightarrow c = a \cdot (k_1 - k_2)$, which proves (since $(k_1 - k_2) \in \mathbb{Z}$) that $a|c$.

(viii) From $a|b \Rightarrow b = a \cdot c$, for some $c \in \mathbb{Z}$. Thus, $k \cdot b = k \cdot (a \cdot c) = a \cdot (k \cdot c)$, which proves that $a|k \cdot b$.

(ix) From $a|b_1, a|b_2, \ldots, a|b_n$ we have the n equations $b_1 = a \cdot c_1$, $b_2 = a \cdot c_2$, ..., $b_n = a \cdot c_n$, for integers $c_1, c_2, \ldots c_n$. Therefore,

$$k_1 b_1 + k_2 b_2 + \ldots + k_n b_n = k_1 \cdot (a \cdot c_1) + k_2 \cdot (a \cdot c_2) + \ldots + k_n (a \cdot c_n);$$

$$k_1 b_1 + k_2 b_2 + \ldots + k_n b_n = (k_1 \cdot c_1 + k_2 \cdot c_2 + \ldots + k_n \cdot c_n) \cdot a,$$

which proves that $a|(k_1 \cdot b_1 + k_2 \cdot b_2 + \ldots + k_n \cdot b_n)$.

(x) If $b|a$, with $a \neq 0$, then (not only $b \neq 0$ by **Definition 2** but also) $c \neq 0$, where c is the unique integer c such that $a = b \cdot c$; and thus $|a| = |b| \cdot |c|$. But $|c| \geq 1$ which implies $|a| = |b| \cdot |c| \geq |b|$. Form the inequality $|a| \geq |b|$, we infer that $-|a| \leq b \leq |a|$. But there are exactly $2 \cdot |a|$ integers which lie between the integers $-|a|$ and $|a|$ (the students should verify this). Thus, if S is the set of divisors of the integer a, the set S contains at most $2 \cdot |a|$ elements (integers); but S is a subset of the set $\{-|a|, (-|a| + 1), \ldots, -1, 1, \ldots, |a| - 1, |a|\}$, the set containing all the $2|a|$ nonzero integers starting with $-|a|$ and ending with $|a|$. therefore, by Theorem 2 (xi), S must be a finite set (apply that Theorem with $n = 2|a|$).

The next theorem we will prove is one that probably many of the readers of this book, have seen before. It is, what is commonly refereed to as the Division Theorem.

Theorem 7 Let a, b be integers with $b \neq 0$. Then there exist unique integers q and r such that $a = b \cdot q + r$ and $0 \leq r < |b|$. this can also be stated more formally as follows: $\forall a, b$, with $b \neq 0$ and $a, b \in \mathbb{Z}$ there exists a unique pair (q, r) in the Cartesian product

$$\mathbb{Z} \times \left\{ \begin{array}{l} \overbrace{\qquad 0, 1, \ldots, |b| - 1 \qquad} \\ \text{integers from zero to } |b| - 1 \\ \text{the set containing all the} \end{array} \right\}, \text{ such that } a = b \cdot q + r.$$

Proof: First we prove the existence of such integers q and r. We proceed by mathematical Induction (in one of the three cases below). In Practice Problem P1 we will give an alternative proof that involves the use of the Well Ordering Property. We distinguish between three cases: $a > 0$, $a = 0, a < 0$.

Case 1: $a > 0$.

Since $a > 0$ and $a \in \mathbb{Z}$, it follows that a is a natural number, so we can use the principle of mathematical Induction (on the natural number a). For $a = 1$: if $|b| > 1$, then $1 = b \cdot 0 + 1$, so the statement is true with $q = 0$, $r = 1$, and $0 \leq r < |b|$. If $|b| = 1$, the statement is again true with $q = 1$ and $r = 0$, if $b = 1$; and with $q = -1$, $r = 0$, if $b = -1$.

The Inductive Step: Assume the statement to be true for some natural number $a = k$. We then have $k = b \cdot q_1 + r_1$, for some integers q_1 and r_1, with $0 \leq r_1 < |b|$. Thus $k + 1 = b \cdot q_1 + (r_1 + 1)$, and since the nonnegative integer r_1 satisfies $r_1 < |b|$, we conclude that $r_1 \leq |b| - 1$ (refer to Example 3, part (i)); so that $r_1 + 1 \leq |b|$. If $r_1 + 1 < |b|$, then the inductive step is complete, because then we simply take $q = q_1$ and $r = r_1 + 1$. If on the other hand $r_1 + 1 = |b|$, then there are two possibilities: either $b = r_1 + 1$ (which happens when $b > 0$) or $b =$ by $-(r_1 + 1)$ (which happens when $b < 0$). In this first case we have $k+1 = b\cdot(q_1+1)+0$, so we can take $q = q_1+1$ and $r = 0$. In the second case we have $k + 1 = b \cdot (q_1 - 1) + 0$, so that we can take $= q_1 - 1$ and $r = 0$. The Inductive step is complete.

Case 2: $a = 0$.

This case is easy because $0 = b \cdot 0 + 0$, so we simply take $q = 0$ and $r = 0$.

Case 3: $a < 0$.

Here we make use of the fact that by (the already proven) **Case 1**, we already know that since $-a$ is a natural number, there exist integers q' and r' such that

$$(-a) = b \cdot q' + r' \text{ and } 0 < r' < |b|$$

If $r' = 0$, we are done, because then $a = b \cdot (-q') + 0$, so we can take $q = -q'$ and $r = 0$. If $r' > 0$ and $b > 0$, then from $a = b \cdot (-q') - r'$; $a = b \cdot (-q' - 1) + (b - r')$, we can take $q = -q' - 1$ and $r = b - r'$; also from $0 \leq r' < |b| = b$, we easily conclude

$$0 \geq -r' > -b \Rightarrow b \geq b - r' > 0$$

But $r' \neq 0$ (because $r' > 0$), hence in fact $0 < b - r' < b$; it is now clear that $0 < r < b$ and so $0 \leq r < b$; finally if $r' > 0$ and $b < 0$, then from $a = b \cdot (-q') - r'$; $a = b \cdot (-q' + 1) + (-b - r')$; we take $q = -q' + 1$ and $r = -b - r'$. These choices for q and $r$do work; one only needs to ensure that $0 \leq r < |b|$, which is equivalent to $0 \leq -b - r' < -b \Leftrightarrow b \leq -r' < 0 \Leftrightarrow 0 < r' \leq -b$ which is true because in fact $0 \leq r' < |b| = -b$; and since $r' \neq 0$ (because $r' < 0$), it follows that $0 < r' < -b$ and so $0 < r' \leq -b$. The proof is **Case 3** is now complete.

Now, that we have established the existence of the integers q and r, we must also show that they are unique. Let (q_1, r_1) and (q_2, r_2) be two pairs satisfying

$$\left.\begin{array}{r} a = b \cdot q_1 + r_1, \quad \text{with} \quad 0 \leq r_1 < |b| \\ \text{and} \quad a = b \cdot q_2 + r_2, \quad \text{with} \quad 0 \leq r_2 < |b| \end{array}\right\}.$$

Subtracting the two equations yields, $r_2 - r_1 = b \cdot (q_1 - q_2)$ which in turn implies

$|r_2 - r_1| = |r_1 - r_2| = |b| \cdot |q_1 - q_2|$. On the other hand the two inequalities above imply

$$\left.\begin{array}{l} 0 \leq r_1 < |b| \\ \\ -|b| < -r_2 \leq 0 \end{array}\right\} \Rightarrow \text{(add)} \ -|b| < r_1 - r_2 < |b| \Rightarrow |r_1 - r_2| < |b|$$

and from the above equations for $|r_1 - r_2|$ we conclude that

$$0 \leq |b| \cdot |q_1 - q_2| < |b|.$$

Since both $|b|$ and $|q_1 - q_2|$ are nonnegative integers, it follows from the last inequality that $|b| \cdot |q_1 - q_2| = 0$; for if it were $|b| \cdot |q_1 - q_2| \neq 0$ we would then have $|b| \geq 1$ and $|q_1 - q_2| \geq 1$, which would imply $|b| \cdot |q_1 - q_2| \geq |b|$, contradicting the previous inequality $|b| \cdot |q_1 - q_2| < |b|$. Hence $|b| \cdot |q_1 - q_2| = 0$; however $b \neq 0$ which allows us to conclude $|q_1 - q_2| = 0 \Rightarrow q_1 - q_2 = 0$; $q_1 = q_2$. From $r_2 - r_1 = b \cdot (q_1 - q_2)$, it now easily follows that $r_1 = r_2$. The proof is complete. ■

Remark 5: If we look at theorem 6 more closely, we will see that theorem 6 gives rise to a mapping from the set $\underbrace{\mathbb{Z} \times (\mathbb{Z} - \{0\})}_{\text{set of all integers except zero}}$ to the set $\mathbb{Z} \times (\mathbb{N} \cup \{0\})$. This is because according to Theorem 6 for every pair (a, b) in $\mathbb{Z} \times (\mathbb{Z} - \{0\})$, there exists a unique pair (q, r) in $\mathbb{Z} \times (\mathbb{N} \cup \{0\})$ such that $a = b \cdot q + r$ and $0 \leq r < |b|$. It is easy to see that this mapping is surjective but not injective; since given a pair (q, r) in the Cartesian product $\mathbb{Z} \times (\mathbb{N} \cup \{0\})$, we can take b to be any integer whose absolute value exceeds r; and then take $a = b \cdot q + r$.

Definition 3: Let a and b be integers with $b \neq 0$. The process of finding the unique integers q and r (as postulated by Theorem 6) such that $a = b \cdot q + r$ and $0 \leq r < |b|$, is called **the division of a with b** and is denoted by $\boldsymbol{a : b}$; the integer q is called the **quotient** of the division, and the integer r is called the **remainder** of the division. We can also say that when the integer a is **divided with** the integer b, it yields a quotient q and a remainder r. When $r = 0$, the division is called **exact**.

Remark 6: Note that from Definitions 2 and 3, it is immediately clear that the nonzero integer b divides the integer a, if and only if, the division of a with b is exact.

Example 3: We give three examples of division

(a) $125 : 15$; $125 = (150 \cdot (8) + 15$; $a = 125$, $b = 15$, $q = 8$, $r = 5$; $0 \leq 5 < |15|$.

(b) $92 : (-14)$; $92 = (-14) \cdot (-6) + 8$; $a = 92$, $b = -14$, $q = -6$, $r = 8$; $0 \leq 8 < |-14|$.

(c) $638 : 29$; $638 = (29) \cdot (22)$; $a = 638$, $b = 29$, $q = 22$, $r = 0$; $0 \leq 0 < |29|$.

Theorem 8 (i) Let $a, b, c \in \mathbb{Z}$, with $c \neq 0$. Then the remainders of the division $a : c$ and $b : c$ are equal if, and only if, the difference $(a - b)$ is a multiple of c; that is if, and only if, the integer $(a - b)$ is divisible by the integer c.

(ii) Let $a_1, \ldots, a_n$ be integers, b a nonzero integer, and $r_1, \ldots r_n$ the remainders of the divisions $a_1 : b, \ldots, a_n : b$, respectively. Then the remainders of the divisions $(a_1 + \ldots + a_n) : b$ and $(r_1 + \ldots + r_n) : b$ are equal.

(iii) Let $a, b \in \mathbb{Z}$, with $b \neq 0$, and k any integer. Then the remainders of the divisions $a : b$ and $(a + k \cdot b) : b$ are equal.

(iv) Let $a_1, \dots, a_n, b \in \mathbb{Z}$, with $b \neq 0$; and $k \in \mathbb{Z}$. then the remainders of the divisions, $(a_1 \dots a_n) : b$ and $[a_n \dots (a_m + k \cdot b) \dots a_n] : b$ are equal.

(v) If $r_1, \dots, r_n$ are the remainders of the divisions $a_1 : b, \dots, a_n : b$ (with $a_1, \dots a_n \in \mathbb{Z}$, b a nonzero integer), then the remainders of the divisions $(a_1 \dots a_n) : b$ and $(r_1 \dots r_n) : b$ are equal.

(vi) For every natural number n, and any integers a and b with $b \neq 0$; if r is the remainder of the division $a : b$ then the remainders of the divisions $a^n : b$ and $r^n : b$ are equal.

Proof: (i) Let r_1 and r_2 be the remainders of $a : c$ and $b : c$ respectively; and q_1, q_2 the quotients of these two divisions. We have $a = c \cdot q_1 + r_1$; with $0 \leq r_1 < |c|$, and $b = c \cdot q_2 + r_2$; with $0 \leq r_2 < |c|$. Subtracting the two equations yields,

$$a - b = c \cdot (q_1 - q_2) + (r_1 - r_2)$$

If $r_1 - r_2 = 0$ (i.e. $r_1 = r_2$), then the last equation clearly shows that $(a - b)$ is a multiple of c, according to Definition 1. Conversely, if $a - b = k \cdot c$, for some integer k, then we obtain $c \cdot [k - (q_1 - q_2)] = r_1 - r_2$, which establishes that the integer $(r_1 - r_2)$ is a multiple of c. On the other hand, form $0 \leq r_1 < |c|$ and $0 \leq r_2 < |c|$ we obtain

$$\left.\begin{array}{l} 0 \leq r_1 < |c|; \text{ and} \\ \\ -|c| < -r_2 \leq 0 \end{array}\right\} \overset{\text{add}}{\Rightarrow} -|c| < r_1 - r_2 < |c|.$$

But the only multiple of c strictly between the integers $-|c|$ and $|c|$, is the number zero as it can be easily established (To the reader: prove that for every natural number n, the only integer in the finite sequence $(-n+1), (-n+2), \dots, -1, 0, 1, \dots, (n-2), (n-1)$, that is divisible by n, is the number zero.) Hence $r_1 - r_2 = 0$; $r_1 = r_2$.

(ii) We have

$$\begin{array}{l} a_1 = b \cdot q_1 + r_1; \quad 0 \leq r_1 < |b| \\ \dots \\ a_n = b \cdot q_n + r_n; \quad 0 \leq r_n < |b| \end{array}$$

By adding the n equations memberwise we obtain,

$$(a_1 + \dots + a_n) = b \cdot (q_1 + \dots + q_n) + (r_1 + \dots + r_n).$$

The result immediately follows by part (i) (In that part, just replace c by b, a by $(a_1 + \dots + a_n)$ and b by $(r_1 + \dots + r_n)$), since $(a_1 + \dots + a_n) - (r_1 + \dots + r_n) = b \cdot (q_1 + \dots + q_n)$, and thus the above difference is a multiple of b.

(iii) This is an immediate consequence of part (i) since $(a + kb) - a = kb$, and thus a multiple of b.

(iv) We have

$$a_1 \dots a_{m-1} \cdot (a_m + k \cdot b) \cdot a_{m+1} \dots a_n =$$

$$= (a_1 \dots a_{m-1} \cdot a_{m+1} \dots a_n) \cdot (a_m + k \cdot b) =$$

$$= (a_1 \dots a_{m-1} \cdot a_m \cdot a_{m+1} \dots a_n) + k \cdot b \cdot (a_1 \dots a_{m-1} \cdot a_{m+1} \dots a_n)$$

Thus

$$[a_1 \dots (a_m + k \cdot b) \dots a_n] - (a_1 \dots a_n) =$$

$$= (a_1 \dots a_{m-1} \cdot a_m \cdot a_{m+1} \dots a_n) + k \cdot b \cdot (a_1 \dots a_{m-1} \cdot a_{m+1} \dots a_n)$$

$$-(a_1 \dots a_{m-1} \cdot a_m \cdot a_{m+1} \dots a_n) =$$

$$= k \cdot b \cdot (a_1 \dots a_{m-1} \cdot a_{m+1} \dots a_n),$$

a multiple of b, and therefore the result readily follows from part (i).

(v) This can be proved by induction on the natural number n, see Exercise 7. Instead we offer an alternative proof, by listing the n equations

$$\begin{aligned} a_1 &= b \cdot q_1 + r_1; \quad 0 \le r_1 < |b| \\ &\dots \\ a_n &= b \cdot q_n + r_n; \quad 0 \le r_n < |b| \end{aligned}$$

If we multiply the n equations memberwise we obtain, $a_1 \dots a_n = (b \cdot q_1 + r_1) \dots (b \cdot q_n + r_n)$. Looking at the right-hand side of the last equation, makes it clear that if we were to completely multiply out the n binomial factors, then among the resulting 2^n terms (each term itself being a product), each term, with the sole exception of $(r_1 \dots r_n)$ would contain a b factor; in other words $(a_1 \dots a_n) = b \cdot k + (r_1 \dots r_n)$, for some $k \in \mathbb{Z}$ and the result follows from part (i).

(vi) This is an immediate consequence of the previous part, part (v); just set $r_1 = \dots = r_n = r$ and $a_1 = \dots = a_n = a$, and the result follows. ■

4.3 The Greatest Common Divisor of two Integers

Definition 4: Let a, b be integers with $|a| + |b| > 0$ (which means that not both a and b are zero; at leat one of the is nonzero)

An integer c is said to be a **common divisor** of a and b if, and only if, $c|a$ and $c|b$; that is, c is a common divisor of a and b, exactly when c divides a and c divides b. From now on, if we say that the integer c is a common divisor of both integers a and b, we will automatically assume that at least one of a and b is nonzero. (In other words we exclude the case $a = b = 0$; of course every integer divides zero, so there is not much else to be said on that.)

Theorem 9 Suppose that a, b are integers with $|a| + |b| > 0$.

(i) The set of all common divisors of a and b, is finite.

(ii) Among all the common divisors of a and b, there exists a greatest one.

Proof: (i) Since at least one of a and b is nonzero, let us assume, without loss of generality, that a is nonzero. If δ is a common divisor of a and b, then $\delta|a$ and $\delta|b$, so in particular $\delta|a$; thus the set of common divisors of a and b, is a subset of the set of all divisors of the integer a; by Theorem 5(x), the set of all divisors of $a \neq 0$ is finite; hence as a subset, by theorem 2(xi), we can conclude that the set of all common divisors of a and b, must be finite.

(ii) By part (i), the set of all common divisors of a and b, is finite. Hence by Theorem 5 (iii), that set has a maximum element; that is, there is a greatest divisor among the, divisors of a and b. ■

Example 4: Let $a = 21- = 2 \cdot 3 \cdot 5 \cdot$ and $b = 364 = 4 \cdot 7 \cdot 13$. The set S of all common divisors of a and b is given by $S = \{-1, 1, 2, -2, 7, -7, 14, -14\}$; the maximum or greatest element of S is the number 14.

Definition 5: Let $a, b \in \mathbb{Z}$, with $|a| + |b| > 0$. The largest among the common divisors of a and b, is called the **greatest common divisor** of a and b, and it is denoted by $\boldsymbol{(a, b)}$ (**note**: Since the same notation is used for (a, b), a member of the Cartesian product $\mathbb{Z} \times \mathbb{Z}$, the distinction should be made in context).

Theorem 10 Let a, b be integers with $|a| + |b| > 0$.

(i) The greatest common divisor (a, b) of a and b, is positive; $(a, b) > 0$.

(ii) There exist integers m and n such that $(a, b) = m \cdot a + n \cdot b$.

(iii) If $\delta = \lambda \cdot a + \mu \cdot b$, for some integers λ and μ, then δ is divisible by every common divisor of a and b.

(iv) If $\delta > 0$, δ is a common divisor of a and b, and δ is divisible by every common divisor of a and b, then $\delta = (a, b)$.

(v) If $\delta > 0$ is a common divisor of a and b, and there exist integers k and l such that $\delta = k \cdot a + l \cdot b$, then $\delta = (a, b)$.

(vi) The integer (a, b) is divisible by every common divisor of a and b.

(vii) A necessary and sufficient condition for an integer δ to be the greatest common divisor of a and b is the following: ($\delta > 0$, δ a common divisor of a and b and $\delta = t \cdot a + u \cdot b$, for some $t, u \in \mathbb{Z}$).

(viii) A necessary and sufficient condition for an integer δ to satisfy $\delta = (a, b)$ is the following: ($\delta > 0$, δ a common divisor of a and b, and δ is divisible by every common divisor of a and b).

Proof: (i) Since (a, b) is a real number, there are exactly three possibilities: $(a, b) > 0$, $(a, b) = 0$, or $(a, b) < 0$. Since $(a, b)|a$ and $(a, b)|b$, the second possibility is immediately excluded because $|a| + |b| > 0$ and hence $|a| + |b| \neq 0$. Now if the third possibility holds we arrive at a contradiction; indeed if $(a, b) < 0$, then $-(a, b) > 0$. but $-(a, b)$ is also a common divisor of a and b, since (a, b) is; this follows from Theorem 5 (ii).

Clearly though $(a, b) < -(a, b)$, in view of $(a, b) < 0$. This is a contradiction, simply because (a, b) is assumed to be the greatest common divisor of a and b; thus, the third possibility above has been eliminated as well; this leaves the first possibility, namely $(a, b) > 0$.

(ii) Let $D = (a, b)$ be the greatest common divisor of a and b. consider the set,

$$S = \{x|\ x \in \mathbb{N} \text{ and } x = t_1 \cdot a + t_2 \cdot b, \text{ for some } t_1,\ t_2 \in \mathbb{Z}\}.$$

The set S contains precisely those natural numbers or positive integers that are of the form (integer) $\cdot a+$ (integer) $\cdot b$. First we show that the set S is nonempty, that is, it contains at least one element. Indeed if $a \neq 0$, take $t_1 = 1$ and $t_2 = 0$, in the case $a > 0$; and in the case $a < 0$, take $t_1 = -1$ and $t_2 = 0$. In both cases, the resulting integer $(t_1 \cdot a + t_2 \cdot b)$ is the natural number $|a|$. Likewise, if $b \neq 0$, take $t_1 = 0$ and $t_2 = 1$; or $t_1 = 0$ and $t_2 = -1$, according to whether b is positive or negative, respectively. The resulting integer is the natural number $|b|$. Keep in mind that by hypothesis $|a| + |b| > 0$; which guarantees that either $|a| > 0$ and $|b| = 0$;or $|a| = 0$ and $|b| > 0$; or $|a| > 0$ and $|b| > 0$.

Thus we see that the set S is never empty: it must contain either the natural number $|a|$; or the natural number $|b|$; or both. According to the Well-Ordered Property, the set S has a least element; let d then be the smallest natural number in the set S. Then by definition we must have $d = m \cdot a + n \cdot b$, for some integers m and n. Next we show that $d|a$ and $d|b$; that is, we show that the integer d is a common divisor of a and b. Consider the division $a : d$; there exist (unique) integers q and r such that $a = d \cdot q + r$ and $0 \leq r < |d| = d$. Hence $a = (m \cdot a + n \cdot b) \cdot q + r \Rightarrow r = a \cdot (1 - mq) + b \cdot (-nq)$, which (by taking $t_1 = 1 - mq$ and $t_2 = -nq$) implies that the integer r is a member of the set S. If $r \neq 0$, then from $0 \leq r < d$, we conclude $0 < r < d$; which says that r is a positive integer less than d, and also a member of S; contrary to the fact that d is the least element in S. Thus r must be zero: $r = 0 \Rightarrow a = d \cdot q + 0 = d \cdot q \Rightarrow d|a$. In a similar way we prove that d is a divisor of b. Thus, the integer d is a common divisor of a and b. On the other hand, $d \leq D = (a, b)$, since D is the greatest common divisor of a and b. Moreover since $D|a$ and $D|b$, there exist by Definition 1, integers k_1 and k_2 such that $a = D \cdot k_1$ and $b = D \cdot k_2$. Now from $d = m \cdot a + n \cdot b$ we obtain $d = D \cdot (m \cdot k_1 + n \cdot k_2)$. Obviously $(m \cdot k_1 + n \cdot k_2)$ must be a positive integer, since both d and D are; thus $d = D \cdot (m \cdot k_1 + n \cdot k_2) \geq D \cdot 1 = D$. We have established $d \leq D$ and $D \geq d$; which proves $d = D$; and from $d = ma + nb \Rightarrow D = (a, b) = m \cdot a + n \cdot b$.

(iii) If $\delta = \lambda \cdot a + \mu b$ and d is a common divisor of a and b, then by setting $a = d \cdot a_1$ and $b = d \cdot b_1$ with $a_1,\ b_1 \in \mathbb{Z}$, we obtain $\delta = d \cdot (\lambda a_1 + \mu b_1)$, which proves that d divides δ.

(iv) Let d be a common divisor of a and b. By hypothesis, $\delta = d \cdot k$, for some integer k. and so, since $\delta > 0$, $\delta = |\delta| = |d| \cdot |k| \geq |d| \cdot 1 = |d| \Rightarrow |d| \leq \delta$. Thus every member of the set S (the set of all common divisors of a and b) has absolute value not exceeding δ; but if $d \in S$, then $-d \in S$ as well; since $-d$ is also a common divisor of a and b, if d is (by Theorem 5 (ii)). This clearly then shows that every member of S, does not exceed δ; and since $\delta \in S$, we conclude that $\delta = (a, b)$.

(v) Suppose $\delta > 0$ and $\delta = k \cdot a + l \cdot b$ for $k, l \in \mathbb{Z}$; and let d be a common divisor of a and b. Put $a = d \cdot a_1,\ b = d \cdot b_1$, with $a_1, b_1 \in \mathbb{Z}$. We have, $\delta = d \cdot (ka_1 + lb_1)$, which establishes that $d|\delta$. Thus δ is divisible by every common divisor of a and b; and since we also have $\delta > 0$, the result follows by part (iv).

(vi) By part (ii), we have $(a, b) = m \cdot a + n \cdot b$, for some integers m and n. Let d be a common divisor of a and b; again, by putting $a = d \cdot a_1, b = d \cdot b_1$ for $a_1, b_1 \in \mathbb{Z}$, we obtain $(a, b) = d \cdot (ma_1 + n \cdot b_1)$ which proves that d divides (a, b).

(vii) This is an immediate consequence of parts (i), (ii) and (v).

(viii) This is an immediate consequence of parts (i), (iv), and (vi).

4.4 The Process of Finding the Greatest Common Divisor of Two Integers

In this Section we will discuss the process or method for finding the greatest common divisor of two integers. This process is commonly referred to as the **Euclidean Algorithm**, in honor of the eminent ancient Greek mathematician Euclid. But first, we need the two Theorems stated below, both of whose proofs are short.

Theorem 11 Let $a, b \in \mathbb{Z}$ with $b \neq 0$.

(i) If $b|a$, then $(a,b) = |b|$.

(ii) If r is the remainder of the division $a : b$, then every common divisor of a and b, is a divisor of r.

(iii) Every common divisor of r and b, is a divisor of a. (Here r stands for the same remainder as in part (ii).)

(iv) $(a,b) = (r,b)$ (again r is the remainder of part (i)).

Proof: (i) Clearly $|b|$ is a common divisor of a and b, since by hypothesis $b|a$ and thus $|b||a$ (obviously because $b = |b|$ or $b = -|b|$ depending on whether $b > 0$ or $b < 0$ respectively); and also $|b||b$. If d is a common divisor of a and b, then d divides b and thus $|d|$ is a divisor of $|b|$; we set $|b| = k \cdot |d|$, for some natural number k; thus $|b| = k \cdot |d| \geq 1 \cdot |d| = |d| \geq d$, which proves that no common divisor of a and b exceeds $|b|$; this coupled with the fact that $|b|$ is a common divisor of a and b, implies $|b| = (a,b)$.

(ii) Let q and r be the quotient and remainder of the division $a : b$; we have $a = b \cdot q + r$, with $0 \leq r < |b|$. If δ is a common divisor of a and b, then $a = \delta \cdot a_1,\ b = \delta \cdot b_1$, for $a_1, b_1 \in \mathbb{Z}$. We have $\delta \cdot a_1 = \delta \cdot b_1 \cdot q + r \Rightarrow \delta \cdot (a_1 - b_1 \cdot q) = r$, which proves that δ is a divisor of r.

(iii) Let $d \in \mathbb{Z}$, $d|r$, and $d|b$. We set $r = d \cdot r_1$ and $b = d \cdot b_1$, for some integers r_1, b_1. From $a = b \cdot q + r$ we then obtain $a = d \cdot b_1 \cdot q + d \cdot r_1 \Rightarrow a = d(b_1 q + r_1)$, which proves that d is a divisor of a.

(iv) By parts (ii) and (iii), it immediately follows that the integers a, b; and the integers r, b; have exactly the same common divisors: if S_1 is the set of all common divisors of a and b; and S_2 is the set of all common divisors b and r, then S_1 is a subset of S_2; and S_2 is a subset of S_1. Consequently $S_1 = S_2 \Rightarrow (a,b) = (r,b)$, since (a,b) is the maximum element of S_1 and (r,b) is the maximum element of S_2. ∎

Theorem 12 Let a and b be natural numbers (or positive integers) with $a > b$. A sequence of nonnegative integers $r_0, r_1, \ldots, r_n, r_{n+1}, \ldots$; can be generated as follows:

(1)	$a = b \cdot q_1 + r_1,$	with $0 \le r_1 < b = r_0$
(2)	$b = r_1 \cdot q_2 + r_2,$	with $0 \le r_2 < r_1,$ if $r_1 \neq 0$
(3)	$r_1 = r_2 \cdot q_3 + r_3,$	with $0 \le r_3 < r_2,$ if$r_2 \neq 0$
$\vdots$	$\vdots$	
(n)	$r_{n-2} = r_{n-1} \cdot q_n + r_n,$	with $0 \le r_n < r_{n-1},$ if $r_{n-1} \neq 0$
$(n+1)$	$r_{n-1} = r_n \cdot q_{n+1} + r_{n+1},$	with $0 \le r_{n+1} < r_n,$ if $r_n \neq 0$
$\vdots$		

(Note that the next term of this sequence is defined, only if the previous term is nonzero; also note that the terms of this sequence are $b = r_0$, the first term; (and starting with the second term), they are the remainders of the divisions $r_0 : r_1, r_1 : r_2, \ldots, r_{n-2} : r_{n-1}, r_{n-1} : r_n, \ldots$.)

Then we have the following:

(i) The above sequence is finite, that is, it terminates. Specifically, there exists a natural number k such that $r_k = 0$ and $r_i \neq 0$ for every integer i with $0 \le i < k$; $i = 0, 1, \ldots, k-1$.

(ii) If k is the natural number of part (i), then $r_{k-1} = (a, b)$ (in other words, the greatest common divisor (a, b), is the last non zero remainder in the finite sequence that starts with $r_0 = b$ and ends with $r_k = 0$; and except for its last term, all of its other k terms, $r_0, \ldots, r_{k-1}$, are nonzero).

Proof: (i) We argue by contradiction: suppose to the contrary that all the remainders $r_1, r_2, \ldots, r_n, \ldots$, are nonzero; this automatically implies that each term r_n, with $n \ge 1$, would then be a natural number (or positive integer). The set S containing all those remainders r_n would be a subset of the set of natural numbers $\mathbb{N}$. Also note that all the terms r_n are distinct from each other, since by the way they are defined each of them is less than the previous one (so this is a decreasing sequence $r_1, r_2, \ldots, r_n, \ldots$). According to the Well Ordering Property, the set S must contain a least element, a smallest natural number r_m, for some positive integer m. Hence $r_m \le r_n$, for every natural number n. But $r_{m+1} < r_m$, by the way this sequence is defined, we have a contradiction. Therefore there exists at least one natural number such that $r_t = 0$. And so the set $T = \{t \in \mathbb{N} | r_t = 0\}$ is nonempty, and thus, as a subset of $\mathbb{N}$, it must contain a least element k; thus k is the smallest value of t such that $r_t = 0$; clearly then $r_k = 0$, and $r_t \neq 0$, for $t = 0, 1, \ldots, k-1$; or equivalently, $r_k = 0$ and $r_i \neq 0$ for every integer i with $0 \le i < k$. Also note that obviously the set T is a singleton set, that it only contains only one element, namely the natural number k; that is because all the remainders r_n are distinct (as remarked above) and hence there is room for only one of them to be zero.

(ii) By Theorem 10(iv) we have, $(a, b) = (b, r_1) = (r_1, r_2) = \ldots = (r_{k-2}, r_{k-1}) = (r_{k-1}, r_k) = (r_{k-1}, 0) = r_{k-1} \neq 0$; thus $(a, b) = r_{k-1}$. ■

Finding (a, b)

(1) First note that since, obviously $(a, b) = (|a|, |b|)$, (because any non zero integer has exactly the same set of divisors as its absolute value does), we can always work with the positive integers $|a|, |b|$, in case at least one of a or b is negative (also, if

either a or b is zero, then (a, b) is equal to $|b|$ or $|a|$ respectively).

(2) We select the largest among $a' = |a|$ and $|b| = b'$. (If $|a| = |b|$, then $(a, b) = |a| = |b|$, by Theorem 10(i)) say a' and we perform the division $a' : b'$, with remainder r_1: if $r_1 = 0$ (i.e. if b' is a divisor of a'), then by Theorem 10(i), we have $(a', b') = b'$. If $r_1 > 0$, we continue by performing the division $b' : r_1$, with remainder r_2. continuing in this manner, Theorem 11 guarantees that at some point we will have $r_k = 0$ and $(a', b') = r_{k-1}$.

Example 5 Find the values of (216,156), (1155,72) and (323,260).

(i)

$$\begin{array}{ll} a = 216,\ b = 156 & \\ 216 = (156) \cdot (1) + 60; & r_1 = 60 \\ 156 = (60) \cdot (2) + 36; & r_2 = 36 \\ 60 = (36) \cdot (1) + 24; & r_3 = 24 \\ 36 = (24) \cdot (1) + 12; & r_4 = 12 \\ 24 = (12) \cdot (2) + 0; & r_5 = 0 \end{array}$$

We see that $k = 5$, $r_k = r_5 = 0$, and $r_{k-1} = r_4 = 12$.

Thus (216,156)=12

(ii)

$$\begin{array}{ll} 1155 = (72) \cdot (16) + 3; & r_1 = 3 \\ 72 = (3) \cdot (24) + 0; & r_2 = 0 \end{array}$$

We see that $k = 2$, $r_k = r_2 = 0$, and $r_{k-1} = r_1 = 3$.

Thus we have (1155,72)=33

(iii)

$$\begin{array}{ll} a = 323,\ b = 260 & \\ 323 = (260) \cdot (1) + 63; & r_1 = 63 \\ 260 = (63) \cdot (4) + 8; & r_2 = 8 \\ 63 = (8) \cdot (7) + 7; & r_3 = 7 \\ 8 = (7) \cdot (1) + 1; & r_4 = 1 \\ 7 = (1) \cdot (7) + 0; & r_5 = 0. \end{array}$$

In this case we have $k = 5, r_k = r_5 = 0$, and $r_{k-1} = r_4 = 1$. Therefore (3233,60)=1.

In the last numerical example, we saw that the greatest common divisor of $a = 323$ and $b = 26$, is the natural number 1. when this occurs we say that the two integers are relatively prime; this is an important concept as we will see later. We have the following definition.

Definition 6: Two non zero integers are called **relatively prime**, if their greatest common divisor is equal to 1. An alternative terminology can also be used: two integers whose greatest common divisor is equal to 1, can also be referred to as **coprime**.

4.5 Properties of the Greatest Common Divisor of Two Integers

The following Theorem lists a number of useful properties of the greatest common divisor of two integers. All statements in the Theorem involve integers which, for simplicity are assumed to be nonzero.

Theorem 13 Let $a, b, c, d,\ a_1, a_2, \ldots, a_m, b_1, b_2, \ldots, b_k$, be nonzero integers.

(i) $(ac, bc) = |c| \cdot (a, b)$.

(ii) If $c|a$ and $c|b$, then $\left(\frac{a}{c}, \frac{b}{c}\right) = \frac{(a,b)}{|c|}$.

(iii) A necessary and sufficient condition for the integer d to be the greatest common divisor of a and b, is $d > 0,\ d|a, d|b$, and $\left(\frac{a}{d}, \frac{b}{d}\right) = 1$.

(iv) If ρ is a rational number, then there exist integers m and l, with $l \neq 0$, and such that $\rho = \frac{m}{l}$ and $(m, l) = 1$.

(v) If $c|a \cdot b$ and $(c, b) = 1$, then $c|a$.

(vi) if $(a, b) = 1$ and $\frac{x}{y} = \frac{a}{b}$, where x and y are integers such that $xy \neq 0$, then there exist an integer t such that $x = t \cdot a,\ y = t \cdot b$, and $t \neq 0$.

(vii) If $d = (a, b)$ and$(c, b) = 1$, then $(a \cdot c, b) = d$.

(vii) If $(a_i, b) = 1$ for $i = 1, 2, \ldots, n$; then $(a_1 \cdot a_2 \ldots a_n, b) = d$.

(ix) If n is a natural number and $(a, b) = 1$, then $(a^n, b) = 1$.

(x) If each of the integers $a_1, a_2, \ldots, a_n$ is relatively prime to each of the integers $b_1, b_2, \ldots, b_k$, then $(a_1 \cdot a_2 \ldots a_n, b_1 \cdot b_2 \ldots b_k) = 1$.

(xi) If n and k are natural numbers and $(a, b) = 1$, then $(a^n, b^k) = 1$.

Proof: (i) By Theorem 9 (ii), there exist integers t_1 and t_2 such that $(a, b) = t_1 \cdot a + t_2 \cdot b \Rightarrow |c| \cdot (a, b) = t_1 \cdot (a|c|) + t_2 \cdot (b|c|)$. Clearly $|c| \cdot (a, b) > 0$ and $|c| \cdot (a, b)$ is a common divisor of $|c| \cdot a$ and $|c| \cdot b$ (just set $\delta = (a, b),\ a = \delta \cdot a_1, b = \delta \cdot b_1$, and multiply by $|c|$); this combined with the last equation above and Theorem 9 (vii), implies that $|c| \cdot (a, b)$ is the greatest common divisor of $|c|a$ and $|c|b$; $|c| \cdot (a, b) = (|c|a, |c|b)$.

(ii) This is proved similarly as part (i), see Exercise 8.

(iii) If $d = (a, b)$, then by part (ii) we must have $\left(\frac{a}{d}, \frac{b}{d}\right) = \frac{(a,b)}{d} = \frac{d}{d} = 1$. Conversely, if we assume that d is a positive common divisor of a and b and that $\left(\frac{a}{d}, \frac{b}{d}\right) = 1$, then by part (ii) we infer that

$$\left(\frac{a}{d}, \frac{b}{d}\right) = \frac{(a, b)}{|d|} \Rightarrow 1 = \frac{(a, b)}{|d|} \Rightarrow (a, b) = |d| = d,$$

since $d > 0$.

(iv) Since ρ is rational, it is of the form $\rho = \frac{m_1}{l_1}$, where m_1, l_1 are integers with $l_1 \neq 0$. Let $\delta = (m_1, l_1)$; then $m_1 = \delta \cdot m$ and $l_1 = \delta \cdot l$, for some integers m and l; thus, $\rho = \frac{m_1}{l_1} = \frac{\delta \cdot m}{\delta \cdot l} = \frac{m}{l}$, with $l \neq 0$ (since $l_1 \neq 0$).

(v) We offer two different proofs for this part.

First Proof: Since $(c, b) = 1$, by part (i) we must have $(ac, ab) = |a| \cdot 1 = |a|$; but $c|ac$ (obviously: $ac = c \cdot a$) and $c|b$ by hypothesis, which implies $c|a \cdot b$ (just set $b = c \cdot b_1 \Rightarrow ab = c \cdot (ab_1)$). Thus c is a common divisor of ac and ab; by Theorem

9 (viii), c must divide the greatest common divisor of ac and ab; but $(ac, ab) = |a|$; hence $c||a|$ which implies $c|a$.

Second Proof: Since $(c, b) = 1$, by Theorem 9 (ii), we must have $1 = x_1 \cdot c + x_2 \cdot b$, for some integers x_1, x_2. By multiplying the last equation by a we obtain, $a = c \cdot (x_1 a) + b \cdot (x_2 a)$. By hypothesis, c divides b; we put $b = c \cdot b_1$ and the last equation yields $a = c \cdot [x_1 a + b_1 \cdot x_2 \cdot a]$, which prove that c divides a.

(vi) From $\frac{x}{y} = \frac{a}{b}$, it follows that $bx = ay \Rightarrow a|bx$; but $(a, b) = 1$ by hypothesis. Applying part (v), we conclude that a must divide x. Set $x = t \cdot a$, for $t \in \mathbb{Z}$. Thus from $bx = ay$ we now have $b \cdot t \cdot a = a \cdot y$; but $a \neq 0$ (in view $xy \neq 0$: if it were $a = 0$, then $\frac{x}{y} = \frac{a}{b}$ would imply $x = 0$). Therefore $b \cdot t \cdot a = ay \Rightarrow bt = y$ and we are done. Clearly $t \neq 0$, since $y \neq 0$.

(vii) By Theorem 9 (ii) we know that there exist integers λ_1, μ_1 and λ_2, μ_2 such that, $\left.\begin{array}{r} \delta = \lambda_1 \cdot a + \mu_1 \cdot b \\ \text{and} \quad 1 = \lambda_2 \cdot c + \mu_2 \cdot b \end{array}\right\}$; multiplying the two equations memberwise yields $\delta = \delta \cdot 1 = (\lambda_1 \cdot a + \mu_1 \cdot b)\,(\lambda_2 \cdot c + \mu_2 \cdot b) \Rightarrow \delta = (\lambda_1 \lambda_2) \cdot (ac) + (\lambda_1 \mu_2 a + \mu_1 \lambda_2 c + \mu_1 \mu_2 b) \cdot b$, and by putting $\lambda_1 \lambda_2 = \lambda$, and $\mu = \lambda_1 \mu_2 a + \mu_1 \lambda_2 c + \mu_1 \mu_2 b$, both integers, we now have, $\delta = \lambda \cdot (ac) + \mu \cdot b$; which couched with the fact that $\delta > 0$ and Theorem 9 (vii), proves that δ is the greatest common divisor of the integers ac and b; $\delta = (ac, b)$.

(viii) We proceed by induction. The statement holds for $n = 1$, by hypothesis. Assume the statement to hold true for $n = k$: $(a_1 \cdot a_2 \ldots a_k, b) = 1$, and suppose that $(a_{k+1}, b) = 1$. From the previous part, part (vii), it then follows that $((a_1 \cdot a_2 \ldots a_k) \cdot a_{k+1}, b) = 1$; $(a_1 \cdot a_2 \ldots a_k \cdot a_{k+1}, b) = 1$.

(ix) This is an immediate consequence of the previous part, part (viii); just take $a_1 = a_2 = \ldots = a_n = a$.

(x) From part (viii), it follows that $(a_1 \cdot a_2 \ldots a_n, b_i) = 1$ for $i = 1, 2, \ldots, k$. Thus by the same part, it also follows that $(a_1 \cdot a_2 \ldots a_n, b_1 \cdot b_2 \ldots b_k) = 1$.

(xi) This is an immediate consequence of the previous part, part (x); just take $a_1 = a_2 \ldots = a_n = a$ and $b_1 = b_2 = \ldots = b_k = b$. ■

The following and last Theorem of this Section provides a foundation for the Fundamental Theorem of Number Theory, which we will prove later in this Chapter.

Theorem 14 Suppose that the nonzero integers $b_1, b_2, \ldots, b_n$ are mutually relatively prime; that is any two of them are relatively prime. Also assume that each of them divides the integer a; $b_i|a$, for $i = 1, 2, \ldots, n$.

Then the product $(b_1 \cdot b_2 \ldots b_n)$ must divide a; $(b_1 \cdot b_2 \ldots b_n)|a$.

Proof: We proceed by Mathematical Induction. Obviously the Theorem holds true for $n = 1$. Now assume the inductive hypothesis: $(b_1 \cdot b_2 \ldots b_k)|a$ which means that $a = (b_1 \cdot b_2 \ldots b_k) \cdot q_k$, for some $q_k \in \mathbb{Z}$. Let b_{k+1} be any integer which is relatively prime to each of the integers $b_1, b_2, \ldots, b_k$; and which is a divisor of a; in other words b_{k+1} satisfies two conditions: $b_{k+1}|a$ and $(b_{k+1}, b_i) = 1$, for $i = 1, 2, \ldots, k$.

By Theorem 12 (viii), the condition $(b_{k+1}, b_i) = 1$, for $i = 1, 2, \ldots, k$; implies that $(b_{k+1}, b_1 \cdot b_2 \ldots b_k) = 1$. On the other hand, $b_{k+1}|a = (b_1 \cdot b_2 \ldots b_k) \cdot q_k$. Since b_{k+1} divides the product $(b_1 \cdot b_2 \ldots b_k) \cdot q_k$, and since it is relatively prime to the integer $(b_1 \cdot b_2 \ldots b_k)$, it follows by Theorem 12(v), that b_{k+1} must be a divisor of q_k: $q_k = b_k \cdot q_{k+1}$; going back to $a = (b_1 \cdot b_2 \ldots b_k) \cdot q_k$, we conclude $a = (b_1 \cdot b_2 \ldots b_k \cdot b_{k+1}) \cdot q_{k+1}$; $(b_1 \cdot b_2 \ldots b_k \cdot b_{k+1}) | a$, the inductive step is now complete. ■

4.6 The Least Common Multiple of Two Integers

Definition 7: (i) Let a and b be nonzero integers. An integer c is said to be a **common multiple of a and b**, if it is divisible by both a and b; that is, if it is a multiple of the integer a and a multiple of the integer b.

(ii) Among all the positive common multiples of a and b (such a natural number exists, by the Well Ordering Property), the least one, is called the **least common multiple of a and b**, and it is denoted by $[a, b]$.

The following Theorem provides the connection between the leat common multiple and the greatest common divisor between two integers.

Theorem 15 Let a and b be non zero integers; that is $ab \neq 0$.

(i) $|ab| = (a, b) \cdot [a, b]$. In other words the product of the greatest common divisor, and the common multiple, is equal to the absolute value of the product of the integers.

(ii) Let l be a positive integer. A necessary and sufficient condition in order that $l = [a, b]$ is the following condition: (l is a positive common multiple of a and b; and every other common multiple of a and b; is divisible by l).

(iii) If a and b are relatively prime, then $[a, b] = a \cdot b$. Conversely, if $[a, b] = a \cdot b$, then $(a, b) = 1$.

Proof: (i) Let $\delta = (a, b)$; then $a = \delta \cdot a_1$ and $b = \delta \cdot b_1$, for some non zero integers a_1 and b_1; by Theorem 12 (iii), the integers a_1 and $_1$ must be relatively prime; $(a_1, b_1) = 1$. Let c be a common multiple of a and b; then by Definition 7 we must have $c = a \cdot k = b \cdot \lambda$, where k and λ are non zero integers. Applying $a = \delta \cdot a_1$, $b = \delta \cdot b_1$ yields $c = \delta \cdot a_1 \cdot k = \delta \cdot b_1 \cdot \lambda \Rightarrow a_1 \cdot k = b_1 \cdot \lambda$; thus b_1 is a divisor of the product $a_1 \cdot k$; however $(a_1, b_1) = 1$ and so by Theorem 12 (v) we conclude that b_1 must divide k; $b_1 | k \Rightarrow k = b_1 \cdot \mu$, for some integer μ. Consequently,

$$c = a \cdot k = a \cdot b_1 \cdot \mu = a \cdot \left(\frac{b}{\delta}\right) \cdot \mu = \frac{ab}{\delta} \cdot \mu,$$

since $b = \delta \cdot b_1$. Note that $\frac{ab}{\delta}$ is an integer since $\frac{b}{\delta}$ is. We have, $|c| = \frac{|ab|}{\delta} \cdot |\mu| \geq \frac{|ab|}{\delta}$, which proves that every common multiple of a and b is greater than or equal to the positive integer $\frac{|ab|}{\delta}$; but $\frac{|ab|}{\delta}$ is a common multiple of a and b) itself since obviously $\frac{|ab|}{\delta} = a \cdot \frac{|b|}{\delta}$ if $a > 0$; and $\frac{|ab|}{\delta} = a \cdot \left(-\frac{|b|}{\delta}\right)$, if $a < 0$; likewise, $\frac{|ab|}{\delta} = b \cdot \left(\frac{|a|}{\delta}\right)$, if $a > 0$; and $\frac{|ab|}{\delta} = b \cdot \left(-\frac{|a|}{\delta}\right)$, if $a < 0$. Thus by Definition 7, the positive integer $\frac{|ab|}{\delta}$ must be the least common multiple of a and b;

$$\frac{|ab|}{\delta} = [a, b] \Rightarrow |ab| = \delta \cdot [a, b] = (a, b) \cdot [a, b].$$

(ii) The condition is necessary, because from the proof of part (i), it is clear that if c is a common multiple of a and b, it must be of the form $c = \left(\frac{ab}{\delta}\right) \cdot \mu$, where $\delta = (a, b)$ and $\mu \in \mathbb{Z} - \{0\}$. So that, $|c| = \frac{|ab|}{\delta} \cdot |\mu| = l \cdot |\mu|$, where $l = \frac{|ab|}{\delta}$. Thus $l||c| \Rightarrow l|c$. The condition is also sufficient because if l is a natural number such that $l|c$; where c can be any positive common multiple of a and b, then $c = l \cdot k$ (for $k \in \mathbb{N}$) which implies $c = l \cdot k \geq l$.

(iii) If $(a, b) = 1$, then it follows from part (i) that $[a, b] = a \cdot b$. Conversely, if $[a, b] = ab$, then it follows from part (i) that $(a, b) = 1$. ∎

Remark 7: The first part of Theorem 14 has a practical usefulness: if we know the greatest common divisor (a, b) of two nonzero integers a and b, then their least common multiple is simply the integer $[a, b] = \frac{|a \cdot b|}{(a,b)}$

Example 6 (i) $a = 18,\ b = 30,\ (a, b) = 6,\ [a, b] = \frac{(18)\cdot(30)}{6} = 90.$

(ii) $a = 27,\ b = 63,\ (a, b) = 9,\ [a, b] = \frac{(27.(63)}{9} = (3) \cdot (63) = 189.$

(iii) $a = 55,\ b = 14,\ (a, b) = 1,\ [a, b] = (55) \cdot (14) = 770.$

4.7 Prime numbers & the Fundamental Theorem of Arithmetic

Definition 8: (i) A natural number greater than 1 is called a **prime number** or simply a **prime**, if its only positive integer divisors are itself and the number 1.

(ii) A negative integer is said to be a **negative prime number** or simply a **negative prime**, if its absolute value is a prime number.

(iii) A nonzero integer whose absolute value is grater than 1 and not a prime, is called a **composite number**.

Example 7 (a) We list here all the prime numbers not exceeding 100: 2, 3, 5, 7, 11, 13, 17, 19, 23, 29, 31, 37, 41, 43, 47, 53, 59, 61, 67, 71, 73, 79, 83, 89, 97.

(b) As the observant student must have noticed, every composite number must be divisible by a prime less than the absolute value of the composite number. Here are some examples: 4, 6, 8, 9. 10, 12, 14, 15, 16, 18, 20, 21, 22, 24, 25, 26, 27, 28, 30.

At the end of this Section, we will prove that there exist infinitely many prime numbers. Since the time of the ancient Greek mathematical Euclid, several ingenious proofs, establishing the infinitude of primes, have been offered.

Remark 8: It follows from Definition 8 that if a natural number n is a composite number then it must have a divisor β such that $1 < \beta < n$; which means that the positive integer $\frac{n}{\beta}$ must also lie strictly between 1 and n (this follows from $1 < \beta < n$: $\frac{1}{\beta} < 1 < \frac{n}{\beta} < n$); and so $n = \beta \cdot \left(\frac{n}{\beta}\right)$; if we set $\gamma = \frac{n}{\beta}$, we have $n = \beta \cdot \gamma$, with the positive integers β, γ satisfying $2 \leq \beta, \gamma \leq n-1$.

Theorem 16 (i) Every integer must be either a prime, or a negative prime, or a composite number; or otherwise it must equal 1, or -1, or zero.

(ii) Let $a, b \in \mathbb{Z}$ such that $b \neq 0, 1, -1$ and $|a|$ is a prime number; and suppose that b divides a. Then $|b| = |a|$; that is, b must be itself a prime or a negative prime.

(iii) If $b \in \mathbb{Z}$, $|b|$ is a prime, and a is not a divisor of the integer b, then a and b must be relatively prime; $(a, b) = 1$.

(iv) If $b \in \mathbb{Z}$, $|b|$ is a prime and b divides the product $a_1 \cdot a_2 \ldots a_n$, where $a_1, a_2, \ldots, a_n$ are nonzero integers, then b must divide at least one of the integers $a_1, a_2, \ldots, a_n$.

(v) Suppose that the integers $b, a_1, a_2, \ldots, a_n$ are such that $|b|, |a_1|, |a_2|, \ldots, |a_n|$ are prime numbers; and that the integer b is a divisor of the product $a_1 \cdot a_2 \ldots a_n$. Then at least one of $a_1, a_2, \ldots, a_n$; say a_k, is such that $|b| = |a_k|$.

Proof: (i) According to Definition 8, if $n \in \mathbb{Z}$ and $|n| > 1$, then either $|n|$ is a prime or otherwise it is composite; thus if $|n| > 1$ (since $|n| = n$ or $n = -|n|$), it follows that n is either a prime, or a negative prime; or a composite number. If on the other hand $|n| \leq 1$, then either $|n| = 1$ or $|n| = 0$; in the first case $n = 1$ or -1; in the second case $n = 0$.

(ii) Since $b \in \mathbb{Z}$ and $b \neq 0, 1, -1$, it follows that $|b| > 1$; on the other hand $b|a$ implies that $|b|$is a divisor of $|a|$; but $|a|$ is a prime: its only divisors are itself and the number 1; in virtue of $|b| > 1$ and $|b| \,|\, |a|$, we can now conclude that $|b| = |a|$.

(iii) Let $(a, b) = \delta$, the greatest common divisor of a and b; then $\delta|a$ and $\delta|b$; since $|b|$ is a prime, the only positive divisors of $|b|$ must be 1 and $|b|$; and since $\delta > 0$, it follows that either $\delta = 1$ or $\delta = |b|$; if $\delta = |b|$, then from $\delta | a$, it follows that $|b|$ must be a divisor of a; and hence b must also be a divisor of a, contrary to the hypothesis that b does not divide a. We therefore conclude that $\delta = 1 = (a, b)$.

(iv) We argue by contradiction: if we assume that the integer b does not divide any of the integers $a_1, a_2, \ldots, a_n$; then by the previous part (iii), it would follow (since $|b|$ is a prime) that b would be relatively prime to each of the integers $a_1, a_2, \ldots, a_n$; $(b, a_1) = (b, a_2) = \ldots = (b, a_n) = 1$. By Theorem 12 (viii), we conclude $(b, a_1 \cdot a_2 \ldots a_n) = 1$ which is a contradiction because (by virtue of the hypothesis $b | a_1 \cdot a_2 \ldots a_n$) we know that $(b, a_1 \cdot a_2 \ldots a_n) = |b| > 1$ (since $|b|$ is a prime). In conclusion, b must divide at least one of the integers $a_1, a_2, \ldots, a_n$.

(v) From the previous part, part (iv), we know that b must divide at least one of the integers $a_1, a_2, \ldots, a_n$ say $b | a_k$ (where k is one of the natural numbers $1, 2, \ldots, n$); thus $|b|$ divides $|a_k|$. Hence, since $|a_k|$ is a prime number and $b \neq 0, 1, -1$ (by the fact that $|b|$ is a prime), we infer by part (ii) that $|b| = |a_k|$. ■

The next theorem is the cornerstone of Arithmetic. It states that every natural number or positive integer can be factored in a unique way as a product of primes; and since every negative integer is the opposite of a natural number, the same can be said about the factorization of negative integers; except that a minus sign must be placed in front of the prime factorization.

Theorem 17 (The Fundamental Theorem of Arithmetic: The Unique Factorization Theorem)

(i) For every natural number $n \geq 2$, there exists a unique natural number m and unique primes $p_1, p_2, \ldots, p_m$ such that,

$$p_1 \leq p_2 \leq \ldots \leq p_m \text{ and } n = p_1 \cdot p_2 \ldots p_m$$

(ii) For every natural number $n \geq 2$, there exists a unique natural number k, unique natural numbers $\alpha_1, \ldots, \alpha_k$ and unique primes $q_1, q_2, \ldots q_k$ such that,

$$q_1 < q_2 < \ldots < q_k \text{ and } n = q_1^{\alpha_1} \cdot q_2^{\alpha_2} \ldots q_k^{\alpha_k}$$

Proof: (i) First we prove the following statement (existence): "For every natural number $n \geq 2$, there exists at least one natural number m and primes $p_1, p_2, \ldots, p_m$ such that, $n = p_1 \cdot p_2 \ldots p_m$ and $p_1 \leq p_2 \leq \ldots \leq p_m$." We proceed by induction on the natural number n, starting with $n = 2$. For $n = 2$, $2 = p_1$, which is a prime, so the statement is true for $n = 2$. We will use the method of complete Mathematical Induction.

The Inductive Hypothesis: Assume the statement to hold true for all natural numbers $n = 2, \ldots, k$. We will show that this assumption implies the truth of the statement for $n = k + 1$. Indeed, if the integer $(k + 1)$ is a prime number, then $k+1 = p_1$, and we are done. If the integer $(k+1)$ is composite then (refer to Remark 8) $k+1 = \beta \cdot \gamma$, where the positive integers β and γ satisfy $1 < \beta < k+1$ and $1 < \gamma < k+1$. We apply the inductive hypothesis to the numbers β and γ: Since1 $< \beta < k + 1$ and $1 < \gamma < k+1$, the integers β and γ are integers somewhere between 2 and k; hence by the inductive hypothesis we must have $\beta = p_1 \cdot p_2 \dots p_m$ and $\gamma = q_1 \cdot q_2 \dots q_l$, for primes $p_1, p_2, \dots, p_m, q_1, q_2, \dots, q_l$ that satisfy $p_1 \leq p_2 \leq \dots \leq p_m$ and $q_1 \leq q_2 \leq \dots \leq q_l$. We can relabel the $(m+l)$ primes $p_1, p_2, \dots p_m, q_1, q_2, \dots q_l$ (since the set of real numbers is totally ordered under $\leq$) as $r_1, r_2, \dots, r_{m+l}$ so that $r_1 \leq r_2 \leq \dots \leq r_{m+l}$. Thus $k+1 = \beta \cdot \gamma = (p_1 \cdot p_2 \dots p_m) \cdot (q_1 \cdot q_2 \dots q_l) =$ (since multiplication is commutative) $= r_1 \cdot r_2 \dots r_{m+l}$, with $r_1 \leq r_2 \leq \dots \leq r_{m+l}$, and the induction step is complete. To complete the proof we must prove the following statement (uniqueness) "The natural number m and the m prime numbers $p_1, p_2, \dots, p_m$ are unique." Suppose that $n = p_1 \cdot p_2 \dots p_m = s_1 \cdot s_2 \dots s_t$, where m, t are natural numbers and the primes $p_1, p_2, \dots, p_m, s_1, s_2, \dots, s_t$ satisfy $p_1 \leq p_2 \leq \dots \leq p_m$ and $s_1 \leq s_2 \leq \dots \leq s_t$. Below we show that we must have $m = t$ and $p_i = s_i$, for all $i = 1, 2, \dots, m = t$. Again we proceed as before (in the case of proving the existence part) by using the Principle of Complete Mathematical Induction on the integer n. For $n = 2$, obviously the statement holds since $n = 2 = p_1$ is a prime. Assume (inductive step) that the statement holds for all natural numbers $n = 2, \dots, k$; we will show that this assumption implies that the statement also holds true for $n = k+1$. Indeed if the number $(k+1)$ is a prime number, then $k+1 = p_1$ and the statement obviously holds true. Now suppose that the integer $(k+1)$ is a composite number. From $k+1 = p_1 \cdot p_2 \dots p_m = s_1 \cdot s_2 \dots s_t$ it follows that the prime p_m divides the product $(s_1 \cdot s_2 \dots s_t)$, and hence by Theorem 15(v) (which applies, because all $s_1, s_2, \dots, s_t$ are primes) it follows that $p_m = s_\rho$, where ρ is one of the natural numbers 1 through t; moreover, since $s_1 \leq s_2 \leq \dots \leq s_t$, it follows that $s_\rho \leq s_k$, for all $k = \rho, \rho+1, \dots, t$. Thus $p_m \leq s_k$, for $k = \rho, \rho+1, \dots, t$. Likewise applying the same thought process to the fact that the prime s_t divides the product of primes $(p_1 \cdot p_2 \dots p_m)$ we conclude that there must exist a natural number v such that $s_t \leq p_\sigma$ for all $\sigma = v, v+1, \dots m$. Since $p_m \leq s_k$ holds for $k = \rho, \rho+1, \dots, t$, it holds in particular for $k = t$; $p_m \leq s_t$. Likewise, since $s_t \leq p_\sigma$ holds for $\sigma = v, v+1, \dots m$, it holds in particular for $\sigma = m$; $s_t \leq p_m$.

The inequalities ($p_m \leq s_t$ and $s_t \leq p_m$) clearly imply $p_m = s_t = q$. From $k+1 = p_1 \cdot p_2 \dots p_m = s_1 \cdot s_2 \dots s_t$ and $p_m = s_t = q$ we infer that $\frac{k+1}{q} = p_1 \cdot p_2 \dots p_{m-1} = s_1 \cdot s_2 \dots s_{t-1}$. The prime number q is a divisor of $(k+1)$ and since the integer $k+1$ is composite it follows that $1 < q < k+1 \Rightarrow q < k+1 < q \cdot (k+1) \Rightarrow 1 < \frac{k+1}{q} < k+1 \Rightarrow 2 \leq \frac{k+1}{q} \leq k$, since $\frac{k+1}{q}$ is an integer. Since the positive integer $\left(\frac{k+1}{q}\right)$ falls between 2 and k, it is now clear that the inductive hypothesis applies to $\left(\frac{k+1}{q}\right)$; which means that $\frac{k+1}{q} = p_1 \cdot p_2 \dots p_{m-1} = s_1 \cdot s_2 \dots s_{t-1}$ implies $m-1 = t-1$ and $p_i = s_i$ for $i = 1, 2, \dots, m-1 = t-1$; by the uniqueness statement of the inductive hypothesis. Thus $m = t$, $q = p_m = s_t$, and $p_i = s_i$, for $i = 1, 2, \dots, m-1 = t-1$. Hence $k+1 = p_1 \cdot p_2 \dots p_{m-1} \cdot p_m = s_1 \cdot s_2 \dots s_{t-1} \cdot s_t$, with $t = i$ and $p_i = s_i$ for $i = 1, 2, \dots m = t$. The inductive step is complete.

(ii) From part (i) we know that for every natural number $n \geq 2$ there exists a unique positive integer m and prime numbers $p_1, p_2, \dots p_m$ such that,

$$n = p_1 \cdot p_2 \dots p_m \text{ and } p_1 \leq p_2 \leq \dots \leq p_m.$$

Because of the inequalities condition that the primes $p_1, p_2, \ldots, p_m$ satisfy, at most m of them can be equal to p_1; let α_1 be the numbers of the primes (among $p_1, p_2, \ldots, p_m$) that are equal to p_1; obviously α_1 is a positive integer not exceeding m:

$$p_1 = p_2 = \ldots = p_{\alpha_1};\ \alpha_1 \in \mathbb{N} \text{ and } \alpha_1 \leq m.$$

The remaining primes (if any at all) $p_{\alpha_1+1}, p_{(\alpha_1+2)}, \ldots, p_m$ must all be greater than p_{α_1} and they satisfy $p_{(\alpha_1+1)} \leq p_{(\alpha_1+2)} \leq \ldots \leq p_m$. We repeat this process by applying it to the prime $p_{(\alpha_1+1)}$ remaining ones) that are equal to $p_{(\alpha_1+1)}$:

$$p_{(\alpha_1+1)} = p_{(\alpha_1+2)} = \ldots = p_{\alpha_2}.$$

Continuing this way we can partition the natural number m into k parts; $\alpha_1, \alpha_2, \ldots, \alpha_k$; so that $m = \alpha_1 + \alpha_2 + \ldots + \alpha_k$. We put,

$$q_1 = p_1 = p_2 = \ldots = p_{\alpha_1};\ q_2 = p_{(\alpha_1+2)} = \ldots = p_{\alpha_2},$$

$$\ldots, q_k = p_{(\alpha_{k-1}+1)} = p_{(\alpha_{k-1}+2)} = \ldots = p_{\alpha_k} = p_m$$

We have,

$$n = p_1 \cdot p_2 \ldots p_m = \underbrace{q_1 \cdot q_1 \ldots q_1)}_{a_1 \text{ times}} \cdot \underbrace{(q_2 \cdot q_2 \ldots q_2)}_{a_2 \text{ times}} \ldots \underbrace{(q_k \cdot q_k \ldots q_k)}_{a_k \text{ times}} = q_1^{\alpha_1} \cdot q_2^{\alpha_2} \ldots q_k^{\alpha_k}$$

with $q_1 < q_2 < \ldots < q_k$ (because of the way the q_i's were defined); clearly the primes $q_1, q_2, \ldots, q_k$ are all unique since the primes $p_1, p_2, \ldots, p_m$ are.

The proof is complete. ■

The following Theorem is am important result that has been in proofs throughout number theory; it says that if two relatively prime integers have a product that is equal to the nth power of an integer, then each of the two integers in the product, must be a nth power itself. Below we prove the result for positive integers; it can be easily extended to just integers.

Theorem 18 Let a, b, c and n be positive integers such that $(a, b) = 1$ and $a \cdot b = c^n$. Then $a = a_1^n$ and $b = b_1^n$, for some positive integers a_1, b_1 such that $(a_1, b_1) = 1$.

Proof: If either of a or b is equal to 1, the result is trivial. Otherwise suppose that $a \geq 2$ and $b \geq 2$. According to the Unique Factorization Theorem, Theorem 16, we must have $a = p_1^{\alpha_1} \ldots p_m^{\alpha_m}$ and $b = q_1^{\beta_1} \ldots q_k^{\beta_k}$, where the distinct primes $p_1, \ldots, p_m$ satisfy $p_1 < \ldots < p_m$; likewise $q_1 < \ldots < q_k$. On account of the fact that a and b are relatively prime, no q_i prime can equal to any of the p_j primes. Also by hypothesis we know that $a \cdot b = c^n$, which implies $p_1^{\alpha_1} \ldots p_m^{\alpha_m} \cdot q_1^{\beta_1} \ldots q_k^{\beta_k} = c^n$. If r is any prime divisor of c, then $r | c^n$ and so r must divide the left-hand side of the last equation, which, according to Theorem 15 (v), shows that r must be one of the primes $p_1, \ldots, p_m, q_1, \ldots, q_k$. Conversely, if t is any of the primes $p_1, \ldots p_m, q_q, \ldots, q_k$; then the above equation shows that t must divide $c^n = \underbrace{c \ldots c}_{n \text{ times}}$; thus by Theorem 15 (iv),

t must divide c. This analysis shows that the distinct prime divisors of the integer c are precisely the primes $p_1, \ldots, p_m, q_1, \ldots, q_k$. Again Theorem 16 implies that, $c = p_1^{\gamma_1} \ldots p_m^{\gamma_m} \cdot q_1^{\delta_1} \ldots q_k^{\delta_k}$; keeping in mind that the prime bases $p_1, \ldots, p_m, q_1, \ldots, q_k$ may not be in increasing order of magnitude (one may have to permute the order of prime powers in the expression for c). Hence

$$p_1^{\alpha_1} \ldots p_m^{\alpha_m} \cdot q_1^{\beta_1} \ldots q_k^{\beta_k} = \left(p_1^{\gamma_1} \ldots p_m^{\gamma_m} \cdot q_1^{\delta_1} \ldots q_k^{\delta_k}\right)^n;$$

$$p_1^{\alpha_1} \ldots p_m^{\alpha_m} \cdot q_k^{\beta_1} \ldots q_k^{\beta_k} = p_m^{(n \cdot \gamma_1)} \ldots p_m^{(n \cdot \gamma_m)} \cdot q_1^{(n \cdot \delta_1)} \ldots q_k^{(n \cdot \delta_k)}$$

We must have $\alpha_i = n \cdot \gamma_i$, for $i = 1, \ldots, m$ and $\beta_j = n \cdot \delta_j$, for $j = 1, \ldots, k$; for if the exponents α_i did not match with the exponents $n \cdot \gamma_i$; or the exponents β_j did not match with the exponents $n \cdot \delta_j$; then according to Theorem 15 (v), one of the primes $p_1, \ldots, p_m, q_1, \ldots, q_k$ would have to equal to one of the remaining primes leading to a contradiction (the student should fill in the details). Therefore,

$$a = p_1^{\alpha_1} \ldots p_m^{\alpha_m} = p_1^{(n \cdot \gamma_1)} \ldots p_m^{(n \cdot \gamma_m)} = (p_1^{\gamma_1} \ldots p_m^{\gamma_m})^n = a_1^n,$$

where $a_1 = p_1^{\gamma_1} \ldots p_m^{\gamma_m}$. Likewise we have,

$$b = q_1^{\beta_1} \ldots q_k^{\beta_k} = q_1^{(n \cdot \delta_1)} \ldots q_k^{(n \cdot \delta_k)} = \left(q_1^{\delta_1} \ldots q_k^{\delta_k}\right)^n = b_1^n,$$

where $b_1 = q_1^{\delta_1} \ldots q_k^{\delta_k}$. By Theorem 12 (x) and Theorem 15 (parts (ii) and (iii)), it clearly follows that $\left(p_1^{\delta_1} \ldots p_m^{\gamma_m}, q_1^{\delta_1} \ldots q_k^{\delta_k}\right) = 1$; $(a_1, b_1) = 1$. The proof is complete. ■

An important concept in number theory is that of a **complete residue system modulo n**, where n is a positive integer. Recall that given a positive integer n, there are exactly n possible remainders of the division $x : n$, for $x \in \mathbb{Z}$; these possible remainders are the numbers $0, 1, \ldots, (n-1)$; that is, the n integers starting (in increasing order) with 0 and ending with $(n-1)$. So, the set $\{0, 1, \ldots, n-1\}$ is the simplest complete residue system modulo n one can find; now imagine any other set of integers, a subset of $\mathbb{Z}$, with the property that every integer in the set, when divided by n, leaves a remainder that is distinct from all the remainders that all the other integers in that set leave when divided by n; such a set would have, by necessity, exactly n elements: n integers $k_1, \ldots, k_n$ such that the divisions $k_1 : n, \ldots k_n : n$ would yield the n distinct remainders $0, 1, \ldots, n-1$; but not necessarily in that order. For example, let $n = 6$; consider the numbers $k_1 = 12$, $k_2 = 8$, $k_3 = 13$, $k_4 = 11$, $k_5 = 22$, $k_6 = 15$; these numbers do form a complete residue system modulo 6, since the remainders of the divisions $k_1 : 6$, $k_2 : 6$, $k_3 : 6$, $k_4 : 6$, $k_5 : 6$, $k_6 : 6$ are the numbers 0,2,1,5,4,3, respectively. The numbers $l_1 = 4$, $l_2 = 6$, $l_3 = 16$, $l_4 = 1$, $l_5 = 9$, $l_6 = 14$; do not form a complete residue system modulo 6, since the six remainders are 4,0,4,1,3,2; here we have a situation where the remainder 5 is missing from the list. If we allow repetitions of remainders, we can still have a complete residue system modulo n, as long as the remainders that are produced cover the entire range from 0 to $(n-1)$; such a general system of a complete residue system modulo n would have more than n elements (if it has at least one repetition of a remainder) and it is called **general**.

Definition 9: (i) Let n be a natural number and $\beta_1, \ldots, \beta_n$ be integers. The set $s = \{\beta_1, \ldots, \beta_n\}$ is called an **exact-complete residue system modulo n**, if $\{r_i | r_i$ is the remainder of the division $\beta_i : n$, for $i = 1, \ldots, n\} = \{0, 1, \ldots n-1\}$ = the set of integers from 0 to $(n-1)$.

(ii) Let n be a natural number and S a subset of $\mathbb{Z}$, containing more than n elements.

The set S is called a **general-complete residue system modulo n**, if $\{r_x | r_x$ is the remainder of the division $x : n$; $x \in S\} = \{0, 1, \ldots, n-1\}$ = the set of integers from 0 to $(n-1)$.

Theorem 19 (i) Let n be a natural number and k an integer relatively prime to n; $(n, k) = 1$; and b an integer. If $\{\beta_1, \ldots, \beta_n\}$ is an exact-complete residue system modulo n, then so is the set $\{k \cdot \beta_1 + b, \ldots, k \cdot \beta_n + b\}$.

(ii) The set $\{0, k, 2k, \ldots, (n-1) \cdot k\}$; where k may be an integer relatively prime to n; is an exact-complete residue system modulo n.

(iii) If $q_1, q_2, \ldots, q_{n-1}$ are the quotients of the $(n-1)$ divisions $k : n$, $2k : n, \ldots, [(n-1) \cdot k] : n$, then

$$q_1 + q_2 + \ldots + q_{n-1} = \frac{(k-1) \cdot (n-1)}{2}$$

(iv) Let p be a prime number, greater than 3; $p > 3$. Then among the consecutive positive integers $2, 3, \ldots, p-3$, $p-2$; there does not exist a number x such that the division $x^2 : p$ leaves remainder 1.

(v) Let p be a prime number. Then for every number x in the sequence (of consecutive positive integers) $1, 2, \ldots, p-2$, $p-1$; there exists a unique number y in the same sequence such that the division $(x \cdot y) : p$, leaves remainder 1. In fact when $x = 1$, then $y = 1$ as well. When $x = p-1$, then $y = p-1$ as well. In all other cases, for $p > 3$ and $2 \leq x \leq p-2$, the number y also falls between 2 and $(p-2)$; and $x \neq y$.

Proof: (i) It only suffices to show the remainders $r_1, \ldots, r_n$ of the n divisions $(k \cdot \beta_1 + b) : n, \ldots, (k \cdot \beta_n + b) : n$, are all distinct. We can give a quick argument by contradiction. Suppose $r_i = r_j$ for some $i \neq j$, with $1 \leq i, j \leq n$. Then,

$$k \cdot \beta_i + b = n \cdot q_i + r_i \quad \text{and} \quad k \cdot \beta_j + b = n \cdot q_j + r_j,$$

where q_i and q_j are the quotients of the divisions $(k \cdot \beta_i + b) : n$ and $(k \cdot \beta_j + b) : n$ respectively. Subtracting the two equations yields,

$$k \cdot (\beta_k - \beta_j) = n \cdot (q_i - q_j) + (r_i - r_j); \quad \text{and since} \quad r_k - r_j = 0,$$

we arrive at $k \cdot (\beta_i - \beta_j) = n \cdot (q_i - q_j)$; the last equation implies that the integer n divides the product $k \cdot (\beta_i - \beta_j)$; however $(n, k) = 1$ and thus by Theorem 12 (v), it follows that n must be divisor of the difference $(\beta_i - \beta_j)$; applying Theorem 7(i) we infer that the remainders of the divisions $\beta_i : n$ and $\beta_j : n$ must be equal, contrary to the fact that the remainders of the n divisions $\beta_1 : n, \dots, \beta_i : n, \dots, \beta_j : n, \dots, \beta_n : n$, must all be distinct since by hypothesis the set $\{\beta_1, \dots, \beta_n\}$ is an exact-complete residue system modulo n. Hence the assumption $r_i = r_j$, $i \neq j$, $1 \leq i, j \leq n$, must be false, which means that $r_1, \dots, r_n$ are all distinct.

(ii) This is an immediate consequence of part (i) with $\beta_1 = 0$;, $\beta_2 = 1, \dots, \beta_n = n-1$ and $b = 0$.

(iii) Let $r_0, r_1, r_1, \dots, r_{n-1}$ be the remainders of the n divisions $0 : n$, $k : n$, $2k : n, \dots, [(n-1) \cdot k] : k$ respectively; by part (ii), the sequence $r_0, r_1, r_2, \dots, r_n$ is just a permutation or reordering of the integers $0, 1, 2, \dots, n-1$;thus

$$r_0 + r_1 + \ldots + r_{n-1} = 0 + 1 + \ldots + (n-1); \quad \text{but } r_0 = 0.$$

Hence $r_1 + r_2 + \ldots + r_{n-1} = 1 + 2 + \ldots + (n-1)$.

In Section 1 we encountered the sum $1 + 2 + \ldots + (n-1)$; it is equal to $\frac{[(n-1)+1]}{2} \cdot (n-1) = \frac{n(n-1)}{2}$. Therefore $r_1 + r_2 + \ldots + r_{n-1} = \frac{n \cdot (n-1)}{2}$. Let $q_1, q_2, \dots q_{n-1}$ be the quotients of the $(n-1)$ divisions $k : n$, $2k : n, \dots, [(n-1) \cdot k] : k$; we have,

$$\left.\begin{array}{l} k = n \cdot q_1 + r_1 \\ 2k = n \cdot q_2 + r_2 \\ \vdots \\ (n-1) \cdot k = n \cdot q_{n-1} + r_{n-1} \end{array}\right\} \begin{array}{l} \Rightarrow \text{ add } k + 2k + \ldots + (n-1) \cdot k = \\ = n \cdot (q_1 + q_2 + \ldots + q_{n-1}) \\ +(r_1 + r_2 + \ldots + r_{n-1}); \end{array}$$

$$[(1 + 2 + \ldots + (n-1)] \cdot k = n \cdot (q_1 + q_2 + \ldots + q_{n-1}) + \frac{n \cdot (n-1)}{2};$$

$$\frac{n \cdot (n-1) \cdot k}{2} - \frac{n \cdot (n-1)}{2} = n \cdot (q_1 + q_2 + \ldots + q_{n-1});$$

$$\frac{n \cdot (n-1) \cdot (k-1)}{2} = n \cdot (q_1 + q_2 + \ldots + q_{n-1}) \Rightarrow$$

$$\Rightarrow q_1 + q_2 + \ldots + q_{n-1} = \frac{(n-1) \cdot (k-1)}{2};$$

the proof is complete.

(iv) If $x \in \mathbb{Z}$ and the division $x^2 : p$ leaves remainder 1, then $x^2 = p \cdot k + 1$, for some integer k (the quotient of the division); $x^2 - 1 = p \cdot k \Rightarrow (x-1) \cdot (x+1) = p \cdot k$. Since the prime number p divides the product $(x-1) \cdot (x+1)$, it must divide (at least) (Note: p actually cannot divide both $(x-1)$ and $(x+1)$ Why?) one of the two factors, according to Theorem 15(iv); if $p|(x-1)$ then $x - 1 = p \cdot \lambda$ for some integer λ; then $x = p \cdot \lambda + 1$; clearly since $0 \leq 1 < p$, the number 1 is the remainder of the division $x : p$; if on the other hand $p|x+1$, then $x + 1 = p \cdot \mu$, for some $\mu \in \mathbb{Z}$; so that

$$x = p \cdot \mu - 1;\ x = p \cdot \mu - p + (p-1) \Rightarrow x = p \cdot (\mu - 1) + (p-1);$$

since $0 \leq p - 1 < p$, it follows that $(p-1)$ is the remainder of the division $x : p$. In summary, we see that whenever $x \in \mathbb{Z}$ is such that the division $x^2 : p$ leaves remainder 1; then the division $x : p$ must leave remainder either 1 or $(p-1)$. Now if we look at the numbers, $2, 3, \ldots, p-3, p-2$, it is obvious that the remainders of the divisions $2 : p, 3 : p, \ldots (p-3); p, (p-2) : p$; are respectively these numbers themselves: $2, 3, \ldots, p-3, p-2$; (all these divisions have quotient equal to zero); none of these division have remainder 1 or $p-1$. We are done.

(v) By part (ii) we know that $\{0, k, 2k, \ldots, (p-1) \cdot k\}$; $(k, p) = 1$; is an exact-complete residue system modulo p. That means the remainders of the divisions $0 : p, k : p, 2k : p, \ldots, [(p-1) \cdot k] : p$; are the numbers $0, 1, 2, \ldots, p-1$ in some order; but the remainder of the division $0 : p$ is obviously zero. This establishes that the remainders of the $(p-1)$ divisions $k : p, 2k : p, \ldots [(p-1)] : p$ are the numbers $1, 2, \ldots, p-2, p-1$ in some order; remember that $(k, p) = 1$. Clearly then if we take k to be any integer between 1 and $(p-1)$; k will be relatively prime to p, because the only divisor (greater than 1) that p may have in common with k is p itself; but none of the integers $1, 2, \ldots, p-1$ is divisible by p. Hence if $x = k$, with k being one of the numbers $1, 2, \ldots, p-1$, it follows that precisely one of the divisions $x : p, 2x : p, \ldots, [(p-1) \cdot x] : p$ must heave remainder 1; say the division $(y \cdot x) : p$ leaves remainder 1. Thus for every x in the sequence $1, 2, \ldots, p-1$; there exists a unique y in the same sequence such that the integer $(x \cdot y)$ leaves remainder 1 when divided with p; the uniqueness of y is clearly guaranteed due to the fact that the set $\{0, x, 2x, \ldots, (p-1) \cdot x\}$ is an exact-complete residue system modulo p, so no repetitions of remainders is possible.

If $x = 1$, then obviously $y = 1$; since $1 \cdot 1 = 1$ leaves remainder 1 when divided with p. When $x = p-1$, $y = p-1$; indeed

$$(p-1) \cdot (p-1) = p^2 - 2p + 1 \Rightarrow (p-1) \cdot (p-1) = p \cdot (p-2) + 1;$$

and since $0 \leq 1 < p$, this proves that the remainder of the division $[(p-1) \cdot (p-1)] : p$ leaves remainder 1. It is now clear that for every x in the sequence $2, 3, \ldots, p-2, p-3$ (for $p > 3$), there exists a unique y in the same sequence such that $(x \cdot y)$ leaves remainder 1 when divided with p. The proof is complete. ■

Remark 9: (i) With regard to Definition 9(i); a perhaps simpler, equivalent definition would be to simply state that the integers $\beta_1, \beta_2, \ldots \beta_n$ form an exact-complete residue system modulo n, if the sequence of the remainders $r_1, r_2, \ldots, r_n$ (of the divisions $\beta_1 = n$, $\beta_2 : n, \ldots, \beta_n : n$ respectively) is a permutation of the sequence of integers from 0 to $n-1$; $0, 1, \ldots, n-1$.

(ii) Only prime numbers have the property of Theorem 18(v); in the language of rings (Chapter 2) this translates into saying that ever nonzero element in the ring $\mathbb{Z}_p = \{0, 1, 2, \ldots, p-1\}$ has a multiplicative inverse (modulo p); that makes the ring $\mathbb{Z}_p$ actually in a field; on the other hand if n is a composite number, $\mathbb{Z}_n$ is never a field; just a ring. This is because among the numbers $1, 2, \ldots, n-1$, there are some that are not relatively prime to n; where as for $n = p$ =prime, all the numbers $1, 2, \ldots, p-1$ are relative prime to p.

Infinitely Many Primes

The first historically known proof was due to Euclid, the ancient-Greek mathematician. We reproduce this proof below. To be able to carry out the proof we first need to show that every integer $n \geq 2$ must have a prime divisor. The argument goes like this: let n be an integer, $n \geq 2$. If n is a prime, then n has a prime divisor, namely itself. If n is not a prime, then it is a composite number: $n = d_1 \cdot k_1$, where d_1 and k_1 are positive integers such that $2 \leq d_1, k_1 < n$. We repeat the process: if d_1 is a prime, we are done. If not, then $d_1 = d_2 \cdot k_2$ with $2 \leq d_2, k_2 < d_1 < n$. Continuing this way, we have a decreasing sequence of positive integers $d_1, d_2, \ldots, d_m, \ldots$; with $d_1 > d_2 > \ldots > d_m > \ldots$. Each term d_i of this sequence must satisfy $2 \leq d_i < n$; or equivalently $2 \leq d_i \leq n-1$. There are exactly $((n-1) - 2 + 1) = n - 2$ integers from 2 to $(n-1)$ (please note: the case $n = 2$ is trivial; so we may assume $n \geq 3$). This shows that the above decreasing sequence can have at most $(n-2)$ members. Therefore the sequence $d_1, d_2, \ldots, d_m, \ldots$ must be finite, with the last term, say d_t, being its least one. Since $2 \leq d_t < n$ and d_t cannot have any divisors other than itself or 1, it follows that d_t must a prime number.

Now that we have shown that every integer grater than 1 must have a prime divisor, we can prove that there must exist infinitely many primes. We argue by contradiction: if the number of primes is finite, say N; then let $2 = p_1, 3 = p_2, \ldots, p_N$ be the list of all primes. Consider the natural number,

$$k = p_1 p_2 \ldots p_N + 1; \text{ clearly } k \geq 2$$

The positive integer k must have a prime divisor; thus one of the primes $p_1, p_2, \ldots, p_N$ must be a prime divisor of k; say p_j is such a divisor of k; then $k = M \cdot p_j$, for some positive integer M. We have

$$M \cdot p_j = p_1 \cdot p_2 \ldots p_n + 1 \Rightarrow$$

$$\Rightarrow M \cdot p_j - (p_1 \cdot p_2 \ldots p_{j-1} \cdot p_j \cdot p_{j+1} \ldots p_N) = 1 \Rightarrow$$

$$\Rightarrow p_j \cdot [M - (p_1 \cdot p_2 \ldots p_{j-1} \cdot p_{j+1} \ldots p_N] = 1,$$

which shows that the prime p_j divides 1; an impossibility. The proof is complete.

Remark 10: (i) We remark here that there exist arbitrary long intervals of real numbers, containing no primes. For example given any positive integer n, consider the n consecutive positive integers,

$$(n+1)! + 2,\ (n+1)! + 3, \ldots, (n+1)! + n, (n+1)! + (n+1),$$

where $(n+1)!$ (pronounced $(n+1)$ factorial) is simply the product $(n+1)! = 1 \cdot 2 \ldots n \cdot (n+1)$; the product of all integers starting with 1 and ending with $(n+1)$. In general the factorial symbol, $k!$ stands for the product $1 \cdot 2 \ldots (k-1) \cdot k$, where k is a positive integer (also 0! is defined to be 1). Now each of the n numbers in the above sequence, is composite; if j is any of the integers $2, 3, \dots, n, n+1$; then we have,

$$(n+1)! + j = 1 \cdot 2 \dots (j-1) \cdot j \cdot (j+1) \dots n \cdot (n+1) + j =$$

$$= j \cdot [1 \cdot 2 \dots (j-1) \cdot (j+1) \dots n \cdot (n+1)] + j;$$

$$(n+1)! + j = j \cdot [(1 \cdot 2 \dots (j-1) \cdot (j+1) \dots n \cdot (n+1)) + 1];$$

We see that the integer j is a divisor of the integer $((n+1)!+j)$; and $2 \leq j < (n+1)!+j$. It is now clear that since the integer $((n+1)!+j)$ has a divisor greater than 1 but less than itself; it must be a composite number.

(ii) The concept of unique factorization is of central importance in number theory. Recall that both $\mathbb{Q}$ and $\mathbb{Z}$ are rings; $\mathbb{Z}$ is a subring of $\mathbb{Q}$; but $\mathbb{Q}$ is not just a commutative ring with identity, or even just an integral domain; it is a field. So one can think of $\mathbb{Q}$ as the field of fractions of $\mathbb{Z}$; and conversely, one can think of $\mathbb{Z}$ as the ring of integers of $\mathbb{Q}$. This concept is generalized in number theory by considering what are called number fields. Examples of number fields are the quadratic number fields, usually denoted by $\mathbb{Q}(\sqrt{d})$; d here stands for a nonzero integer that is not an integer square or minus an integer square (actually it suffices for d to plus or minus a square-free integer: an integer that is the product of distinct primes). Recall from Chapter 2 (in the practice problems Section) that $\mathbb{Q}(\sqrt{d})$ consists of all real numbers of the form $a + b\sqrt{d}$, where a and b are rational numbers. It turns out that when d is of the form, $d = 4k+1$, then the ring of integers of $\mathbb{Q}(\sqrt{d})$ consists of all reals of the form $a + b\sqrt{d}$, where $a, b \in \mathbb{Z}$; together with all reals of the form $\frac{a}{2} + \frac{b}{2} \cdot \sqrt{d}$, where a and b are both odd integers. When d is of the form, $d = 4k+3$, then only the reals of the form $a + b\sqrt{d}$, with a and $b \in \mathbb{Z}$, comprise the ring of $\mathbb{Q}(\sqrt{d})$. The ring of integers of a number field is said to have the unique factorization property if each of its elements can be uniquely factored as a product of prime elements; uniqueness in this context means uniqueness up to order of prime factors and units (recall the concept of a unit from Chapter 2). Even though the concept of a prime element in the ring of integers of a number field bears similarity to that of a prime number in $\mathbb{Z}$, a precise definition of such a concept is beyond the scope of this book. The eminent German mathematician of the eighteenth century, Frederick Gauss, extensively investigated the number field $\mathbb{Q}(\sqrt{-1}) = \mathbb{Q}(i)$ (the purely imaginary number i from complex analysis), and discovered that the prime elements in the ring of integers $\mathbb{Z}[i] = \mathbb{Z}[\sqrt{-1}]$, are precisely the elements of the form $a + b\sqrt{-1}$; where a and b are integers such that the sum $a^2 + b^2$ is a prime number; for example $2 + 3\sqrt{-1}$ is a prime element since $2^2 + 3^2 = 13$; and so is its conjugate $2 - 3\sqrt{-1}$. It should be noted here that the unique factorization property is quite a rare phenomenon among number fields.

4.8 The Language of congruences

The notion of congruence is of key importance in number theory. Among other things, it helps reduce the number of symbols needed to carry out proofs in number theory. In a nutshell, two integers will be congruent modulo a positive integer n, exactly when they belong to the same residue class modulo n; that is to say, they leave the same remainder when divided by n.

Definition 10: Let n be a positive integer. Two integers a and b are said to be **congruent modulo n** if, and only if, the integer $(a-b)$ is divisible by n. If a is congruent to b modulo n, we then write $a \equiv b \pmod{n}$.

Definition 10: Let n be a positive integer. We say that the **integer a is congruent to the integer b modulo n** if, and only if, the integer $(a-b)$ is divisible by n.

Notation: If a is congruent to b modulo n, we write $\boldsymbol{a \equiv b (\text{mod } n)}$. If a is not congruent to b, we write $a \not\equiv b \pmod{n}$.

Theorem 20 (i) Let n be a (fixed) positive integer. The relation in the set $\mathbb{Z}$ defined by "integer x is congruent to integer y", is an equivalence relation.

(ii) Let $n \in \mathbb{N}$, a and $b \in \mathbb{Z}$; then $a \equiv b \pmod{n}$ if, and only if, the divisions $a : n$ and $b : n$ leave the same remainder.

For parts (iii) - (vii), let $a, b \in \mathbb{Z}, n \in \mathbb{N}$, and such that $a \equiv b \pmod{n}$. We then have the following:

(iii) $a + c \equiv b + c (\text{mod } n)$, for any integer c.

(iv) $a \cdot c \equiv b \cdot c (\text{mod } n)$, for any integer c.

(v) If $c, d \in \mathbb{Z}$ such that $c \equiv d \pmod{n}$, then $a + c \equiv b + d \pmod{n}$.

(vi) If $c, d \in \mathbb{Z}$ such that $c \equiv d \pmod{n}$, then $a \cdot c \equiv b \cdot d \pmod{n}$.

(vii) If $k \in \mathbb{N}$, then $a^k \equiv b^k \pmod{n}$.

(viii) If the integers a, b, c satisfy the congruence, $a \cdot c \equiv b \cdot c \pmod{n}$ (where $n \in \mathbb{N}$), then $a \equiv b \,[\text{mod } (\frac{n}{d})]$, where $d = (c, n)$; the greatest common divisor of c and n.

(ix) If $c \in \mathbb{Z}$, $n \in \mathbb{Z}$, and $(c, n) - 1$; and the integers a and b satisfy the congruence $a \cdot c \equiv b \cdot c (\text{mod } n)$, then $a \equiv b (\text{mod } n)$.

(x) If $n_1, n_2 \in \mathbb{N}$ and the integers a, b satisfy $a \equiv b(\text{mod } n_1)$ and $a \equiv b(\text{mod } n_2)$; then $a \equiv b(\text{mod } [n_1, n_2])$, where $[n_1, n_2]$ stands for the least common multiple of n_1 and n_2.

Proof: (i) The relation is reflexive: $a \equiv a(\text{mod } n)$, since $a - a = 0$ which is obviously divisible by any positive integer n. It is also symmetric: if $a \equiv b(\text{mod } n)$ then $(a-b)$ is divisible by n; by Theorem 5(ii) n must also divide the integer $-(a-b) = b-a$; hence $b \equiv a(\text{mod } n)$, which shows that the congruence relation is also symmetric. Finally it is transitive: suppose $a \equiv b(\text{mod } n)$ and $b \equiv c(\text{mod } n)$; then $n|(a-b)$ and $n|(b-c)$; by Theorem 5(ix) n must divide the sum $(a-b) + (b-c) = a - c$; hence $a \equiv c(\text{mod})$; the relation is also transitive.

(ii) Since by Definition 10, $a \equiv b(\text{mod } n)$ if, and only if, n is a divisor of$(a-b)$;

it follows by Theorem 7(i) that $a \equiv b(\text{mod } c)$ if, and only if, the divisions $a : n$ and $b : n$ have the same remainders.

(iii) We have $(a+c)-(b+c) = a-b$; but $a-b$ is divisible by n since by hypothesis $a \equiv b(\text{mod } n)$. Thus $(a+c)-(b+c)$ is divisible by n; which implies by Definition 10 that $a+c \equiv b+c(\text{mod } n)$.

(iv) We have $a \cdot c - b \cdot c = (a-b) \cdot c$; but $n|a-b$, since $a \equiv b(\text{mod})$; thus by Theorem 5(viii), it follows that n must divide that product's equal, namely the difference $a \cdot c - bc$; again from Definition 10 we conclude $a \cdot c \equiv b \cdot c(\text{mod } n)$.

(v) From $a \equiv b(\text{mod } n)$ and $c \equiv d(\text{mod } n)$, we know that n divides the differences $(a-b)$ and $(c-d)$. Hence by Theorem 5(ix), n must divide the sum $(a-b)+(c-d) = (a+c)-(b+d)$; again, Definition 10 implies $a+c = b+d(\text{mod } n)$.

(vi) Here we use a little trick:

$$a \cdot c - b \cdot d = a \cdot c - b \cdot c + b \cdot c - b \cdot d = (a-b) \cdot c + b \cdot (c-d);$$

on the other hand we know that $n|(a-b)$ and $n|(c-d)$ because of the assumptions $a \equiv b(\text{mod } n)$ and $c \equiv d(\text{mod } n)$; and Definition 10. Therefore Theorem 5(ix) tells us that the integer n must divide the sum $(a-b) \cdot c + b \cdot (c-d) = a \cdot c - b \cdot d$, which proves that $a \cdot c \equiv b \cdot d(\text{mod } n)$.

(viii) This part can reproved by using part (vi) (with $a = c$ and $b = d$) and mathematical induction on the natural number k. We urge the reader to do the proof. An alternative proof may be offered though through the use of the identity,

$$a^k - b^k = (a-b) \cdot \left(a^{k-1} + a^{k-2} \cdot b^1 + \ldots + a^1 \cdot b^{k-2} + b^{k-1}\right).$$

On account of $n|(a-b)$ (since $a \equiv b(\text{mod } n)$), it follows that n divides the right-hand side of the last equation (by Theorem 5(viii)) and hence $n|(a^k - b^k)$; $a^k \equiv b^k(\text{mod } n)$. Or one can evoke the transitivity of Divisibility, Theorem 5(iv): since $n|(a-b)$ and $(a-b)|a^k - b^k$ (from the last equation), it follows that $n|(a^k - b^k)$; $a^k \equiv b^k(\text{mod } n)$.

(viii) The hypothesis $a \cdot c \equiv b \cdot c(\text{mod } n)$ implies by Definition 10, that n is a divisor of the difference $a \cdot c - b \cdot c = (a-b) \cdot c$; we can set $(a-b)c = n \cdot t$, where t is some integer. Now $d = (c, n)$ is the greatest common divisor of c and n; we have $c = d \cdot \left(\frac{c}{d}\right)$ and $n = d \cdot \left(\frac{n}{d}\right)$, where the integers $\frac{c}{d}$ and $\frac{n}{d}$ are relatively prime by Theorem 12(iii); $\left(\frac{c}{d}, \frac{n}{d}\right) = 1$, The equation $(a-b)c = nt$ yields

$$(a-b) \cdot d \cdot \left(\frac{c}{d}\right) = d \cdot \left(\frac{n}{d}\right) \cdot t \Rightarrow (a-b) \cdot \left(\frac{c}{d}\right) = \left(\frac{n}{d}\right) \cdot t;$$

The last equation clearly shows that the integer $\frac{n}{d}$ is a divisor of the product $(a-b) \cdot \left(\frac{c}{d}\right)$; but it is also relatively prime to the integer $\frac{c}{d}$; consequently, according to Theorem 12(v), $\frac{n}{d}$ must divide the difference $(a-b)$; which means that,

$$a \equiv b \left(\text{mod } \frac{n}{d}\right)$$

(ix) This is an immediate consequence of the previous part, part (viii), with $d = 1$.

(x) The congruences $a \equiv b(\text{mod } n_1)$ and $a \equiv b(\text{mod } n_2)$ imply that positive integers n_1 and n_2 are both divisors of the difference $(a-b)$; put it equivalently the difference $(a-b)$ is a common multiple of the integers n_1 and n_2; by Theorem 14(ii) it then follows that the integer $(a-b)$ is divisible by $[n_1, n_2]$ the least common multiple of n_1 and n_2; thus,

$$a \equiv b\,(\text{mod } [n_1, n_2])\,. \blacksquare$$

Remark 11: (i) If we apply Theorem 19(iii) with $c = -b$, we obtain $a - b \equiv 0(\text{mod } n)$; which is not surprising since the assumption $a \equiv b(\text{mod } n)$ implies that $(a-b)$ is divisible by n; which means that the remainder of the division $(a-b) : n$ is zero.

(ii) The congruence relation modulo n, is an equivalence relation with exactly n equivalence classes. Each class contains all the (infinitely many) integers which leave the same remainder r when divided with n; r being one of the integers $0, 1, \ldots, n-1$; thus, the rth equivalence class consists of all integers of the form $q \cdot n + r$. We have already seen this back in Chapter 2.

(iii) Most of the properties of congruences listed in Theorem 19, are reminiscent of the usual properties of real numbers. That is because, working with congruences modulo n is tantamount to working in the ring $n\mathbb{Z}$ which we first encountered in Chapter 2. Both $n\mathbb{Z}$ and $\mathbb{R}$ are commutative rings with identity; and so, really, properties (i) through (vii) are really a consequence of the commutative ring structure. Properties (viii) and (ix) are where the differentiation occurs: in the set of reals $\mathbb{R}$, we can cancel any nonzero real number factor from both sides of an equation, we can do so since $\mathbb{R}$ is a field. However, unless n is a prime, $n\mathbb{Z}$ is not a field, but just a commutative ring with identity which has zero divisors. We can cancel a factor from both sides of a congruence only if it is relatively prime to the modulus n (Theorem 19(ix)). Otherwise, we must resort to property (viii). Also, note that when n is a prime, we can cancel any factor from both sides of a congruence modulo n, as long as the common factor c is not a multiple of the prime modulus n.

We close this Section with Theorem 20; the first part of which is historically referred to as Fermat's Little Theorem; and the second pat is known as Wilson's Theorem. Pierre Fermat was a highly ingenious mathematician who lived in seventeenth century France, and who made major contributions to number theory, as well as to other branches of mathematics and physics. In reference to the work "Little" to Fermat's Little Theorem: the reason for this is that Fermat was also known for another, "big" Theorem which was really a conjecture until 1993. Fermat, inspired by many solved and unsolved problems in diophantine analysis (The branch of number theory that deals with algebraic equations to be solved in integers), proposed the following problem: "For every natural number $n \geq 3$, the equation, in unknowns x, y, z; $x^n + y^n = z^n$ has no solution in nonzero integers". Interestingly, the case $n = 2$ has been well understood for some time: this is the case of Pythagorean triples; the equation $x^2 + y^2 = z^2$ not only has nonzero solutions, but it has infinitely many such solutions; the set of solutions to the Pythagorean equation $x^2 + y^2 = z^2$ is completely understood and the solutions can be described in terms of two independently integer valued parameters.

We remark here that the word "Diophantine" refers to the great Greek mathematician Diophantus who lived in Alexandria, Egypt in the late second and third century A.D. the word "Pythagorean" refers to Pythagoras, the other great Greek mathematician and philosopher of the sixth century B.C.

Fermat claimed to have discovered a marvelous proof to his problem/conjecture, which he never revealed though. Many mathematicians in the years that ensued Fermat's life, were highly skeptical of Fermat's claim; especially since for nearly three and a half centuries and despite the efforts of many highly ingenious and prodigious mathematicians, the problem remained unsolved. It was only in 1993 that Fermat's "Big Theorem" was finally proved by Andrew Wiles, a mathematician at Princeton University who had worked intensively in almost complete isolation and secrecy on the Fermat problem for nearly seven years. Wiles was able to effectively use some highly advanced tools and sophisticated results found in both number theory and arithmetic geometry, a kind of hybrid branch (of mathematics) between number theory and algebraic geometry; Professor Wiles was able to combine some powerful results, which had already been established by other researchers in an area of mathematics entitled "elliptic curves". That combined effort of Wiles, finally did the trick and solved the Fermat problem to the satisfaction of the experts. There is a plethora of difficult and unsolved problems in number theory to keep future generations of Ph.D mathematicians occupied for many years to come.

Theorem 21 (i) (Fermat's Little Theorem) Let p be a prime number. If a is any integer not divisible by p (which is equivalent to saying $(a, p) = 1$), then $a^{p-1} \equiv 1 (\text{mod } p)$.

(ii) If p is a prime and a any integer, then $a^p \equiv a (\text{mod } p)$.

(iii) (Wilson) If p is a prime then,

$$(p-1)! \equiv -1 (\text{mod } p)$$

Conversely, if a positive integer $n \geq 2$ satisfies the congruence $(n-1)! \equiv -1 (\text{mod } n)$, then n must be a prime.

Here the symbol $k!$ stands for the product $(1 \cdot 2 \ldots k)$, the product of all positive integers from 1 to k.

Proof: (i) Since a is not divisible by the prime p, it follows that a must be relatively prime to p; $(a, p) = 1$. Now consider the p divisions $0 : p, a : p, 2a : p, \ldots [(p-1) \cdot a : p]$; and let $r_0, r_1, \ldots, r_{p-1}$ be the respective remainders. Clearly $r_0 = 0$. By Theorem 18(ii), the set, $\{0, a, 2a, \ldots, (p-1) \cdot a\}$ is an exact-complete residue system modulo p; in view of $r_0 = 0$, it follows the sequence $r_1, r_2, \ldots, r_{p-1}$ must be simply a permutation of the sequence of numbers $1, 2, \ldots, (p-1)$. Hence,

$$r_1 \cdot r_2 \ldots r_{p-1} = 1 \cdot 2 \ldots (p-1)$$

Furthermore since $(j \cdot a - r_j)$ (for $j = 1, 2, \ldots, p-1$) is divisible by p (by the very definition of the division $(j \cdot a) : p$), it follows from Definition 10 that,

$$j \cdot a - r_j \equiv (\text{mod } p); \text{ for } j = 1, 2, \ldots, p-1$$

Therefore,

$$\left.\begin{array}{l} a \equiv r_1(\text{mod } p) \\ 2a \equiv r_2(\text{modp}) \\ \vdots \\ (p-1)a \equiv r_{p-1}(\text{mod } p) \end{array}\right\} \quad \begin{array}{l} 1st \text{ congruence} \\ 2nd \text{ congruence} \\ \vdots \\ (p-1)th \text{ congruence} \end{array}$$

By Theorem 19(vi) (which can be easily extended to more than two congruences, by mathematical induction), we can multiply memberwise the $(p-1)$ congruences listed above to obtain,

$$[1 \cdot 2 \ldots (p-1)] \cdot a^{p-1} \equiv (r_1 \cdot r_2 \ldots r_{p-1})(\text{mod } p);$$

but $r_1 \cdot r_2 \ldots r_{p-1} \equiv 1 \cdot 2 \ldots (p-1)$, as we have already seen above. Thus, we now have,

$$[1 \cdot 2 \ldots (p-1)] \cdot a^{p-1} \equiv [1 \cdot 2 \ldots (p-1)](\text{mod } p);$$

However the product $[1 \cdot 2 \ldots (p-1)]$ is relatively prime to p; because obviously $(1,p) = (2,p) = \ldots = (p-1,p) = 1$ and by Theorem 12(viii). Hence by (the cancellation Law) Theorem 19(ix) we conclude that $a^{p-1} \equiv 1(\text{mod } p)$.

(ii) If a is a multiple of p; then $a \equiv 0(\text{mod } p)$ and also (trivially) $\underbrace{a \cdot a \ldots a}_{(p-1) \text{ times}} \equiv 0(\text{mod } p)$; $a^{p-1} \equiv 0(\text{mod } p)$. Thus in this case, we have $a^{p-1} \equiv 0(\text{mod } p)$. If a is not a multiple of p, then by the previous part, part (i), weknot that $a^{p-1} \equiv 1(\text{mod } p) \Rightarrow$ (by Theorem 19(iv)) $a^{p-1} \cdot a \equiv 1 \cdot a(\text{mod } p)$; $a^p \equiv a(\text{mod } p)$; we are done.

(iii) There are $(p-1)$ factors in the product $[1 \cdot 2 \ldots (p-1)]$; namely, the positive integers from 1 to $(p-1)$; if we consider only the $(p-3)$ consecutive integers $2, 3, \ldots, (p-2)$, we can apply Theorem 18(v): for each integer x between 2 and $(p-2)$, there exists a unique integer y, also between 2 and $(p-2)$ such that $x \neq y$ and $x \cdot y \equiv 1(\text{mod } p)$. We can list exactly $(\frac{p-3}{2})$ such pairs (x_i, y_i) with $x_i \cdot y_i \equiv 1(\text{mod } p)$; for $i = 1, \ldots, \frac{p-3}{2}$. We have

$$[2 \cdot 3 \ldots (p-2] \equiv (x_1 \cdot y_1) \cdot (x_2 \cdot y_2) \ldots (x_{\frac{p-3}{2}} \cdot y_{\frac{p-3}{2}})(\text{mod } p).$$

$$[2 \cdot 3 \ldots (p-2)] \equiv \underbrace{1 \ldots 1}_{(\frac{p-3}{2}) \text{ times}} \equiv 1(\text{mod} p).$$

Now multiply both sides of the last congruence by $1 \cdot (p-1)$:

$$1 \cdot 2 \cdot 3 \ldots (p-2) \cdot (p-1) \equiv 1 \cdot 1 \cdot (p-1)(\text{mod} p);$$
$$(p-1)! \equiv (p-1)(\text{mod } p)$$

However $p - 1 \equiv -1(\text{mod} p)$ so that we arrive at, $(p-1)! \equiv -1(\text{mod} p)$. ∎

4.9 Practice Problems

P2. Give an alternative proof for the existence part of Theorem 6, (for $a, b, \in \mathbb{Z}$, with $b > 0$, there exist integers q and r such that $a = b \cdot q + r$, with $0 \le r < b$), using the Well Ordering Property.

Solution: Let $a, b, \in \mathbb{Z}$, with $b > 0$. consider the set, $S = \{x | x = a - bt$ such that $t \in \mathbb{Z}$ and $a - bt \ge 0\}$. By definition, S consists of nonnegative integers of a special kind. The set S is nonempty: for $t = -|a|$, we have $x = a - b(-|a|) = a + b \cdot |a| \ge 0$, since in virtue of $b > 0$ we have $b \ge 1$ (because $b \in \mathbb{Z}$) and so $b \cdot |a| \ge 1 \cdot |a| = |a| \ge -a \Rightarrow b \cdot |a| + a \ge 0$, which clearly shows that S is nonempty. Since S contains only nonnegative integers, it follows by the Well Ordering Property (extended to the set of nonnegative integers), that S contains a least element, some nonnegative integer r. Since $r \in S \Rightarrow r = a - b \cdot q$, for some integer q. We claim that $r < b$; if it were $r \ge b$ then we would have $r - b \ge 0 \Rightarrow (a - bq) - q \ge 0 \Rightarrow a - b(q+1) \ge 0$; but then the integer $a - b \cdot (q+1)$ would be a member of S (for $t = q+1$); however $r - b < r$, in view of $b > 0$; since $r - b = a - b \cdot (q+1) \in S$, this contradicts the fact that r is the smallest elements of S. Thus, we have proved $a = b \cdot q + r$, with $0 \le r < b$; this establishes the existence part of Theorem 6.

P3. If $q_1, q_2, \ldots, q_{n-1}, q_n$ are the quotients of the divisions $a : b_1, q_1 : b_2, \ldots, q_{n-2} : b_{n-1}, q_{n-1} : b_n$, respectively ($b_1, b_2, \ldots, b_{n-1}, b_n$ are positive integers; a an integer), then the quotient of the division $a : (b_1 \cdot b_2 \ldots b_n)$ is the integer q_n.

Solution: We proceed by the way of mathematical induction. The result obviously holds true for $n = 1$. Suppose it holds for $n = k$, that is, $a = (b_1 \cdot b_2 \ldots b_k) \cdot q_k + r_k$ with $0 \le r_k < |b_1 \cdot b_2 \ldots b_k|$. Consider the division $q_k : b_{k+1}$; we have $q_k = q_{k+1} \cdot b_{k+1} + r_{k+1}$, with $0 \le r_{k+1} < |b_{k+1}|$; or equivalently, $0 \le r_{k+1} \le |b_{k+1}| - 1$. Combining the above two equations yields.

$$a = (b_1 \cdot b_2 \ldots b_k)(q_{k+1} \cdot b_{k+1} + r_{k+1}) + r_k;$$
$$a = (b_1 \cdot b_2 \ldots b_k \cdot b_{k+1}) \cdot q_{k+1} + (b_1 \cdot b_2 \ldots b_k) \cdot r_{k+1} + r_k \tag{1}$$

On account of $0 \le r_k < |b_1 \cdot b_2 \ldots b_k|$ and $r_{k+1} \le |b_{k+1}| - 1$ we decide that,

$$0 \le b_1 \cdot b_2 \ldots b_k| \cdot r_{k+1} + r_k \le |b_1 \cdot b_2 \ldots b_k| \cdot (|b_{k+1}| - 1) + |b_1 \cdot b_2 \ldots b_k|$$
$$\Rightarrow 0 \le |b_1 \cdot b_2 \ldots b_k| \cdot r_{k+1} + r_k \le |b_1 \cdot b_2 \ldots b_k \cdot b_{k+1}|$$
$$-|b_1 \cdot b_2 \ldots b_k| + |b_1 \cdot b_2 \ldots b_k|; 0 \le |b_1 \cdot b_2 \ldots b_k| \cdot r_{k+1} + r_k \le |b_1 \cdot b_2 \ldots b_k \cdot b_{k+1}| \tag{2}$$

Let $R = (b_1 \cdot b_2 \ldots b_k) \cdot r_{k+1} + r_k$. By the problem's hypothesis $b_1, b_2, \ldots, b_k, b_{k+1}$ are positive integers; since also $r_k, r_{k+1} \ge 0$, we conclude that $R \ge 0$ as well as, by (2),

$$0 \le R \le b_1 \cdot b_2 \ldots b_k \cdot b_{k+1} \tag{3}$$

Also by (1)

$$\Rightarrow a = (b_1 \cdot b_2 \dots b_k \cdot b_{k+1}) \cdot q_{k+1} + R \tag{4}$$

According to Theorem 6, inequalities (3) and equation (4) clearly shows that R and q are the unique remainder and quotient respectively of the division $a : (b_1 \cdot b_2 \dots b_k \cdot b_{k+1})$; the inductive step is complete.

P3. If each of the integers a and b is a sum of two integer squares, prove that the product $a \cdot b$ is also a sum of two integer squares.

Solution: Suppose that $a = m_1^2 + n_1^2$ and $b = m_2^2 + n_2^2$, where m_1, n_1, m_2, n_2 are integers. We have

$$a \cdot b = (m_1^2 + n_1^2) \cdot (m_2^2 + n_2^2) = (m_1 m_2)^2 + (m_1 n_2)^2 + (n_1 m_2)^2 + (n_1 n_2)^2 =$$

$$= [(m_1 m_2 + n_1 n_2)^2 + 2(m_1 m_2)(n_1 n_2) + (n_1 n_2)^2] + [(m_1 n_2)^2$$

$$-2(m_1 n_2)(n_1 m_2) + (n_1 m_2)^2] = (m_1 m_2 + n_1 n_2)^2 + (m_1 m_2 - n_1 m_2)^2],$$

and the proof is complete since obviously both numbers $(m_1 m_2 + n_1 n_2)$ and $(m_1 n_2 - n_1 m_2)$ are integers.

P4. If an even integer is the sum of two integer squares, then so is the half of that integer.

Solution: An even integer is a multiple of 2. Let $n = 2k$ be the sum of two integer squares: $n = 2k = \alpha^2 + \beta^2$, where $\alpha, \beta \in \mathbb{Z}$. First observe that either both α and β are even integers; or they are both odd integers. Indeed from the identity $\alpha^2 + \beta^2 = (\alpha + \beta)^2 - 2\alpha\beta$; and since $\alpha^2 + \beta^2 = 2k$, it follows that $(\alpha + \beta)^2 = 2 \cdot (k + \alpha\beta)$, which proves that the square $(\alpha + \beta)^2$ is an even integer; this implies that $(\alpha + \beta)$ must be an even integer; for if $(\alpha + \beta)$ were odd, then in the language of congruences we would have $\alpha + \beta \equiv 1(\text{mod } 2)$ and by Theorem 19(ii) this would imply $(\alpha + \beta)^2 \equiv 1(\text{mod } 2)$, contrary to the fact that $(\alpha + \beta)^2 \equiv 0(\text{mod } 2)$ (see above). We conclude that $\alpha + \beta \equiv 0(\text{mod } 2) \Rightarrow$ (either $\alpha \equiv \beta \equiv 0(\text{mod } 2)$ or $\alpha \equiv \beta \equiv 1(\text{mod } 2)$) (since the other two possibilities: $\alpha \equiv 1, \beta \equiv 0(\text{mod } 2)$; or $\alpha \equiv 0$, $\beta \equiv 1(\text{mod } 2)$ obviously imply $\alpha + \beta \equiv 1(\text{mod } 2)$).

Case 1: $\alpha \equiv \beta \equiv 0\text{mod } 2)$; set $\alpha = 2\alpha_1$ and $\beta = 2\beta_1$, with $\alpha_1, \beta_1 \in \mathbb{Z}$. We have $n = 2k = \alpha^2 + \beta^2 \Rightarrow \frac{n}{2} = k = 2(\alpha_1^2 + \beta_1^2) \Rightarrow \frac{n}{2} = (\alpha_1 + \beta_1)^2 + (\alpha_1 - \beta_1)^2$, and we are done.

Case 2: $\alpha \equiv \beta \equiv 1(\text{mod } 2)$; set $\alpha = 2\alpha_1 + 1$ and $\beta = 2\beta_1 + 1$, for integers α_1 and β_1.

We have

$$n = 2k = (2\alpha_1 + 1)^2 + (2\beta_1 + 1)^2 \Rightarrow$$

$$\Rightarrow \frac{n}{2} = k = \frac{4\alpha_1^2 + 4\alpha_1 + 4\beta_1^2 + 4\beta_1 + 2}{2};$$

$$\frac{n}{2} = k = 2\alpha_1^2 + 2\alpha_1 + 2\beta_1^2 + 2\beta_1 + 1;$$

$$\frac{n}{2} = k = (\alpha_1^2 - 2\alpha_1\beta_1 + \beta_1^2) + (\alpha_1^2 + \beta_1^2 + 1^2 + 2\alpha_1\beta_1 + 2\alpha_1 + 2\beta_1);$$

$$\frac{n}{2} = k = (\alpha_1 - \beta_1)^2 + (\alpha_1 + \beta_1 + 1)^2;$$

the proof is complete.

P5. If the integers α, β are both even or both odd, the rational number $\frac{\alpha^2+\beta^2}{2}$ is actually the sum of two integer squares (and thus it is itself an integer).

Solution: From the identity $(\alpha+\beta)^2 + (\alpha-\beta)^2 = 2 \cdot (\alpha^2+\beta^2)$ we infer that

$$\frac{2(\alpha^2+\beta^2)}{4} = \frac{(\alpha+\beta)^2 + (\alpha-\beta)^2}{4} \Rightarrow \frac{\alpha^2+\beta^2}{2} = \left(\frac{\alpha+\beta}{2}\right)^2 + \left(\frac{\alpha-\beta}{2}\right)^2;$$

however both $\left(\frac{\alpha+\beta}{2}\right)$ and $\left(\frac{\alpha-\beta}{2}\right)$ are integers in view of the fact $\alpha+\beta \equiv 0(\text{mod } 2)$ and $\alpha - \beta \equiv 0(\text{mod } 2)$ (since we know by hypothesis that either $\alpha \equiv \beta \equiv 0(\text{mod } 2)$ or $\alpha \equiv \beta \equiv 1(\text{mod } 2)$). We are done.

P6. If the integers α, β, γ satisfy the condition $\alpha^2 + 2\beta = \gamma^2$, then $\alpha \equiv \gamma(\text{mod } 2)$ and the integer $(\alpha^2 + \beta)$ is a sum of two integer squares.

Solution: From $\alpha^2 + 2\beta = \gamma^2$, it follows $\alpha^2 \equiv \gamma^2(\text{mod } 2)$, since $2\beta \equiv 0(\text{mod } 2)$. The last congruence easily implies $\alpha \equiv \gamma(\text{mod } 2)$ since if $\alpha \equiv 0$ and$\gamma \equiv 1(\text{mod } 2)$; or vice versa; obviously in that case $\alpha^2 \not\equiv \gamma^2(\text{mod } 2)$. Also note that the same conclusion can be drawn (namely that $\alpha \equiv \gamma(\text{mod } 2)$), if the congruence $\alpha^k \equiv \gamma^m(\text{mod } 2)$, where k, m are positive integers (the argument is the same). We have thus shown that $\alpha \equiv \gamma(\text{mod } 2)$, which is to say that either both α and γ are even integers or they are both odd. To show that $(\alpha^2 + \beta)$ is a sum of two integer squares we use the given condition:

$$\alpha^2 + 2\beta = \gamma^2 \Rightarrow 2\alpha^2 + 2\beta = \gamma^2 + \alpha^2 \Rightarrow 2(\alpha^2 + \beta) = \gamma^2 + \alpha^2 \Rightarrow \alpha^2 + \beta = \frac{\gamma^2 + \alpha^2}{2};$$

since γ and α are either both even or both odd (as shown above), it follows by the previous practice problem, P5, that $\frac{\gamma^2+\alpha^2}{2}$ is the sum of two integer squares; consequently its equal, $(\alpha^2 + \beta)$, must also be a sum of two integer squares.

P7. If $\alpha, \beta, \gamma \in \mathbb{Z}$ and 3 does not divide the product $\alpha \cdot \beta \cdot \gamma$, then (a) $\alpha^2 \equiv \beta^2 \equiv \gamma^2 \equiv 1(\text{mod } 3)$ (b) $\alpha^2 \equiv \beta^6 \equiv \gamma^6 \equiv 1(\text{mod } 9)$ (c) $\alpha^2 + \beta^2 + \gamma^2 \equiv 0(\text{mod } 3)$.

Solution: Since $3 \dagger \alpha\beta\gamma$ or equivalently $\alpha\beta\gamma \not\equiv 0 (\mathrm{mod}\ 3)$, it follows that each of α, β, γ is not divisible by $3 : \alpha \not\equiv 0.\ \beta \not\equiv 0,\ \gamma \not\equiv 0 (\mathrm{mod}\ 3)$. Consider $\alpha^2 - 1 = (\alpha - 1)(\alpha + 1)$; there are three congruence classes modulo $3 : 0, 1$, and 2. Since $a \not\equiv 0(\mathrm{mod}\ 3)$, $\alpha \equiv 1$ or $2(\mathrm{mod}\ 3)$; if $\alpha \equiv 2(\mathrm{mod}\ 3)$, then $\alpha + 1 \equiv 0(\mathrm{mod}\ 3)$ and if $\alpha \equiv 1(\mathrm{mod}\ 3)$ then $\alpha - 1 \equiv 0(\mathrm{mod}\ 3)$. In either case the product $\alpha^2 - 1 = (\alpha-1)(\alpha+1)$ is divisible by $3 : \alpha^2 - 1 \equiv 0(\mathrm{mod}\ 3) \Rightarrow \alpha^2 \equiv 1(\mathrm{mod}\ 3)$. Similarly, $\beta^2 \equiv \gamma^2 \equiv (\mathrm{mod}\ 3)$.

(b) By part (a) we know that $\alpha^2 \equiv 1(\mathrm{mod}\ 3) \Rightarrow \alpha^2 = 3k + 1$, where k is a nonnegative integer. Cubing both sides we obtain,

$$(\alpha^2)^3 = \alpha^2 = (3k+1)^3 = (3k)^3 + 3(3k)^2 \cdot (1) + 3(3k) \cdot (1)^2 + (1)^3 \Rightarrow$$

$$\alpha^6 = 27k^3 + 27k^2 + 9k + 1 \Rightarrow \alpha^6 = 9 \cdot (3k^3 + 3k^2 + k) + 1 \Rightarrow \alpha^6 \equiv 1(\mathrm{mod}\ 9).$$

Another method would be to factor the difference

$$\alpha^6 - 1 = (\alpha^2)^3 - 1 = (\alpha^2 - 1)[(\alpha^2)^2 + (\alpha^2) + 1] = (\alpha^2 - 1) \cdot (\alpha^4 + \alpha^2 + 1).$$

By part (a) we know that $a^2 - 1 = 0(\mathrm{mod}\ 3)$; but also $\alpha^2 \equiv 1(\mathrm{mod}\ 3)$ implies (by Theorem 19(vii)) $(\alpha^2)^2 \equiv (1)^2(\mathrm{mod}\ 3) \Rightarrow \alpha^4 \equiv 1(\mathrm{mod}\ 3)$. By Theorem 19(v), the three congruences $\alpha^2 \equiv 1, \alpha^4 \equiv 1, 1 \equiv 1(\mathrm{mod}\ 3)$ imply $\alpha^4 + \alpha^2 + 1 \equiv 1 + 1 + 1(\mathrm{mod}\ 3)$; $\alpha^4 + \alpha^2 + 1 \equiv 0(\mathrm{mod}\ 3)$. Since each of the factors $(\alpha^2 - 1)$ and $(\alpha^4 + \alpha^2 + 1)$ is divisible by 3 it follows that the product $\alpha^6 - 1 = (\alpha^2 - 1)(\alpha^4 + \alpha^2 + 1)$ is divisible by 9: $\alpha^6 - 1 \equiv 0(\mathrm{mod}\ 9) \Rightarrow \alpha^6 \equiv 1(\mathrm{mod}\ 9)$; likewise $\beta^6 \equiv \gamma^6 \equiv 1(\mathrm{mod}\ 9)$.

(c) This follows at once from part (a):

$$\alpha^2 \equiv 1, \beta^2 \equiv 1, \gamma^2 \equiv 1(\mathrm{mod}\ 3) \Rightarrow \alpha^2 + \beta^2 + \gamma^2 \equiv 1 + 1 + 1 \equiv 0(\mathrm{mod}\ 3).$$

P8. If α, β, x, y are integers and $(x - y)$ is a divisor of $(\alpha x + \beta y)$, then $(x - y)$ must also be a divisor of the product $(\alpha + \beta)(x + y)$.

Solution: Since $(x - y) | (\alpha x + \beta y)$, we have $\alpha x + \beta y = k \cdot (x - y)$, for $k \in \mathbb{Z}$. A bit of algebra yields

$$x \cdot (k - \alpha) = y \cdot (k + \beta). \qquad (1)$$

Let $\delta = (x, y)$, the greatest common divisor of x and y. We set $x = \delta \cdot x_1, y = \delta \cdot y_1$ with $(x_1, y_1) = 1$ (by Theorem 12(iii)). Substituting for x and y back in equation (1) gives

$$\delta \cdot x_1 \cdot (k - \alpha) = \delta \cdot y_1 \cdot (k + \beta) \Rightarrow x_1 \cdot (k - \alpha) = y_1 \cdot (k + \beta). \qquad (2)$$

Now we apply Theorem 12(v): Since $x_1 | y_1 \cdot (k+\beta)$, $(x_1, y_1) = 1$, it follows that x_1 must divide $(k+\beta)$; by the same argument y_1 must divide $(k - \alpha)$; we put $k + \beta = x_1 \cdot k_1$ and

$k-\alpha = y_1 \cdot k_2$; gong back to equation (2) we obtain, $x_1 \cdot y_1 \cdot k_2 = y_1 \cdot x_1 \cdot k_1 \Rightarrow k_2 = k_1 = \lambda$. From the equations $k - \alpha = y_1 \cdot k_2$, $k + \beta = x_1 \cdot k_1$, and with $k_1 = k_2 = \lambda$, we infer

$$(k+\beta) - (k-\alpha) = \lambda \cdot (x_1 - y_1) \Rightarrow \beta + \alpha = \lambda \cdot (x_1 - y_1).$$

Therefore,

$$(\alpha+\beta)\cdot(x+y) = \lambda \cdot (x_1 - y_1) \cdot (\delta x_1 + \delta y_1) = \lambda \cdot (x_1 - y_1) \cdot \delta \cdot (x_1 + y_1) \Rightarrow$$

$$(\alpha+\beta)\cdot(x+y) = \lambda \cdot (\delta x_1 - \delta y_1) \cdot (x_1 + y_1);$$

$$(\alpha+\beta)\cdot(x+y) = \lambda \cdot (x-y) \cdot (x_1 + y_1),$$

which proves that $(x-y)$ is a divisor of $(\alpha+\beta)\cdot(x+y)$.

P9. Let α, β, n be natural numbers such that $\alpha > 1, \beta > 1$, and $\alpha \cdot \beta = 2^n + 1$. Prove that the (positive) integers $(\alpha - 1)$ and $(\beta - 1)$ must be divisible by the same highest power of 2.

Solution: From $\alpha \cdot \beta = 2^n + 1, \alpha, \beta, n \in \mathbb{N}$ and $\alpha, \beta > 1$ we infer that both $(\alpha - 1)$ and $(\beta - 1)$ are odd positive integers and hence by the Fundamental Theorem of arithmetic $\alpha = 2^k \cdot \alpha_1$, for $k \in \mathbb{N}$ and α_1 an odd positive integer. Likewise $\beta = 2^m \cdot \beta_1$, for $m \in \mathbb{N}$ and β_1 an odd positive integer. Apply

$$\alpha \cdot \beta = 2^n + 1 : (2^k \cdot \alpha_1 + 1) \cdot (2^m \cdot \beta_1 + 1) = 2^n + 1 \Rightarrow$$

$$\Rightarrow 2^{k+m} \cdot \alpha_1 \beta_1 + 2^k \cdot \alpha_1 + 2^m \cdot \beta_1 + 1 = 2^n + 1 \Rightarrow \tag{1}$$

$$\Rightarrow 2^{k+m} \cdot \alpha_1 \cdot \beta_1 + 2^k \cdot \alpha_1 + 2^m \cdot \beta_1 = 2^n;$$

We claim that unless $k = m$ a contradiction will ensue. Indeed if $k \neq m$ then either $k > m$ or $k < m$. If $k > m$, then $k - m > 0$, which means that $(k - m)$ is a positive integer. Equation (1) implies,

$$2^m \cdot]2^k \cdot \alpha_1 \cdot \beta_1 + 2^{k-m} \cdot \alpha_1 + \beta_1] = 2^n \tag{2}$$

Note that the integer $2^k \cdot \alpha_1 \cdot \beta_1 + 2^{k-m} \cdot \alpha_1 + \beta_1 = 2^{k-m} \cdot (2^m \cdot \alpha_1 \cdot \beta_1 + \alpha_1) + \beta_1$ is an odd integer since $(k - m) \in \mathbb{N}$ and β_1 is odd; if we look at equation (2) we see that it is contradictory: if $m < n$, then (2) implies $2^k \cdot \alpha_1 \cdot \beta_1 + 2^{k-m} \cdot \alpha_1 + \beta_1 = 2^{n-m}$, an impossibility in view of $(n - m) \in \mathbb{N}$; which means that the left-hand side of the last equation is odd but the right-hand side is even. If $m \geq n$, again a contradiction results from (2) since in that case the left-hand side will be greater than the right-hand side.

The argument that equation (1) leads to a contradiction when $k < m$, is similar. We conclude that $k = m$; that is the positive integers $(\alpha - 1)$ and $(\beta - 1)$ are divisible by the same highest power of 2.

P10. If the integers α, β are even and each of them is of the form $x^2 + k \cdot y^2$, where k is a fixed odd integer, the number $\frac{\alpha \cdot \beta}{4}$ (which is an integer) must also be of the same form.

Solution: Suppose that $\alpha = x_1^2 + k \cdot y_1^2$, $\beta = x_1^2 + k \cdot y_2^2$, for integers x_1, y_1, x_2, y_2. We compute the product

$$\alpha \cdot \beta = (x_1^2 + k \cdot y_1^2) \cdot (x_2^2 + k \cdot y_2^2) = (x_1x_2)^2 + (ky_1y_2)^2 + k \cdot [(y_1x_2)^2 + (x_1y_2)^2] =$$

$$= (x_1x_2)^2 + (ky_1y_2)^2 + 2kx_1x_2y_1y_2 + k \cdot [(y_1x_2)^2 + (x_1y_2)^2 - x_1x_2y_1y_2]$$

$$= (x_2x_2 + ky_1y_2)^2 + k \cdot (y_1x_2 - x_1y_2)^2$$

$$\Rightarrow \frac{\alpha \cdot \beta}{4} = \left(\frac{x_1x_2 + ky_1y_2}{2}\right)^2 + k \cdot \left(\frac{y_1x_2 - x_1y_2}{2}\right)^2. \tag{1}$$

To finish the proof, all we have to do, is show that each of the rational numbers $\left(\frac{x_1x_2+ky_1y_2}{2}\right)$ and $\left(\frac{y_1x_2-x_1y_2}{2}\right)$ is actually an integer. By hypothesis we know that $\alpha \equiv \beta \equiv 0 (\text{mod } 2)$; since $\alpha = x_1^2 + k \cdot y_1^2$, $\beta = x_2^2 + k \cdot y_2^2$ and k is odd, it is easy to see that $x_1 \equiv y_1 (\text{mod } 2)$ and $x_2 \equiv y_2 (\text{mod } 2)$.

There are exactly four possibilities:

1) x_1, y_1, x_2, y_2 are all even.

2) x_1, y_1 are both even while x_2, y_2 are both odd.

3) x_1, y_1 are both odd; while x_2, y_2 are both even.

4) x_1, y_1, x_2, y_2 are all odd.

Since k is odd, it is easy to see that in all four possibilities above, both integers $(x_1x_2 + ky_1y_2)$ and $(y_1x_2 - x_1y_2)$ are even; thus $\left(\frac{x_1x_2+ky_1y_2}{2}\right)$ and $\left(\frac{y_1x_2-x_1y_2}{2}\right)$ are both integers and since equation (1) holds true, we are done.

P11. In number theory courses students learn about the concept of the **order** of an integer a modulo n. This is defined as follows: if n is a positive integer and a an integer relatively prime to n; $(a, n) = 1$; then the order of a modulo n is defined to be the least positive integer or natural number k such that $a^k \equiv 1 (\text{mod } n)$. This concept is of central importance in number theory; in an involved and long proof in number theory courses, it is proved that for every $n \in \mathbb{N}$ and every $a \in \mathbb{Z}$, such that $(n, a) = 1$, such a positive integer k always exists. In this practice problem, prove that if $m \in \mathbb{N}$ and $a^m \equiv 1 (\text{mod } n)$, then the order k must be a divisor of m.

Solution: Since k is the order, $a^k \equiv 1 (\text{mod } n)$. Also, by assumption, $a^m \equiv 1 (\text{mod } n)$. Consider the division $m : k$; $m = q \cdot k + r$, with $0 \le r < k$. We have

$$a^m \equiv 1 (\text{mod } n) \Rightarrow a^{q \cdot k + r} \equiv 1 (\text{mod } n) \Rightarrow (a^k) q \cdot a^r \equiv 1 (\text{mod } n).$$

Now apply $a^k \equiv 1 (\text{mod } n) \Rightarrow (a^k)^q \equiv 1^q \equiv 1 (\text{mod } n)$. Therefore from $(a^k)^q \cdot a^r \equiv 1 (\text{mod } n)$ we obtain $1 \cdot a^r \equiv 1 (\text{mod } n) \Rightarrow a^r \equiv 1 (\text{mod } n)$. The remainder r satisfies $0 \le r < k$; if $1 \le r < k$, then this together with the last congruence violates the fact that k is the least natural number t such that $a^t \equiv 1 (\text{mod } n)$. We conclude that $r = 0 \Rightarrow m = q \cdot k$; k is a divisor of m.

P12. If p is a prime number, $a \in \mathbb{Z}$ with $(a, p) = 1$, show that the order of a modulo p must divide $(p-1)$.

Solution: By Fermat's Little Theorem, Theorem 20(i), we know that $a^{p-1} \equiv 1(\text{mod } p)$; thus, by the previous practice problem, P11, it follows that the order of a modulo p, must be a divisor of the integer $(p-1)$.

P13. Use the concept of the order modulo a positive integer, and practice problem P12, to show that for every natural number n not divisible by 3, the integer $(3^{2n} + 3^n + 1)$ is a multiple of 13.

Solution: By practice problem P12, for $p = 13$, we know that the order of 3 modulo 13, must be a divisor of $p-1 = 13-1 = 12$; the positive integer divisors of 12 are $1, 2, 3, 4, 6$, and 12. We compute $3^1 \equiv 3$, $3^2 \equiv 9$, $3^3 \equiv 27 \equiv 1(\text{mod } 13)$. No need to compute further, 3 is the smallest natural number k such that $3^k \equiv 1(\text{mod } 13)$. Thus, 3 is the order of 3 modulo 13. Next apply the identity $x^3-1 = (x-1)(x^2+x+1)$, with $x = 3^n$, n a natural number.

$$(3^n)^3 - 1 = (3^n - 1)[(3^n)^2 + 3^n + 1] \Rightarrow (3^3)^n - 1 = (3^n - 1)(3^{2n} + 3^n + 1).$$

Clearly $(3^3)^n - 1 = (27)^n - 1 \equiv (1)^n - 1 \equiv 1 - 1 \equiv 0(\text{mod } 13)$, since $27 \equiv 1(\text{mod } 13)$. Therefore $(3^n)^3 - 1$ is divisible by 13, and hence 13 must divide the product $(3^n - 1) \cdot (3^{2n} + 3^n + 1)$. Now if n is not a multiple of 3, the integer $(3^n - 1)$ cannot be divisible by 13.(because if $13|3^n - 1$, then by practice problem P12 the order of 3 (mod 13), which is 3, would have to divide n). Since the prime13 does not divide $3^n - 1$ (for $n \not\equiv 0(\text{mod } 3)$) and it divides the product $(3^n - 1) \cdot (3^{2n} + 3^n + 1)$ it must divide the second factor, $3^{2n} + 3^n + 1$, according to Theorem 12(v).

P14. In each case below, use mathematical induction to prove the given statement.

(a) For every natural number n, the integer $(3^{2n+2} - 2^{n+1})$ is divisible by 7.

(b) For every natural number n, the integer $(2^{2n-1} \cdot 3^{n+2} + 1)$ is divisible by 11.

(c) For every natural number n, the integer $(3^{4n+1} + 10 \cdot 3^{2n} - 13)$ is divisible by 64.

Solution: (a) For $n = 1 : 3^4 - 2^4 = 81 - 4 = 77 \equiv (\text{mod } 7)$, the statement holds true.

Inductive hypothesis: Assume $3^{2k+2} - 2^{k+1} \equiv 0(\text{mod } 7)$. Multiply both sides by $3^2 : 3^2 \cdot (3^{2k+2} - 2^{k+1}) \equiv 3^2 \cdot 0 \equiv 0(\text{mod } 7) \Rightarrow 3^{2(k+1)+2} - 3^2 \cdot 2^{k+1} \equiv (\text{mod } 7)$; substituting $9 = 3^2 = 2 + 7$ in the last congruence gives,

$$3^{2(k+1)+2} - (7+2) \cdot 2^{k+1} \equiv 0(\text{mod } 7) \Rightarrow 3^{2(k+1)+2} - 2^{(k+1)+1} - 7 \cdot 2^{k+1} \equiv 0(\text{mod } 7);$$

but $-7.2^{k+1} \equiv 0(\text{mod } 7)$, and so we obtain, $3^{2(k+1)+2} - 2^{(k+1)+1} \equiv 0(\text{mod } 7)$; the statement also holds for $n = k+1$.

(b) For $n = 1 : 2^{2-1} \cdot 3^{1+2} + 1 = 2 \cdot 3^3 + 1 = 54 + 1 = 55 \equiv 0(\text{mod } 11)$. Assume the statement to hold true for $n = k$:

$$2^{2k-1} \cdot 3^{k+2} + 1 \equiv 0(\text{mod } 11).$$

Multiply the last congruence by $(2^2 \cdot 3)$ to obtain,

$$2^{2(k+1)-1} \cdot 3^{(k+1)+2} + 2^2 \cdot 3 \equiv ((\text{mod } 11) \Rightarrow$$

$$\Rightarrow 2^{2(k+1)-1} \cdot 3^{(k+1)+2} + 1 + 11 \equiv 0(\text{mod } 11;$$

$$2^{2(2+1)-1} \cdot 3^{(k+1)+2} + 1 \equiv 0(\text{mod } 11);$$

which proves that the statement also holds true for $n = k + 1$.

(c) For $n = 1: \quad 3^5 + 10 \cdot 3^2 - 13 = 243 + 90 - 13 = 320 = 5 \cdot 64 \equiv 0(\text{mod } 64)$. Inductive hypothesis: for $n = k$, assume the statement to hold true: $3^{4k+1} + 10 \cdot 3^{2k} - 13 \equiv 0(\text{mod } 64)$; multiply both sides of this congruence by 3^4 to obtain, $3^{4(k+1)+1} + 10 \cdot 3^2 \cdot 3^{2(k+1)} - 13 \cdot 3^4 \equiv (\text{mod } 64)$; substitute for $3^2 = 9 = 8+1$ in the last congruence

$$\left(3^{4(k+1)+1} + 10 \cdot 3^{2(k+1)} - 13\right) + \left(80 \cdot 3^{2(k+1)} - 13 \cdot 3^4 + 13\right) \equiv (\text{mod } 64). \quad (1)$$

Consider the quantity $80 \cdot 3^{2(k+1)} - 13 \cdot 3^4 + 13$; we reduce it modulo 64, by using $80 \equiv 16(\text{mod } 64)$ and $3^4 = 81 \equiv 17(\text{mod } 64)$. We have,

$$\begin{aligned} 80 \cdot 3^{2(k+1)} - 13 \cdot 3^4 + 13 &\equiv 16 \cdot 3^{2(k+1)} - 13 \cdot 17 + 13(\text{mod } 64); \\ 80 \cdot 3^{2(k+1)} - 13 \cdot 3^4 + 13 &\equiv 16 \cdot \left[3^{2(k+1)} - 13\right] (\text{mod } 64). \end{aligned} \quad (2)$$

However, $3^{2(k+1)} - 13 \equiv 0(\text{mod } 4)$, since $3^2 = 0 \equiv 1(\text{mod } 4)$ and $13 \equiv 1(\text{mod } 4)$; and so $3^{2(k+1)} - 13 \equiv (3^2)^{k+1} - 1 \equiv 1 - 1 \equiv 0(\text{mod } 4)$. Therefore $16 \cdot [3^{2(k+1)} - 13] \equiv 0(\text{mod } 64)$, since the number $\left(3^{2(k+1)} - 13\right)$ is divisible by 4. Going back to (2) we see that

$$80 \cdot 3^{2(k+1)} - 13 \cdot 3^4 + 13 \equiv 0(\text{mod } 64) \quad (3)$$

From congruences (1) and (3) it is now clear that, $3^{4(k+1)+1} + 10 \cdot 3^{2(k+1)} - 13 \equiv 0(\text{mod } 64)$; the statement also holds true for $n = k + 1$.

P15. Show that the integers, which, when divided with 6 or 9 respectively, leave remainder 4 or 7 respectively must be of the form $(18k - 2)$, for $k \in \mathbb{Z}$; or equivalently, of the form $(18m + 16)$; $m \in \mathbb{Z}$.

Solution: If n is such an integer, then $n = 6q_1 + 4$ and simultaneously $n = 9q_2 + 7$; for integers q_1, q_2. As a result we must have $6q_1 + 4 = 9q_2 + 7 \Rightarrow 6q_1 = 9q_2 + 3 \Rightarrow$ (dividing both sides by 3) $2q_1 = 3q_2 + 1$; performing the simple division $3 : 2$, we have $3 = 2 \cdot 1 + 1$; substituting for $3 = 2 + 1$ in the last equation yields $2q_1 = 2q_2 + (q_2 + 1)$; which shows that the integer $(q_2 + 1)$ must be divisible by 2 : set $q_2 + 1 = 2k$; $q_2 = 2k - 1 \Rightarrow 2q_1 = 3(2k - 1) + 1 = 6k - 2$; $q_1 = 3k - 1$.

Check: $n = 6q_1 + 4 = 6(3k - 1) + 4 = 18k - 2$

And $n = 9q_2 + 7 = 9(2k - 1) + 7 = 18k - 2$.

From $n = 18k - 2$; by replacing -2 by $(16 - 18)$ we see that, $n = 18k + 16 - 18$; $n = 18 \cdot (k-1) + 16$; set $m = k - 1$ to obtain $n = 18m + 16$.

We remark here that an alternative may be employed in this problem: once we write $n = 6q_1 + 4$; then in view of $(6,9) = 3$, we may just consider the classes of q_1 modulo 3: $q_1 = 3m$, $q_2 = 3m+1$, or $q_2 = 3m+2$; so that respectively, $n = 6q_1 + 4 = 18m + 4$, $18m + 10$, or $18m + 16$. Obviously only $(18m+16)$ leaves remainder 7 when divided by 9, since $18m + 16 = 9(2m+1) + 7$.

P16. Find the remainder of the division, $[(3869)^{37} \cdot (5637)^{31}] : 8$.

First we reduce each of the numbers 3869, 5637, modulo 8.

$$\begin{aligned} 3869 &= (8) \cdot (483) + 5; \quad \text{thus } 3869 \equiv 5 (\text{mod } 8) \\ \text{And} \quad 5637 &= (8) \cdot (704) + 3; \quad \text{thus } 5637 \equiv 3 (\text{mod } 8). \end{aligned}$$

Next we find the orders of 5 and 3 modulo 8: $5^1 \equiv 5$; $5^2 \equiv 25 \equiv 1 (\text{mod } 8)$; thus the order of 5 modulo 8 is 2. And $3^1 \equiv 3$, $3^2 \equiv 9 \equiv 1 (\text{mod } 8)$; likewise the order of 3 modulo 8 is 2. We have

$$\begin{aligned} (3869^{37} &\equiv 5^{37} \equiv 4^{(2 \cdot 18)+1} \equiv (5^2)^{18} \cdot 5 \equiv 1^{18} \cdot 5 \equiv 1 \cdot 5 \equiv 5 (\text{mod } 8). \\ \text{And} \quad (5637)^{31} &\equiv 3^{31} \equiv 3^{(2 \cdot 15)+1} \equiv (3^2)^{15} \cdot 3 \equiv 1^{15} \cdot 3 \equiv 1 \cdot 3 \equiv 3 (\text{mod } 8). \end{aligned}$$

Hence $(3869)^{37} \cdot (5637)^{31} \equiv 5 \cdot 3 \equiv 15 \equiv 7 (\text{mod } 8)$; therefore $(3869)^{37} \cdot (5637)^{31} = 8 \cdot q + 7$, for some positive integer q and since $0 \leq 7 < 8$, it follows that q is the quotient and 7 the remainder of the division $[(3869)^{37} \cdot (5637)^{31}] : 8$.

P17. If α, β, γ, n, are nonzero integers, prove each of the following:

(a) $(\alpha, \beta) = (5\alpha + 3\beta, 13\alpha + 8\beta)$.

(b) $(\alpha, \beta) = (\alpha + \beta, [\alpha, \beta])$.

(c) $\left(2n+1, \frac{n(n+1)}{2}\right) = 1$.

Solution: (a) It suffices to show that every common divisor d of α and β is also a common divisor of the integers $(5\alpha + 3\beta)$ and $(13\alpha + 8\beta)$; and conversely. Indeed by Theorem 5(ix) if $d|\alpha$ and $d|\beta$, then $d|(5\alpha + 3\beta)$ and $d|(13\alpha + 8\beta)$. Let us prove the converse: if $\delta|(5\alpha + 3\beta)$ and $\delta|(13\alpha + 8\beta)$, then $\delta|\alpha$ and $\delta|\beta$.

We apply Theorem 5(ix) again: the assumption $\delta|(5\alpha + 3\beta)$ and $\delta|(13\alpha + 8\beta)$ implies that $\delta|[-13 \cdot (5\alpha + 3\beta) - 3 \cdot (13\alpha + 8\beta)] = \beta$. Thus $\delta|\alpha$ and $\delta|\beta$. We have shown that the set of all common divisors of α and β, is equal to the set of all common divisors of $(5\alpha + 3\beta)$ and $(13\alpha + 8\beta)$. Therefore $(\alpha, \beta) = (5\alpha + 3\beta, 13\alpha + 8\beta)$.

(b) It is obvious that if $d|\alpha$ and $d|\beta$, then $d|(\alpha + \beta)$ and $d|[\alpha, \beta] =$ least common multiple of α and β. Now suppose that $\delta|(\alpha + \beta)$ and $\delta|[\alpha, \beta]$. By Theorem 14(i), $|\alpha \cdot \beta| = (\alpha, \beta) \cdot [\alpha, \beta]$; we set $\alpha = D \cdot \alpha_1$, and $\beta = D \cdot \beta_1$, with $(\alpha_1, \beta_1) = 1$. The last equation gives $|\alpha_1 \cdot \beta_1| \cdot D^2 = D \cdot [\alpha, \beta] \Rightarrow [\alpha, \beta] = D \cdot |\alpha_1 \beta_1|$. Now look at $\delta|(\alpha + \beta) = D \cdot (\alpha_1 + \beta_1)$ and $\delta|[\alpha, \beta] = D \cdot |\alpha_1 \beta_1|$. We apply Theorem 16(i): the integer δ factors into a product of primes uniquely, as postulated by that Theorem. If we can show that every prime divisor of δ, also divides D, then it will follow $\delta|D$. Pick p, a prime divisor of δ. Since $p|\delta$ and $\delta| \cdot D \cdot |\alpha_1 \beta_1|$, we must have $p|D \cdot |\alpha_1 \beta_1|$.

If $p|D$, we are done. If $p||\alpha_1\beta_1|$, then $p|(\alpha_1\beta_1)$ and by Theorem 15(iv), p must divide α_1 or β_1 but not both, since (α_1, β_1). If $p|\alpha_1$, then from $\delta|D(\alpha_1+\beta_1) = D\alpha_1 + D\beta_1$, and $p|\delta$, we infer that $p|[(D\alpha_1 + D\beta_1) - D\alpha_1] = D\beta_1$. However $p|\alpha_1$ and in view of $(\alpha_1, \beta_1) = 1$, we conclude that $(p_1, \beta_1) = 1$. Thus the fact that $p|D\beta_1$ implies by Theorem 12(v), that $p|D$. In a similar fashion we can show that the assumption $p|\beta_1$ implies $p|D$. In all cases, $p|D \Rightarrow \delta|D = (\alpha, \beta) \Rightarrow \delta$ is a common divisor of α and β; thus the set of common divisors of α and β is the same as the set of common divisors of $(\alpha+\beta)$ and $[\alpha, \beta]$. Therefore $(\alpha, \beta) = (\alpha+\beta, [\alpha, \beta])$.

(c) Note that $n(n+1)$ is always an even integer. If $n = 2k$, $k \neq 0$, then $\frac{n(n+1)}{2} = k \cdot (2k+1)$ and $2n+1 = 4k+1$. Note that unless $k = -1$, the integer $k \cdot (2k+1)$ is different from 1 or -1; so it has a prime divisor, say p. Since $p|k \cdot (2k+1) \Rightarrow p|k$ or $p|(2k+1)$. We claim that p cannot also divide $(4k+1)$: If it did, then either $p|k$ and $p|4k+1$; or $p|(2k+1)$ and $p|4k+1$; in the first case, $p|[(4k+1) - 4k] = 1$, a contradiction. In the second case, $p|[2(2k+1) - (4k+1)]$; $p|1$, a contradiction again.

Finally suppose that $n = 2k+1$; then we have $\frac{n(n+1)}{2} = (2k+1)(k+1)$ and $2n+1 = 2(2k+1)+1 = 4k+3$. Again if $p|[(2k+1)\cdot(k+1)]$ and $p|4k+3$ we will have a contradiction since this implies $p|(2k+1)$ and $p|(4k+3)$; or $p|(k+1)$ and $p|(4k+3)$. In the first case we conclude $p|[(4k+3) - 2(2k+1)] = 1$; and in the second case we conclude $p|[4(k+1) - (4k+3)] = 1$; again a contradiction. Hence the integers $\frac{n(n+1)}{2}$ and $(2n+1)$ cannot have a prime in common; which is to say,

$$\left(\frac{n(n+1)}{2}, 2n+1\right) = 1$$

P18. If $\epsilon = [\alpha, \beta]$, the least common multiple of two nonzero integers α, β; and $\gamma \in \mathbb{Z}$ such that $\gamma \neq 0$ and $(\gamma, \beta) = 1$, then $[\alpha\gamma, \beta] = \epsilon \cdot |\gamma|$.

Solution: By Theorem 14(i); or by Remark 7 we know that

$$[\alpha\gamma, \beta] = \frac{|(\alpha\gamma)\cdot\beta|}{(\alpha\gamma, \beta)} = \frac{|\alpha|\cdot|\beta|\cdot|\gamma|}{(\alpha\gamma, \beta)}$$

Since $(\gamma, \beta) = 1$, then in view of $(\alpha\gamma, \beta)|(\alpha\gamma)$ and $(\alpha\gamma, \beta)|\beta$, it follows that $(\alpha\gamma, \beta) = (\alpha, \beta)$. This requires some explanation: for simplicity let $d = (\alpha\gamma, \beta)$. Clearly, d cannot have a factor greater than 1 in common with γ, because of the condition $(\gamma, \beta) = 1$. Thus we conclude that $(d, \gamma) = 1$. By Theorem 12(v), the condition $(d, \gamma) = 1$ together with $d|(\alpha\gamma)$ imply $d|\alpha$. Hence $d|\alpha$ and $d|\beta$; which means that d is a common divisor of α and β and so $d|\delta = (\alpha, \beta)$. But since $\delta|\alpha$ and $\delta|\beta$ it follows that $\delta|(\alpha\gamma)$ and $\delta|\beta$; which means that δ is a common divisor of $(\alpha\gamma)$ and β; δ must divide $d = (\alpha\gamma, \beta)$, by Theorem 9(viii). We have $(d|\delta$ and $\delta|d) \Rightarrow$ (since both d and δ are positive integers) $d = \delta$; $(\alpha\gamma, \beta) = (\alpha, \beta)$. Going back to

$$[\alpha\gamma, \beta] = \frac{|\alpha|\cdot|\beta|\cdot|\gamma|}{(\alpha\gamma, \beta)} = \frac{|\alpha|\cdot|\beta|}{(\alpha, \beta)}\cdot|\gamma| = \epsilon\cdot|\gamma|,$$

by Remark 7.

P20. If α, β, γ are nonzero integers such that $(\gamma, \beta) = 1$ and $\gamma | \alpha$; then $\left[\frac{\alpha}{\gamma}, \beta\right] = \frac{\epsilon}{|\gamma|}$, where $\epsilon = [\alpha, \beta]$.

Solution: Since $\gamma | \alpha$, the number $\frac{\alpha}{\gamma}$ is an integer. We write $\alpha = \left(\frac{\alpha}{\gamma}\right) \cdot \gamma$; because of the condition $(\gamma, \beta) = 1$ we can apply the previous problem P18:

$$\epsilon = [\alpha, \beta] = \left[\left(\frac{\alpha}{\gamma}\right) \cdot \gamma, \beta\right] \underset{\text{by P18}}{=} \left[\frac{\alpha}{\gamma}, \beta\right] \cdot |\gamma| \Rightarrow \frac{\epsilon}{|\gamma|} = \left[\frac{\alpha}{\gamma}, \beta\right].$$

We are done.

P20. (i) Show that if the positive integer k is not a multiple of 3, that is, if $k \not\equiv 0 (\text{mod } 3)$, then the positive integer $(2^{2k} + 2^k + 1)$ is divisible by 7; $2^{2k} + 2^k + 1 \equiv 0 (\text{mod } 7)$.

(ii) For every positive integer α, with $\alpha \not\equiv 0 (\text{mod } 3)$, show that $2^{4\alpha} + 2^{2\alpha} + 1 \equiv 0 (\text{mod } 7)$.

(iii) For every natural number n, $2^{2^{n+1}} + 2^{2^n} + 1 \equiv 0 (\text{mod } 7)$.

Solution: (i) We first find the order of 2 modulo 7 (refer to practice problem P11). We have $2^1 = 2$, $2^2 \equiv 4$, $2^3 \equiv 8 \equiv 1 (\text{mod } 3)$; thus, 3 is the order of 2 modulo 7. We apply the identity $a^3 - 1 = (a-1)(a^2 + a + 1)$ with $a = 2^k$: $(2^k)^3 - 1 = (2^k - 1)(2^{2k} + 2^k + 1)$; note that $2^k - 1 \not\equiv 0 (\text{mod } 7)$, by Problem P11: If it were $2^k - 1 \equiv 0 (\text{mod } 7)$, then $2^k \equiv 1 (\text{mod } 7)$ and thus 3 (the order of 2 modulo 7) would have to divide k, contrary to the hypothesis of the problem. Thus 7 does not divide $2^k - 1$ and since 7 is a prime, $(7, 2^k - 1) = 1$. On the other hand, $(2^k)^3 - 1 = (2^3)^k - 1 \equiv (1)^k - 1 \equiv 1 - 1 \equiv 0 (\text{mod } 7)$, which shows that prime 7 divides the product $(2^k - 1)(2^{2k} + 2^k + 1)$; but $(7, 2^k - 1) = 1$. According to Theorem 12(v), 7 must divide $2^{2k} + 2^k + 1$; $2^{2k} + 2^k + 1 \equiv 0 (\text{mod } 7)$.

(ii) This follows immediately from part (i): just put $k = 2\alpha$; clearly $\alpha \equiv 0 (\text{mod } 3)$ implies $k \not\equiv 0 (\text{mod } 3)$.

(iii) Note that $2^{n+1} = 2 \cdot 2^n$; if we set $k = 2^n$, then k is not a multiple of 3, for any value of the positive integer exponent $n : k \not\equiv 0 (\text{mod } 3)$, for all $n \in \mathbb{N}$; the result follows from part (i).

P21. Let a, b, c be integers with $|a| + |b| + |c| > 0$ (which means that at least one of a, b, c is nonzero). An integer δ is called a **common divisor** of a, b, c if δ is a divisor of each of the three integers a, b, and c.

(a) Show that the set of common divisors of the integers a, b, and c, has a maximum element, that being a positive integer denoted by $\boldsymbol{(a, b, c)}$ and called the **greatest common divisor** of a, b, and c.

(b) Show that $(a, b, c) = ((a, b), c) = (a, (b, c)) = ((a, c), b)$.

(c) An integer ρ is called a **common multiple** of the nonzero integers a, b, and c, if it is divisible by each of the integers a, b, and c. Show that among all positive common multiples of a, b, c, there is at least one, called the **least common multiple** of a, b and c; and it is denoted by $\boldsymbol{[a, b, c]}$ (Here we actually assume that $abc \neq 0$.)

(d) Show that $[a, b, c] = [[a, b], c] = [a, [b, c]] = [[a, c], b]$; with a, b, c, being nonzero integers.

(e) Show that if a, b, c are nonzero integers then,

$$(a,b,c) \cdot [a,b,c] \leq |a \cdot b \cdot c|$$

furthermore show that the equal sign holds exactly (or only if) when, $(a, b) = (a, c) = (b, c) = 1$; in other words the equal sign holds precisely when the integers a, b, c are mutually relatively prime.

Solution: (a) The condition $|a| + |b| + |c| > 0$ implies at least one of a, b, c must be nonzero. Without loss of generality, assume $a \neq 0$. Then $|a| + |b| > 0$ as well. Clearly, every common divisor δ of a, b, c, is also a common divisor of the integers a and b. Since $|a| + b| > 0$, Theorem 9(i) applies: the greatest common divisor (a, b) of a and b, is a positive integer: $(a, b) \geq 1$. Now, consider the set of all positive common divisors of the three integers a, b, c: each such positive divisor δ, is also a positive integer divisor of a and b; thus $1 \leq \delta \leq (a, b)$; which proves that there are at most (a, b) positive common divisors of a, b, and c. Also note that the set of all positive common divisors of a, b, c, is nonempty, since it contains the number 1. We can now conclude that since $1 \leq \delta \leq (a, b)$; among all positive common divisors of a, b, c, there exists a greatest one, denoted by (a, b, c); and (a, b, c) cannot exceed $(a, b) : 1 \leq (a, b, c) \leq (a, b)$.

(b) We will show that $(a, b, c) = ((a, b), c)$. Establishing the other two equalities, is done is a similar fashion. It suffices to show that the set of common divisors of the three integers a, b, c; is equal to the set of common divisors of the two integers (a, b) and c. Indeed, let $\delta \in \mathbb{Z}$ and $\delta | a,\ \delta | b,\ \delta | c$; since $\delta | a,\ \delta | b$, it follows by Theorem 9(vi) that $\delta | (a, b)$; hence δ is common divisor of (a, b) and c. Conversely, suppose d is a common divisor of (a, b) and c; since $d | (a, b)$, it follows, again by Theorem 9(viii), that $d | a$ and $d | b$; and so $d | a, d | b, d | c$. We have proved that every common divisor of a, b, c, is also a common divisor of (a, b) and c; and conversely every common divisor of (a, b) and c, is also a common divisor of a, b, and c. Therefore,

$$(a, b, c) = ((a, b), c)$$

(c) Among all positive common multiples of the three nonzero integers a, b, c, there has to be a least one: this is guaranteed by the Well Ordering Property, since the set of all positive common multiples of a, b, and c, is a subset of the set of natural numbers $\mathbb{N}$.

(d) We prove that $[a, b, c] == [[a, b], c]$. The other two equalities are established in a similar way. If l is a common multiple of a, b, and c; then obviously l is a common multiple of a and b; by Theorem 14(ii), l must be divisible by $[a, b]$, which shows that l is a common multiple of the integers $[a, b]$ and c. Conversely let λ be a multiple of $[a, b]$ and c; by the transitivity of divisibility (Theorem 5(iv)) and the very definition of $[a, b]$, it follows that since λ is divisible by $[a, b]$ and $[a, b]$ is divisible by both a and b; λ must be divisible by both a and b; that is, λ is common multiple of a and b; and since by hypothesis λ is a multiple of c as well, we conclude that λ is a common multiple of a, b, and c. We have in effect proved that the set of all common multiples of a, b, and c, is equal to the set of all common multiples of $[a, b]$ and c. Hence the set of all positive common multiple of a, b, c, must be the same as the set of all positive common multiples of $[a, b]$, and c. Therefore,

$$[a, b, c] = [[a, b], c].$$

NOTE: We note here that a formula for $[a, b, c]$ in terms of $[a, b]$ can be obtained: by Remark 7 we know that $[[a, b], c] = \frac{[a,b] \cdot |c|}{([a,b],c)}$. Thus, $[a, b, c] = \frac{[a,b] \cdot |c|}{([a,b],c)}$.

(e) According to the **NOTE** at the end of the proof for part (d), we know that

$$[a, b, c] = \frac{[a, b] \cdot |c|}{([a, b], c)} \tag{1}$$

Because of (1) the statement we are asked to prove in this part is equivalent to

$$(a, b, c) \cdot \left(\frac{[a, b] \cdot |c|}{([a, b], c)} \right) \leq |a| \cdot |b| \cdot |c| \tag{2}$$

$$\begin{gathered} \Leftrightarrow (\text{since } abc \neq 0) \frac{(a, b, c) \cdot [a, b]}{([a, b], c)} \leq |a| \cdot |b| \\ \Leftrightarrow (\text{since } ([a, b], c) > 0)(a, b, c) \cdot [a, b] \leq |ab| \cdot ([a, b], c) \end{gathered} \tag{3}$$

(Since all steps are reversible back to (2) and hence from (1); to the statement we are trying to prove) if we establish (3), then the proof will be complete. First observe that,

$$1 \leq [a, b] \leq |a| \cdot |b| \tag{4}$$

Inequalities (4) should be clear, in view of the fact that $[a, b]$ is the least common multiple of a and b; and $|a||b|$ is a positive common multiple of a and b. Next we have, $(a, b, c) = ((a, b), c)$ by part (b) of this problem. On the other hand, the positive integer (a, b) is a divisor of $[a, b]$, since $(a, b)|a$ and $(a, b)|b$; and $a, b|[a, b]$. This implies that

$$1 \leq ((a, b), c) \leq ([a, b], c). \tag{5}$$

Since $((a, b), c)$ divides both (a, b) and c; and thus $((a, b), c)$ is a common divisor of $[a, b]$ and c (in view of $(a, b)|[a, b]$ – see above). Multiplying (4) and (5) memberwise yields.

$$1 \leq [a, b] \cdot ((a, b), c) \leq |ab|.([a, b], c);$$

Using $((a, b), c) = (a, b, c)$ finally gives,

$$1 \leq [a, b] \cdot (a, b, c) \leq |ab| \cdot ([a, b], c),$$

which clearly establishes (3).

From the above proof it is evident that if either of the inequalities $[a, b] \leq |a| \cdot |b|$ and $((a, b), c) \leq ([a, b], c)$, holds with a strict inequality sign then (3) will also hold with a strict inequality sign. Thus, the equal sign will hold in (3) if, and only if both equalities hold below:

$$[a, b] = |a| \cdot |b| \text{ and } ((a, b), c) = ([a, b], c) \tag{6}$$

The first equality in (6) implies, by Theorem 14(iii), that $(a, b) = 1$; then, from the second equality in (6) we obtain $(1, c) = ([a, b], c)$; but $(1, c) = 1$ and so $1 = ([a, b], c)$; making use of $|a| \cdot |b| = [a, b]$ we arrive at $(|a| \cdot |b|, c) = 1$ which implies $(c, |a|) = 1 = (c, |b|)$. Altogether $(a, b) = (a, c) = (b, c) = 1$. The proof is complete.

We note here that the last step should be clear: if it were $\delta = (c, |a|) > 1$(or $(c, |b|) > 1$), then δ would divide the product $|a| \cdot |b|$ as well; so that $(|a| \cdot |b|, c) \geq \delta > 1$, contrary to $(|a| \cdot |b|, c) = 1$. ∎

P22. This problem goes back to Chapter 2. Show that if m, n are positive integers which are relatively prime, then the ring $\mathbb{Z}_{mn}$ is isomorphic to the ring $\mathbb{Z}_m \times \mathbb{Z}_n$.

Solution: Recall that $\mathbb{Z}_m$ is the ring whose elements are the integers $0, 1, \ldots, (m-1)$; and $\mathbb{Z}_n$ has elements $0, 1, \ldots, (n-1)$; both of these rings are commutative rings with identity (the number 1). The operations of addition and multiplication in those rings, are simply the addition and multiplication modulo m and n respectively. As for the ring $\mathbb{Z}_m \times \mathbb{Z}_n$, this is simply the Cartesian product ring that has $(m \cdot n)$ elements, each element being a pair (i, j); with $0 \leq i \leq m-1$ and $0 \leq j \leq n-1$. Addition and multiplication in $\mathbb{Z}_m \times \mathbb{Z}_n$ is simply the coordinate-wise addition and multiplication modulo m and modulo n. Finally the elements of $\mathbb{Z}_{mn}$ are the integers $0, 1, \ldots, (mn-1)$; the two operations are addition and multiplication modulo (mn).

Consider the mapping $f : \mathbb{Z}_{mn} \to \mathbb{Z}_m \times \mathbb{Z}_n$, defined by $f(x) = (x_m, x_n)$, where x_m is the remainder of the division $x : m$; and x_n is the remainder of the division $x : n$. Back in Chapter 2, we showed that this mapping is actually a homorphism. Here we show that it is also a bijection: a one-to-one and onto or surjective mapping; this will complete the proof, since a bijective homomorphism is by definition an isomorphism.

First we show that the above mapping is onto or surjective. Let (i, j) be an element of $\mathbb{Z}_m \times \mathbb{Z}_n$; with i being one of the integers $0, 1, \ldots, m-1$; j one of the integers $0, 1, \ldots, n-1$. Below, we will show that there is an integer α such that,

$$\left.\begin{array}{r} \alpha \equiv i (\text{mod } n) \\ \text{and} \quad \alpha \equiv j (\text{mod } n) \end{array}\right\}$$

Indeed: by Theorem 9(ii), there exist integers k andλ such that,

$$\begin{array}{l} k \cdot m + \lambda \cdot n = (m, n); \\ k \cdot m + \lambda \cdot n = 1 \end{array}$$

Multiply both sides by

$$(i-j): \quad (i-j)km + (i-j)\lambda n = i - j \Rightarrow j + (i-j)\lambda n = i + (j-i)km.$$

Let $\alpha = j + (i-j)\lambda n = i + (j-i)km$. By definition 10, $\alpha \equiv i(\text{mod } m)$, since $m|(\alpha - i)$. Likewise $\alpha \equiv j(\text{mod } n)$; we have established the existence of an integer α, that satisfies the system of congruences,

$$\left.\begin{array}{r} x \equiv i(\text{mod } m) \\ \text{and} \quad x \equiv (\text{mod } n) \end{array}\right\} \qquad (1)$$

Next we show that the integer α is unique modulo (mn): which is to say that if another integer β also satisfies the system of congruence (1), then $\alpha \equiv \beta(\text{mod } mn)$. To see why, suppose that β satisfies (1); then

$$\alpha \equiv \beta \equiv i(\text{mod } m) \text{ and } \alpha \equiv \beta \equiv j(\text{mod } n) \Rightarrow \alpha - \beta \equiv 0(\text{mod } m) \text{ and}$$
$$\alpha - \beta \equiv 0(\text{mod } n)$$

The last two congruences say that both m and n divide the integer $(\alpha - \beta)$; however, since $(m, n) = 1$, Theorem 13 applies. According to Theorem 13, since m and n are relatively prime and each divides $(\alpha - \beta)$, their product must also divide $(\alpha - \beta)$. Thus $\alpha - \beta \equiv 0(\text{mod } mn) \Rightarrow \alpha \equiv \beta(\text{mod } mn)$, which proves the claim. Among all integers which are congruent to α modulo (mn), there exists a smallest nonnegative one; one of the nonnegative integers $0, 1, 2, \ldots, (mn-1)$. Let l be that smallest nonnegative integer; $\alpha \equiv l(\text{mod } mn)$. Obviously the last congruence implies $\alpha \equiv l(\text{mod } m)$ and $\alpha \equiv l(\text{mod } n)$; thus l satisfies the system (1) and since l is an element of $\mathbb{Z}_{mn}$, it follows that the mapping f is surjective. Also note that the integer l is none other than the remainder of the division $\alpha : (mn)$.

Finally l is the only integer x among $0, 1, 2, \ldots, (mn-1)$, such that $f(x) = (i, j)$; that is true because we have already proved that any integer x that satisfies $f(x) = (i, j)$; must also satisfy the system (1) and hence it must be congruent to α modulo (mn); and hence congruent to l modulo (mn). This proves that the mapping f is also one-to-one (since l is the unique integer x in $\mathbb{Z}_{mn}$ such that $f(x) = (i, j)$). ■

Remark 12: In the proof that was given above (practice problem P22), we in effect proved the **Chinese Remainder Theorem** for $k = 2$; the Chinese Remainder Theorem says that given a natural number k, and positive integers $n_1, n_2, \ldots, n_k$ which are mutually relatively prime (i.e. $(n_i, n_j) = 1$, for $i \neq j$; $1 \leq i, j \leq k$), and any integers $r_1, r_2, \ldots r_k$; then the system of simultaneous congruences,

$$\left.\begin{array}{l} x \equiv r_1(\text{mod } n_1) \\ x \equiv r_2(\text{mod } n_2) \\ \vdots \\ x \equiv r_k(\text{mod } n_k) \end{array}\right\},$$

has a unique solution modulo $(n_1 \cdot n_2 \ldots n_k)$. This is proved in undergraduate number theory courses; the proof not only establishes the existence and uniqueness (modulo $(n_1 \cdot n_2 \ldots n_k)$) but it actually provides a computational method or algorithm to actually find that solution.

4.10 Exercise set

1. Prove Theorem 1, part (ii).

2. Prove Theorem 1, part (iv).

3. Prove Theorem 1, part (v).

4. Prove Theorem 1, part (vii).

5. Prove Theorem 3.

6. Prove part (iii) of Example 3.

7. Given a proof that uses mathematical induction for Theorem 7(v).

8. Give a proof that uses mathematical induction for Theorem 7(v).

9. Use mathematical induction to prove that the given statement holds for every natural number n, for parts (i)- (ii); for parts (iii)-(iv), you may want to use known results such as Theorem 1.

(i) $\frac{3}{1^2 \cdot 2^2} + \frac{5}{2^2 \cdot 3^2} + \ldots + \frac{2n+1}{2^n \cdot (n+1)^2} = \frac{n \cdot (n+2)}{(n+1)^2}$.

(ii) $1 \cdot 2^2 + 2 \cdot 3^2 + \ldots + (n-1) \cdot n^2 = \frac{n \cdot (n+1)(3n^2 - n - 2)}{12}$, for $n \in \mathbb{Z}$ and $n \geq 2$.

(iii) $\frac{n^2}{2} < 1 + 2 + \ldots + n < \frac{(n+1)^2}{2}$.

(iv) $\frac{n^3}{3} < 1^2 + 2^2 + \ldots + n^2 < \frac{(n+1)^3}{3}$.

(v) $\frac{n^4}{4} < 1^3 + 2^3 + \ldots + n^3 < \frac{(n+1)^4}{4}$.

10. Let $\alpha, \beta \in \mathbb{Z}$. Prove each of the following:

(i) If α and β are both even, so are the integers $(\alpha + \beta)$, $(\alpha - \beta)$, and $(\alpha \cdot \beta)$.

(ii) If α and β are both odd, then the integers $(\alpha + \beta)$ and $(\alpha - \beta)$ are both even; but $(a \cdot b)$ is odd.

(iii) If one of α, β, is even and the other is odd, then both integers $(\alpha + \beta)$ and $(\alpha - \beta)$ are odd, whereas $(\alpha \cdot \beta)$ is even.

11. If the integer α is a sum of two integer squares, then so is the integer 2α.

Hint: $2 = 1^2 + 1^2$.

12. Let m and n be fixed integers.

(i) Find the number of even integers that lie between the integers $2n$ and $2(n+m)$.

(ii) Find the number of odd integers that lie between $2n$ and $2(n+m)$.

1. Prove each of the following:

   (i) The product of two consecutive integers is an even integer.

   (ii) The product of two successive even integers is a multiple of 8.

   (iii) The only divisors of the integers 1 and -1, are the integers 1 and -1.

13. If $\alpha, \beta \in \mathbb{Z}$ and 5 does not divide the product ab, then

   (i) The remainder of the division $\alpha^4 : 5$, is equal to 1.

   (ii) The remainder of the division $\alpha^2 : 5$, is either equal to 1, or equal to 4.

   (iii) $5|\alpha^4 - \beta^4$.

14. If $\alpha, \beta \in \mathbb{Z}$ then (i) $3|(\alpha^3 - \alpha)$, (ii) If $3|(\alpha \pm \beta)$, then $3|(\alpha^3 \pm \beta^3)$, (iii) If $7|(10\alpha + \beta)$, then $7|(\alpha^3 - \beta^3)$.

15. If $\alpha, \beta \in \mathbb{Z}$, then (i) $600|\alpha^2 \cdot (\alpha^4 - 1) \cdot (\alpha^4 - 16)$, (ii) $30|\alpha\beta \cdot (\alpha^4 - \beta^4)$.

16. If $\alpha_1\alpha_2, \dots, a_n$ are positive integers; each of them greater than 1; n and x are both positive integers, and $\alpha = x^{n-1} + x^{n-2} + \dots + x + 1$, $\beta = x^{(na_1-1)} + x^{(na_2-2)} + \dots + x^{(na_n-n)}$, then $\alpha|\beta$. Note that $na_i - i \geq 1$, since $a_i \geq 2$ and $n \geq i$).

Hint: Let $y = x^n$; note that $(x-1) \cdot \alpha = y - 1$ and $(x-1) \cdot \beta = (x-1) \cdot \left[x^{(na_1-1)} + \dots + x^{(na_i-i)} + \dots + x^{(na_n-n)}\right]$; write the ith term, $x^{(na_i-i)}$, as follows:

$$x^{(na_i-i)} = x^{(na_i-i)} - x^{(n-i)} + x^{(n-i)} = x^{n-i} \cdot \left(x^{n(a_i-1)} - 1\right) + x^{n-i}$$

Note that $x^{n(a_i-1)} - 1 = (x^n)^{(a_i-1)} - 1 = y^{(a_i-1)} - 1$, which shows that since $a_i - 1 \geq 1$, the last expression is divisible by $(y-1)$. From there conclude that the sum the above brackets contains, is a sum of two quantities, one of which is divisible by $(y-1)$. (Because it is the sum of terms of the form $y^{(a_i-1)} - 1$; for $i = 1, 2, \dots, n$). The other quantity is really the sum $\sum_{i=1}^{i=n} x^{n-i} = x^{n-1} + \dots + x + 1$. But then, $(x-1)(x^{n-1} + \dots + x + 1) = x^n - 1 = y - 1$. From all the above conclude that $(x-1) \cdot \beta$ is divisible by $(y-1)$; finally conclude that $\alpha|\beta$.

17. Show that each of the natural numbers $1, 2, \dots, n$ must divide at least one of the natural numbers $n+1, n+2, \dots 2n$.

18. In each of the following use either mathematical induction or the concept of the order of an integer modulo a prime number (refer to practice problems P11-P13); or a combination of these two methods to show that the given number b divides the given expression (n is a natural number)

   (i) $b|(2^{2n} + 15n - 1)$; $b = 9$.

(ii) $b|(3 \cdot 5^{2n+1} + 2^{3n+1})$; $b = 17$

(iii) $b|(7^{2n+1} - 4n - 7)$; $b = 288$.

(iv) $b|(n^{13} - n)$; $b = 2730$.

(v) $b|n(n+1)(2n+1)(3n^2+3n-1)$; $b = 30$.

19. Show that there does not exist an integer α such that the remainder of the division $\alpha : 15$ is equal to 10, while the remainder of the division $\alpha : 10$ is 7.

20. Describe all the integers with the property that if they are divided by 3,4,5,6, they produce remainders 2,3,4, and 5 respectively.

21. If $\alpha\beta \in \mathbb{Z}$ and $\alpha > \beta$, prove that the remainders of the divisions $\alpha : (\alpha - \beta)$ and $\beta : (\alpha - \beta)$ are equal; whereas the two quotients differ by 1.

22. When the numbers 688 and 486 are divided by the positive integer δ, they leave remainders 17 and 13 respectively. Find the value(s) of δ.

23. Prove each of the following for any two integers a and b such that $(a, b) = 1$.

(i) $(a + kb, b) = 1$, for any integer k.

(ii) $(a+b, a-b) = 1$ or 2; depending on whether $a+b \equiv 1$ or $a+b \equiv 0 \pmod 2$ respectively.

24. Suppose that $A = 11\alpha + 2\beta$, $B = 18\alpha + 5\beta (\alpha, \beta \in \mathbb{Z})$. Show that either $(A, B) = 1$ or $(A, B) = 19$.

25. If $n \in \mathbb{N}$, $a \in \mathbb{Z}$, and $(a, 5) = 1$, prove that $a^{8n} + 3a^{4n} - 4 \equiv 0 \pmod 5$.

26. Suppose that $\alpha, \beta, m \in \mathbb{N}$ and $(\alpha, \beta) = 1$. Moreover assume $(\alpha - \beta)$ to be an odd integer. Show that $((\alpha+\beta)^m, (\alpha-\beta)^m) = 1$.

27. Let α, β, γ be nonzero integers. Prove each of the following:

(i) $[\alpha\gamma, \beta\gamma] = |\gamma| \cdot [\alpha, \beta]$.

(ii) If $\gamma|\alpha$ and $\gamma|\beta$, then $\left[\frac{\alpha}{\gamma}, \frac{\beta}{\gamma}\right] = \frac{[\alpha,\beta]}{|\gamma|}$.

28. If $\alpha, \beta, \gamma, \delta, \alpha\beta\gamma\delta \neq 0$, and $\frac{\alpha}{\beta} = \frac{\gamma}{\delta}$, show that $\frac{(\alpha,\delta)}{(\beta,\gamma)} = \frac{[\beta,\gamma]}{[\alpha,\delta]}$.

29. Find each of the following in terms of the natural number n:

(i) $[n, n+1]$.

(ii) $[n, n+2]$.

(iii) $\left[\frac{(n-1)n(n+1)}{6}, \frac{n(n+1)(n+2)}{6}\right]$.

30. Let a and b be positive integers. Let $p_1, \ldots, p_n$ be the distinct primes that a and b have in common; if a and b have no primes in common, that is, if $(a, b) = 1$, assume $p_1 = \ldots = p_n = 1$. Furthermore let $q_1, \ldots, q_k$ be those distinct primes (if any) that divide a but do not divide b; and let $r_1, \ldots, r_m$ be the distinct primes (if any) that divide b but do not divide a. (If no such primes q_i exist, take $q_1 = \ldots = q_k = 1$. Likewise, take $r_1, = \ldots = r_m = 1$, if no such primes r_j exist.) According to Theorem 16(ii), we must have

$$a = p_1^{\alpha_1} \ldots p_n^{\alpha_n} \cdot q_1^{\beta_1} \ldots q_k^{\beta_k} \text{ and } b = p_1^{\delta_1} \ldots p_n^{\delta_n} \cdot r_1^{\gamma_1} \ldots r_m^{\gamma_m},$$

with the exponents $\alpha_1, \ldots \alpha_n, \delta_1, \ldots, \delta_n$ being positive integers, if $(a, b) > 1$. If $(a, b) = 1$, then take $\alpha_1 = \ldots = \alpha_n = \delta_1 = \ldots = \delta_n = 0$; and with the exponents $\beta_1, \ldots, \beta_k$ being positive integers if $k \geq 1$; if no q_i's exist, assume $\beta_1 = \ldots = \beta_k = 0$. Likewise, assume $\gamma_1, \ldots, \gamma_m$ to be positive integers if $m \geq 1$; if no r_j's exist, assume $\gamma_1 = \ldots == \gamma_m = 0$. Prove each of the following:

(i) $(a, b) = p_1^{m_1} \cdot p_2^{m_2} \ldots p_n^{m_n}$; where $m_i = \min\{\alpha_i, \delta_i\}$, for $i = 1, 2, \ldots n$.

(ii) $[a, b] = p_1^{M_1} \cdot p_2^{M_2} \ldots p_n^{M_n} \cdot q_1^{\beta_1} \ldots q_k^{\beta_k} \cdot r_1^{\gamma_1} \ldots r_m^{\gamma_m}$, where $M_j = \max\{\alpha_j, \delta_j\}$, for $j = 1, 2, \ldots n$.

31. Recall from Theorem 9(ii) that if δ is the greatest common divisor of the integers a and b (with $|a| + |b| > 0$), then $\delta = m \cdot a + n \cdot b$, for some integers m and n.

(i) Show, by giving a simple example (or counterexample actually) that if an integer k satisfies $k = t_i \cdot a + t_2 \cdot b$, for $t_1, t_2 \in \mathbb{Z}$, that by itself, does not necessarily make k to be the greatest common divisor of a and b. But if, as Theorem 9(v) postulates, k is a common divisor of a and b, which also satisfies $k = t_1 \cdot a + t_2 \cdot b$, then $k = (a, b)$.

(ii) Prove that the integers m and n of part (ii) of Theorem 9 are not unique; in fact, show that there exist infinitely many integer pairs (t_1, t_2) that satisfy $(a, b) = t_1 \cdot a + t_2 \cdot b$, by the following procedure outlined below. Let (m, n) be such a pair, which exists by Theorem 9(ii). to find all the other such pairs (t_1, t_2) we must solve

$$m \cdot a + n \cdot b = t_1 \cdot a + t_2 \cdot b \Leftrightarrow a \cdot (m - t_1) = b \cdot (t_2 - n) \Leftrightarrow \frac{a}{\delta} \cdot (m - t_1) = \frac{b}{\delta} \cdot (t_2 - n);$$

where $\delta = (a, b)$. Next use the fact that $\left(\frac{a}{\delta}, \frac{b}{\delta}\right) = 1$ (Theorem 12(iii)) and Theorem 12(v) to proceed further.

32. Perform each of the following divisions; find the quotient and the remainder of each division.

(a) $312 : 19$

(b) $578 : 17$

(c) $525 : 12$

(d) $1,007 : 33$

33. In each case use the Euclidean Algorithm to find the greatest common divisor (a, b) of the given integers a and b.

(i) $a = 204,\ b = 492$ **(ii)** $a = 511,\ b = 371$

(iii) $a = 495,\ b = 264$ **(iv)** $a = 355,\ b = 183$

34. Find the least common multiple $[a, b]$ of the given integers a and b.

(a) $a = 324,\ b = 1701$ **(b)** $a = 539,\ b = 121$

35. In each case below we practice problem P30, to find the greatest common divisor and the greatest common multiple of the given integers a and b

(i) $a = p^3,\ b = p^7 \cdot q$, where p and q are distinct primes.

(ii)] $a = p^3 \cdot q^5 \cdot r^7,\ b = p^4 \cdot q^5 \cdot r^8$, where p, q, and r are distinct primes.

(iii) $a = p^{20} \cdot q^{12} \cdot r^7,\ b = p^{17} \cdot q^{14} \cdot r^4$, where p, q, and r are distinct primes.

36. Let p and q be distinct primes. Prove that for every integer a such that $(a, pq) = 1$, the congruence $a^{[p-1,q-1]} \equiv 1 (\text{mod } pq)$ holds true.

Hint: First apply Fermat's Little Theorem for each prime p, q. Then use the fact that $[p-1, q-1]$ is a multiple of p; and a multiple of q..

Chapter 5

Powers and Roots

5.1 Powers with Integer Exponents

Of the six sections in this chapter, this section will be undoubtedly the most familiar to most readers of this text, given their precalculus and calculus background.

Definition 1: (i) For each nonzero real number a, we define $\boldsymbol{a^{\circ} = 1}$

(ii) For each nonzero real number a and each positive integer or natural number n, we define,

$$a^n = \underbrace{a \cdot a \ldots a}_{n \text{ times}}$$

(iii) We call a^n the **nth power of (the real number)a**, if a is nonzero and n a natural number.

(iv) We call 0 the nth power of a, if $a = 0$ and $n \in \mathbb{N}$.

(v) We call $\frac{1}{a^{-n}}$, the nth power of a, if a is a nonzero real number and n is a negative integer. We write, $a^n = \frac{1}{a^{-n}}$.

Terminology: The real number a is called the **base** of the nth power a^n; and the integer n is called the **exponent** of the nth power a^n.

Remark 1: Note that for every nonzero real number a and any integer n, we must have $a^n = \frac{1}{a^{-n}}$. If $n \leq 0$, this is of course Definition 1(v) and 1(i). If n is a positive integer, then $(-n)$ is a negative integer, so by Definition 1(v) again we must have $a^{(-n)} = \frac{1}{a^{-(-n)}} = \frac{1}{a^n} \Rightarrow$ (multiply both sides by $\frac{a^n}{a^{-n}}$) $\Rightarrow \frac{a^n \cdot a^{-n}}{a^{-n}} = \frac{a^n}{a^{-n} \cdot a^n} \Rightarrow a^n = \frac{1}{a^{-n}}$.

Theorem 1 Let a, b, be nonzero real numbers and m, n be integers, $m, n \in \mathbb{Z}$. Then,

(i) $a^m \cdot a^n = a^{m+n}$

(ii) $(a^m)^n = a^{mn}$

(iii) $(ab)^n = a^n \cdot b^n$

(iv) $\left(\frac{a}{b}\right)^n = \frac{a^n}{b^n}$

(v) $a^n > 0$ if either $a > 0$; or $a < 0$ and n being an even integer. And, $a^n < 0$ when $a < 0$ and n being an odd integer.

(vi) If a and b are both positive; then in order that $a > b$ it is necessary and sufficient that the following condition holds true: (either $a^n > b^n$ and $n > 0$; or $a^n < b^n$ and $n < 0$).

(vii) If $a^n = b^n$, then $|a| = |b|$ (which means $a = \pm b$) or $n = 0$.

Proof: (i) If both m and n are positive integers, we can offer a proof by double induction as follows: first we show that the statement is true for $n = 1$ and every natural number m. Indeed, for $n = 1$ and $m = 1$ the statement is true: $a^2 = a \cdot a$, by Definition 1(ii). Suppose that the statement is true $n = 1$ and $m = k$: $a^{k+1} = a^k \cdot a$; multiply both sides by a to obtain $a^{1+(k+1)} = a^{k+1} \cdot a$, which shows that the statement is also true for $n = 1$ and $m = k + 1$. We have thus shown that the statement is true for $n = 1$ and every natural number m. Next we use induction on n. Indeed, the statement is true for $n = 1$ and every natural number m. Inductive hypothesis: Assume that the statement is true for $n = k$ and every natural number $m : a^{m+k} = a^m \cdot a^k \Rightarrow a^{m+k} \cdot a = a^m \cdot a^k \cdot a \Rightarrow a^{m+(k+1)} = a^m \cdot a^{k+1}$, which proves that the statement is also true for every natural m and $n = k + 1$. The inductive step is complete; we conclude that the statement holds true for every natural number m and every natural number n.

Next we prove the statement for m and n both being negative integers.
Since $m < 0$ and $n < 0$, it follows $m + n < 0$; $(m + n)$ is a negative integer. By Definition (1)v we have $a^{m+n} = \frac{1}{a^{-(m+n)}}$; however, both $(-m)$ and $(-n)$ are positive integers and so, as we have already established, $a^{-(m+n)} = a^{(-m-n)} = a^{-m} \cdot a^{-n}$; so that, $a^{m+n} = \frac{1}{a^{-(m+n)}} = \frac{1}{a^{-m} \cdot a^{-n}} = a^{-(-m)} \cdot a^{-(-n)} = a^m \cdot a^n$, by Definition 1(v).
Now, suppose that $m > 0$, $n < 0$ (or vice-versa) and $m + n > 0$. From Definition 1(v) we have $a^m \cdot a^n = \frac{a^m}{a^{-n}} = \frac{a^{[(m+n)+(-n)]}}{a^{-n}}$. But $m+n > 0$ and $-n > 0$; so that both $(m + n)$ and $(-n)$ are natural numbers; as we have already proved, we must have $a^{[(m+n)+(-n)]} = a^{m+n} \cdot a^{-n}$ and so by going back we obtain, $a^m \cdot a^n = \frac{a^{[(m+n)+(-n)]}}{a^{-n}} = \frac{a^{m+n} \cdot a^{-n}}{a^{-n}} = a^{m+n}$.
Next, suppose that $m > 0$, $n < 0$, and $m + n < 0$. By Remark 1, we know that $a^m = \frac{1}{a^{-m}}$ and $a^n = \frac{1}{a^{-n}}$. Hence, $a^m \cdot a^n = \frac{1}{a^{-m}} \cdot \frac{1}{a^{-n}} = \frac{1}{a^{-(m+n)}} = a^{m+n}$, by Definition 1(v).
What about the case $m > 0$, $n < 0$ and $m+n = 0$? If that is the case then $n = -m$. by Definition 1(i) we have $a^{m+n} = a^0 = 1 \Rightarrow a^{m+(-m)} = 1 \Rightarrow a^m \cdot a^{-m} = 1 \Leftrightarrow a^{-m} = \frac{1}{a^m}$, which is true according to Definition 1(v) and since $m \in \mathbb{N}$.

Finally, if either $m = 0$ or $n = 0$ or both m and n are zero; the statement $a^{m+n} = a^m \cdot a^n$, is trivially true since $a^0 = 1$ (by Definition 1(i)).

(ii) Exercise 1 (may use double induction).

(iii) Exercise 2 (may use induction).

(iv) This is an immediate consequence of parts (ii) and (iii): $\left(\frac{a}{b}\right)^n = (a \cdot b^{-1})^n =$

$a^n \cdot (b^{-1})^n = a^n \cdot b^{(-1) \cdot n} = a^n \cdot b^{-n} = \frac{a^n}{b^n}$, by Definition 1(v).

(v) First suppose that $a > 0$. We prove $a^n > 0$ for every $n \in \mathbb{N}$, by using mathematical induction: for $n = 1$ the statement is true since $a^1 = a > 0$. Assume the statement to hold true for $n = k : a^k > 0$. Since both a^k and a are positive, so must be their product (recall back from Chapter 3 that if $\alpha > 0$ and $\beta > 0$, then $\alpha\beta > 0$; this is a consequence of **Axiom 15** of the system of real numbers): $a^k \cdot a^1 > 0 \Rightarrow a^{k+1} > 0$, and the induction is complete. Next if $n = 0$, then $a^0 = 1 > 0$ (back in Chapter 3, we actually proved that $1 > 0$). If n is a negative integer then $(-n)$ is a natural number: since $a > 0$, we have already shown that we must have $a^{-n} > 0$; by part (i), we have $a^{(n-n)} = a^n \cdot a^{-n}$; $a^0 = a^n \cdot a^{-n} \Rightarrow 1 = a^n \cdot a^{-n}$ (or we could simply evoke Definition 1(v)) and since $1 > 0$, we conclude $a^n \cdot a^{-n} > 0$. As we have already proved in Chapter 3, either both a^n and a^{-n} are positive; or they are both negative; but $a^{-n} > 0$ (see above) and thus $a^n > 0$. Now suppose that $a < 0$ and n is an even integer. We write $-a = (-1)a > 0$, since $a < 0$. By our analysis above, we must have $(-a)^n = [(-1)a]^n > 0$; and by part (iii) it follows that $(-a)^n = [(-1)a]^n > 0$; and by part (iii) it follows that $(-a)^n = (-1)^n a^n > 0$. Again, $(-1)^n$ and a^n must either be both positive or both negative. Since n is even we must have $(-1)^n > 0$: indeed, if $n = 0$, $(-1)^0 = 1$. If n is a positive integer, set $n = 2\ell$, $\ell \in \mathbb{N}$; an easy induction on ℓ (start with $(-1)^2 = (-1) \cdot (-1) > 0$) proves that $(-1)^n = (-1)^{2\ell} > 0$, for all $\ell \in \mathbb{N}$. When n is a negative even integer, simply write $(-1)^n \cdot (-1)^{-n} = (-1)^0 = 1 > 0$; since $(-n)$ is a positive integer, we already know that $(-1)^{-n} > 0$; in fact, $(-1)^n = 1$. We conclude that $(-1)^n > 0$ and from (see above) $(-a)^n = (-1)^n a^n > 0$, we conclude that $a^n > 0$.

Lastly, the proof that $a^n < 0$, for $a < 0$ and n an odd integer is left to the student - see Exercise 3. Again, one can start with $-a > 0$; $-a = (-1) \cdot a > 0$. then set $n = 2\ell - 1$, $\ell \in \mathbb{N}$, and proceed with an induction on ℓ, by proving that $(-1)^{2\ell-1} = -1 < 0$. Finally, by employing the familiar trick $(-1)^n \cdot (-1)^{-n} > 0$, also prove that $(-1)^n < 0$, when n is a negative odd integer. And then use $(-a)^n = (-1)^n a^n > 0$, to draw the conclusion that $a^n < 0$.

(vi) Exercise 3.

(vii) Suppose $a^n = b^n$. If $a = 0$, then $0 = b^n$; according to Definition 1, $b = 0$ and n must be a positive integer (the expression b^n is undefined for $b = 0$ and n being a nonpositive integer; also clearly if $b \neq 0$ and n is any integer b^n cannot equal 0, as it easily follows from Definition 1). Likewise if $b = 0$, $0 = a^n$ will imply $a = 0$ and $n \in \mathbb{N}$.

Next assume $ab \neq 0$. First assume n to be a natural number. We have, $a^n = b^n \Rightarrow |a^n| = |b^n| \Rightarrow |a|^n = |b|^n$ (note: an easy induction can establish $|c^n| = |c|^n$ for every natural n and any nonzero real number c). We can apply the by now familiar identity for the difference of two nth powers: $|a|^n = |b|^n \Rightarrow |a|^n - |b|^n = 0 \Rightarrow (|a| - |b|) \cdot (|a|^{n-1} + |a|^{n-2} \cdot |b|^1 + \ldots + |a|^1 \cdot |b|^{n-2} + |b|^{n-1}) = 0$. Since $|a| \neq 0$ and $|b| \neq 0$; $|a| > 0$ and $|b| > 0$, which clearly shows that the second factor of the last product above is positive (if $n = 1$, the second factor is simply equal 1). Hence, we conclude that $|a| - |b| = 0 \Rightarrow |a| = |b|$. Next, if n is a negative ineger, then $(-n)$ is a positive integer or natural number. Set $m = -n$: $a^n = b^n \Rightarrow a^{-m} = b^{-m} \Rightarrow$ (by

Definition 1) $\frac{1}{a^m} = \frac{1}{b^m} \Rightarrow \frac{a^m b^m}{a^m} = \frac{a^m b^m}{b^m} \Rightarrow a^m = b^m \Rightarrow |a^m| = |b^m| \Rightarrow |a|^m = |b|^m$, the rest of the argument is identical to the one given above for n. We conclude $|a| = |b|$. Finally, if $n = 0$, then $a^n = b^n$ holds true since $a^0 = 1 = b^0$ (by Definition 1(i)), regardless of the values of the nonzero real numbers a and b. ■

Remark 2: The following are immediate consequences of Theorem 1. Below we offer a proof for part (i); parts (ii), (iii), and (iv) are to be proved by the reader in Exercises 15, 16, 17. Let k be a positive integer; $k \in \mathbb{N}$

(i) If $\beta < \alpha < 0$, then $\beta^{2k} > \alpha^{2k}$, $\beta^{2k+1} < \alpha^{2k+1}$, $\beta^{-2k} < \alpha^{-2k}$, and $\beta^{-(2k+1)} > \alpha^{-(2k+1)}$.

(ii) If $\beta < 0 < \alpha$ and $|\beta| < \alpha$, then $\beta^{2k} < \alpha^{2k}$, $\beta^{2k+1} < \alpha^{2k+1}$, $\beta^{-2k} > \alpha^{-2k}$, and $\beta^{-(2k+1)} < \alpha^{-(2k+1)}$.

(iii) If $\beta < 0 < \alpha$ and $|\beta| > \alpha$, then $\beta^{2k} > \alpha^{2k}$, $\beta^{2k+1} < \alpha^{2k+1}$, $\beta^{-2k} < \alpha^{-2k}$, and $\beta^{-(2k+1)} < \alpha^{-(2k+1)}$.

(iv) If $\beta < 0 < \alpha$ and $|\beta| = \alpha$, then $\beta^{2k} = \alpha^{2k}$, $\beta^{2k+1} < \alpha^{2k+1}$, $\beta^{-2k} = \alpha^{-2k}$, $\beta^{-(2k+1)} < \alpha^{-(2k+1)}$.

Let us prove (i). From $\beta < \alpha < 0$ we obtain by multiplying the first inequality, $\beta < \alpha$, with $\dfrac{1}{\alpha} < 0$;

$$\frac{\beta}{\alpha} > \frac{\alpha}{\alpha};$$

$$\frac{\beta}{\alpha} > 1.$$

We apply part (vi) of Theorem 1 with $a = \frac{\beta}{\alpha}$, $b = 1$, and $n = 2k > 0$: we have $\frac{\beta}{\alpha} > 1 \Rightarrow \left(\frac{\beta}{\alpha}\right)^{2k} > 1^{2k} = 1 \Rightarrow \frac{\beta^{2k}}{\alpha^{2k}} > 1$; but $\alpha^{2k} > 0$, since $\alpha^{2k} = (\alpha^k)^2 \geq 0$ and since $\alpha < 0$, we actually have $\alpha^{2k} > 0$; $\frac{\beta^{2k}}{\alpha^{2k}} > 1 \Rightarrow \alpha^{2k} \cdot \left(\frac{\beta^{2k}}{\alpha^{2k}}\right) > \alpha^{2k} \cdot 1 \Rightarrow \beta^{2k} > \alpha^{2k}$. Next, from $\beta^{2k} > \alpha^{2k} \Rightarrow \frac{\beta^{2k}}{\alpha^{2k} \cdot \beta^{2k}} > \frac{\alpha^{2k}}{\alpha^{2k} \cdot \beta^{2k}} \Rightarrow \frac{1}{\alpha^{2k}} > \frac{1}{\beta^{2k}}$; $\alpha^{-2k} > \beta^{-2k}$; or equivalently $\beta^{-2k} < \alpha^{-2k}$. We go back to $\frac{\beta}{\alpha} > 1$ and apply Th. 1(vi), with $n = 2k + 1$ this time: $\left(\frac{\beta}{\alpha}\right)^{2k+1} > 1 - -^{(2k+1)} = 1 \Rightarrow \frac{\beta^{2k+1}}{\alpha^{2k+1}} > 1$. However, $\alpha^{2k+1} < 0$ by part (v) of Theorem 1, since $\alpha < 0$ and $(2k + 1)$ is an odd integer. Hence,

$$\frac{\beta^{2k+1}}{\alpha^{2k+1}} > 1 \Rightarrow \alpha^{2k+1} \cdot \left(\frac{\beta^{2k+1}}{\alpha^{2k+1}}\right) < \alpha^{2k+1} \cdot 1 = \alpha^{2k+1} \Rightarrow \beta^{2k+1} < \alpha^{2k+1}.$$

Note that both powers α^{2k+1} and β^{2k+1} are negative real numbers since $\alpha < 0$, $\beta < 0$ and $(2k+1)$ is odd, according to Theorem 1(vi). Therefore $\alpha^{2k+1} \cdot \beta^{2k+1} > 0$; so that,

$$\beta^{2k+1} < \alpha^{2k+1} \Rightarrow \left(\frac{1}{\alpha^{2k+1} \cdot \beta^{2k+1}}\right) \cdot \beta^{2k+1} < \left(\frac{1}{\alpha^{2k+1} \cdot \beta^{2k+1}}\right) \cdot \alpha^{2k+1}$$

$$\Rightarrow \frac{1}{\alpha^{2k+1}} < \frac{1}{\beta^{2k+1}}; \; \alpha^{-(2k+1)} < \beta^{-(2k+1)}$$

or equivalently, $\beta^{-(2k+1)} > \alpha^{-(2k+1)}$. We are done.

5.2 Bionomial Coefficients

Undoubtedly, every calculus student must have seen the expansions of the binomial expressions $(a+b)^0$, $(a+b)^1$, $(a+b)^2$, and $(a+b)^3$:

$$
\begin{aligned}
(a+b)^0 &= 1 \\
(a+b)^1 &= 1\cdot a + 1\cdot b \\
(a+b)^2 &= 1\cdot a^2 + 2ab + 1\cdot b^2 \\
(a+b)^3 &= 1\cdot a^3 + 3a^2b + 3ab^2 + 1\cdot b^3
\end{aligned}
$$

We list two more binomial expansions, that some students of calculus may have not seen before:

$$
\begin{aligned}
(a+b)^4 &= 1\cdot a^4 + 4a^3b + 6a^2b^2 + 4ab^3 + 1\cdot b^4 \\
(a+b)^5 &= 1\cdot a^5 + 5a^4b + 10a^3b^2 + 10a^2b^3 + 5ab^4 + 1\cdot b^5
\end{aligned}
$$

For each value of the nonnegative integer n, we can arrange the coefficients of each binomial expansion $(a+b)^n$ horizontally, to create a triangular configuration, commonly called **Pascal's Triangle** (which has interesting properties):

$$
\begin{array}{lccccccccccc}
n=0 & & & & & & 1 & & & & & \\
n=1 & & & & & 1 & & 1 & & & & \\
n=2 & & & & 1 & & 2 & & 1 & & & \\
n=3 & & & 1 & & 3 & & 3 & & 1 & & \\
n=4 & & 1 & & 4 & & 6 & & 4 & & 1 & \\
\vdots & 1 & & 5 & & 10 & & 10 & & 5 & & 1
\end{array}
$$

Pascal's Triangle extends indefinitely as n increases. The numbers on each row of Pascal's Triangle are what are called the **binomial coefficients**. Note that given a nonnegative integer n, there are exactly $(n+1)$ binomial coefficients. If n is even, then $(n+1)$ is odd, and the highest binomial coefficient occurs exactly in the middle of Pascal's Triangle nth row. If n is odd, $(n+1)$ is even, and in that case there are two equal highest middle binomial coefficients. Also observe that binomial coefficients that are on the same row in Pascal's Triangle and which are equidistant from the middle of that row, are equal; this will be proved in Theorem 2.

For the definition of a binomial coefficient, we need to recall the notion of $n!$ (n factorial), where n is a nonnegative integer. Recall from Chapter 4 that,

$$n! = \begin{cases} 1, \text{ if } n = 0 \\ 1 \cdot 2 \cdots n = \text{ product of the first } n \text{ positive integers, if } n \in \mathbb{N}. \end{cases}$$

We have the following definition.

Definition 2: Let n, k be nonnegative integers such that $0 \leq k \leq n$. The **binomial coefficient** $\begin{pmatrix} n \\ k \end{pmatrix}$ is defined to be the following rational number:

$$\begin{pmatrix} n \\ k \end{pmatrix} = \frac{n!}{k!(n-k)!}$$

Remark 3: (i) It follows immediately from Definition 2 that if $k = 0$, then $\begin{pmatrix} n \\ k \end{pmatrix} = \begin{pmatrix} n \\ 0 \end{pmatrix} = 1$. Indeed, $\begin{pmatrix} n \\ 0 \end{pmatrix} = \frac{n!}{0!(n-0)!} = \frac{n!}{(0!)\cdot(n!)} = \frac{n!}{1\cdot(n!)} = \frac{1}{1} = 1$.

(ii) For $n = k$, we have $\begin{pmatrix} n \\ k \end{pmatrix} = \begin{pmatrix} n \\ n \end{pmatrix} = \frac{n!}{(n!)(0!)} = 1$.

(iii) For $n \geq 1$ and $k = 1$ we have,

$$\begin{pmatrix} n \\ k \end{pmatrix} = \begin{pmatrix} n \\ 1 \end{pmatrix} = \frac{n!}{1!(n-1)!} = \frac{1 \cdot 2 \ldots (n-1) \cdot n}{1 \cdot 2 \ldots (n-1)} = n$$

Likewise, for $n \geq 1$ and $k = n - 1$ we have,

$$\begin{pmatrix} n \\ k \end{pmatrix} = \begin{pmatrix} n \\ n-1 \end{pmatrix} = \frac{n!}{(n-1)![n-(n-1)]!} = \frac{n!}{(n-1)!1!} = n$$

The next Theorem presents two fundamental properties of the binomial coefficients; it also shows that each binomial coefficient must actually be a positive integer.

Theorem 2: (i) If n and k be nonnegative integers such that $n \geq k$, then $\begin{pmatrix} n \\ k \end{pmatrix} = \begin{pmatrix} n \\ n-k \end{pmatrix}$.

(ii) If n and k are nonnegative integers such that $n \geq k + 1$, then $\begin{pmatrix} n \\ k \end{pmatrix} + \begin{pmatrix} n \\ k+1 \end{pmatrix} = \begin{pmatrix} n+1 \\ k+1 \end{pmatrix}$.

(iii) If n and k are nonnegative integers such that $n \geq k$, then $\begin{pmatrix} n \\ k \end{pmatrix}$ is a positive integer.

Proof: (i) We have, by Definition 2,

$$\begin{pmatrix} n \\ k \end{pmatrix} = \frac{n!}{k!(n-k)!} = \frac{n!}{(n-k)![n-(n-k)]!} = \begin{pmatrix} n \\ n-k \end{pmatrix},$$

again by Definition 2.

(ii) We have,

$$\binom{n}{k}+\binom{n}{k+1}=\frac{n!}{k!(n-k)!}+\frac{n!}{(k+1)![n-(k+1)]!}$$

$$=\frac{n!(k+1)}{\underbrace{k!(k+1)\cdot}_{(k+1)!}(n-k)!}+\frac{n!(n-k)}{(k+1)!\underbrace{[[n-(k+1)]!(n-k)}_{\text{this is equal to }(n-k)!}}$$

$$=\frac{n!(k+1)+n!(n-k)}{(k+1)!(n-k)!}=\frac{n![(k+1)+(n-k)]}{(k+1)!((n+1)-(k+1))!}$$

$$=\frac{n!(n+1)}{(k+1)!((n+1)-(k+1))!}$$

$$=\frac{(n+1)!}{(k+1)!((n+1)-(k+1))!}=\binom{n+1}{k+1}$$

Note that in the above proof we used the fact (by the definition of the factorial) that $m!(m+1)=(m+1)!$, in three different instances. In the first instance with $m=k$; in the second instance with $m=n-(k+1)$, so that $m+1=n-k$; and in the third instance with $m=n$.

(iii) We prove the following statement: "For every nonnegative integer n, the $(n+1)$ binomial coefficients $\binom{n}{0},\binom{n}{1},\ldots,\binom{n}{n}$, are all positive integers". We use mathematical induction, starting with $n=0$: for $n=0$, there is only one binomial coefficient in the above list, namely $\binom{0}{0}=\frac{0!}{0!0!}=1$, which is a positive integer. Suppose for $n=k$ (nonnegative integer) the statements hold true: $\binom{k}{0},\binom{k}{1},\ldots,\binom{k}{k-1},\binom{k}{k}$ are all positive integers. Now consider the binomial coefficients $\binom{k+1}{r}$, where $r=0,1,\ldots,k+1$. If $r=0$, $\binom{k+1}{0}=\frac{(k+1)!}{(0!)(k+1)!}=1$, which is a positive integer. Next, assume r to be any of the numbers $1,\ldots,k,k+1$. Since $1\le r\le k+1$, it follows that $0\le r-1\le k$; that is, $(r-1)$ is a nonnegative integer: thus, by part (ii) or this theorem, we have $\binom{k+1}{r}=\binom{k}{r}+\binom{k}{r-1}$.

Suppose that $1\le r\le k$; $0\le r-1\le k-1<k$. By the inductive hypothesis both binomial coefficients $\binom{k}{r}$ and $\binom{k}{r-1}$ are positive integers, since they are among $\binom{k}{0},\binom{k}{1},\ldots,\binom{k}{k-1},\binom{k}{k}$. Thus their sum is also a positive integer, and so, the equality $\binom{k+1}{r}=\binom{k}{r}+\binom{k}{r-1}$ implies that the bino-

mial coefficient $\binom{k+1}{r}$ must also be an integer for $r = 1, \ldots, k$. Next we have $\binom{k+1}{k+1} = \frac{(k+1)!}{0!(k+1)!} = 1 = \binom{k+1}{0}$. Thus, under $k \geq 1$, we have shown that all the binomial coefficients $\binom{k+1}{0}, \binom{k+1}{1}, \ldots, \binom{k+1}{k}, \binom{k+1}{k}$ must be positive integers. Finally, if $k = 0$, there are two choices for r, namely $r = 0, 1$. When $r = 0$, $\binom{k+1}{r} = \binom{1}{0} = 1$; and when $r = 1$, likewise $\binom{k+1}{r} = \binom{1}{1} = 1$. In all cases, we have shown that the inductive hypothesis implies that the statement must also hold true for $n = k+1$. We are done. ■

Theorem 3: (Binomial Theorem) For every natural number or positive integer n and any real numbers a and b the following expansion holds true:

$$(a+b)^n = \binom{n}{0} a^n + \binom{n}{1} a^{n-1} \cdot b^1 + \binom{n}{2} a^{n-2} \cdot b^2 + \ldots +$$
$$+ \binom{n}{i} \cdot a^{n-i} \cdot b^i + \ldots + \binom{n}{n-2} a^2 \cdot b^{n-2} +$$
$$+ \binom{n}{n-1} a^1 b^{n-1} + \binom{n}{n} b^n = \sum_{i=0}^{n} \binom{n}{i} a^{n-i} \cdot b^i$$

(Note that the first and last coefficients in the above expansion are always equal to $1: \binom{n}{0} = \binom{n}{n} = 1$; and the second and second from last coefficients are equal to $n: \binom{n}{1} = \binom{n}{n-1}$).

Proof: By using mathematical induction. For $n = 1$, $(a+b)^1 = a^1 + b^1 = \binom{1}{0} a^1 + \binom{1}{1} b^1$, since trivially $\binom{1}{0} = \binom{1}{1} = 1$.

Inductive Hypothesis: Suppose that for $n = k \in \mathbb{N}$

$$(a+b)^k = \binom{k}{0} a^k + \binom{k}{1} a^{k-1} \cdot b^1 + \ldots + \binom{k}{i} a^{k-i} \cdot b^i$$
$$+ \ldots + \binom{k}{k-1} a^1 b^{k-1} + \binom{k}{k-1} a^1 b^{k-1} + \binom{k}{k} b^k \tag{1}$$

Multiply equation (1) first by a and secondly by b to obtain respectively,

$$
\begin{aligned}
(a+b)^k \cdot a = & \binom{k}{0} a^{k+1} + \binom{k}{1} a^k b^1 + \ldots + \binom{k}{i} a^{(k+1-i)} \cdot b^i \\
& + \ldots + \binom{k}{k-1} a^2 b^{k-1} + \binom{k}{k} a^1 b^k
\end{aligned} \tag{2}
$$

and,

$$
\begin{aligned}
(a+b)^k \cdot b = & \binom{k}{0} a^k \cdot b^1 + \binom{k}{1} a^{k-1} \cdot b^2 + \ldots + \binom{k}{i} a^{k-1} \cdot b^{i+1} \\
& + \ldots + \binom{k}{k-1} a^1 \cdot b^k + \binom{k}{k} b^{k+1}
\end{aligned} \tag{3}
$$

Now we add equations (2) and (3) memberwise, noting that $(a+b)^k a + (a+b)^k \cdot b = (a+b)^k \cdot (a+b) = (a+b)^{k+1}$. We obtain,

$$
\begin{aligned}
(a+b)^{k+1} = & \binom{k}{0} a^{k+1} + \left[\binom{k}{0} + \binom{k}{1}\right] a^k b^1 \\
& + \ldots + \left[\binom{k}{i-1} + \binom{k}{i}\right] a^{(k+1-i)} \cdot b^i \\
& + \ldots + \left[\binom{k}{k-1} + \binom{k}{k}\right] a^1 \cdot b^k + \binom{k}{k} b^{k+1}
\end{aligned} \tag{4}
$$

By Theorem 2(ii), we know that for $1 \leq i \leq k$,

$$
\binom{k}{i-1} + \binom{k}{i} = \binom{k+1}{i} \tag{5}
$$

Combining equations (4) and (5) together with the fact that $\binom{k}{0} = 1 = \binom{k+1}{0} = \binom{k}{k} = \binom{k+1}{k+1}$, we arrive at the following expansion:

$$
\begin{aligned}
(a+b)^{k+1} = & \binom{k+1}{0} a^{k+1} + \binom{k+1}{1} a^k b^1 \\
& + \ldots + \binom{k+1}{i} a^{(k+1-i)} \cdot b^i \\
& + \ldots + \binom{k+1}{k} a^1 b^k + \binom{k+1}{k+1} b^{k+1} \\
= & \sum_{i=0}^{k+1} \binom{k+1}{i} a^{(k+1-i)} \cdot b^i
\end{aligned}
$$

The inductive step is complete. ■

Remark 4: (i) If we look at Pascal's Trianle more closely and we pick two adjacent numbers (binomial coefficients) on a row, we observe that their sum is equal to the number directly beneath the two numbers on the row directly below: for example,

$$\begin{matrix} 4 & & 6 \\ & 10 & \end{matrix}$$ so that $4 + 6 = 10$.

This simply illustrates Theorem 2(ii): $\binom{n}{k} + \binom{n}{k+1} = \binom{n+1}{k+1}$. In our example $\binom{4}{1} + \binom{4}{2} = \binom{5}{2}$. Because the three numbers involved can be thought as the vertices of a triangle, often, Theorem 2(ii) is referred to as Pascal's Triangle Equality.

(ii) The binomial coefficients have actually another significance. Let n be a natural number and k a nonnegative integer, and S any set containing n elements. In introducing set theory or discrete mathematics courses, it is shown that the number of distinct subsets of S, each containing k elements, is equal to the binomial coefficient $\binom{n}{k}$. When $k = 0$, $\binom{n}{0} = 1$, not surprising, since there is only one subset of S with no elements: the empty set. For example, out of any set containing 6 elements, we can only form $\binom{6}{3} = \frac{6!}{3!3!} = \frac{4 \cdot 5 \cdot 6}{1 \cdot 2 \cdot 3} = 4 \cdot 5 = 20$ distinct subsets, each containing 3 elements.

(iii) Sometimes it is useful to simplify the expression for the binomial coefficient $\binom{n}{k}$:

$$\begin{aligned} \binom{n}{k} &= \frac{n!}{k!(n-k)!} \\ &= \frac{1 \cdot 2 \ldots (n-k)[(n-k)+1] \ldots n}{(1 \cdot 2 \ldots k) \cdot [1 \cdot 2 \ldots (n-k)]} \\ &= \frac{[(n-k)+1] \ldots n}{1 \cdot 2 \ldots k} \end{aligned}$$

(iv) Because of the set theoretic property of $\binom{n}{k}$, sometimes the binomial coefficient $\binom{n}{k}$ is read "n choose k".

5.3 Radicals with positive integer indices

In this section we will prove the existence and uniqueness of the nth root of a non-negative real number. A host of properties will subsequently follow. We begin this section with an example that at first may seem of marginal significance and of purely technical nature. However, this example will be instrumental in establishing Theorem 4, the fundamental theorem of this section; it will be used in its proof.

Example 1: (i) Prove that if k and ϵ are real numbers such that $0 < k < 1$ and $\epsilon \geq 0$, then $(\epsilon + k)^n \leq \epsilon^n + k \cdot [(1+\epsilon)^n - \epsilon^n]$, for every natural number n.

We proceed by the use of mathematical induction. For $n = 1 : (\epsilon + k)^1 \leq \epsilon^1 + k \cdot [(1+\epsilon) - \epsilon] \Leftrightarrow \epsilon + k \leq \epsilon + k$, which is true since $\epsilon + k = \epsilon + k$. The inductive hypothesis: the statement holds true for $n = m$:

$$(\epsilon + k)^m \leq \epsilon^m + k \cdot [(1+\epsilon)^m - \epsilon^m] \tag{6}$$

Since $0 < k$ and $0 \leq \epsilon$, it follows $0 < k + \epsilon$. Multiplying both sides of inequality (5) by $(\epsilon + k)$ we obtain,

$$\begin{aligned}
(\epsilon + k)^{m+1} &\leq (\epsilon + k) \cdot \{\epsilon^m + k[(1+\epsilon)^m - \epsilon^m]\}; \\
&= \epsilon^{m+1} - k\epsilon^{m+1} + k\epsilon(1+\epsilon)^m + k\{\epsilon^m \\
&\quad +k(1+\epsilon)^m - k\epsilon^m\} \\
&= \epsilon^{m+1} - k\epsilon^{m+1} + k\epsilon(1+\epsilon)^m + k \cdot \{\epsilon^m(1-k) \\
&\quad -(1-k)(1+\epsilon)^m + (1+\epsilon)^m\} \\
&= \epsilon^{m+1} - k\epsilon^{m+1} + k\epsilon(1+\epsilon)^m + k \cdot \{(1+\epsilon)^m \\
&\quad +(1-k)[\epsilon^m - (1+\epsilon)^m]\}
\end{aligned} \tag{7}$$

$$(\epsilon + k) \leq \epsilon^{m+1} - k\epsilon^{m+1} + k\epsilon(1+\epsilon)^m + k(1+\epsilon)^m + k(1-k)[\epsilon^m - (1+\epsilon)^m]$$

By hypothesis we know that $0 < k < 1$; hence $k \cdot (1-k) < 0$; also $\epsilon^m - (1+\epsilon)^m < 0$, since $0 \leq \epsilon$ and so $0 \leq \epsilon < 1 + \epsilon$; and by Theorem 1(vi), with $a = 1+\epsilon$, $b = \epsilon$ (if $\epsilon = 0$ obviously, $\epsilon^m - (1+\epsilon)^m < 0$) and $n = m$ (positive since $m \in \mathbb{N}$), it follows that $(1+\epsilon)^m > \epsilon^m$; $\epsilon^m - (1+\epsilon)^m < 0$.

From the above analysis we conclude that $k(1-k)[\epsilon^m - (1+\epsilon)^m] < 0$. Applying this to inequality (7), we see that if we remove the negative term $k(1-k)[\epsilon^m - (1+\epsilon)^m]$ from the right-hand side of (7); the right-hand side of (7) becomes larger. Hence,

$$(\epsilon + k)^{m+1} < \epsilon^{m+1} - k\epsilon^{m+1} + k\epsilon(1+\epsilon)^m + k(1+\epsilon)^m;$$

$$(\epsilon + k)^{m+1} < \epsilon^{m+1} - k\epsilon^{m+1} + k(1+\epsilon)^m \cdot [\epsilon + 1];$$

$$(\epsilon + k)^{m+1} < \epsilon^{m+1} - k\epsilon^{m+1} + k(1+\epsilon)^{m+1} \Rightarrow$$

$$\Rightarrow (\epsilon + k)^{m+1} < \epsilon^{m+1} + k \cdot \left[(1+\epsilon)^{m+1} - \epsilon^{m+1}\right],$$

which is the statement we wanted to prove for $n = m + 1$. The induction proof is complete. ■

(vi) Suppose that ϵ and k are real numbers such that $0 \leq \epsilon$ and $0 < \epsilon - k$; and either $k > 0$ or $k < -1$. Prove that for every natural number n,

$$(\epsilon - k)^n \geq \epsilon^n - k \cdot \left[(1+\epsilon)^n - \epsilon^n\right].$$

We invite the reader to write a proof for this statement. The proof is similar to the one given for part (i) of this example.

Theorem 4: For every natural number n and every nonnegative real number a, there exists a unique nonnegative real number such that $x^n = a$.

Proof: Let n be a natural number and a, a nonnegative real number. Consider the set $S = \{y \mid y \in \mathbb{R}, y \geq 0, \text{ and } y^n \leq a\}$. The number 0 is a member of S, since $0^n - 0 \leq a$. Hence, S is nonempty. Now consider the set,

$$U = \{t \mid t \in \mathbb{R} \text{ and } y \leq t, \text{ for every } y \in S\}.$$

In other words, the set U consists of all of those real numbers (upperbounds), each of which is greater than or equal to every member of the set S. First, note that U is a nonempty set: if $0 < a < 1$, then $1^n = 1 > a > y^n > 0, \forall y \in S$. Now $1^n > y^n > 0$ implies, by Theorem 1(vi) that $1 > y \geq 0$. So we see that if $0 \leq a < 1$, the number 1 is a member of the set U. If, on the other hand, $a \geq 1$, then by virtue of $n - 1 \geq 0$ (because $n \in \mathbb{N}$) and Theorem 1(vi) we must have $a^{n-1} \geq 1^{n-1} = 1 \Rightarrow a^{n-1} \cdot a \geq 1 \cdot a$, since $0 \leq a$. Therefore, $a^n \geq a$ and since $a \geq y^n, \forall y \in S$, we conclude $a^n \geq y^n \geq 0$ and by Theorem 1(vi) once more, we must have $a \geq y$. Thus, if $a \geq 1$, we see that the number a is actually a member of the set U.

In both cases, we see that the set U is nonempty. According to **Axiom 16** of the real number system $\mathbb{R}$, the set U must have a least or minimum element ϵ. Note that $\epsilon > 0$, since S consists only of nonnegative inegers and no element of U is less than any element of S. We write this observation:

(i) $\epsilon \geq y \geq 0, \ \forall y \in S$

We make two more observations:

(ii) If $\epsilon > 0$, and $\gamma \in \mathbb{R}$ such that $0 \leq \gamma < \epsilon$, then γ must belong to the set S: indeed, if it were otherwise, we would have $\gamma^n > a$; but $a \geq \epsilon^n$, since ϵ belongs to S; hence, we would have $\gamma^n > \epsilon^n \geq 0$, and by Theorem 1(vi), $\gamma > \epsilon$, contradicting $0 \leq \gamma < \epsilon$ above.

(iii) Every real number δ such that $\delta > \epsilon$, is not a member of the set S. Because $\delta > \epsilon \geq y,\ \forall y \in S$; and and so $\delta > y,\ \forall y \in S$, so δ cannot be a member of S.

Below, we show that $\epsilon^n = a$. According to the system of real numbers $\mathbb{R}$, which is totally ordered under $\leq$ and $<$, there are exactly three possibilities: either $\epsilon^n < a$, or $\epsilon^n = a$, or $\epsilon^n > a$. We will show that the assumption that either the first or third possibility holds true, leads to a contradiction.

Indeed, let us first assume $\epsilon^n < a$. Then, $a - \epsilon^n > 0$ and since $(1+\epsilon)^n > \epsilon^n$ (by virtue of $1+\epsilon > \epsilon \geq 0$ and Theorem 1(vi)), it follows that $0 < \frac{a-\epsilon^n}{(1+\epsilon)^n - \epsilon^n}$.

Now the real number $(1+\epsilon)$ cannot belong to S because $\epsilon < 1+\epsilon$, and so, by observation (iii) above, the real number $\delta = 1+\epsilon$ is not a member of S. Hence, by the way S is defined, we must have $(1+\epsilon)^n > a$. This implies, $0 < \dfrac{a-\epsilon^n}{(1+\epsilon)^n - \epsilon^n} < \dfrac{(1+\epsilon)^n - \epsilon^n}{(1+\epsilon)^n - \epsilon^n} = 1$. Pick any real number k strictly between 0 and $\dfrac{a-\epsilon^n}{(1+\epsilon)^n - \epsilon^n}$. (Recall that there are infinitely many reals between any two given reals; just take their average and repeat the process); $0 < k < \dfrac{a-\epsilon^n}{(1+\epsilon)^n - \epsilon^n} < 1$. The middle inequality implies

$$\begin{aligned} & k \cdot [(1+\epsilon)^n - \epsilon^n] < a - \epsilon^n \\ \Rightarrow\ & k \cdot [(1+\epsilon)^n - \epsilon^n] + \epsilon^n < a. \end{aligned} \tag{8}$$

On the other hand, according to Example 1(i), we know that

$$(\epsilon + k)^n \leq \epsilon^n + k \cdot [(1+\epsilon)^n - \epsilon^n]. \tag{9}$$

By combining (8) and (9) we arrive at $(\epsilon + k)^n < a$, which is a contradiction since the real number $\delta = \epsilon + k$ is greater than ϵ (because $k > 0$); and so, by observation (iii) above, δ cannot be a member of the set S, which means that $\delta = (\epsilon + k)^n > a$, contrary to $(\epsilon + k)^n < a$. This concludes the argument that the (first possibility)$\epsilon^n < a$ cannot hold. Next we show that the third possibility $\epsilon^n > a$ cannot hold either. Pick any r such that, (note that $\epsilon > 0$) $0 < r < \min\left\{\epsilon, \dfrac{\epsilon^n - a}{(1+\epsilon)^n - \epsilon^n}\right\}$; then we obtain, in particular,

$$\begin{aligned} & 0 < r < \frac{\epsilon^n - a}{(1+\epsilon)^n - \epsilon^n} \Rightarrow r \cdot [(1+\epsilon)^n - \epsilon^n] < \epsilon^n - a \\ \Rightarrow\ & a < \epsilon^n - r \cdot [(1+\epsilon)^n - \epsilon^n] \end{aligned} \tag{10}$$

Also by Example 1(ii) (with $r = k$) we know that ,

$$\epsilon^n - r \cdot [(1+\epsilon)^n - \epsilon^n] \leq (\epsilon - r)^n \tag{11}$$

From (10) and (11) we infer that $a < (\epsilon - r)^n$ which is a contradiction because $\epsilon - r > 0$ (see above) and so, by observation (ii), the positive real $(\epsilon - r)$ must belong to the set

S. But the inequality $a < (\epsilon - r)^n$ clearly shows that $(\epsilon - r)$ cannot be a member of S, according to the way S is defined. We have a contradiction, which shows that the (third) possibility $\epsilon^n > a$ cannot hold either. We conclude that $\epsilon^n = a$. Thus the set, $E = \{x \mid x \in \mathbb{R}, x \geq 0, \text{ and } x^n = a\}$ is nonempty, since ϵ is one of its members. Next and lastly, we demonstrate that ϵ is the only element of E, establishing uniqueness. Indeed, suppose that ϵ_1 is some element of E. Then, we must have

$$\epsilon_1^n = a = \epsilon^n.$$

The very fact that $\epsilon_1^n = \epsilon^n$, $n \in \mathbb{N}$, $\epsilon_1 > 0$, and $\epsilon > 0$, implies according to Theorem 1(vii), $| \mid \epsilon_1 \mid = \mid \epsilon \mid \Rightarrow \epsilon_1 = \epsilon$, since $\epsilon_1, \epsilon > 0$ (trivially if either ϵ or ϵ_1 is zero, so must be the other one since the nth power, $n \in \mathbb{N}$, of a positive number, must be positive). This concludes the proof: $x = \epsilon$ is the unique nonnegative real such that $x^n = a$. ■

Theorem 5: Let a be a real number and n a natural number.

(i) If $a > 0$ and n is even, there exists a unique negative real number v such that $v^n = a$.

(ii) If $a > 0$ and n is odd, then there does not exist any negative real number u such that $u^n = a$.

(iii) If $a < 0$, then there does not exist a nonnegative number x such that $x^n = a$.

(iv) If $a < 0$ and n is odd, then there exists a unique negative real number w such that $w^n = a$.

(v) If $a < 0$ and n is even, then there does not exist a negative real number t such that $t^n = a$.

Proof: (i) By Theorem 4, there exists a unique real number x such that $x \geq 0$ and $x^n = a$; but $(-x)^n = x^n$ because n is even. Indeed, put $n = 2k, k \in \mathbb{N}$, so that $(-x)^n = (-x)^{2k} = [(-x)^2]^k = [(-1)^2 \cdot x^2]^k = (x^2)^k = x^{2k}$; we see that $(-x)^n = x^n = a$. Note that x must actually be positive, since $x \geq 0, x^n = a$, and $a > 0$ (if it were $x = 0$, then $x^n = 0$). Therefore, $v = -x < 0$ is a negative real number such that $v^n = a$. The real number v is unique: if $v_1 < 0$ is such that $v_1^n = a$, then $-v_1 > 0$ and $(-v_1)^n = v_1^n = a$, because n is even. But $(-v_1)$ must be unique according to Theorem 4, which means (since also $(-v)^n = x^n = a$) that $-v_1 = -v$; $v_1 = v$. The uniqueness of v is established.

(ii) If u is a negative real, then $-u > 0 \Rightarrow (-u)^n > 0$, by Theorem 1(v); but $(-u)^n = (-1)^n \cdot u^n = (-1) \cdot u^n$, since n is odd. Thus $(-u)^n = (-1) \cdot u^n > 0 \Rightarrow u^n < 0$, which proves that u^n cannot equal a, since $0 < a$ and so (from $u^n < 0$) we must have $u^n < a$.

(iii) If $x \geq 0$, then either $x = 0$ or $x > 0$. If $x = 0 \Rightarrow x^n = 0^n = 0$ (since $n \in \mathbb{N}$); if on the other hand, $x > 0 \Rightarrow$ (by Theorem 1(v)) $x^n > 0$; and since $0 > a$, we conclude $x^n > a$; x^n cannot equal a.

(iv) By Theorem 4, and since $-a > 0$, there exists a unique positive real number x (obviously x cannot be zero) such that $x^n = -a$ which implies $(-1) \cdot x^n = a$; because n is odd we also have $(-1)^n = -1$. Thus, $(-1)x^n = a \Rightarrow (-1)^n \cdot x^n = a \Rightarrow (-x)^n - a$.

The existence of the real number w is established by simply taking $w = -x$. The uniqueness of w follows from the fact that x is unique. (The reader should fill in a couple of details: if w_1 is another such negative real, show that $-w - -w_1 \rightarrow w - w_1$.)

(v) If t is a negative real and n even; set $n = 2k, k \in \mathbb{N}$. We have $-t > 0 \Rightarrow (-t)^n > 0$; $(-1)^n \cdot t^n > 0$; but $(-1)^n = (-1)^{2k} = [(-1)^2]^k = 1^k = 1$; hence $(-1)^n t^n > 0 \Rightarrow t^n > 0$; but $0 > a$; so that $t^n > a$; t^n cannot equal a. We are done. ∎

Now that we have Theorem 5 in place we can give a precise definition of the nth root of a real number. It should be pointed out here that the definition goes as far as it can, as long as we require the nth root of a real number to be also a real number. Once we extend this notion to the field of complex numbers, the nth root of a number acquires a broader meaning. Students of Complex Analysis learn that every complex number (and hence every real number, since the set of reals $\mathbb{R}$ is a subset of the set of compelx numbers $\mathbb{C}$) possesses actually exactly n nth roots.

Definition 3: Let n be a natural number greater than 1; that is, $n \geq 2$; and let a be a real number. We call the ***n*th root of *a***:

(i) If $a \geq 0$, the unique nonnegative number x such that $x^n = a$.

(ii) If $a < 0$ and n is odd, the unique negative number is x such that $x^n = a$.

Notation: In both cases, the nth root of a is denoted by $\sqrt[n]{a}$; the symbol $\sqrt{a}$ is called a **radical**; the under the radical quantity a is called the **radicant** of the nth root; the natural number n is called the **square root** of a; it is customary to omit the index $n = 2$. Then we simply write $\sqrt{a}$ and we mean $\sqrt[2]{a}$; in all other cases, the index n must appear in its proper place. Also, when $n = 3$, $\sqrt[3]{a}$ is called the **cubic or cube root** of a.

Remark 5: From Definition 3 and Theorem 5, we deduce that all real numbers x such that $x^n = a$, with $n \geq 2$ and $n \in \mathbb{N}$ are the following:

(i) $x_1 = \sqrt[n]{a} \geq 0$, if $a \geq 0$, and for every natural number $n \geq 2$. For example, if $a = 64$ and $n = 2$, then $x_1 = \sqrt{64} = 8$. If $a = 64$ and $n = 3$, then $x_1 = \sqrt[3]{64} = 4$.

(ii) $x_2 = -\sqrt[n]{a}$, if $a \geq 0$, and for every even natural number $n \geq 2$. For example, if $a = 64$ and $n = 2$, then $x_2 = -\sqrt{64} = -8$.

(iii) $x_3 = \sqrt[n]{a}$, if $a < 0$, and for every odd natural number $n \geq 3$. For example, if $a = -64$ and $n = 3$, then $x_3 = \sqrt[3]{-64} = -4$.

The following theorem establishes some of the basic properties of nth roots.

Theorem 6: Let a and b be a real number, k and n integers.

(i) If $a \geq 0$, and $k, n \geq 2$, then $\left(\sqrt[n]{a}\right)^k = \sqrt[n]{a^k}$.

(ii) If $a \geq 0$ and $n \geq 2$; or if $a < 0$ and n is an odd natural number, then $\sqrt[n]{a^n} = a$.

(iii) If $a \geq 0$ and $n \geq 2$, then $\left(\sqrt[n]{a}\right)^n = \sqrt[n]{a^n} = a$.

(iv) If $a < 0$ and n is an even natural number, then $\sqrt[n]{a^n} = \sqrt[n]{|a|^n} = |a| = -a$.

If $a \geq 0$ and $b \geq 0$, $k, n \in \mathbb{N}$, with $k, n \geq 2$; or if k, n are odd natural numbers (with a, b having any combination of signs: both nonnegative, both negative, or one nonnegative, the other negative) then

(v) $\sqrt[k]{\sqrt[n]{a}} = \sqrt[kn]{a}$.

(vi) $\sqrt[n]{a} \cdot \sqrt[n]{b} = \sqrt[n]{ab}$

(vii) If $a > b > 0$ and $n \in \mathbb{N}$ with $n \geq 2$; or if $b < a < 0$, $n \in \mathbb{N}, n \geq 3$, and n odd, then $\sqrt[n]{a} > \sqrt[n]{b}$.

Proof: (i) Put $\sqrt[n]{a} = x$; then by Definition 3(i), x is the unique nonnegative number such that $x^n = a$; $(x^n)^k = a^k \Rightarrow x^{nk} = a^k$; $(x^k)^n = a^k$. Since $x^k \geq 0$ (because $x \geq 0$) from the last equation it follows that x^k is the unique nth root of a^k; $x^k = \sqrt[n]{a^k}$; and therefore, since $x = \sqrt[n]{a}$, we conclude $\left(\sqrt[n]{a}\right)^k = \sqrt[n]{a^k}$.

(ii) If $y = a^n$, then according to Definition 3, we must have $a = \sqrt[n]{y}$; and since $y = a^n$, we obtain $a = \sqrt[n]{a^n}$.

(iii) We apply part (i) with $n = k$ we have, $\left(\sqrt[n]{a}\right)^n = \sqrt[n]{a^n}$; moreover, by part (ii) we know that $\sqrt[n]{a^n} = a$. Hence, $\left(\sqrt[n]{a}\right)^n = \sqrt[n]{a^n} = a$.

(iv) Since n is even and $a < 0$; we have $|a| = -a$ and $|a|^n = (-a)^n = (-1)^n a^n = a^n \Rightarrow \sqrt[n]{|a|^n} = \sqrt[n]{a^n}$. Furthermore, since $|a| \geq 0$, it follows by part (iii) that $\sqrt[n]{|a|^n} = |a|$; combining this with $|a| = -a$ and what we have above, we see that $\sqrt[n]{a^n} = \sqrt[n]{|a|^n} = |a| = -a$.

(v)] Observe that $\left(\sqrt[k]{\sqrt[n]{a}}\right)^{kn} = \left[\left(\sqrt[k]{\sqrt[n]{a}}\right)^k\right]^n = \left(\sqrt[n]{a}\right)^n = a$, by part (iii). This proves that the real number $\sqrt[k]{\sqrt[n]{a}}$, is the (kn)th root of a, according to Definition 3. Hence $\sqrt[k]{\sqrt[n]{a}} = \sqrt[kn]{a}$.

(vi) We have $\left(\sqrt[n]{a} \cdot \sqrt[n]{b}\right)^n = \left(\sqrt[n]{a}\right)^n \cdot \left(\sqrt[n]{b}\right)^n = ab$, which proves that $\sqrt[n]{a} \cdot \sqrt[n]{b} = \sqrt[n]{ab}$.

(vii) If $a > b > 0$, then $\sqrt[n]{a}$ and $\sqrt[n]{b}$ are also positive. The inequality $a > b$ can be written as $\left(\sqrt[n]{a}\right)^n > \left(\sqrt[n]{b}\right)^n$, and so, Theorem 1(vi) implies $\sqrt[n]{a} > \sqrt[n]{b}$.

Next, suppose that $b < a < 0$ and $n \in \mathbb{N}$, with $n =$ odd ≥ 3. We have $-b > -a > 0$; and thus, $\sqrt[n]{-b}$, $\sqrt[n]{-a}$ are both positive. By Theorem 1(vi) once more, we deduce that $-b > -a$; $\left(\sqrt[n]{-b}\right)^n > \left(\sqrt[n]{-a}\right)^n$ implies $\sqrt[n]{-b} > \sqrt[n]{-a}$; however, $\sqrt[n]{-b} = \sqrt[n]{(-1) \cdot b} =$ (by part (vi)) $= \sqrt[n]{-1} \cdot \sqrt[n]{b} = (-1) \cdot \sqrt[n]{b}$, since n is odd. Likewise $\sqrt[n]{-a} = -\sqrt[n]{a}$. Therefore from the inequality $\sqrt[n]{-b} > \sqrt[n]{-a}$ we obtain $-\sqrt[n]{b} > -\sqrt[n]{a}$. Multiplying both sides of the last inequality by (-1) produces $\sqrt[n]{b} < \sqrt[n]{a}$; or $\sqrt[n]{a} > \sqrt[n]{b}$. We are done. ∎

Remark 6: From part (vi) we can obtain the quotient version of that property: under the same assumptions as in part (vi) but with the added assumption $b \neq 0$, we can replace b with $b^{-1} = \frac{1}{b}$ to obtain $\sqrt[n]{\frac{a}{b}} = \frac{\sqrt[n]{a}}{\sqrt[n]{b}}$.

Having established Theorem 6, we are now in a position to introduce the concept of a rational power of a positive real number. We have the following definition.

Definition 4: Let a be a positive real number, $a > 0$; and $r = \frac{m}{n}$ a rational number (so that $m, n \in \mathbb{Z}; n \neq 0$). We call **power of the real number a with exponent the rational number r**,

(i) The real number $\sqrt[|n|]{a^{|m|}}$, if $r > 0$.

(ii) The real number $\dfrac{1}{\sqrt[|n|]{a^{|m|}}}$, if $r < 0$.

(iii) The real number 1, if $r = 0$.

Notation: We write $a^r = \sqrt[|n|]{a^{|m|}}$, $\frac{1}{\sqrt[|n|]{a^{|m|}}}$, or 1; according to whether $r > 0, r < 0$, or $r = 0$ respectively.

Remark 7: (i) Note that the above definition of the power a^r does not depend on the choice of m and n. To see why, let us suppose $r > 0$; if $r = \frac{m}{n}$ and $\delta = (m, n)$, the greatest common divisor of m and n; then $m = \delta \cdot m_1$ and $n = \delta \cdot n_1$, with $(m_1, n_1) = 1$ (as we have seen in Chapter 4). The integers m_1 and n_1 are in lowest terms. It is also clear that the positive integers $|m_1|$ and $|n_1|$ are unique: for every nonzero rational r, there exists a unique pair $(|m_1|, |n_1|)$, such that $r = \frac{m_1}{n_1}$, and we have $\left(\sqrt[|n_1|]{a^{|m_1|}}\right)^{|n|} = \left(\sqrt[|n_1|]{a^{|m_1|}}\right)^{\delta \cdot |n_1|} = \left[\left(\sqrt[|n_1|]{a^{|m_1|}}\right)^{|n_1|}\right]^{\delta} = \left(a^{|m_1|}\right)^{\delta} = a^{|m_1| \cdot \delta} = a^{|m|}$, which shows that the real number $\sqrt[|n_1|]{a^{|m_1|}}$ is the $|n|$th root of $a^{|m|}$; which proves that a^r is well-defined, regardless of the choice of m and n such that $r = \frac{m}{n}$. So, for example, if $a > 0$, the meaning of $a^{\frac{4}{6}}$ is exactly the same as that of $a^{\frac{2}{3}}$: $a^{\frac{4}{6}}$ is exactly the same as that of $a^{\frac{2}{3}}$: $a^{\frac{4}{6}} = a^{\frac{2}{3}}$.

(ii) Note from Definition 4, it is immediately evident that if $a > 0$ and $r \in \mathbb{Q}$, then $a^{-r} = \frac{1}{a^r}$.

(iii) Note the importance of $a > 0$ in Definition 4: If one allowed $a < 0$, there would be problems. For example, if r is of the form, $r = \frac{\text{odd}}{\text{even}}$, the definition of a^r does depend on the choice of m and n: for instance, take $r = \frac{3}{2}$. We have, $a^{\frac{3}{2}} = \sqrt{a^3}$, which is undefined (by Theorem 5, parts (iii) and (v)), since $a^3 < 0$. On the other hand, $\frac{3}{2} = \frac{4}{6}$; and $a^{\frac{4}{6}} = \sqrt[6]{a^4}$, which is perfectly defined. Of course, if $r = \frac{\text{odd}}{\text{odd}}$, the above problem would not exist, and one could extend the definition in this case (likewise, for $r = \frac{\text{even}}{\text{odd}}$). But when combining powers with rational exponents, some of which have negative bases (or a single negative base a), problems could still arise when we have to add or multiply the exponents. That explains why we work only with positive bases a. Also, the case $a = 0$, obviously poses problems when $r \leq 0$, since 0^r would be undefined; but for $r > 0$, $0^r = 0$.

The proof of the following theorem, is left as a series of exercises. The properties listed in Theorem 7, will undoubtedly be familiar to the reader of this book.

Theorem 7: Let a, b, be positive real numbers; $a > 0, b > 0$. Also let r, r_1, r_2, be rational numbers. We have the following:

(i) $a^{r_1} \cdot a^{r_2} = a^{r_1 + r_2}$

(ii) $(a^{r_1})^{r_2} = a^{r_1 r_2}$

(iii) $(ab)^r = a^r b^r$

(iv) $\left(\frac{a}{b}\right)^r = \frac{a^r}{b^r}$

(v) $a^r > 0$

(vi) In order that $a > b$, it is necessary and sufficient that:

$$\begin{cases} a^r > b^r, & \text{if } r > 0; \\ a^r < b^r, & \text{if } r < 0 \end{cases}$$

(vii) If $a^r = b^r$, then $a = b$ or $r = 0$; and conversely.

Proof:

(i) Exercise 5

(ii) Exercise 6

(iii) Exercise 7

(iv) Exercise 8

(v) Exercise 9

(vi) Exercise 10

(vii) Exercise 11

5.4 Practice Problems

P1. Back in Chapter 4, we came across the sum $(1+2+\ldots+n)$; the sum of the first n natural numbers. As an example of the use of mathematical induction we showed that $1+2+\ldots+n=\frac{n(n+1)}{2}$. Note that if we set $S_k=1^k+2^k+\ldots+n^k$; where k is a natural number, and n is fixed, then the above mentioned sum is simply S_1. In this practice problem we prove that,

$$(n+1)^{k+1}=\binom{k+1}{1}\cdot S_k+\binom{k+1}{2}\cdot S_{k-1}+\ldots+\binom{k+1}{k}S_1+(n+1).$$

This shows, that if all the previous sums $S_1,\ldots,S_{k-1}$ are known, one can compute the next sum, S_k.

Solution: We apply the binomial theorem, Theorem 3, n times; where the natural number n is regarded fixed. We apply Theorem 3 with $b=1$ and $a=1,2,\ldots,n$, and with the exponent being $(k+1)$:

$$\begin{aligned}
(1+1)^{k+1} &= 1^{k+1}+\binom{k+1}{1}1^k+\binom{k+1}{2}1^{k+1} \quad (1)\\
&\quad +\ldots+\binom{k+1}{k}1^1+1\\
(2+1)^{k+1} &= 2^{k+1}+\binom{k+1}{1}2^k+\binom{k+1}{2}2^{k-1} \quad (2)\\
&\quad +\ldots+\binom{k+1}{k}2^1+1\\
&\vdots \qquad\qquad\qquad\qquad\qquad\qquad \vdots\\
(n+1)^{k+1} &= n^{k+1}+\binom{k+1}{1}n^k+\binom{k+1}{2}n^{k-1} \quad (n)\\
&\quad +\ldots+\binom{k+1}{k}n^1+1
\end{aligned}$$

If we add the n equalities memberwise we obtain,

$$\begin{aligned}
2^{k+1}+3^{k+1}+\ldots+(n+1)^{k+1} &= \left[1^{k+1}+2^{k+1}+\ldots+n^{k+1}\right]\\
&+\binom{k+1}{1}\cdot\left[1^k+2^k+\ldots+n^k\right]\\
&+\binom{k+1}{2}\cdot\left[1^{k-1}+2^{k-1}+\ldots+n^{k-1}\right]\\
&+\ldots+\binom{k+1}{k}\cdot[1^1+2^1+\ldots+n^1]+n;
\end{aligned}$$

hence,

$$(n+1)^{k+1} + S_{k+1} - 1 = S_{k+1} + \binom{k+1}{1} \cdot S_k + \binom{k+1}{2} \cdot S_{k-1} + \ldots + \binom{k+1}{k} \cdot S_2 + n$$

which implies,

$$(n+1)^{k+1} = \binom{k+1}{1} \cdot S_k + \binom{k+1}{2} \cdot S_{k-1} + \ldots + \binom{k+1}{k} \cdot S_1 + (n+1).$$

We are done. As a practicing example, let us try $k = 2$:

$$(n+1)^3 = \binom{3}{1} \cdot S_2 + \binom{3}{2} \cdot S_1 + (n+1);$$

$$(n+1)^3 = 3S_2 + 3S_1 + (n+1);$$

and since $S_1 - 1 + 2 + \ldots + n - \frac{n(n+1)}{2}$, we can solve for the last equation for S_2 to obtain, $S_2 = \dfrac{n(n+1)(2n+1)}{6} = 1^2 + 2^2 + \ldots + n^2$, which is familiar from Chapter 4. Likewise, if the student tries $k = 3$ and solves for S_3, she or he will arrive at $S_3 = \left[\frac{n(n+1)}{2}\right]^2$.

P2. Let n and k be natural numbers and a a nonnegative real number; $a \geq 0$. Prove that $\sqrt[nk]{a^k} = \sqrt[n]{a}$.

Solution: By Theorem 6(iii), we have $\sqrt[k]{a^k} = a$. Since the two nonnegative real numbers $\sqrt[k]{a^k}$ and a are equal, their nth roots (which exist and are unique by Theorem 4) must be equal: $\sqrt[n]{\sqrt[k]{a^k}} = \sqrt[n]{a}$. On the other hand, by Theorem 6(v), we must have $\sqrt[n]{\sqrt[k]{a^k}} = \sqrt[nk]{a^k}$. Therefore, it follows that $\sqrt[nk]{a^k} = \sqrt[n]{a}$.

P3. Suppose that the real numbers α, β, γ satisfy the condition $\alpha\beta + \alpha\gamma + \beta\gamma = 1$. Prove that $(1+\alpha^2)(1+\beta^2)(1+\gamma^2) = [\alpha\beta\gamma - (\alpha+\beta+\gamma)]^2$.

Solution: From the identify, $(\alpha+\beta+\gamma)^2 = \alpha^2 + \beta^2 + \gamma^2 + 2(\alpha\beta + \alpha\gamma + \beta\gamma)$, it follows in view of the problem's hypothesis that,

$$(\alpha+\beta+\gamma)^2 = \alpha^2 + \beta^2 + \gamma^2 + 2 \quad (1)$$

Next, square both sides of $\alpha\beta + \alpha\gamma + \beta\gamma = 1$ to obtain

$$(\alpha\beta)^2 + (\alpha\gamma)^2 + (\beta\gamma)^2 + 2\alpha\beta\gamma(\alpha+\beta+\gamma) = 1 \quad (2)$$

Now consider,

$$\begin{aligned}(1+\alpha^2)(1+\beta^2)(1+\gamma^2) &= (1+\alpha^2\beta^2+\alpha^2+\beta^2)(1+\gamma^2)\\ &= 1+\alpha^2\beta^2+\alpha^2+\beta^2+\gamma^2+\alpha^2\beta^2\gamma^2\\ &\quad +\alpha^2\gamma^2+\beta^2\gamma^2;\end{aligned}$$

so we have,

$$(1+\alpha^2)(1+\beta^2)(1+\gamma^2) = 1+(\alpha^2+\beta^2+\gamma^2) + [(\alpha\beta)^2+(\alpha\gamma)^2+(\beta\gamma)^2]+(\alpha\beta\gamma)^2 \quad (3)$$

Combining (1), (2), and (3) yields,

$$\begin{aligned}(1+\alpha^2)(1+\beta^2)(1+\gamma^2) &= 1+[(\alpha+\beta+\gamma)^2-2]\\ &\quad +[1-2\alpha\beta\gamma(\alpha+\beta+\gamma)]+(\alpha\beta\gamma)^2;\\ (1+\alpha^2)(1+\beta^2)(1+\gamma^2) &= (\alpha+\beta+\gamma)^2-2(\alpha\beta\gamma)(\alpha+\beta+\gamma)\\ &\quad +(\alpha\beta\gamma)^2;\\ (1+\alpha^2)(1+\beta^2)(1+\gamma^2) &= [\alpha\beta\gamma-(\alpha+\beta+\gamma)]^2.\end{aligned}$$

We are done.

P4. If α, β are positive rational numbers such that $\alpha\beta > 1$, show that for any rational exponents r_1, r_2,

$$\left(\alpha^2-\frac{1}{\beta^2}\right)^{r_1}\cdot\left(\alpha-\frac{1}{\beta}\right)^{r_2-r_1} : \left(\beta^2-\frac{1}{\alpha^2}\right)^{r_2}\cdot\left(\beta+\frac{1}{\alpha}\right)^{r_1-r_2} = \left(\frac{\alpha}{\beta}\right)^{r_1+r_2}$$

Solution: The hypothesis $\alpha, \beta > 0$ and $\alpha\beta > 1$, ensures that real numbers $\alpha^2-\frac{1}{\beta^2}, \alpha-\frac{1}{\beta}, \beta^2-\frac{1}{\alpha^2}$, and $\beta+\frac{1}{\alpha}$, are all positive. Hence, when they are raised to any rational exponent, all these powers are defined and they are positive real numbers. We have,

$$\frac{\left(\alpha^2-\frac{1}{\beta^2}\right)^{r_1}\cdot\left(\alpha-\frac{1}{\beta}\right)^{r_2-r_1}}{\left(\beta^2-\frac{1}{\alpha^2}\right)^{r_2}\cdot\left(\beta+\frac{1}{\alpha}\right)^{r_1-r_2}}$$

$$= \frac{\left[\frac{(\alpha\beta)^2-1}{\beta^2}\right]^{r_1}\cdot\left[\frac{\alpha\beta-1}{\beta}\right]^{r_2-r_1}}{\left[\frac{(\alpha\beta)^2-1}{\alpha^2}\right]^{r_2}\cdot\left[\frac{\alpha\beta+1}{\alpha}\right]^{r_1-r_2}}$$

$$= \frac{\left[(\alpha\beta)^2-1\right]^{r_1}\cdot\left(\beta^2\right)^{-r_1}\cdot(\alpha\beta-1)^{r_2-r_1}\cdot\beta^{-(r_2-r_1)}}{\left[(\alpha\beta)^2-1\right]^{r_2}\cdot(\alpha^2)^{-r_2}\cdot(\alpha\beta+1)^{r_1-r_2}\cdot\alpha^{-(r_1-r_2)}}$$

$$\begin{aligned}
&= [(\alpha\beta)^2-1]^{r_1-r_2}\cdot\beta^{-2r_1}\cdot\beta^{r_1-r_2}\cdot\alpha^{2r_2}\cdot\alpha^{r_1-r_2} \\
&\quad\cdot(\alpha\beta-1)^{r_2-r_1}\cdot(\alpha\beta+1)^{r_2-r_1}
\end{aligned}$$

$$\begin{aligned}
&= [(\alpha\beta+1)(\alpha\beta-1)]^{r_1-r_2}\cdot(\alpha\beta-1)^{r_2-r_1} \\
&\quad\cdot(\alpha\beta+1)^{r_2-r_1}\cdot\beta^{-(r_1+r_2)}\cdot\alpha^{r_1+r_2}
\end{aligned}$$

$$\begin{aligned}
&= (\alpha\beta+1)^{r_1-r_2}\cdot(\alpha\beta+1)^{r_2-r_1}\cdot(\alpha\beta-1)^{r_1-r_2} \\
&\quad\cdot(\alpha\beta-1)^{r_2-r_1}\cdot\left(\frac{\alpha}{\beta}\right)^{r_1+r_2}
\end{aligned}$$

$$= (\alpha\beta+1)^0\cdot(\alpha\beta-1)^0\cdot\left(\frac{\alpha}{\beta}\right)^{r_1+r_2} = \left(\frac{\alpha}{\beta}\right)^{r_1+r_2}$$

P5. (i) Let α, β be real numbers such that $\beta > 0$ and either $\beta^2 \leq \alpha \leq 2\beta^2$ or $\alpha > 2\beta^2$. Show that the quantity $A = \sqrt{\alpha + 2\beta\sqrt{\alpha-\beta^2}} + \sqrt{\alpha - 2\beta\sqrt{\alpha-\beta^2}}$ is equal to 2β when $\beta^2 \leq \alpha \leq 2\beta^2$; otherwise, when $\alpha > 2\beta^2$, $A = 2\sqrt{\alpha-\beta^2}$.

(ii) Show that if $A = \sqrt{\alpha+2\sqrt{\alpha-1}} + \sqrt{\alpha-2\sqrt{\alpha-1}}$ and either $1 \leq \alpha \leq 2$ or $\alpha > 2$; then $A = 2$ or $A = 2\sqrt{\alpha-1}$ respectively.

Solution: (i) Square both sides to obtain,

$$\begin{aligned}
A^2 &= \alpha + 2\beta\sqrt{\alpha-\beta^2} \\
&\quad +2\sqrt{(\alpha+2\beta\sqrt{\alpha-\beta^2})(\alpha-2\beta\sqrt{\alpha-\beta^2})} \\
&\quad +\alpha - 2\beta\sqrt{\alpha-\beta^2};
\end{aligned}$$

$$A^2 = 2\alpha + 2\sqrt{\alpha^2 - (2\beta\sqrt{\alpha-\beta^2})^2};$$

$$\begin{aligned}
A^2 &= 2\alpha + 2\sqrt{\alpha^2 - 4\beta^2(\alpha-\beta^2)} \\
&= 2\alpha + 2\sqrt{\alpha^2 - 2(2\beta^2)\alpha + (2\beta^2)^2}
\end{aligned}$$

$$A^2 = 2\alpha + 2\sqrt{(\alpha-2\beta^2)^2}.$$

If $\alpha - 2\beta^2 > 0$, then by Theorem 6(iii) we must have, $\sqrt{(\alpha-2\beta^2)^2} = \alpha - 2\beta^2 \Rightarrow A^2 = 2\alpha + 2(\alpha-2\beta^2) = 4(\alpha-2\beta^2) \Rightarrow \sqrt{A^2} = \sqrt{4(\alpha-2\beta^2)}$; but $A \geq 0$ (by the way A is defined, as the sum of two square roots - refer to Remark 5), which implies

$\sqrt{A^2} = A$ and hence $A = 2\sqrt{\alpha - 2\beta^2}$. If, on the other hand, $\beta^2 \leq \alpha \leq 2\beta^2$, then $\alpha - 2\beta^2 \leq 0$, in particular, and so $\sqrt{(\alpha - 2\beta^2)^2} = -(\alpha - 2\beta^2) = 2\beta^2 - \alpha$, by Theorem 6(iv) (note that if $\alpha - 2\beta^2 = 0$, this step is still valid since $-0 = 0$). So, in this case we obtain $A^2 = 2\alpha + 2(2\beta^2 - \alpha) \Rightarrow A^2 = 4\beta^2 \Rightarrow \sqrt{A^2} = \sqrt{4\beta^2}$; and since $A \geq 0$, $\beta \geq 0$ (by hypothesis), we obtain (by Theorem 6(iii) $A = 2\beta$. Note that the part $\beta^2 \leq \alpha$ of the problem's hypothesis is actually needed; $0 \leq \alpha - \beta^2$, so $\sqrt{\alpha - \beta^2}$ is defined as a (nonnegative) real number. Otherwise, if it were $\alpha - \beta^2 < 0$, $\sqrt{\alpha - \beta^2}$ would not be a real number. Recall that $\sqrt{\alpha - \beta^2}$ is part of the formula that defines the nonnegative real number A.

(ii) This is an immediate consequence of part (ii), with $\beta = 1$.

P6. (i) Show that for every natural number $n \geq 2$, $\left(1 + \frac{1}{n}\right)^{n-1} < n$.

(ii) Use part (i) to prove that for every natural number $n \geq 2$, $(n!)^2 > n^n$.

Solution: (i) We proceed by mathematical induction. For $n = 2$, $\left(1 + \frac{1}{2}\right)^1 < 2$; $\frac{3}{2} < 2$, which is true. The inductive hypothesis: assume that for $n = k \geq 2$, $\left(1 + \frac{1}{k}\right)^{k-1} < k$. Multiply both sides by $(1 + \frac{1}{k})$ to obtain $\left(1 + \frac{1}{k}\right)^k < k \cdot \left(1 + \frac{1}{k}\right) = k + 1$. Thus, $\left(1 + \frac{1}{k}\right)^k < k + 1$. On the other hand, $1 + \frac{1}{k+1} < 1 + \frac{1}{k}$ (obviously, since $k > 0$ and so $k + 1 > k$). Therefore, $\left(1 + \frac{1}{k+1}\right)^k < \left(1 + \frac{1}{k}\right)^k$, by Theorem 1(vi). We conclude that $\left(1 + \frac{1}{k+1}\right)^k < \left(1 + \frac{1}{k}\right)^k < k + 1 \Rightarrow \left(1 + \frac{1}{k+1}\right)^k < k + 1$, which proves that the statement also holds for $n = k + 1$.

(ii) We apply mathematical induction. For $n = 3$, $(3!)^2 > 3^3 \Leftrightarrow (1 \cdot 2 \cdot 3)^2 > 3^3 \Leftrightarrow 36 > 27$, which is true. Assume the inductive hypothesis: the statement holds for $n = k$, $k \geq 3$; $(k!)^2 > k^k \Rightarrow (1 \cdot 2 \ldots k)^2 > k^k$; multiply both sides of the last inequality by $(k+1)^2$ to obtain,

$$(1 \cdot \ldots k)^2 \cdot (k+1)^2 > (k+1)^2 \cdot k^k;$$

$$[1 \cdot 2 \ldots k \cdot (k+1)]^2 > (k+1)^2 \cdot k^k; \qquad (1)$$

$$[(k+1)!]^2 > (k+1)^2 \cdot k^k$$

By part (i) of this problem we know (since $k \geq 3$) that

$$k > \left(1 + \frac{1}{k}\right)^{k-1};$$

$$k > \left(\frac{k+1}{k}\right)^{k-1};$$

$$k > \frac{(k+1)^{k-1}}{k^{k-1}};$$

multiply both sides by k^{k-1} to obtain, $k \cdot k^{k-1} > \frac{k^{k-1} \cdot (k+1)^{k-1}}{k^{k-1}} = (k+1)^{k-1}$ (this last step is valid since $k > 0$ and thus $k^{k-1} > 0$); $k^k > (k+1)^{k-1}$. Multiply both sides by $(k+1)^2$:

$$(k+1^2 \cdot k^k > (k+1)^2 \cdot (k+1)^{k-1} = (k+1)^{k+1}. \qquad (2)$$

By (1) and (2) it follows that $[(k+1)!]^2 > (k+1)^{k+1}$; the inductive proof is complete.

P7. Let a, b be distinct real numbers: $a \neq b$. Consider the sequence of real numbers defined as follows: $c_1 = a+b, c_2 = c_1 - \frac{ab}{c_1}, c_3 = c_1 - \frac{ab}{c_2}, \ldots, c_n = c_1 - \frac{ab}{c_{n-1}}, c_{n+1} = c_1 - \frac{ab}{c_n}, \ldots$. Prove that for every natural number n, the following statement holds true: $c_n = \frac{a^{n+1}-b^{n+1}}{a^n-b^n}$.

Solution: We proceed by mathematical induction. For $n=1$, $c_1 = \frac{a^2-b^2}{a-b} = \frac{(a-b)(a+b)}{a-b} = a+b$, which is true by the problem's hypothesis. The inductive hypothesis: assume that for $n=k$, the statement holds true: $c_k = \frac{a^{k+1}-b^{k+1}}{a^k-b^k}$. By the problem's assumption we know that, $c_{k+1} = c_1 - \frac{ab}{c_k} = (a+b) - \frac{ab}{c_k}$. Applying the inductive hypothesis yields,

$$c_{k+1} = (a+b) - \frac{ab}{\left(\frac{a^{k+1}-b^{k+1}}{a^k-b^k}\right)} = (a+b) - \frac{(ab)\cdot(a^k-b^k)}{a^{k+1}-b^{k+1}}$$

$$\Rightarrow \quad c_{k+1} = \frac{(a+b)(a^{k+1}-b^{k+1})-(ab)(a^k-b^k)}{a^{k+1}-b^{k+1}};$$

$$c_{k+1} = \frac{a^{k+2}+b\cdot a^{k+1}-a\cdot b^{k+1}-b^{k+2}-b\cdot a^{k+1}+a\cdot b^{k+1}}{a^{k+1}-b^{k+1}}$$

$$c_{k+1} = \frac{a^{k+2}-b^{k+2}}{a^{k+1}-b^{k+1}}.$$

The inductive step is complete.

P8. Let $a_1, a_2, \ldots, a_n, \ldots$ be a sequence of positive real numbers; $a_n > 0$ for every $n \in \mathbb{N}$. Suppose that the sequence has the property that $a_1 = 1$ and $a_{n+1}^2 > a_n \cdot a_{n+2}$, for every natural number. Prove that

(i) $\sqrt[n+1]{a_{n+2}} > \sqrt[n+2]{a_{n+3}}$; and

(ii) $2 > \sqrt{3} > \sqrt[3]{4} > \ldots > \sqrt[n+1]{n+2} > \sqrt[n+2]{n+3} > \ldots$

Solution: (i) First note that since all the terms of the sequence are positive, and according to Theorem 1(vi), the statement we are asked to prove is equivalent to another, easier to prove statement:

$$\sqrt[n+1]{a_{n+2}} > \sqrt[n+2]{a_{n+3}}$$

$$\Leftrightarrow \left(\sqrt[n+1]{a_{n+2}}\right)^{(n+1)(n+2)} > \left(\sqrt[n+2]{a_{n+3}}\right)^{(n+1)(n+2)}$$

$$\Leftrightarrow \left[\left(\sqrt[n+1]{a_{n+2}}\right)^{n+1}\right]^{n+2} > \left[\left(\sqrt[n+2]{a_{n+3}}\right)^{n+2}\right]^{n+1} \qquad (1)$$

$$\Leftrightarrow (a_{n+2})^{n+2} > (a_{n+3})^{n+1}$$

So, if we prove (1), we will have proven the original statement in part (i). We use mathematical induction. For $n=1$, we must prove $(a_3)^3 > (a_4)^2$. By the problem's hypothesis for $n=1$ we know that,

$$a_2^2 > a_1 \cdot a_3; \quad \text{but } a_1 = 1; \text{ we have,}$$

$$a_2^2 > a_3 \tag{2}$$

Again, by the problem's assumption for $n=2$ we have, $(a_3)^2 > a_2 a_4 \Leftrightarrow$ (since both sides are positive) $\Leftrightarrow (a_3)^4 > (a_2 a_4)^2 = (a_2)^2 \cdot (a_4)^2 > a_3 \cdot (a_4)^2$, by (2). Therefore,

$$(a_3)^4 > a_3 \cdot (a_4)^2, \tag{3}$$

Since $a_3 > 0$, multiplying both sides of (3) by $\frac{1}{a_3}$ produces $(a_3)^3 > (a_4)^2$ which establishes (1) for $n=1$. Next, we assume the inductive hypothesis: we assume the statement (1) to hold for $n=k$:

$$(a_{k+2})^{k+2} > (a_{k+3})^{k+1} \tag{4}$$

We apply the problem's hypothesis (with $n = k+2$): $(a_{k+3})^2 > a_{k+2} \cdot a_{k+4} \Rightarrow$ (by Theorem 1(v), since both sides are positive)

$$\begin{aligned} \Rightarrow \quad \left[(a_{k+3})^2\right]^{(k+2)} &> (a_{k+2})^{k+2} \cdot (a_{k+4})^{k+2} \\ \Rightarrow \quad (a_{k+3})^{2k+4} &> (a_{k+2})^{k+2} \cdot (a_{k+4})^{k+2} \end{aligned} \tag{5}$$

Since all the terms/factors of (4) and (5) are positive, by combining (4) and (5) we deduce that (this step would be invalid if $(a_{k+4})^{k+2}$ of (4) were negative), $(a_{k+3})^{2k+4} > (a_{k+3})^{k+1} \cdot (a_{k+4})^{k+2}$; multiplying both sides of the last inequality with the positive real number $\frac{1}{(a_{k+3})^{k+1}}$ yields,

$$(a_{k+3})^{[(2k+4)-(k+1)]} > (a_{k+4})^{k+2};$$

$$(a_{k+3})^{k+3} > (a_{k+4})^{k+2};$$

the induction process is complete.

(ii) This is an application of part (i). Let $a_n = n$, for $n = 1, 2, 3, \ldots$; $a_n > 0, \forall n \in \mathbb{N}$. Clearly $a_1 = 1$ and $(a_{n+1})^2 = (n+1)^2$, while $a_n \cdot a_{n+2} = n \cdot (n+2)$. Note that the condition $(a_{n+1})^2 > a_n \cdot a_{n+2}$ is indeed satisfied since $(n+1)^2 = n^2 + 2n + 1 > n^2 + 2n = n \cdot (n+2)$. Therefore by part (i) of this problem,

$$\sqrt[n+1]{a_{n+2}} > \sqrt[n+2]{a_{n+3}};$$

$$\sqrt[n+1]{n+2} > \sqrt[n+2]{n+3}.$$

We are done: $\sqrt{3} > \sqrt[3]{4} > \sqrt[4]{5} > \ldots > \sqrt[n+1]{n+2}$; and since $2 > \sqrt{3}$ we have,

$$2 > \sqrt{3} > \sqrt[3]{4} > \sqrt[4]{5} > \ldots > \sqrt[n+1]{n+2} > \sqrt[n+2]{n+3} > \ldots.$$

5.5 Exercise Set

1. Prove part (ii) of Theorem 1.

2. Prove part (iii) of Theorem 1.

3. Prove part (vi) of Theorem 1.

4. Give a proof for part (iii) of Example 1(ii).

5. Prove part (i) of Theorem 7.

6. Prove part (ii) of Theorem 7.

7. Prove part (iii) of Theorem 7.

8. Prove part (iv) of Theorem 7.

9. Prove part (v) of Theorem 7.

10. Prove part (vi) of Theorem 7.

11. Prove part (vii) of Theorem 7.

12. Let $a, b \in \mathbb{R}$ with $a > 0, b > 0$. If $x = \frac{1}{2}\left[\sqrt{\frac{a}{b}} - \sqrt{\frac{b}{a}}\right]$, show that $\frac{2a \cdot \sqrt{1+x^2}}{x+\sqrt{1+x^2}} = a+b$.

13. If $A = \sqrt{1 + a + \sqrt{1 + 2a}} + \sqrt{1 + a - \sqrt{1 + 2a}}$, show that if $a \geq 0$, then $A = \sqrt{2(1 + 2a)}$; but if $a < 0$, then $A = \sqrt{2}$.

14. Use mathematical induction to prove the **Bernoulli** inequality: if a is a real number such that $a \geq -1$, then $(1 + a)^n \geq 1 + na$, for every natural number n. Is it true that the equal holds only for $n = 1$? Show that the answer is in the affirmative.

15. Prove part (ii) of Remark 2.

16. Prove part (iii) of Remark 2.

17. Prove part (iv) of Remark 2.

18. (i) Show that for every natural number $n \geq 2$, $1 < \sqrt[n]{n}$.
    (ii) Prove that for every natural number $n \geq 4$, $\sqrt[n]{n} < \sqrt[3]{3}$.
    Hint: The statement is equivalent to $n^3 < 3^n$, for $n \geq 4$.

19. Use the binomial theorem (Theorem 3) to prove that if $a > 0$, then $(1 + a)^n > 1 + na + \frac{n(n-1)}{2} \cdot a^2$.

20. Show that for every natural number n, $2 + 2^2 + 2^3 + \ldots + 2^n = 2(2^n - 1)$.

21. Prove if a is a real number, $a \neq 1$, then $1 + 2a + 3a^2 + \ldots + na^{n-1} = \frac{1-(n+1)a^n+na^{n+1}}{(1-a)^2}$, for every $n \in \mathbb{N}$.

22. Show that for every natural number n,

(i) $\frac{1}{2} + \frac{2}{2^2} + \frac{3}{2^3} + \ldots + \frac{n}{2^n} = 2 - \left(\frac{n+2}{2^n}\right)$.

(ii) $1 + 9 + 45 + \ldots + (2n-1) \cdot 3^{n-1} = (n-1) \cdot 3^n + 1$.

23. Show that for every $n \in \mathbb{N}$, $(n+5)! < \frac{(n+5)^{n+5}}{2^{n+5}}$.

24. Let $a_1, a_2, \ldots, a_n, a_{n+1}, \ldots$ be a sequence such that $a_{n+1} = 2a_n + 1$, for every natural number n. Prove that $a_n = 2^{n-1} \cdot (a_1 + 1) - 1$, for every $n \in \mathbb{N}$.

25. Prove that for every natural number n,

(i) $\frac{1 \cdot 3 \cdot 5 \ldots (2n-1)}{2 \cdot 4 \cdot 6 \ldots 2n} < \sqrt{\frac{1}{2n+1}}$.

(ii) $\frac{3 \cdot 7 \cdot 11 \ldots (4n-1)}{5 \cdot 9 \cdot 13 \ldots (4n+1)} < \sqrt{\frac{3}{4n+3}}$

26. Show that if $0 < \alpha < 1$, then $(1-\alpha)^n \geq 1 - n\alpha$, for every $n \in \mathbb{N}$; with the equal sign being the case only when $n = 1$.

27. Show that for every natural number n,

(i) $\left(1 - \frac{1}{n^2}\right)^n \geq 1 - \frac{1}{n}$.

(ii) $\left(1 + \frac{1}{n}\right)^n \geq 2$.

(iii) $\left(1 + \frac{1}{n+1}\right)^{n+1} > \left(1 + \frac{1}{n}\right)^n$.

Hint:

(i) Use Exercise 26.

(ii) Apply the binomial theorem with $a = 1$ and $b = \frac{1}{n}$. Show that on the right-hand side of the expansion, the term 1 and the term $\binom{n}{1} \cdot \frac{1}{n} = 1$ both appear, with the remaining of the terms being positive.

(iii) Use part (i) as follows: by part (i) we know that $\left(1 - \frac{1}{m^2}\right)^m \geq 1 - \frac{1}{m}$, for $m \in \mathbb{N}$; in particular for $m \geq 2$. Factor the left-hand side (by factoring the quantity $1 - \frac{1}{m^2}$) and manipulate it algebraically in order to obtain $\left(\frac{m-1}{m}\right)^{m-1} \geq \left(\frac{m}{m+1}\right)^m$; now let $m = n+1$, with $n \in \mathbb{N}$, since $m \in \mathbb{N}$ and $m \geq 2$. Algebraically manipulate the last inequality in order to arrive at the desired result.

28. Prove that if $0 < \alpha < 1$, then $(1+\alpha)^n \cdot (1+\alpha^n) > 2^{n+1} \cdot \alpha^n$, for every $n \in \mathbb{N}$.

29. Show that if $\alpha > 0$, then $\alpha^{2n} + \alpha^{2n-1} + \alpha^{2n-2} + \ldots + \alpha + 1 \geq (2n+1) \cdot a^n$, $\forall n \in \mathbb{N}$.

30. Show that if $0 < \alpha < 1$, then $1 - \alpha^{2n+1} > (2n+1) \cdot \alpha^n \cdot (1 - \alpha)$, for every natural number n.

31. Suppose that α, β are real numbers such that $\alpha \geq \beta > 0$. Show that $\alpha^{n+1} + \beta^{n+1} \geq \alpha^n \cdot \beta + \alpha \cdot \beta^n$, $\forall n \in \mathbb{N}$. When does the equal sign hold?

32. Prove that if $n \in \mathbb{N}$, then $(n+2)^{n+2} < 2^{n+2} \cdot (n+1)^{n+1}$.

33. If $m, n \in \mathbb{Z}$, $m > n$, and a is a real number greater than 1, then $a^m > a^n$.

34. If α, β are positive real numbers, then $\sqrt{\alpha} - \sqrt{\beta} < \sqrt{\alpha + \beta} < \sqrt{\alpha} + \sqrt{\beta}$.

35. If $n \in \mathbb{N}$ and $n > 2$, then $2 \cdot 4 \cdot 6 \ldots (2n+2) < (n+2)^{n+1}$.

Hint: Note that the given statement is equivalent to $2^{n+1} \cdot (n+1)! < (n+2)^{n+1}$; use mathematical induction and Exercise 27(ii).

Chapter 6

Irrational Numbers

Undoubtedly, the reader of this book must have come across, in his/her mathematical life, real numbers that are not rational. That is, real numbers that cannot be represented as a ratio of two integers. Such numbers are called **irrational** numbers. Probably the most famous of all irrational numbers is $\sqrt{2}$. This is probably the most well-known irrational number to a wide mathematical audience. There are several interesting and ingenious proofs that establish $\sqrt{2}$ as an irrational numbers. Some of these proofs are more geometric in nature. Some are more purely number-theoretic in nature. On the other hand, the student of this book, has most likely come across irrational numbers such as π, $e, ln3$, $\log 2$, e. Such numbers are frequently encountered in a first Calculus course. There important differences between irrational numbers like $\sqrt{2}$ or $\sqrt{3}$ and irrational numbers like π, e, or $ln2$. The former numbers are called **algebraic** because they are root of zeros of polynomial functions with integer coefficients ($\sqrt{2}$ is a root of the polynomial function $f(x) = x^2 - 2$). The latter numbers like π, e, or $ln2$are not roots or zeros to any polynomial equation with integer coefficients. Such irrational numbers are known as **transcendental** numbers. We note here that the set of algebraic real numbers contain the set -real numbers, as pointed out above.

6.1 Discovering Irrational Numbers

As we know, the rational numbers form a ring, and what is more they actually form a field. If $a, b \in \mathbb{Q}$, the all four combinations $(a+b)$, $(a-b)$, $a \cdot b$, and $\frac{a}{b}(b \neq 0)$ are also rational numbers. However, this is not the case with irrational numbers. The sum of two irrationals may or may not be an irrational number. For example, the sum of the two irrationals $(\sqrt{2}-3)$ and $(\sqrt{2}+3)$ is the irrational number $2\sqrt{2}$. The sum of the two irrationals $(3-\sqrt{2})$ and $(3+\sqrt{2})$ is the rational number 6. The same is true for products. The product of the irrational numbers $\sqrt{2}-3$ and $\sqrt{2}-3$ is a rational number since, $(\sqrt{2}-3)(\sqrt{2}+3) = (\sqrt{2})^2 - (3)^2 = 4 - 9 = -5$. The product of $\sqrt{2}$ and $(\sqrt{2}+3)$ is irrational: $\sqrt{2} \cdot (\sqrt{2}+3) = (\sqrt{2})^2 + 3 \cdot \sqrt{2} = 2 + 3\sqrt{2}$. Another example: $\sqrt{3} \cdot \sqrt{12} = \sqrt{36} = 6$ is a rational number but both $\sqrt{3}$ and $\sqrt{12} = 2\sqrt{3}$ are irrational numbers.

Example 1: Assuming $\sqrt{6}$ to be irrational, we will show that the number $\sqrt{2}+\sqrt{3}$ is irrational as well. We argue by contradiction. Suppose, to the contrary that $\sqrt{2}+\sqrt{3} = \rho$, where ρ is a rational number. Squaring both sides, $2+2(\sqrt{2})(\sqrt{3})+3 = \rho^2 \Rightarrow 2\sqrt{6} = \rho^2 - 5 \Rightarrow \sqrt{6} = \frac{\rho^2-5}{2}$, which is a contradiction, since $\frac{\rho^2-5}{6}$ is a rational number (in view of the fact that ρ is rational) while $\sqrt{6}$ is irrational.

In the following example, we find out that an expression involving radicals may temptingly appear to be an irrational number but in actuality, it maybe a rational number. The sample example also illustrates a technique that can be employed to tackle similar problems.

Example 2: Show that the number $\sqrt[3]{45+29\sqrt{2}} + \sqrt[3]{45-29\sqrt{2}}$ is an integer. If we set, $a = \sqrt[3]{45+29\sqrt{2}}$ and $b = \sqrt[3]{45-29\sqrt{2}}$, from the Bionomial Theorem we know that,

$$(a+b)^2 = a^3 + 3a^2b + 3ab + b^3;$$

$$(a+b)^3 = a^3 + b^3 + 3ab(a+b);$$

the number we are trying to show to be an integer is simply the sum $S = a + b$. Also, if $P = ab$ then $P = \sqrt[3]{(45+29\sqrt{2})(45-20\sqrt{2}} = \sqrt[3]{(45)^2 - (2)(29)^2} = \sqrt[3]{(2025)-(1682)} = \sqrt[3]{343} = \sqrt[3]{(7)^3} = 7$. Also $a^3 = 45 + 29\sqrt{2}$, $b^3 = 45 - 29\sqrt{2}$. Going back to the identity $(a+b)^3 = a^3 + b^3 + 3ab(a+b)$, we now have $S^3 = (45+29\sqrt{2}) + (45-29\sqrt{2}) + 3 \cdot P \cdot S$; $S^3 = 90 + 3 \cdot (7) \cdot S$ (since $P = 7$); $S^3 - 21S - 90 = 0$. Thus S is one of the three roots of the cubic equation $^3 - 21x - 90 = 0$. Accordingly to pre calculus methods if this last equation has a rational root p/q (p and q integers whose greatest common divisor is 1), then p must be a divisor of -90 and q must be a divisor of 1. That creates a lot of possible rational roots $p/q = \pm1, \pm2, \pm3, \pm5, \pm6, \pm9, \pm10, \pm15, \pm18, \pm30$. It should be noted here, that calculus methods can eliminate most of these possibilities. Even if one is determined to check all of these possible rational roots, before long one finds that $6^3 - (21)(6) - 90 = 216 - 126 - 90 - 0$. Thus the number 6 is one of the three roots of the equation $x^3 - 21x - 90 = 0$. Performing the division (or synthetic division) of the polynomial function $x^3-21x-90$ with the binomial $x-6$ we find that $x^3-21x-90=(x-6)(x^2-15x+15)$. The other two roots of the equation $x^3-21x-90=0$ are simply the two zeros of the trinomial $x^2-15x+15$. These two roots can be found by applying the quadratic formula: $r_1=\frac{15+\sqrt{(-15)^2-4(15)}}{2}=\frac{15+\sqrt{165}}{2}$ and $r^2=\frac{15-\sqrt{165}}{2}$. Thus the three roots of the cubic equation $x^3-21x-90=0$ are the numbers 6, $\frac{15+\sqrt{165}}{2}$, and $\frac{15-\sqrt{165}}{2}$. Since we have shown that our sum S is a root to the above cubic, it must be one of these numbers. Now if we look at our number $S=\sqrt[3]{45+29\sqrt{2}}+\sqrt[3]{45-29\sqrt{2}}$, we can find an upper and a lower bound: from

$$\begin{aligned}
&1<\sqrt{2}<2\Rightarrow 29<29\sqrt{2}<58\Rightarrow\\
&\Rightarrow 29+45<45+29\sqrt{2}<45+58;\\
&74<45+29\sqrt{2}<103;\\
&\sqrt[3]{74}<\sqrt[3]{45+29\sqrt{2}}<\sqrt[3]{103}
\end{aligned}\tag{1}$$

On the other hand, it is easily seen that

$$0<\sqrt[3]{45-29\sqrt{2}}<2 \tag{2}$$

(the reader should show this by cubing each side of the two simultaneous inequalities.) Adding (1) and (2) memberwise we arrive at,

$$4<\sqrt[3]{74}<S<\sqrt[3]{103}+2<5+2=7;\quad 4<S<7.$$

Thus, our sum S must be a real number greater than four but less than 7. On the other hand, it is easily seen that $1<\frac{15-\sqrt{165}}{2}<2$ and $13<\frac{15+\sqrt{165}}{2}<14$. Clearly then, none of the two roots $r_1=\frac{15-\sqrt{165}}{2}$ or $r_2=\frac{15+\sqrt{165}}{2}$ falls in the open interval $(4,7)$. Therefore since $4<6<7$, we conclude that $S=6$; $\sqrt[3]{45+29\sqrt{2}}+\sqrt[3]{45-29\sqrt{2}}=6$.

Theorem 1 If a is an irrational number and r a rational number, then the numbers $(a+r)$, $(a-r)$, $a\cdot r$ (for $r\neq 0$), $\frac{a}{r}$ (for $r\neq 0$), and $\frac{r}{a}$ (for $r\neq 0$), are all irrational numbers.

Proof: By contradiction, if we assume $a+r=\rho$ to be a rational number then $a=\rho-r$ will be rational, since the difference of two rationals is rational; hence a contradiction. The argument is similar for proving the irrationality of the difference $(a-r)$. Next, if we assume $a\cdot r=\rho=$ rational then since $r\neq 0$, we have $a=\frac{\rho}{r}=$ rational. Since the ratio of two rationals is rational, we have a contradiction again. Finally, if we assume $\frac{a}{r}=\rho=$ rational, then $a=r\cdot\rho=$ rational, since the product of two rational numbers is rational, a contradiction once more. Likewise if $\frac{r}{a}$ were rational, say $\frac{r}{a}=t=$ rational then (since $t\neq 0$ because $r\neq 0$), we have $a=\frac{r}{t}=$ ratio; a contradiction. The proof is complete.

The next Theorem establishes a whole class of irrational numbers in an explicit way.

Theorem 2 (i) If a natural number k is the nth power of a natural number, than $\sqrt[n]{k}$ is a rational number.

(ii) If a natural number k is not the nth power of a natural number, then $\sqrt[n]{k}$ is an irrational number.

(iii) Let k and m be natural numbers which are relatively prime. That is $(k, m) = 1$ (where (k, m) denotes the greatest common divisor of k and m). Let n be a natural number. If at least one of the numbers $\sqrt[n]{k}$ and $\sqrt[n]{m}$ is irrational, then so is the product $\sqrt[n]{km}$. In particular if both $\sqrt[n]{k}$ and $\sqrt[n]{m}$ are irrational numbers, so is $\sqrt[n]{km}$.

(iv) Let $k_1, k_2, \dots, k_l$ be natural numbers which are mutually relatively prime. That is $(k_i, k_j) = 1$, for all i and j such that $i \neq j$ and $1 \leq i, j \leq l$. Let n be a natural number. If at least one of the l numbers, $\sqrt[n]{k_1}, \dots, \sqrt[n]{k_l}$ is irrational so is the product $\sqrt[n]{k_1 \dots k_l}$. In particular if all the numbers $\sqrt[n]{k_l}, \dots, \sqrt[n]{k_l}$ are irrational, so is their product $\sqrt[n]{k_1 \dots k_l}$.

(v) If k, m, n are natural numbers with $(k, m) = 1$ and at least one of the numbers $\sqrt[n]{k}$ and $\sqrt[n]{m}$ is irrational then so is the ration $\sqrt[n]{\frac{k}{m}}$.

(vi) The nth root of a positive rational number r is an irrational number if, and only if, r is not the nth power of a rational number.

Proof: (i) Suppose that $k = t^n$, where t is a natural number. Then $\sqrt[n]{k} = \sqrt[n]{t^n} = t$, which since it is an integer, it is, in particular, a rational number. ∎

(ii) We prove the contrapositive: if $\sqrt[n]{k}$ is a rational number, then k must be the nth power of a natural number. Indeed suppose that $\sqrt[n]{k} = \frac{p}{q}$, where p and q are positive integers with $(p, q) = 1$ (obviously $\sqrt[n]{k}$ is positive, since the principal nth root of a positive real must be positive). We have $k = \frac{p^n}{q^n} \Rightarrow k \cdot q^n = p^n$; which shows that q^n divides p^n. Now q cannot have any prime divisors, since if r is a prime divisor of q, then the above equations hows that r must divide p^n and consequently (as we have seen in Chapter 4) since r is a prime, r must divide p. This would violate the relative primeness condition $(p, q) = 1$. Hence q cannot have any prime divisors, which implies, since $q \geq 1$ (remember q is a natural number), that $q = 1$. Hence $k \cdot q^n = p^n$ implies $k = p^n$ and we are done. ∎

(iii) First we show that the natural number $\sqrt[n]{k \cdot m}$ cannot be the nth power of a natural number. We do this by arguing by contradiction: if it were $k \cdot m = t^n$, for some natural number t, then in view of $(k, m) = 1$ and according to Theorem 17 in Chapter 4 each of k and m would have to be the nth power of a natural number; $k = t_1^n$ and $m = t_2^n$, for $t_1, t_2 \in \mathbb{N}$. Then by part (i) both real numbers $\sqrt[n]{k}$ and $\sqrt[n]{m}$ would be rational numbers, contrary to hypothesis. Hence $k \cdot m$ cannot be the nth power of a natural number. Applying part (ii) we conclude that the number $\sqrt[n]{km}$ is an irrational number.

(iv) We proceed by induction starting with $l = 2$: the statement is true for $l = 2$, simply by part (iii). Assume that the statement holds true for some natural number $l = r \geq 2$. We will show that this assumption implies that the statement is also true for $l = r + 1$. So the inductive hypothesis is that if $k_1, \dots, k_r$ are mutually relatively prime natural numbers with at least one of the real numbers $\sqrt[n]{k_1}, \dots, \sqrt[n]{k_r}$ being irrational, then the product $\sqrt[n]{k_1 \dots k_r}$ is also irrational. Now, let $k_1, \dots, k_r, k_{r+1}$ be any $(r + 1)$ natural numbers which are mutually relatively prime and such that at least one of the numbers $\sqrt[n]{k_1}, \dots, \sqrt[n]{k_r}, \sqrt[n]{k_{r+1}}$ is irrational.

Clearly $k_1, \ldots, k_r$ are mutually relatively prime since the numbers $k_1, \ldots k_r, k_{r+1}$ are. Furthermore if there is an irrational number among $\sqrt[n]{k_1}, \ldots, \sqrt[n]{k_r}$, then the inductive hypothesis applies: the number $\sqrt[n]{k_1 \ldots k_r}$ must also be irrational; but $\sqrt[n]{k_1 \ldots k_r} \cdot \sqrt[n]{k_{r+1}} = \sqrt[n]{k_1 \ldots k_r \cdot k_{r+1}}$. If $\sqrt[n]{k_{r+1}}$ is also irrational, then in view of $(k_1 \ldots k_r, k_{r+1}) = 1$. (This follows from the fact that k_{r+1} is relatively prime to each of the numbers $k_1, \ldots k_r$). We conclude by part (iii) (the case $l = 2$) that the product of the two irrational numbers $\sqrt[n]{k_1 \ldots k_r}$ and $\sqrt[n]{k_{r+1}}$ must also be irrational. If on the other hand $\sqrt[n]{k_{r+1}}$ is rational, then again this product $(\sqrt[n]{k_1, \ldots k_r}) \cdot (\sqrt[n]{k_{r+1}})$ will be irrational since by Theorem 1, the product of a nonzero rational number and an irrational number, must be irrational. Finally, to compete the proof, if all the r numbers $\sqrt[n]{k_1}, \ldots, \sqrt[n]{k_r}$ are rational, then $\sqrt[n]{k_{r+1}}$ must be irrational due to the assumption that at least one of $\sqrt[n]{k_1}, \ldots, \sqrt[n]{k_r}, \sqrt[n]{k_{r+1}}$ is irrational. Again, Theorem 1 applies since $\sqrt[n]{k_1 \ldots k_r}$ is rational. In that case (because it is the product of r rational numbers), and $\sqrt[n]{k_{r+1}}$ is irrational; hence $\sqrt[n]{k_1 \ldots k_r + k_{r+1}}$ is irrational. The inductive step is now complete. ■

(v) We prove the contrapositive statement: if, under the assumption $(k, m) = 1$,, the number $\frac{\sqrt[n]{k}}{\sqrt[n]{m}}$ is rational, then both $\sqrt[n]{k}$ and $\sqrt[n]{m}$ must be rational. Indeed assume $\frac{\sqrt[n]{k}}{\sqrt[n]{m}} = \rho$, is a rational number. We have $\frac{k}{m} = \rho^n$. If we set $\rho = \frac{t}{l}$, where l and t are positive integers with $(t, l) = 1$, we must then have $\frac{k}{m} = \frac{t^n}{l^n} \Rightarrow k \cdot l^n = m \cdot t^n$. Since $(k, m) = 1$, the last equation shows that k must divide t^n and for the same reason, m must divide l^n; set $t^n = k \cdot t_1$ and $l^n = m \cdot l_1$. Form $k \cdot l^n = m \cdot t^n$ we obtain $k \cdot m \cdot l_1 = m \cdot k \cdot t_1 \Rightarrow$ (since km is positive and hence nonzero) $l_1 = t_1$ which implies $l_1 = t_1 = 1$, in view of$(t, l) = 1$, and hence $(t^n, l^n) = 1$. Thus, $k = t^n$ and $m = l^n$ which implies $t = \sqrt[n]{k}$ and $l = \sqrt[n]{m}$, proving that both numbers $\sqrt[n]{k}$ and $\sqrt[n]{m}$ are integers and hence rationals. This proof is complete. ■

(vi) Clearly, the given statement is equivalent to the following statement. The nth root of a positive rational r is a rational number if, and only if, r is the nth power of a rational number. Indeed, if $r = \rho^n$, where ρ is rational, then $\sqrt[n]{r} = \sqrt[n]{\rho^n} = \rho$, a rational number. Conversely if $\sqrt[n]{r} = \rho$ for some rational number ρ, then $(\sqrt[n]{r})^n = \rho^n$; $r = \rho^n$. We are done. ■

Remark 1: (i) Recall the contrapositive of an inference statement $(p \Rightarrow q)$. The contrapositive is the statement $(\rceil q \Rightarrow \rceil p)$, where $\rceil$ is the negation symbol. In other words $(p \Rightarrow q) \Leftrightarrow (\rceil q \Rightarrow \rceil p)$.

(ii) With regard to part (iv) of the theorem, we are asked to prove $(p \Leftrightarrow q)$, where p is the statement, "The nth root of a positive rational number r is an irrational number" and q the statement, "r is not the nth power of a rational number." The statement $(p \Leftrightarrow q)$ is equivalent to $(\rceil p \Leftrightarrow \rceil q)$. It is easier to prove the latter, so the reader can see why we chose that approach.

(iii) Both parts (iii) and (iv) are false if the natural numbers k and m (and $k_1 \ldots, k_l$) are replaced by positive rationals. One can see this by using a very simple counter example. Take any number N so that $\sqrt{N}$ is irrational (e.g. $N = 2, 3, 5, \ldots$). N can be any natural number that is not a perfect square. Then both numbers $\sqrt{k} = \frac{1}{\sqrt{N}}$ and $\sqrt{m} = \sqrt{N}$ are irrational, but the product $\sqrt{k \cdot m} = \frac{1}{\sqrt{N}} \cdot \sqrt{N} = 1$, is obviously a rational number.

Theorem 3 Suppose that m and n are natural numbers (and hence positive),d the greatest common divisor of m and n; so that, $m = dm_1, n = dn_1$; where m_1 and n_1 natural numbers which are relatively prime; $(m_1, n_1) = 1$. Consider the real number, $\rho = k_1\sqrt{m} + k_2\sqrt{n}$, where k_1, k_2 are integers (which means they can be negative), with at least one of them being nonzero. If, and only if, (at least) one of the following five conditions holds true, will the number ρ be a rational number.

Condition (0): $k_1 = k_2 = 0$; $\rho = 0$.

Condition (a): $k_2 \neq 0, k_1 = 0$ and n is a perfect square (i.e. integer square).

Condition (b): $k_1 \neq 0, k_2 = 0$ and m is a perfect square.

Condition (c): $k_1k_2 \neq 0$, d **not** a perfect square, and both m_1 and n_1 being perfect squares: $m_1 = M^2$ and $n_1 = N^2$; M, N natural numbers and the integers k_1, k_2, M, N satisfying $k_1M + k_2N = 0$. When condition (c) holds, then $\rho = 0$.

Condition (d): Both m and n are perfect squares, k_1 and k_2 can be any integers. If all five conditions fail, and only then, will the number ρ be an irrational number. In particular if at least one of k_1 or k_2 is not zero, mn is not a perfect square and neither m nor n is a perfect square, then ρ must be an irrational number. (Note that both conditions (c) and (d) imply that mn is a perfect square.)

Remark 2: Note that conditions (a), (b), and (c) are mutually exclusive; but condition (d) is not mutually exclusive with any of conditions (a), (b), or (c). Also note that in condition (d), one k_1, k_2 must be positive, the other negative (since M, and N are natural numbers or positive integers).

Proof: Our method of proof will be to assume that ρ is rational and show that (at least) one of conditions (0), (a), (b), (c), or (d) must be satisfied. Conversely we will show that if any of these five conditions is held, ρ must be rational.

So first, assume that ρ is rational. Clearly the possibility $k_1 = k_2 = 0$ is consistent with that assumption; $\rho = 0$ in the case $k_1 = k_2 = 0$. Now assume that at least one of the integers k_1, k_2 is nonzero. We square both sides to obtain,

$$m \cdot k_1^2 + 2k_1k_2\sqrt{mn} + k_2n^2 = \rho^2 \Rightarrow 2k_1k_2\sqrt{mn} = \rho^2 - (mk_1^2 + k_2n^2) \qquad (1)$$

Since at lest on of k_1, k_2 is nonzero, there are precisely three possibilities: $k_1 = 0$ and $k_2 \neq 0$; $k_1 \neq 0$ and $k_2 = 0$; or $k_1k_2 \neq 0$. If $k_1 = 0$ and $k_2 \neq 0$ and equation (1) implies,

$$\rho^2 = nk_2^2 \qquad (2)$$

Since n and k_2^2 are positive integers, equation (2) shows that ρ^2 is a positive integer and hence $|\rho|$ is the square root of a positive integer or natural number. But $|\rho|$ is a rational number (because ρ is); but according to Theorem 1 (parts (i) and (ii)), the square root of a natural number is rational precisely when the natural number is a perfect square. Hence, we conclude that $|\rho|$ is the square root of a perfect square, when means that $|\rho|$ is a natural number; and hence so is ρ^2.

Thus, equation (2) says that nk_2^2 is equal to an integer square. Then n must be a perfect square (we leave this as an exercise to the student: show that if x, y, z are natural numbers such that $x \cdot y^2 = z^2$, then x must be a perfect square; hint: $x = \left(\frac{z}{y}\right)^2$ = rational square – apply Theorem 1). Hence condition (a) is satisfied. In the same way, the possibility $k_1 \neq 0,\ k_2 \neq 0$ leads (from equation (1)) to the equation $\rho^2 = mk_1^2$ and a similar argument shows that m must be a perfect square. Hence, in that case, condition (b) must hold. Next assume $k_1k_2 \neq 0$. In that case equation (1) implies $\sqrt{mn} = \frac{\rho^2-(mk_1^2+nk_2^2)}{2k_1k_2}$, which means that $\sqrt{mn}$ is a rational number. Now we make use of $m = d \cdot m_1,\ n = d \cdot n_1$, with $(m_1, n_1) = 1$. We have, $\sqrt{mn} = \sqrt{d^2 \cdot m_1n_1} = d \cdot \sqrt{m_1n_1}$. Since $d \cdot \sqrt{m_1n_1}$ is rational, so must be the number $\frac{d \cdot \sqrt{m_1n_1}}{d} = \sqrt{m_1n_1}$. Then according to Theorem 1 (parts (i) and (ii)), the integer m_1n_1 must be a perfect square, say $m_1n_1 = t^2;\ t \in \mathbb{N}$; now we use the fact that $(m_1, n_1) = 1$. According to Theorem 17 in Chapter 4, each number m_1, n_1, must be a perfect square; so $m_1 = M^2$ and $n_1 = N^2$, where M, N are natural numbers. Now, let us go back to $\rho = k_1\sqrt{m} + k_2\sqrt{n}$; we have,

$$\begin{aligned} \rho &= k_1 \cdot \sqrt{d \cdot m_1} + k_2 \cdot \sqrt{d \cdot n_1} = k_1 \cdot \sqrt{d \cdot M^2} + k_2 \cdot \sqrt{d \cdot N^2} = \\ &= k_1 \cdot M \cdot \sqrt{d} + k_2 \cdot N \cdot \sqrt{d};\ \rho = (k_1 \cdot M + k_2 \cdot N) \cdot \sqrt{d}. \end{aligned} \tag{3}$$

If d is not a perfect square, then by Theorem 1 (parts (i) and (ii)), the number $\sqrt{d}$ must be irrational, and thus (3) implies in that case that $k_1M + k_2N = 0$; condition (c) is established.

Finally, if d is a perfect square, say $d = D^2$; we have $m = dm_1 = D^2M^2 = (DM)^2$ and $n = dn_1 = D^2 \cdot N^2 = (DN)^2$ which proves that both m and n are perfect squares; hence, condition (d) is established. Note that $\rho = (k_1M + k_2N)D$, when condition (d) holds. To conclude, we must prove the converse of this theorem. Namely, that if any of the conditions (0) through (d) holds, then the number $\rho = k_1\sqrt{m} + k_2\sqrt{n}$ must be rational. The proof of this is easy and it is left to the student (just assume condition (0), and compute ρ to show that it is rational; then assume condition (a) and so on). To close, the last statement in the Theorem says that if, not both k_1, k_2 are zero, mn is not a perfect square, and if neither m nor n is a perfect square then ρ must be irrational. Indeed, if mn is not a perfect square then neither condition (d) nor condition (c) can take hold: because each of these two conditions implies that mn is a perfect square (specifically $mn = (dM^2)(dN^2) = (dMN)^2$). Also, since neither m nor n is a perfect square, neither condition (a) nor condition (b) can be satisfied; and condition (0) cannot be satisfied either if not both k_1, k_2 are zero. Since none of the conditions (0) through (d) is satisfied, ρ must be irrational; since these conditions are necessary and sufficient for the rationality of ρ. End of proof. ∎

The following Theorem is an application of Theorem 3.

Theorem 4 (i) If α, γ, are rational numbers, β and δ are positive rational numbers, $\sqrt{\beta}$ and $\sqrt{\delta}$ both irrationals and if the equality, $\alpha + \sqrt{\beta} = \gamma + \sqrt{\delta}$ holds true then $\alpha = \gamma$ and $\beta = \delta$.

(ii) There do not exist rationals α and γ, and positive rationals β and δ such that

$\sqrt{\beta}$ and $\sqrt{\delta}$ are irrationals and the equality $\alpha+\sqrt{\beta}=\gamma-\sqrt{\delta}$ holds true.

Proof. (i) We have $\alpha+\sqrt{\beta}=\gamma+\sqrt{\delta} \Rightarrow \sqrt{\beta}-\sqrt{\delta}=\gamma-\alpha$. Let D be the least common denominator of the three rational numbers β, δ, and $\gamma-\alpha$. In other words the positive integer D is simply the least common multiple of the three denominators of the rational numbers β, δ, and $\gamma-\alpha$, when they are written in lowest terms. So we can write, $\beta=\frac{m}{D}$, $\delta=\frac{n}{D}$, $\gamma-\alpha=\frac{t}{D}$, where m, n, D are positive integers or natural numbers; and t is an integer. Thus the equation $\sqrt{\beta}-\sqrt{\delta}=\gamma-\alpha$ implies $\frac{\sqrt{m}}{\sqrt{D}}-\frac{\sqrt{n}}{\sqrt{D}}=\frac{\gamma-\alpha}{D}$; multiplying both sides by D we obtain $\sqrt{mD}-\sqrt{nD}=t$; since t is an integer, it is a rational number. Now we apply Theorem 3 with the coefficients k_1 and k_2 being 1 and -1, respectively; $k_1=1$, $k_2=-1$ (and mD and nD in place of m and n in that Theorem). According to Theorem 3, since t is rational and mD, nD natural numbers, one of the five conditions (0), (a), (b), (c), or (d) must hold. Conditions (0), (a), (b) are immediately eliminated since both k_1 and k_2 are nonzero. Now, condition (d) cannot hold either: indeed if mD and nD are both perfect squares then $mD=t_1^2$, $nD=t_2^2$, for some natural numbers t_1 and t_2; but then $\beta=\frac{m}{D}=\frac{mD}{D^2}=\frac{t_1^2}{D^2}=\left(\frac{t_1}{D}\right)^2$; and likewise $\delta=\frac{n}{D}=\frac{nD}{D^2}=\frac{t_2^2}{D^2}=\left(\frac{t_2}{D}\right)^2$, which proves that both numbers $\sqrt{\beta}=\frac{t_1}{D}$, $\sqrt{\delta}=\frac{t_2}{D}$ are rational numbers violating the hypothesis of the theorem. Therefore, there is room for only one condition to hold, namely (c): write $Dm=d\cdot m_1$, $Dn=d\cdot n_1$, with $(m_1,n_1)=1$, $m_1=M^2$, $n_1=N^2$ and the greatest common divisor d of m and n, being a nonperfect square; now applying $Dm=d\cdot M^2$ and $Dn=d\cdot N^2$ back to equation $\sqrt{mD}-\sqrt{nD}=t$ we now have $\sqrt{dM^2}-\sqrt{dN^2}=t \Rightarrow (M-N)\sqrt{d}=t$; and since d is not a perfect square, it follows (by Theorem 1 (i) and (ii)) that $\sqrt{d}$ is irrational; and so the last equality is possible only when $M-N=0=t$; $M=N$ which implies, $Dm=Dn \Rightarrow \beta=\frac{mD}{D^2}=\frac{nD}{D^2}=\delta$; and since $\beta=\delta$, it is now clear from $\sqrt{\beta}-\sqrt{\delta}=\gamma-\alpha$, that $\gamma-\alpha=0$; $\alpha=\gamma$. ■

(ii) The proof is similar to that given in part (i). The only difference is that because the resulting equation is $\sqrt{mD}+\sqrt{nD}=t$ (instead of $\sqrt{mD}-\sqrt{nD}=t$) one easily sees that condition (c) cannot be satisfied either, since in condition (c), $k_1k_2<0$; whereas here we have $k_1=1=k_2$. End of proof. ■

Remark 3: For part (ii) an alternative, shorter proof may be offered; an easy proof indeed. Starting with $\sqrt{\beta}-\sqrt{\delta}=\gamma-\alpha$, if we assume that $\beta\neq\delta$, then we must also have $\alpha\neq\gamma$. Also $\left(\sqrt{\beta}-\sqrt{\delta})(\sqrt{\beta}+\sqrt{\delta}\right)=\beta-\delta \Rightarrow \sqrt{\beta}-\sqrt{\delta}=\frac{\beta-\delta}{\sqrt{\beta}-\sqrt{\delta}}=\frac{\beta-\delta}{\gamma-\alpha}$. Adding the two equations $\sqrt{\beta}-\sqrt{\delta}=\gamma-\alpha$ and $\sqrt{\beta}+\sqrt{\delta}=\frac{\beta-\delta}{\gamma-\alpha}$ memberwise we obtain $\sqrt[2]{\beta}=\gamma-\alpha+\frac{\beta-\delta}{\gamma-\alpha}$, a rational number, contrary to the hypothesis that $\sqrt{\beta}$ is irrational. Thus the assumption that $\beta\neq\delta$ must be false; we conclude $\beta=\delta$ and from there $\alpha=\gamma$ as well.

Theorem 5 Let k_1, k_2, k_3 be integers with $k_3\neq 0$; and let m, n be natural numbers such that mn is not a perfect square. Let $\rho=k_1\cdot\sqrt{m}+k_2\cdot\sqrt{n}+k_3\cdot\sqrt{mn}$.

(i) If the number ρ is rational, then the integer k_3 must be a divisor of the product k_1k_2, and $\rho=-\frac{k_1k_2}{k_3}$. Equivalently, if k_3 does not divide k_1k_2, then the number ρ is an irrational number.

(ii) If $k_3=1$, the number ρ is rational only when either n is a perfect square (but

m is not) and $k_1 = -\sqrt{n}$; or m is a perfect square (but n is not) and $k_2 = -\sqrt{m}$; and $\rho = -k_1k_2$; otherwise, if these conditions are not satisfied, ρ is irrational.

(iii) If $k_3 = -1$, the number ρ is rational only when either n is a perfect square (but m is not) and $k_1 = \sqrt{n}$; or m is a perfect square (but n is not) and $k_2 = \sqrt{m}$. Otherwise, if these conditions are not satisfied, ρ is irrational.

Proof: (i) From

$$\begin{aligned} &k_1\sqrt{m} + k_2\sqrt{n} + k_3\sqrt{mn} = \rho \Rightarrow (k_1\sqrt{m} + k_2\sqrt{n})^2 = (\rho - k_3\sqrt{mn})^2 \Rightarrow \\ &\Rightarrow m \cdot k_1^2 + 2k_1k_2\sqrt{,m} + nk_2^2 = \rho^2 - 2\rho k_3\sqrt{mn} + mnk_3^2 \Rightarrow \\ &\Rightarrow 2(k_1k_2 + k_3\rho)\sqrt{mn} = \rho^2 + mnk_3^2 - (mk_1^2 + nk_2^2) \end{aligned} \tag{1}$$

Since mn is not a perfect square, Theorem 1 (parts (i) and (ii)) says that $\sqrt{mn}$ must be an irrational number; but ρ is rational, which means that according to (1) we must have $2(k_1k_2 + \rho k_3) = 0$; hence by (1),

$$\left(k_1k_2 + \rho k_3 = 0 \quad \text{and} \quad \rho^2 + mnk_3^2 - (mk_1^2 + nk_2^2) = 0\right) \tag{2}$$

But $k_3 \neq 0$: so that $\rho = -\frac{k_1k_2}{k_3}$ and substituting for ρ in the other equation we obtain

$$k_1^2k_2^2 = k_3^2 \cdot (mk_1^2 + nk_2^2 - mnk_3^2) \tag{3}$$

Equation (3) clearly shows that the integer k_3^2 divides the integer $k_1^2k_2^2$; thus k_3 must be a divisor of k_1k_2 (for the student: fill in the details for this: show that if the a and b are integers and a^2 divides b^2, then a must divide b). The proof is complete. ■

(ii) If $k_3 = 1$ and ρ is rational, equations (2) and (3) imply $\rho = -k_1k_2$ and $k_1^2k_2^2 + mn - mk_1^2 - nk_2^2 = 0$; thus, $m \cdot (n - k_1^2) - k_2^2(n - k_1^2) = 0$; $(m - k_2^2) \cdot (n - k_1^2) = 0$. If $m - k_2^2 = 0$; then $k_2^2 = m$; which proves that the natural number m is a perfect square and $|k_2| = \sqrt{m}$. Going back to the initial equation $k_1\sqrt{m} + k_2\sqrt{n} + k_3\sqrt{mn} = \rho$ we now have, $k_1|k_2| + k_2\sqrt{n} + |k_2|\sqrt{n} = -k_1k_2$;

$$(k_2 + |k_2|)\sqrt{n} = -k_1k_2 - k_1|k_2| \tag{4}$$

Now, $\sqrt{n}$ must be irrational since m is a perfect square and mn is not a perfect square, n cannot be a perfect square (if it were, the product mn of two perfect squares would also be a perfect square). Hence, since $\sqrt{n}$ is irrational, (4) clearly implies $k_2 + |k_2| = 0 = -k_1k_2 + k_1|k_2|$; which means that $k_2 = -|k_2|$; and in virtue of $|k_2| = \sqrt{m}$, we conclude that $k_2 = -\sqrt{m}$. In a similar way one prove that if the second possibility holds, namely $n - k_1^2 = 0$, n must be a perfect square and $k_1 = -\sqrt{n}$. The converse is easy to establish; if either $k_2 = -\sqrt{m}$ or $k_1 = -\sqrt{n}$, one easily sees that $\rho = k_1\sqrt{m} - \sqrt{mn} + \sqrt{mn} = k_1\sqrt{m} = -k_1k_2 =$ rational; or $\rho = -\sqrt{nm} + k_2\sqrt{n} + \sqrt{mn} = k_2\sqrt{m} = -k_1k_2 =$ rational; respectively. ■

(iii) The proof of this is similar to that of part (ii).. The details are left to the student (verify!).

We finish this section by showing that between any two (distinct) real numbers, there exists an irrational number (so if one fixes two real numbers a and b, $a < b$;

one by using this result can show that there exist infinitely many irrational numbers between a and b).

Theorem 6 For every pair (a, b) of real numbers a and b, with $a < b$, there exists an irrational number t strictly between a and b; $a < t < b$.

Proof: Let a and b be real numbers with $a < b$. Pick any irrational number ρ; according to Theorem 4 in Chapter 4 strictly between the real numbers $a - \rho$ and $b - \rho$ there exists a rational number r: $a - \rho < r < b - \rho \Rightarrow a < r + \rho < b$; but the number $t = r + \rho$ is irrational, by Theorem 1, since it is the sum of a rational and irrational number. ∎

6.2 The integer part of a real number

According to Theorem 4 of Chapter 4, for every real number a there exists a unique integer k such that $k \leq a < k+1$; in other words, k is the largest integer not exceeding a.

Definition 1: Given a real number a, the unique integer k such that $k \leq a < k+1$, is called the **integer part** of a, and it is denoted by $[\![a]\!]$.

Example 2: We give some numerical examples:

$$[\![-10.3]\!] = -11,\ [\![0.5]\!] = 0,\ [\![-4]\!] = -4, [\![-0.7]\!] = -1,\ [\![7.2]\!] = 7.$$

Theorem 7 (i) For every real number a there exists a unique real number ϵ such that $0 \leq \epsilon < 1$ and $a = [\![a]\!] + \epsilon$.

(ii) If the real number a is an integer, then $[\![a]\!] + [\![-a]\!] = 0$.

(iii) If the real number a is not an integer, then $[\![a]\!] + [\![-a]\!] = -1$.

(iv) For every real number a and every integer m, the equality $[\![a+m]\!] = [\![a]\!]+[\![m]\!]$, holds true.

Proof: (i) If $[\![a]\!] = k$, then by definition $k \leq a < k+1$. Let $\epsilon = a - [\![a]\!]$; the inequality $k \leq a < k+1$ implies $k - k \leq a - k < (k+1) - k$; $0 \leq a - k < 1$; but $\epsilon = a - [\![a]\!] = a - k$. Hence $0 \leq \epsilon < 1$; the uniqueness of ϵ is immediate, merely by the fact that $\epsilon = a - [\![a]\!]$ and the uniqueness of $k = [\![a]\!]$.

(ii) If $[\![a]\!]$ is an integer, since $k \leq a < k+1$ and since there is no integer lying strictly between k and $(k+1)$ (by Theorem 2(v) in Chapter 4), it follows that $a = k = [\![a]\!]$. Then $-a = -k = -[\![a]\!]$; and from $-k \leq -a < -k+1$ (this clearly holds ture since $-a = -k$) and definition 1, we conclude that $[\![-a]\!] = -k$; consequently,

$$[\![a]\!] + [\![-a]\!] = k + (-k) = 0.$$

(iii) By part (i), we have $a = [\![a]\!]+\epsilon$, with $a \leq \epsilon < 1$; but because a is not an integer, $\epsilon \neq 0$, and so we must have $0 < \epsilon < 1$. So that $-a = -[\![a]\!]-\epsilon = -[\![a]\!]-1+(1-\epsilon)$. Note that since $0 < \epsilon < 1$, we have $-1 < -\epsilon < 0 \Rightarrow 0 < 1-\epsilon < 1$. Now from $0 < 1-\epsilon < 1$ and $-a = -[\![a]\!] - 1 + (1-\epsilon)$ we conclude $-[\![a]\!] - 1 < -a < \underbrace{-[\![a]\!] - 1 + 1} = -[\![a]\!]$. The last inequality says that the real number $-a$ is strictly between the integers $(-[\![a]\!]-1)$ and $-[\![a]\!]$; this, couched with definition 1, allows to conclude that $[\![-a]\!] = -[\![a]\!] - 1$. Therefore, $[\![a]\!] + [\![-a]\!] = [\![a]\!] + (-[\![a]\!] - 1) = -1$. ■

(iv) By definition 1, we have $[\![a]\!] \leq a < [\![a]\!] + 1$; adding m to both sides we find that, $[\![a]\!]+m \leq a+m < ([\![a]\!]+m)+1$; but the real numbers $([\![a]\!]+m)$ and $(([\![a]\!]+m)+1$ are successive integers. Thus the last double-inequality combined with definition 1 imply that $[\![a+m]\!] = [\![a]\!] + m$. End of proof. ■

We conclude this section with the following Theorem.

Theorem 8 (i) For any real numbers a and b,

$$[\![a]\!] + [\![b]\!] \leq [\![a+b]\!] \leq [\![a]\!] + [\![b]\!] + 1.$$

(ii) For any real numbers a and b,

$$[\![a]\!] - [\![b]\!] - 1 \leq [\![a - b]\!] \leq [\![a]\!] - [\![b]\!].$$

(iii) For every natural number n and any real numbers $a_1, \ldots, a_n$,

$$[\![a_1]\!] + \ldots + [\![a_n]\!] \leq [\![a_1 + \ldots + a_n]\!] \leq [\![a_1]\!] + \ldots + [\![a_n]\!] + n.$$

Proof: (i) According to Theorem 7(i), we have $a = [\![a]\!] + \epsilon_1$ and $b = [\![b]\!] + \epsilon_2$, where $0 \leq \epsilon_1, \epsilon_2 < 1$. Thus,

$$a + b = ([\![a]\!] + [\![b]\!]) + (\epsilon_1 + \epsilon_2) \tag{1}$$

From $0 \leq \epsilon_1 < 1$ and $0 \leq \epsilon_2 < 1$ it follows $0 \leq \epsilon_1 + \epsilon_2 < 2$. Also, since $[\![a]\!]$ and $[\![b]\!]$ are both integers, so is the number $([\![a]\!] + [\![b]\!])$ and by applying Theorem 7 (iv) we find that

$$[\![a + b]\!] = [\![a]\!] + [\![b]\!] + [\![\epsilon_1 + \epsilon_2]\!] \tag{2}$$

From

$$0 \leq \epsilon_1 + \epsilon_2 < 2, \text{ we have } 0 \leq [\![\epsilon_1 + \epsilon_2]\!] \leq 1 \tag{3}$$

Combining (1), (2), and (3) we find that

$$[\![a]\!] + [\![b]\!] \leq \underbrace{[\![a]\!] + [\![b]\!] + [\![\epsilon_1 + \epsilon_2]\!]}_{[\![a+b]\!]} \leq [\![a]\!] + [\![b]\!] + 1$$

the proof is complete. ■

(ii) A little trick is involved here: if we write $a = (a - b) + b$ and applying part (i) with $(a - b)$ and b in place of a and b respectively, we have,

$$[\![a - b]\!] + [\![b]\!] \leq [\![a]\!] \leq [\![a - b]\!] + [\![b]\!] + 1$$

Adding $(-[\![b]\!] - 1)$ across the double-inequality gives

$$[\![a - b]\!] - 1 \leq [\![a]\!] - [\![b]\!] - 1 \leq [\![a - b]\!] \tag{4}$$

The second of the inequalities in (4) says that

$$[\![a]\!] - [\![b]\!] - 1 \leq [\![a - b]\!] \tag{5}$$

While the first of the inequalities in (4) says that $[\![a - b]\!] - 1 \leq [\![a]\!] - [\![b]\!] - 1$; or that

$$[\![a - b]\!] \leq [\![a]\!] - [\![b]\!]; \tag{6}$$

combining (5) with (6) we deduce that,

$$[\![a]\!] - [\![b]\!] - 1 \le [\![a - b]\!] \le [\![a]\!] - [\![b]\!]$$

(iii) By induction on the natural number n: For $n = 1$, the statement simply says that $[\![a_1]\!] \le [\![a_1]\!] \le [\![a_1]\!] + 1$ which is true. Assume the statement to hold for some $n = k > 1$; we will show that this assumption implies the truth of the statement for $n = k + 1$. If the statement holds for $n = k$ then we must have,

$$[\![a]\!] + \ldots [\![a_k]\!] \le [\![a_1 + \ldots + a_k]\!] \le [\![a_1]\!] + \ldots + [\![a_k]\!] + k; \tag{7}$$

Let a_{k+1} be a real number. We apply part (i) with $a = a_1 + \ldots + a_k$ and $b = a_{k+1}$:

$$[\![a_a + \ldots + a_k]\!] + [\![a_{k+1}]\!] \le [\![a_1 + \ldots + a_k + a_{k+1}]\!] \le [\![a_1 + \ldots + a_k]\!] + [\![a_{k+1}]\!] + 1 \tag{8}$$

From the first equality of the inductive hypothesis (7) it follows that (by adding $[\![a_{k+1}]\!]$ to both sides)

$$[\![a_1]\!] + \ldots [\![a_k]\!] + [\![a_{k+1}]\!] \le [\![a_1 + \ldots + a_k]\!] + [\![a_{k+1}]\!]. \tag{9}$$

By the second inequality of the inductive hypothesis (7) it follows that (by adding $[\![a_{k+1}]\!] + 1$ to both sides),

$$[\![a_1 + \ldots + a_k]\!] + [\![a_{k+1}]\!] + 1 \le [\![a_1]\!] + \ldots + [\![a_k]\!] + [\![a_{k+1}]\!] + k + 1. \tag{10}$$

Combining (8), (9) and (10) we arrive at,

$$[\![a_1]\!] + \ldots + [\![a_k]\!] + [\![a_{k+1}]\!] \le [\![a_1 + \ldots + a_k + a_{k+1}]\!] \le [\![a_1]\!] + \ldots + [\![a_k]\!] + [\![a_{k+1}]\!] + k + 1$$

which completes the induction process. ■

6.3 Practice Problems

P1. Show that there are no rational numbers α, β, γ such that $\frac{\alpha}{\sqrt{3}} + \frac{\beta}{\sqrt{5}} + \frac{\gamma}{\sqrt{7}} = 1$.

Solution: Suppose to the contrary that such a triple (α, β, γ) is possible:

$$\begin{aligned} &\tfrac{\alpha}{\sqrt{3}} + \tfrac{\beta}{\sqrt{5}} + \tfrac{\gamma}{\sqrt{7}} = 1 \\ &\alpha\sqrt{5}\sqrt{7} + \beta\sqrt{3}\sqrt{7} + \gamma\sqrt{3}\sqrt{5} = \sqrt{3}\sqrt{5}\sqrt{7}; \\ &\sqrt{5}\cdot(\alpha\sqrt{7} + \sqrt{3}\gamma) = \sqrt{3}\sqrt{7}(\sqrt{5} - \beta); \end{aligned} \qquad (1)$$

squaring both sides we obtain

$$\begin{aligned} &5\cdot(7\alpha^2 + 2\alpha\gamma\sqrt{21} + 3\gamma^2) = 21\cdot(5 - 2\beta\sqrt{5} + \beta^2); \\ &(35\alpha^2 + 15\gamma^2 - 21\beta^2 - 105) + 10\alpha\gamma\sqrt{21} - 42\beta\sqrt{5} = 0 \end{aligned} \qquad (2)$$

Equation (1) clearly implies that the number, $10\alpha\gamma\sqrt{21} - 42\beta\sqrt{5}$ is rational; we put

$$10\alpha\gamma\sqrt{21} - 42\beta\sqrt{5} = r \in \mathbb{Q} \qquad (3)$$

Since α, γ, and β are rationals we can set $\alpha = \frac{I_1}{D}$, $\beta = \frac{I_2}{D}$, $\gamma = \frac{I_3}{D}$, where I_1, I_2, I_3 are integers and D is a natural number. Substituting for α, β, and γ in (3) we obtain

$$10I_1I_3\sqrt{21} - 42I_2\cdot D\cdot\sqrt{5} = r\cdot D^2 = \text{rational} \qquad (4)$$

Now we can apply Theorem 3 with $m = 21, n = 5, k_1 = 10I_1I_3$, and $k_2 = -42I_2\cdot D$. Since (4) holds, Theorem 3 applies: one of the five conditions (0) through (d) in Theorem 3 must hold: if condition (0) holds, we must have, $10I_1I_3 = 0 = -42I_2D$; but $D \neq 0$, which means that $I_1 = I_2 = 0$; or $I_3 = I_2 = 0$; in the first case we obtain $\alpha = \beta = 0$, in the second case $\beta = \gamma = 0$. Both lead to a contradiction according to equation (1). The first case ($\alpha = \beta = 0$) implies $\gamma = \sqrt{7}$ which is impossible since $\sqrt{7}$ is irrational according to Theorem 2. The second case ($\beta = \gamma = 0$) implies $\alpha = \sqrt{3}$ which is impossible for the same reason.

Next, note that all of conditions (a) through (d) are immediately ruled out since neither $m = 21$ nor $n = 5$ is a perfect square; we see that the assumption $\frac{\alpha}{\sqrt{3}} + \frac{\beta}{\sqrt{5}} + \frac{\gamma}{\sqrt{7}} = 1$, where $\alpha, \beta, \gamma \in \mathbb{Q}$, leads to a contradiction. Thus we conclude that there simply do not exist three such rationals with the above property. The proof is complete.

P2. Let a, b, c be natural numbers such that $a \neq b, a \neq c$, and the product $b\cdot c$ is not a perfect square. Prove that the number $\frac{\sqrt{a}-\sqrt{b}}{\sqrt{a}-\sqrt{c}}$ is irrational.

Solution:

$$\frac{\sqrt{a}-\sqrt{b}}{\sqrt{a}-\sqrt{c}} = \frac{(\sqrt{a}-\sqrt{b})(\sqrt{a}+\sqrt{c})}{(\sqrt{a}-\sqrt{c})(\sqrt{a}+\sqrt{c})} = \frac{a+\sqrt{ac}-\sqrt{ab}-\sqrt{bc}}{a-c} =$$

$$\frac{1}{a-c}\cdot\left[a+\sqrt{ac}-\sqrt{ab}-\frac{\sqrt{(ab)(ac)}}{a}\right] = \frac{1}{a(a-c)}\cdot\left[a^2+a\sqrt{ac}-a\sqrt{ab}-\sqrt{(ab)(ac)}\right]$$

$$\frac{\sqrt{a}-\sqrt{b}}{\sqrt{a}-\sqrt{c}} - \frac{a^2}{a(a-c)} + \frac{1}{a(a-c)}\cdot\left[a\sqrt{ac}-a\sqrt{ab}-\sqrt{(ab)(ac)}\right] \tag{1}$$

Now consider the number $(a\sqrt{ac}-a\sqrt{ab}-\sqrt{(ab)(bc)}) = \rho$. We claim that it is irrational: we can apply Theorem; indeed, first we see that the hypothesis of Theorem 5 is satisfied with $m = ac$, $n = ab$, since the product $mn = (ab)(ac) = a^2\cdot(bc)$ is not a perfect square. This follows from the fact that bc is not a perfect square (by assumption in our problem); so if bc is not a perfect square then $a^2\cdot(bc)$ cannot be a perfect square either (verify!); thus Theorem 5 applies: in the notation of that Theorem we have $k_1 = a, k_2 = -a$, and $k_3 = -1$; then according to part (iii) of Theorem 5, the number ρ will be rational when either n is a perfect square, $k_1 = \sqrt{n}$ (and m not a perfect square) or m is a perfect square, $k_2 = \sqrt{m}$ (and n not a perfect square). In our case $k_2 = -a$, a negative integer since a is a natural number; so the second possibility, namely that $k_2 = -a = \sqrt{m}$ cannot occur. Thus we must have (first possibility) $k_1 = a = \sqrt{n}$; $k_1 = a = \sqrt{ab}$, with ab being a perfect square; but $a = \sqrt{ab}$ obviously implies $a^2 = ab$ and since $a \neq 0$, we conclude that $a = b$ contrary to the assumption of the problem that $a \neq b$; so we have a contradiction. Therefore the number $\rho = a\sqrt{ac}-a\sqrt{ab}-\sqrt{(ab)(bc)}$ must be irrational; and hence, since a and $a-c$ are rational numbers, and obviously $\frac{1}{a(a-c)} \neq 0$, it follows by equation (1) and Theorem 1 that the number $\frac{\sqrt{a}-\sqrt{b}}{\sqrt{a}-\sqrt{b}}$ must be irrational.

A final note: the assumption $a \neq c$ is needed for the above fraction to be defined (i.e. to have nonzero denominator).

P3. If a is an irrational number and the numbers r_1, r_2, r_3, r_4 are all rational, find the necessary and sufficient conditions for the number $\rho = \frac{a\cdot r_1+r_2}{a\cdot r_3+r_4}$ to be rational.

Solution. First observe that if $r_3 = 0$, then $r_4 \neq 0$ and the given fraction ρ simply becomes $\rho = a\cdot\frac{r_1}{r_4}+\frac{r_2}{r_4}$; and since both $\frac{r_1}{r_4}$ and $\frac{r_2}{r_4}$ are rational, ρ will be rational only when $\frac{r_1}{r_4} = 0; r_1 = 0$ (otherwise by Theorem 1, ρ will be irrational). So, one possibility for the rationality of ρ is $r_3 = r_1 = 0$; when this happens, $\rho = \frac{r_2}{r_4}$. Next suppose that $r_3 \neq 0$: note that $(a\cdot r_3 + r_4)$ cannot be zero under this condition, for if this were the case we would have $a = -\frac{r_4}{43}$ = rational, contrary to the assumption that a is irrational. So under the assumption $r_3 \neq 0$, we have

$$\rho = \frac{ar_1+r_2}{ar_3+r_4} \Leftrightarrow \rho\cdot(ar_3+r_4) = ar_1+r_2 \Leftrightarrow a(\rho r_3 - r_1) = r_2 - \rho r_4 \tag{1}$$

Since a is irrational, we see that (1) clearly shows that if ρ is rational, we must have $\rho r_3 - r_1 = 0$, because both numbers $(\rho r_3 - r_1)$ and $(r_2 - \rho r_4)$ would be rational if ρ is rational. Hence, $\rho r_3 - r_1 = 0 = r_2 - \rho r_4$. From $\rho r_3 - r_1 = 0$, it follows (since $r_3 \neq 0)\rho = \frac{r_1}{r_3}$. From $r_2 - \rho r_4 = 0$ it follows that $\rho = \frac{r_2}{r_4}$, provided that $r_4 \neq 0$; if however $r_4 = 0$, then also $r_2 = 0$. In the first possibility, we have $\rho = \frac{r_1}{r_3} = \frac{r_2}{r_4}$. In the second possibility, we have $\rho = \frac{r_1}{r_3}$ and $r_2 = 0 = r_4$. Conversely, if $r_3 r_4 \neq 0$ and $\rho = \frac{r_1}{r_3} = \frac{r_2}{r_4}$, it is easy to see that indeed $\rho = \frac{ar_1+r_2}{ar_3+r_4}$; since, on account of $r_1 - \rho r_3, r_2 - \rho r_4$ we have $\frac{ar_1+r_2}{ar_3+r_4} - \frac{a\rho r_3+\rho r_4}{ar_3+r_4} - \frac{\rho(ar_3+r_4)}{ar_3+r_4} - \rho$. Conversely, in the second case, when $\rho = \frac{r_1}{r_3}$ and $r_2 = r_4 = 0$ we easily see that again $\rho = \frac{ar_1+r_2}{ar_3+r_4}$. We summarize as follows: The necessary and sufficient conditions for the rationality of ρ are

(a) $r_3 = 0 = r_1$ and $r_4 \neq 0$, in which case $a = \frac{r_2}{r_4}$

or (b) $r_3 \neq 0, r_4 \neq 0$, and $\frac{r_1}{r_3} = \frac{r_2}{r_4}$ in which case $\rho = \frac{r_1}{r_3}$

or (c) $r_3 \neq 0, r_2 = r_4 = 0$, in which case $\rho = \frac{r_1}{r_3}$.

P4. Let $\alpha, \beta, \gamma, \delta$ be rational numbers such that neither β nor δ is the cube of a rational number, and $\alpha + \sqrt[3]{\beta} = \gamma + \sqrt[3]{\delta}$. Prove that $\alpha = \gamma$ and $\beta = \delta$.

Solution: According to Theorem 2(vi), since neither β nor δ is a rational cube, both numbers $\sqrt[3]{\beta}$ and $\sqrt[3]{\delta}$ must be irrational. Form $\alpha + \sqrt[3]{\beta} = \gamma + \sqrt[3]{\delta}$ we have,

$$\gamma - \alpha = \sqrt[3]{\beta} - \sqrt[3]{\delta} \tag{1}$$

We will proceed by showing that the negation of the statement ($\alpha = \gamma$ and $\beta = \delta$) leads to a contradiction. So, let us assume the negation of ($\alpha = \gamma$ and $\beta = \delta$). That means that either ($\alpha = \gamma$ and $\beta \neq \delta$) or ($\alpha \neq \gamma$ and $\beta = \delta$) or ($\alpha \neq \gamma$ and $\beta \neq \delta$). According to (1), there is room for only one of these three possibilities: namely $\alpha \neq \gamma$ and $\beta \neq \gamma$. So now we have three assumptions: two of them are from the hypothesis of the problem: these are (1) and the assumption that neither β nor δ is a rational cube (which means that both $\sqrt[3]{\beta}$ and $\sqrt[3]{\delta}$ are irrational). The third assumption is ($\alpha \neq \gamma$ and $\beta \neq \delta$). We will show that these three assumptions combined lead to a contradiction.

Using the identity

$$(\sqrt[3]{\beta})^3 - (\sqrt[3]{\delta})^3 = \left[\sqrt[3]{\beta} - \sqrt[3]{\delta}\right] \cdot \left[(\sqrt[3]{\beta})^2 + (\sqrt[3]{\beta})(\sqrt[3]{\delta}) + (\sqrt[3]{\delta})^2\right]$$

in conjunction with equation (1) and the assumption $\alpha - \gamma \neq 0$, we find that

$$\sqrt[3]{\beta^2} + \sqrt[3]{(\beta\delta)^2} + \sqrt[3]{\delta^2} = \frac{\beta - \delta}{\gamma - \alpha} \tag{2}$$

Now apply another identity, the expansion of the cube::

$$(\sqrt[3]{\beta} - \sqrt[3]{\delta})^3 = (\sqrt[3]{\beta})^3 - 3(\sqrt[3]{\beta}\sqrt[3]{\delta}) \cdot [\sqrt[3]{\beta} - \sqrt[3]{\delta}] - (\sqrt[3]{\delta})^3;$$

Using equation (1) we obtain,

$$(\gamma-\alpha)^3=\beta-3\sqrt[3]{\beta\delta}\cdot(\gamma-\alpha)-\delta \Rightarrow \sqrt[3]{\beta\delta}=-\frac{\delta-\beta-(\gamma-\alpha)^3}{3(\gamma-\alpha)}=\rho \qquad (3)$$

where ρ is obviously a rational number since α,γ,δ, and β are all rationals.

Since neither of β or δ is a rational cube, it follows that neither of them can be zero: solving for $\sqrt[3]{\delta}$ in (3) we obtain, $\sqrt[3]{\delta}=\frac{\rho}{\sqrt[3]{\beta}}$. Substituting for $\sqrt[3]{\delta}$ in equation (1) implies $\left(\sqrt[3]{\beta}\right)^2+(\alpha-\gamma)\left(\sqrt[3]{\beta}\right)-\rho=0$ (4). According to (4) and the quadratic formula, it follows that $\sqrt[3]{\beta}$ must be one of the two roots of (4), namely one of the numbers $\frac{(\gamma-\alpha)+\sqrt{(\alpha-\gamma)^2+4\rho}}{2}$, $\frac{(\gamma-\alpha)-\sqrt{(\alpha-\gamma)^2+4\rho}}{2}$; but α,γ, and ρ are rational numbers; hence the number $D=(\alpha-\gamma)^2+4\rho$ is also rational; if D is the square of a rational number the we have a contradiction since both numbers $\frac{(\gamma-\alpha)\pm\sqrt{d}}{2}$ will be rational, while $\sqrt[3]{\beta}$ is assumed to be irrational; if D is not a rational square we still have a contradiction because by cubing both sides $\sqrt[3]{\beta}=\frac{(\gamma-\alpha)\pm\sqrt{D}}{2}$ we obtain

$$\beta=\left[\frac{(\gamma-\alpha)^2+3(\gamma-\alpha)D}{8}\right]\pm\left[\frac{3(\gamma-\alpha)^2+D}{8}\right]\sqrt{D};$$

however $\sqrt{D}$ is irrational (as we have already seen, the nth root of a rational number is itself rational, if and only if, that rational number is the nth power of some rational number); hence a contradiction. Thus the negative of the assumption ($\alpha-\gamma$ and $\beta-\delta$) must be false. We conclude that $\alpha=\gamma$ and $\beta=\delta$.

P5. If the numbers $\alpha,\beta,\gamma,\delta$ are rational, both β and δ are positive, $\sqrt{\beta}$ and $\sqrt{\delta}$ are both irrational, and the condition $\sqrt[3]{\alpha+\sqrt{\beta}}=\gamma+\sqrt{\delta}$ holds true, prove that

(a) $\sqrt[3]{\alpha-\sqrt{\beta}}=\gamma-\sqrt{\delta}$ and

(b) $\sqrt[3]{\alpha^2-\beta}$ is a rational number.

Solution: (a) From $\sqrt[3]{\alpha+\sqrt{\beta}}=\gamma+\sqrt{\delta}$ by cubing both sides we obtain

$$\alpha+\sqrt{\beta}=\gamma^3+3\gamma\delta+3(\gamma^2+\delta)\cdot\sqrt{\delta}\Rightarrow\sqrt{\beta}-(3\gamma^2+\delta)\cdot\sqrt{\delta}=\gamma^3+\gamma\delta-\alpha \qquad (1)$$

Equation (1) says that the number $\left(\sqrt{\beta}-(3\gamma^2+\delta)\sqrt{\delta}\right)$ is rational; if we write the rationals β,γ, and δ in lowest terms and if D is the least common multiple of their (positive) denominators, then the number $D\beta$, $D\gamma$, and $D\delta$ must be integers. On the other hand, since the number $(\sqrt{\beta}-(3\gamma^2+\delta)\sqrt{\delta})$ is rational, so must be the number $D^3(\sqrt{\beta}-(3\gamma^2+\delta)\sqrt{\delta})=D^3\cdot\sqrt{\beta}-D^3(3\gamma^2+\delta)\sqrt{\delta}=D^3\cdot\sqrt{\beta}-(3D^2\gamma^2+D^2\delta)\cdot D\cdot\sqrt{\delta}=\sqrt{D^6\cdot\beta}-(3D^2\gamma^2+D^2\delta)\cdot\sqrt{D^2\delta}$. Thus the number, $\rho=\sqrt{D^6\cdot\beta}-(3D^2\gamma^2+D^2\delta)\cdot\sqrt{D^2\delta}$ (2), must be rational. The numbers $m=D^6\cdot\beta$, $n=D^2\delta$, $k_1=1$, and $k_2=-(3D^2\gamma^2+D^2\delta)$ are all integers since D is an integer and so are the numbers $D\beta, D\gamma$, and $D\delta$. We can now apply Theorem 3. According to Theorem 3, one of the five conditions (0) through (d) must hold if (and only if) the number ρ of equation (2) above is rational. However, conditions (0) through (c) are immediately eliminated, since $k_1 = 1 \neq 0$ and since k_2 cannot be zero either: indeed, if $k_2 = 0$, then $(3D^2\gamma^2 + D^2\delta) = 0 \Rightarrow 3D^2\gamma^2 + D^2\delta = 0$, which is impossible since $\delta > 0$ by hypothesis and also $D > 0$ (D is the least common denominator of the rationals β, γ, δ: since β and δ are positive, D must be positive even if $\gamma = 0$), and so $D^2\gamma^2 + D^2\delta > 0$. Thus, since $k_1 k_2 \neq 0$, Theorem 3 says that either condition (c) or condition (d) must hold. If condition (d) holds, then both integers $D^6 \cdot \beta$ and $D^2\delta$ must be perfect squares: $D^6 \cdot \beta = l_1^2$ and $D^2\delta = l_2^2$, where l_1 and l_2 are natural numbers; but then $\sqrt{\beta} = \frac{l_1}{D^3}$ and $\sqrt{\delta} = \frac{l_2}{D}$, contradicting the hypothesis that both $\sqrt{\beta}$ and $\sqrt{\delta}$ are irrationals. So condition (d) cannot hold either. Therefore, the only condition left, namely condition (c) must hold true: according to condition (c), if d is the greatest common divisor of the integers $m = D^6 \cdot \beta$ and $n = \delta \cdot D^2$ then there must exist natural numbers M and N such that, $D^6 \cdot \beta = d \cdot M^2$, $D^2 \cdot \delta = d \cdot N^2$, with $(M, N) = 1$, d not a perfect square, and $k_1 M + k_2 N = 0$; in our case $k_1 = 1$ and $k_2 = -(3D^2\gamma^2 + D^2\delta)$. We have, $M - (3D^2\gamma^2 + D^2\delta)N = 0 \Rightarrow M = D^2(3\gamma^2 + \delta)N \Rightarrow$ $M^2 = D^4 \cdot (3\gamma^2 + \delta)^2 N^2$; substituting for $M^2 = \frac{D^6 \cdot \beta}{d}$ and $N^2 = \frac{D^2 \cdot \delta}{d}$ in the last equation gives, $D^6 \cdot \frac{\beta}{d} = \frac{D^4 \cdot (3\gamma^2 + \delta)^2 \cdot D^2 \cdot \delta}{d} \Rightarrow \beta = (3\gamma^2 + \delta^2)^2 \delta$, which is equivalent (since β and δ are both positive) to,

$$\sqrt{\beta} = (3\gamma^2 + \delta)\sqrt{\delta} \qquad (3)$$

Next, consider the identity

$$(\gamma - \sqrt{\delta})^3 = \gamma^3 + 3\gamma\delta - (3\gamma^2 + \delta)\sqrt{\delta} \qquad (4)$$

Furthermore according to (3), the left-hand of equation (1) must be zero. Thus, so must be the right-hand side of (1), that is,

$$\gamma^3 + 3\gamma\delta - \alpha = 0; \;\; \alpha = \gamma^3 + 3\gamma\delta \qquad (5)$$

Combining identity (4) with equations (5) and (3) we obtain, $(\gamma - \sqrt{\delta})^3 = \alpha - \sqrt{\beta} \Rightarrow \gamma - \sqrt{\delta} = \sqrt[3]{\alpha - \sqrt{\beta}}$; the proof is complete.

(b) By the hypothesis of this problem, $\sqrt[3]{\alpha + \sqrt{\beta}} - \gamma + \sqrt{\delta}$. Also, by part (a), $\sqrt[3]{\alpha - \sqrt{\beta}} = \gamma - \sqrt{\delta}$. Multiplying these two equations memberwise,

$$\sqrt[3]{(\alpha - \sqrt{\beta})(\alpha + \sqrt{\beta})} = (\gamma + \sqrt{\delta})(\gamma - \sqrt{\delta}) \Rightarrow \sqrt[3]{\alpha^2 - \beta} = \gamma^2 - \delta,$$

which proves that $\sqrt[3]{\alpha^2 - \beta}$ is rational.

P6. Suppose that $\alpha, \beta, \gamma, \delta$ are rationalnumbers such that β and δ are positive, both $\sqrt{\beta}$ and $\sqrt{\delta}$ are irrational numbers and they satisfy the condition $\sqrt{\alpha + \sqrt{\beta}} =$

(a) $\sqrt{\alpha - \sqrt{\beta}} = |\gamma - \sqrt{\delta}|$.

(b) $\sqrt{\alpha^2 - \beta}$ is rational and

(c) $a > 0$.

Solution: (a) From $\sqrt{\alpha + \sqrt{\beta}} = \gamma + \sqrt{\delta}$ we infer (by squaring both sides),

$$\alpha + \sqrt{\beta} = \gamma^2 + 2\gamma\sqrt{\delta} + \delta \Rightarrow \sqrt{\beta} - 2\gamma\sqrt{\delta} = \gamma^2 + \delta - \alpha \tag{1}$$

Putting $\beta = \frac{B}{N}, \gamma = \frac{C}{N}, \delta = \frac{D}{N}$ where B, C, and D, and N are positive integers and multiplying equation (1) by N^2 gives,

$$\sqrt{B \cdot N^3} - 2C\sqrt{N \cdot D} = N^2 \cdot (\gamma^2 + \delta - \alpha) = \rho, \tag{2}$$

where ρ is a rational number. Now we apply Theorem 3, with $m = B \cdot N^3, n = DN, k_1 = 1, k_2 = -2C$. According to Theorem 3, conditions (0) and (a) obviously cannot hold since $k_1 = 1 \neq 0$. If condition (b) holds, then $C = 0$, and the integer $B \cdot N^3$ must be a perfect square. But $\beta = \frac{B}{N} \Rightarrow \beta N^4 = B \cdot N^3$; so $B \cdot N^4$ must be a perfect square which means $B \cdot N^4 = t^2$, t a natural number. Thus $B = \frac{t^2}{N^4} \Rightarrow \sqrt{B} = \frac{t}{N^2}$ contradicting the fact that $\sqrt{B}$ is irrational by assumption.

Condition (d) cannot hold either because both integers $B \cdot N^3$ and $D \cdot N$ would have to be perfect squares: this implies both β and δ to be rational squares contrary to the hypothesis that $\sqrt{\beta}$ and $\sqrt{\delta}$ are both irrational. Therefore, we conclude that condition (c) must hold true: note that when condition (c) holds true we must have $\rho = 0$. Going back to equation (2) and substituting for $\rho = 0$ we now have, $N^2 \cdot (\gamma^2 + \delta - \alpha) = 0 \Rightarrow \gamma^2 + \delta - \alpha = 0$; $\alpha = \delta + \gamma^2$ and, from equation (1) we deduce that $\sqrt{\beta} - 2\gamma\sqrt{\delta} = 0 \Rightarrow \sqrt{\beta} = 2\gamma\sqrt{\delta}$. Putting these two together we have,

$$\left.\begin{array}{l} \alpha = \delta + \gamma^2 \\ \\ \sqrt{\beta} = 2\gamma\sqrt{\delta} \end{array}\right\} \tag{3}$$

Apply the identity $(\gamma - \sqrt{\delta})^2 = \gamma^2 - 2\gamma\sqrt{\delta} + \delta$; and by (3) we now have

$$\begin{array}{rl} & (\gamma - \sqrt{\delta})^2 = \gamma^2 + \delta - 2\gamma\sqrt{\delta}; \\ 0 \leq & (\gamma - \sqrt{\delta})^2 = \alpha - \sqrt{\beta} \\ \Rightarrow & \sqrt{(\gamma - \sqrt{\delta})^2} = \sqrt{\alpha - \sqrt{\beta}};\ |\gamma - \sqrt{\delta}| = \sqrt{\alpha - \sqrt{\beta}} \end{array}$$

we are done.

(b) By hypothesis we have $\sqrt{\alpha + \sqrt{\beta}} = \gamma + \sqrt{\delta}$ and by part (a) we have $\sqrt{\alpha - \sqrt{\beta}} = \gamma - \sqrt{\delta}$.

By multiplying these two equations memberwise we arrive at

$$\sqrt{\alpha + \sqrt{\beta}} \cdot \sqrt{\alpha - \sqrt{\beta}} = (\gamma + \sqrt{\delta})(\gamma - \sqrt{\delta}) \Rightarrow$$

$$\Rightarrow \sqrt{(\alpha + \sqrt{\beta})(\alpha - \sqrt{\beta})} = \gamma^2 - (\sqrt{\delta})^2;\ \sqrt{\alpha^2 - \beta} = \gamma^2 - \delta,$$

which proves that $\sqrt{\alpha^2 - \beta}$ is rational.

(c) Since $\alpha = \delta + \gamma^2$ (see equation (3) above) and since $\gamma^2 \geq 0$ and $\delta > 0$ (by hypothesis), it is clear that $\alpha > 0$.

P7. If $\alpha, \beta, \gamma, \delta$ are rational numbers, $\sqrt[3]{\delta}$ irrational, and $\alpha + \beta\sqrt[3]{\delta} + \gamma \cdot \sqrt[3]{\delta^2} = 0$, then $\alpha = \beta = \gamma = 0$.

Solution: We have

$$\alpha + \beta \cdot \sqrt[3]{\delta} + \gamma\sqrt[3]{\delta^2} = 0 \tag{1}$$

So that $\alpha + \beta \cdot \sqrt[3]{\delta} = -\gamma\sqrt[3]{\delta^2}$; cubing both sides yields,

$$\alpha^3 + 3\alpha^2\beta \cdot \sqrt[3]{\delta} + 3\alpha\beta^2 \cdot \sqrt[3]{\delta^2} + \beta^3 \cdot \delta = -\gamma^3 \cdot \delta^2 \tag{2}$$

Below we show that the assumption $\alpha\beta\gamma \neq 0$ leads to a contradiction; indeed suppose $\alpha \cdot \beta \cdot \gamma \neq 0$, then (1) implies $\sqrt[3]{\delta^2} = -\left(\frac{\alpha+\beta \cdot \sqrt[3]{\delta}}{\gamma}\right)$; substituting for $\sqrt[3]{\delta^2}$ in equation (2) we obtain $\alpha^3 + 3\alpha^2\beta \cdot \sqrt[3]{\delta} + 3\alpha\beta^2 \cdot \left(\frac{\alpha+\beta\sqrt[3]{\delta}}{\gamma}\right) + \beta^3\delta = -\gamma^3\delta^2$. We have (by multiplying the last equation by $\gamma \neq 0$)

$$3\alpha\beta \cdot (\alpha\gamma - \beta^2) \cdot \sqrt[3]{\delta} = -\gamma \cdot (\beta^3\delta + \gamma^2\delta^2 + \alpha^3) + 3\alpha^2\beta^2 \tag{3}$$

Since both the right-hand side of (3) and the number $3\alpha\beta(\alpha\gamma - \beta^2)$ are rational, while $\sqrt[3]{\delta}$ is irrational, equation (3) implies $3\alpha\beta(\alpha\gamma - \beta^2) = 0$ (otherwise $\sqrt[3]{\delta}$ would be rational); and since $\alpha\beta \neq 0$, the last equation implies

$$\alpha\gamma = \beta^2 = 0; \qquad \beta^2 = \alpha\gamma \tag{4}$$

We now go back to the original equation (1) and we write $\alpha + \gamma\sqrt[3]{\delta^2} = -\beta \cdot \sqrt[3]{\delta}$; cubing both sides again we have,

$$\alpha^3 + 3\alpha^2\gamma\sqrt[3]{\delta^2} + 3\alpha\gamma^2 \cdot \delta \cdot \sqrt[3]{\delta} + \gamma^3\delta^2 = -\beta^3 \cdot \delta;$$

using (from equation (1)) $\gamma^3\sqrt{\delta^2} = -(a + \beta\sqrt[3]{\delta})$ again as we did before we arrive at

$$\alpha^3 + 3\alpha(\gamma^2\delta - \alpha\beta)\sqrt[3]{\delta} - 3\alpha^3 + \gamma^3\delta^2 = -\beta^3\delta;$$

multiplying the last equation by γ gives,

$$3\alpha\gamma \cdot (\gamma^2 \cdot \delta - \alpha\beta) \cdot \sqrt[3]{\delta} = \gamma \cdot (3\alpha^3 - \alpha^3 - \gamma^3\delta^2 - \beta^3\delta). \tag{5}$$

Again as in the case of equation (3) the irrationality of $\sqrt[3]{\delta}$ implies $3\alpha\gamma(\gamma^2\delta - \alpha\beta) = 0$ and since $\alpha\beta\gamma \neq 0$ we conclude

$$\gamma^2\delta - \alpha\beta = 0 \Rightarrow \qquad \gamma^2\delta = \alpha\beta \tag{6}$$

Now we combine equations (4) and (6) to obtain (since $\alpha\beta\gamma \neq 0$ and $\delta \neq 0$ because $\sqrt[3]{\delta}$ is irrational)

$$\frac{\beta^2}{\gamma^2\delta} = \frac{\alpha\gamma}{\alpha\beta} \Rightarrow \frac{\beta^3}{\gamma^3} = \delta; \quad \delta = (\beta/\gamma)^3 \Rightarrow \sqrt[3]{\delta} = \frac{\beta}{\gamma},$$

a rational number, contrary to the fact that $\sqrt[3]{\delta}$ is irrational by hypothesis. Therefore the assumption $\alpha\beta\gamma \neq 0$ must be false, since under the existing hypothesis of the problem, it has led to a contradiction. We conclude $\alpha\beta\gamma - 0$. Moreover note that the assumption that $\sqrt[3]{\delta}$ is irrational implies that $\sqrt[3]{\delta^2}$ must also be irrational: indeed,, from $\sqrt[3]{\delta^2} \cdot \sqrt[3]{\delta} = \delta$; but $\sqrt[3]{\delta}$ is irrational and if $\sqrt[3]{\delta^2}$ were rational their product would be rational contrary to the fact that their product is a rational number, namely δ. Now, we know that $\alpha\beta\gamma = 0$: if $\alpha = 0$, then equation (1) implies $\beta \cdot \sqrt[3]{\delta} = -\gamma\sqrt[3]{\delta^2} \Rightarrow \beta = -\gamma \cdot \sqrt[3]{\delta}$ (since $\delta \neq 0$). But $\sqrt[3]{\delta}$ is irrational and so the last equation is possible only when $\beta = \gamma = 0$.

Likewise if $\beta = 0$, we obtain $\alpha = -\gamma\sqrt[3]{\delta^2}$; since $\sqrt[3]{\delta^2}$ is irrational, we must have $\alpha = \gamma = 0$.

Finally, if $\gamma = 0$, we obtain $\alpha = -\beta\sqrt[3]{\delta}$ which once more applies $\alpha = \beta = 0$, since $\sqrt[3]{\delta}$ which once more applies $\alpha = \beta = 0$, since $\sqrt[3]{\delta}$ is irrational.

In all cases, we have $\alpha = \beta = \gamma = 0$.

P8. Prove that if the numbers a, b, c, r and s are all rational and the numbers $\sqrt{r}$, $\sqrt{s}$, and $\sqrt{r \cdot s}$ are irrational and if $a + b\sqrt{r} + c \cdot \sqrt{s} = 0$, then $a = b = c = 0$.

Solution: Start with

$$a + b \cdot \sqrt{r} + c \cdot \sqrt{s} = 0 \tag{1}$$

Equation (1) implies $a + b \cdot \sqrt{r} = -c \cdot \sqrt{s}$; squaring both sides gives

$$a^2 + 2ab\sqrt{r} + b^2 \cdot r = c^2 \cdot s \Rightarrow 2ab\sqrt{r} = c^2 \cdot s - (a^2 + b^2 \cdot r) \tag{2}$$

Since a, b, c, and s are all rational, equation (2) clearly shows that $ab = 0$. Otherwise $\sqrt{r}$ would be rational (by dividing both sides of (2)by $2ab$), contrary to the hypothesis that $\sqrt{r}$ is irrational. So we conclude,

$$ab = 0 \tag{3}$$

Going back to (1), we write $b \cdot \sqrt{r} + c \cdot \sqrt{s} = -a$; squaring both sides of the last equation we obtain,

$$b^2 \cdot r + 2bc \cdot \sqrt{rs} + c^2 \cdot s = a^2 \Rightarrow 2bc\sqrt{rs} = a^2 - (b^2 \cdot r + c^2 \cdot s)$$

and since $\sqrt{rs}$ is irrational the last equation shows (as above, in the case of (2)) that we must have $bc = 0$ (4). since (3) and (4) hold true simultaneously there are two possibilities (not mutually exclusive):

$$b = 0 \text{ or } (a = c = 0)$$

If $a = c = 0$, then (1) implies $b\sqrt{r} = 0 \Rightarrow$ (since $r \neq 0$ because $\sqrt{r}$ is irrational) $b = 0$. Thus $a = b = c = 0$ in this case. If the first possibility occurs, $b = 0$, then equation (1) implies $a + c\sqrt{s} = 0 \Rightarrow c\sqrt{s} = -a$; since $\sqrt{s}$ is irrational we must have $c = 0$, otherwise $\sqrt{s}$ would equal $-\frac{a}{c}$, a rational number. Thus $c = 0$ and so $a = 0$ as well. Therefore, once more, we conclude $b = 0 = c = a$.

P9. Find rational numbers $\alpha, \beta, \gamma, \delta$, such that

$$\sqrt{2}-\sqrt{6}+3=\frac{\alpha\cdot(\sqrt{2}+\sqrt{3})+\beta}{\gamma(\sqrt{2}+\sqrt{3})+\delta}$$

Solution. Obviously not both γ and δ can be zero. The required condition is equivalent to,

$$\left.\begin{array}{l}[\gamma(\sqrt{2}+\sqrt{3})+\delta]\cdot[\sqrt{2}-\sqrt{6}+3]=\alpha(\sqrt{2}+\sqrt{3})+\beta \\ \text{and with not both } \gamma \text{ and } \delta \text{ zero.}\end{array}\right\} \quad (1)$$

We have (algebra steps)

$$2\gamma-2\gamma\sqrt{3}+3\gamma\sqrt{2}+\gamma\sqrt{6}-3\gamma\sqrt{2}+3\gamma\sqrt{3}+\delta\sqrt{2}-\delta\sqrt{6}+3\delta=\alpha\sqrt{2}+\alpha\sqrt{3}+\beta$$

$$\Leftrightarrow(\delta-\alpha)\sqrt{2}+(\gamma-\alpha)\sqrt{3}+(\gamma-\delta)\sqrt{6}+(2\gamma+3\delta-\beta)=0. \quad (2)$$

Now apply exercise 20. (You are supposed to work this exercise out, there is an elaborate hint given) with

$$a=2\gamma+3\delta-\beta,\ b=\delta-\alpha,\ c=\gamma-\alpha,\ d=\gamma-\delta,\ r_1=2,\ r_2=3,\ \text{and } r_1r_2=6;$$

note that $\sqrt{2}$, $\sqrt{3}$, and $\sqrt{6}$ are all irrational according to Theorem 2(ii). Thus from exercise 20 we conclude that we must have

$$2\gamma+3\delta-\beta=0,\quad \delta-\alpha=0,\quad \gamma-\alpha=0,\quad \gamma-\delta=0$$

which implies $\alpha=\gamma=\delta=t$ and $\beta=5t$; where t is some rational number. It is now clear that the complete solution to the problem is given by $\alpha\gamma=\delta=t$ and $\beta=5t$, where t can be any nonzero rational number.

Check:

$$\frac{\alpha(\sqrt{2}+\sqrt{3})+\beta}{\gamma(\sqrt{2}+\sqrt{3})+\delta}=\frac{t(\sqrt{2}+\sqrt{3})+5t}{t(\sqrt{2}+\sqrt{3})+t}=\frac{\sqrt{2}+\sqrt{3}+5}{\sqrt{2}+\sqrt{3}+1}=$$

$$\frac{(\sqrt{2}+\sqrt{3}+5)(\sqrt{2}-\sqrt{3}+1)}{(\sqrt{2}+\sqrt{3}+1)(\sqrt{2}-\sqrt{3}+1}=\frac{2-\sqrt{6}+\sqrt{2}+\sqrt{6}-3+\sqrt{3}+5\sqrt{2}-5\sqrt{3}+5}{(\sqrt{2}+1)^2-(\sqrt{3})^2}$$

$$=\frac{6\sqrt{2}-4\sqrt{3}+4}{2+2\sqrt{2}+1-3}=$$

$$=\frac{6\sqrt{2}-4\sqrt{3}+4}{2\sqrt{2}}=\frac{(6\sqrt{2}-4\sqrt{3}+4)(2\sqrt{2})}{(2\sqrt{2})^2}=$$

$$=\frac{24-8\sqrt{6}+8\sqrt{2}}{8}=3-\sqrt{6}+\sqrt{2}$$

P10. Show that for any two real numbers b and c with $b < c$, the inequality $[\![b]\!] \leq [\![c]\!]$ holds true, (where $[\![x]\!]$ stands for the integer part of x.

Solution: By definition 1, $[\![b]\!]$ is the largest integer not exceeding b : $[\![b]\!] \leq b < [\![b]\!]+1$. Since $b < c$, conclude $[\![b]\!] \leq b < c$. By definition 1 again $[\![c]\!]$ is the largest integer not exceeding c. Now we cannot have $[\![c]\!] < [\![b]\!]$, because we already have $[\![b]\!] \leq b < c$, and this would imply that $[\![c]\!]$ is not the largest integer not exceeding b($[\![b]\!]$ would be larger). Since we cannot have $[\![c]\!] < [\![b]\!]$ we conclude that $[\![b]\!] \leq [\![c]\!]$. (Recall that by Axiom 13 of the real numbers that given any two reals a_1 and a_2, exactly one of three possibilities must occur: $a_1 < a_2$, $a_1 = a_2$, or $a_1 > a_2$.) Note that what the statement of this problem really says is that $[\![x]\!]$ is a nondecreasing function.

P11. Show that if k is an integer and n a natural number then $[\![\frac{k}{n}]\!] = q$, where q, is the quotient of the division of k with n.

Solution: From Chapter 4 we know that there exist unique integers q (quotient) and r (remainder) such that $k = q \cdot n + r$, and $0 \leq r \leq n-1$; (r is one of the integers 0 through $(n-1)$). So that $\frac{k}{n} = q + \frac{r}{n}$; and since $0 \leq r \leq n-1$ it follows that $0 \leq \frac{r}{n} \leq \frac{n-1}{n} < 1$ which in turn gives, $q \leq \frac{k}{n} = q + r/n < q+1$. By definition 1, the fact that q is an integer, and $q \leq \frac{k}{n} < q+1$, we conclude that q is the integer part of k/n, that is $q = [\![\frac{k}{n}]\!]$.

P12. Show that for any natural number n and any real number a,

$$[\![\frac{[\![a]\!]}{n}]\!] = [\![\frac{a}{n}]\!]$$

Solution: (a) By definition 1 we have

$$[\![a]\!] \leq a < [\![a]\!]+1 \tag{1}$$

$$\Rightarrow \frac{[\![a]\!]}{n} \leq \frac{a}{n} < \frac{[\![a]\!]+1}{n}$$

Let $k = [\![a]\!]$. According to practice problem P11 if a and r are the quotient and remainder of the division of k with n, we must have $[\![\frac{k}{n}]\!] = q$; and of course, $k = q \cdot n + r$; with $0 \leq r \leq n-1$.

According to (1), $\frac{k}{n} \leq \frac{a}{n} < \frac{k+1}{n}$; combining this with $k = q \cdot n + r$ and $0 \leq r \leq n-1$ we now have,

$$q \leq q + \frac{r}{n} \leq \frac{a}{n} < q + \frac{r+1}{n} \leq q+1;$$

and thus $q \leq \frac{a}{n} < q+1$, which proves by definition 1 that $q = [\![\frac{a}{n}]\!]$. But $q = [\![\frac{k}{n}]\!]$; hence

$$[\![\frac{k}{n}]\!] = [\![\frac{a}{n}]\!]; [\![\frac{[\![a]\!]}{n}]\!] = [\![\frac{a}{n}]\!].$$

6.4 Exercise Set

1. Which of the following numbers are rational? Which are irrational? In each case justify your answer.

   (i) $\sqrt[3]{4}$

   (ii) $(\sqrt{3}) \cdot (\sqrt{48})$

   (iii) $\sqrt[4]{p^2}$, where p is a prime number

   (iv) $\sqrt[6]{k^2}$, where k is a natural number which is a perfect cube

   (v) $\sqrt{\frac{p_1}{p_2}}$, where p_1 and p_2 are distinct primes.

   (vi) $\sqrt[3]{27abc}$, where a, b, c are integers whose product is the cube of an integer.

   (vii) $\sqrt[n]{p \cdot k}$, where p and k are natural numbers such that p is a prime that does not divide k.

2. Show that the number $a = \sqrt{2} + \sqrt[3]{2}$ is irrational.

   Hint: Write $a - \sqrt{2} = \sqrt[3]{2}$; the cube both sides.

3. If a and b are distinct primes, c and d are rational numbers, and $c \cdot \sqrt{a} = d \cdot \sqrt{b}$, prove that $c = d = 0$.

4. If r_1, r_2 are rational numbers, show that in order that the number $(\sqrt{r_1} + \sqrt{r_2})$ be rational, it is necessary and sufficient that both numbers $\sqrt{r_1}$ and $\sqrt{r_2}$ are rational numbers.

5. If $\alpha_1, \beta_1, \gamma_1, \alpha_2, \beta_2, \gamma_2$, and δ are rational and $\sqrt[3]{\delta}$ is irrational, show that the equality

$$\alpha_1 + \beta_1 \sqrt[3]{\delta} + \gamma_1 \sqrt[3]{\delta^2} = \alpha_2 + \beta_2 \sqrt[3]{\delta} + \gamma_2 \sqrt[3]{\delta^2},$$

   implies $\alpha_1 = \alpha_2$, $\beta_1 = \beta_2$, $\gamma_1 = \gamma_2$.

   Hint: Practice problem P7.

6.

   (i) If α, β, γ are rational numbers and $\alpha + \beta\sqrt{2} + \gamma\sqrt[3]{2} = 0$, show that $\alpha = \beta = \gamma = 0$.

   (ii) If 2 is replaced by another prime p, is the same conclusion still valid?

   (iii) If 2 is replaced by a natural number n which is neither a perfect square nor a perfect cube, is the same conclusion still valid?

7. Suppose that the rational numbers $\alpha, \beta, \gamma, \delta$ satisfy $\alpha\sqrt{3} + \beta\sqrt{12} + \gamma\sqrt{27} = \delta$. What conditions must $\alpha, \beta, \gamma, \delta$ satisfy for this to be true?

8. Describe all the rational numbers a, b, c, d, such that $a\sqrt{2}+b\sqrt{3}+c\sqrt{6} = d$. Is it true that the only values of the rationals a, b, c, d are $a = 0, b = 0, d = 0$? Or are there other quadruples (a, b, c, d) of rational numbers that satisfy the equation above?

9. Prove or disprove: if $\sqrt{a}+\sqrt{b} = \sqrt{\gamma}+\sqrt{\delta}$ where $a, b, c, d \in \mathbb{Q}$, and $\sqrt{a}, \sqrt{b}, \sqrt{\gamma}, \sqrt{\delta}$, all being irrational numbers, then ($a = \gamma$ and $b = \delta$) or ($a = \delta$ and $b = \gamma$). If you think this statement is false, then give a counterexample to disprove it. If it is false, can you find a family of quadruples (a, b, γ, δ) that satisfy the above equation?

Hint: Take $a = d \cdot k_1^2$, $b = d \cdot k_2^2$, $c = d \cdot k_3^2$, $\delta = d \cdot k_4^2$, where d is a natural number that is not a perfect square and k_1, k_2, k_3, k_4 are rationals (you may take them to be integers);; choose k_1, k_2, k_3, k_4 in a manner that a, b, γ, δ satisfy the above condition.

1. If n is a natural number, then the numbers

   (i) $\sqrt{5(n^2+2)}$

   (ii) $\sqrt{n(n+1)(n+2)}$

   (iii) $\sqrt[3]{n(n+1)(n+2)}$

   are irrational.

10. Suppose m, n are natural numbers, $m > n$, $(m, n) = 1$, and one of m, n is even, the other odd. Prove that the number $\sqrt{m^2 - n^2}$ is rational if and only if both numbers $\sqrt{m+n}$ and $\sqrt{m-n}$ are rational.

11. If $\sqrt{a}$ and $\sqrt{b}$ are irrational numbers, and a, b are natural numbers, find the necessary and sufficient conditions so that $k\sqrt{a} + \lambda\sqrt{b} = \mu$, for some integers k, λ, μ.

12. Show that for every real number a, $[\![-a]\!] \leq -[\![a]\!]$. When does the equal sign hold and when does the strict inequality hold?

13. Show that for every real number x,

$$[\![x]\!] + [\![-x]\!] = \begin{cases} -1, & \text{if } x \text{ is not an integer} \\ 0, & \text{if } x \in \mathbb{Z} \end{cases}$$

14. Let m and n be positive integers. Show that the number of positive integer multiples of n that do not exceed m, is equal to $[\![\frac{m}{n}]\!]$.

15. According to Theorem 8(i), for any two real numbers a and b,

$$[\![a]\!] + [\![b]\!] \leq [\![a+b]\!] \leq [\![a]\!] + [\![b]\!] + 1.$$

In the first inequality $[\![a]\!] + [\![b]\!] \leq [\![a+b]\!]$. When does the equal sign hold and when does the strict inequality hold? Answer the same question for the second inequality $[\![a+b]\!] \leq [\![a]\!] + [\![b]\!] + 1$ as well.

16. Let n be a natural number and assume that $a - [\![a]\!] \leq \frac{1}{n}$, where a is a real number. Prove that $[\![na]\!] - n[\![a]\!] = 0$.

17. Let b and c be any two real numbers such that $b < c$. According to practice Problem P10, $[\![b]\!] \leq [\![c]\!]$. Show that $[\![b]\!] = [\![c]\!]$ holds only when $(0 < c - b < 1$ and either b is an integer or otherwise there is no integer k such that $b < k \leq c)$. In all other cases, $[\![b]\!] < [\![c]\!]$.

18. If n is a natural number, a is a real number, and $\frac{n-1}{n} \leq a - [\![a]\!]$, show that $[\![na]\!] - n[\![a]\!] = n - 1$

19. Suppose that $\alpha, \beta, \gamma, \delta$ are rational numbers, and r_1 and r_2 are positive rational numbers such that the numbers $\sqrt{r_1}$, $\sqrt{r_2}$ and $\sqrt{r_1 r_2}$ are all irrational. In addition to this hypothesis, assume that the six numbers $\alpha, \beta, \gamma, \delta, r_1, r_2$ satisfy the condition $\alpha + \beta\sqrt{r_1} + \gamma\sqrt{r_2} + \delta \cdot \sqrt{r_1 r_2} = 0$.

Prove that $\alpha = \beta = \gamma = \delta = 0$ by the following procedure below:

i) First observe that if one of $\alpha, \beta, \gamma, \delta$ is zero, then the other three must also be zero according to practice problem P8.

ii) Assume $\alpha\beta\gamma\delta \neq 0$ and show that this assumption combined with the problem hypothesis, leads to a contradiction (thus we cannot have $\alpha\beta\gamma\delta \neq 0$; which means $\alpha\beta\gamma\delta = 0$ and as explained above, this in turn implies $\alpha = \beta = \gamma = \delta = 0$). Do this by the following either method below (it is a good idea to use each method in order to gain a different perspective on the problem).

Method 1: From the given condition we have, $\beta\sqrt{r_1} + \gamma\sqrt{r_2} = -\alpha - \delta \cdot \sqrt{r_1 r_2}$; square both sides, use the fact that $\sqrt{r_1 r_2}$ is irrational, in order to conclude $\beta\gamma - \alpha\delta = 0$; $\beta\gamma = \alpha\delta$; now use the assumption that $\alpha\beta\gamma\delta \neq 0$, in order to write $\frac{\alpha}{\beta} = \frac{\gamma}{\delta} = k$ (a rational number); so that $\alpha = \beta k$ and $\gamma = \delta k$. Go back to your previous equation (obtained after squaring both sides of the equation involving $\sqrt{r_1}$, $\sqrt{r_2}$ and $\sqrt{r_1 r_2}$, listed above), substitute for $\alpha = \beta k$ and $\gamma = \delta \cdot k$ in order to arrive at $(r_1 - k^2)(\beta^2 - \delta^2 \cdot r_2) = 0$; by setting each factor to zero conclude that there is a contradiction.

Method 2: Put

$$\beta = \frac{B}{D}, \; \gamma = \frac{C}{D}, \; \delta = \frac{d}{D}, \; r_1 = \frac{R_1}{D}, \; r_2 = \frac{R_2}{D};$$

where D is a positive integer and B, C, d, R_1, and R_2 integers. Go back to the problem's condition and conclude that,

$$B \cdot \sqrt{R_1 \cdot D} + C \cdot \sqrt{R_2 \cdot D} + d \cdot \sqrt{R_1 R_2} = -D^2 \cdot \alpha;$$

$$B \cdot D \cdot \sqrt{R_1 \cdot D} + C \cdot D \cdot \sqrt{R_2 \cdot D} + d \cdot \sqrt{(R_1 D)(R_2 D)} = -D^3 \cdot \alpha$$

Apply Theorem 5(i) with $\rho = -D^3 \cdot \alpha$, $k_1 = BD$, $k_2 = CD$, $k_3 = d$, $m = R_1 D$, $n = R_2 D$ (note that $k_3 \neq 0$ and $mn = R_1 R_2 D^2$ is not a perfect square because $R_1 R_2 \cdot D^2 = r_1 \cdot r_2 \cdot D^4$; if $R_1 R_2 D^2$ were a perfect square then its equal, $r_1 r_2 D^4$ would be a perfect square, say $r_1 r_2 D^4 = t^2$, where r is a natural number, but then $\sqrt{r_1 r_2} = \frac{t}{D^2} =$ rational, contrary to the problem's assumption that $\sqrt{r_1 r_2}$ is irrational). According to Theorem 5(i) we must have $\rho = -D^3 \cdot \alpha = -\frac{k_1 k_2}{k_3}$; conclude that $\alpha\delta = \beta\gamma$ and from there proceed as we did in Method 1.

Chapter 7

Progressions

In this chapter we will study arithmetic and geometric progressions, as well as mixed progressions. All three kinds of progressions are examples of sequences. Almost every student who has studied mathematics, at least through a first calculus course, has come across the concept of sequence. Such a student has usually seen some examples of sequences so the reader of this book has quite likely at least some informal understanding of what the term sequence means. We start with a formal definition of the term sequence.

Definition 1: (a) A **finite sequence** of k elements, (k a fixed positive integer) and whose terms are real numbers, is a mapping f from the set $\{1, 2, \ldots, k\}$ (the set containing the first k positive integers) to the set of real numbers $\mathbb{R}$. Such a sequence is usually denoted by $a_1, \ldots, a_n, \ldots, a_k$. If n is a positive integer between 1 and k, the **nth term a_n**, is simply the value of the function f at n; $a_n = f(n)$.

(b) **An infinite sequence** whose terms are real numbers, is a mapping f from the set of positive integers or natural numbers to the set of real numbers $\mathbb{R}$, we write $F : \mathbb{N} \to \mathbb{R};\ f(n) = a_n$.

Such a sequence is usually denoted by $a_1, a_2, \ldots a_n, \ldots$. The term a_n is called the nth term of the sequence and it is simply the value of the function at n.

Remark 1: Unlike sets, for which the order in which their elements do not matter, in a sequence the order in which the elements are listed does matter and makes a particular sequence unique. For example, the sequences 1, 8, 10, and 8, 10, 1 are regarded as different sequences. In the first case we have a function f from $\{1, 2, 3\}$ to $\mathbb{R}$ defined as follows: $f := \{1, 2, 3\} \to \mathbb{R};\ f(1) = 1 = a_1,\ f(2) = 8 = a_2$, and $f(3) = 10 = a_3$. In the second case, we have a function $g : \{1, 2, 3\} \to \mathbb{R};\ g(1) = b_1 = 8,\ g(2) = b_2 = 10$, and $g(3) = b_3 = 1$.

Only if two sequences are **equal as functions**, are they regarded one and the same sequence.

7.1 Arithmetic Progressions

Definition 2: A sequence $a_1, a_2, \ldots, a_n, \ldots$ with at least two terms, is called an **arithmetic** progression, if, and only if there exists a (fixed) real number d such that $a_{n+1} = a_n + d$, for every natural number n, if the sequence is infinite. If the sequence if finite with k terms, then $a_{n+1} = a_n + d$ for $n = 1, \ldots, k-1$. The real number d is called the **difference** of the arithmetic progression.

Remark 2: What the above definition really says, is that starting with the second term a_2, each term of the sequence is equal to the sum of the previous term plus the fixed number d.

Definition 3: An arithmetic progression is said to be **increasing** if the real number d (in Definition 2) is positive, and **decreasing** if the real number d is negative, and **constant** if $d = 0$.

Remark 3: Obviously, if $d > 0$, each term will be greater than the previous term, while if $d < 0$, each term will be smaller than the previous one.

Theorem 1: Let $a_1, a_2, \ldots, a_n, \ldots$ be an arithmetic progression with difference d, m and n any natural numbers with $m < n$. The following hold true:

(i) $a_n = a_1 + (n-1)d$

(ii) $a_n = a_{n-m} + md$

(iii) $a_{m+1} + a_{n-m} = a_1 + a_n$

Proof: (i) We may proceed by mathematical induction. The statement obviously holds for $n = 1$ since $a_1 = a_1 + (1-1)d$; $a_1 = a_1 + 0$, which is true. Next we show that if the statement holds for some natural number t, then this assumption implies that the statement must also hold for $(t+1)$. Indeed, if the statement holds for $n = t$, then we have $a_t = a_1 + (t-1)d$, but we also know that $a_{t+1} = a_t + d$, since a_t and a_{t+1} are successive terms of the given arithmetic progression. Thus, $a_t = a_1 + (t-1)d \Rightarrow a_t + d = a_1(t-1)d + d \Rightarrow a_t + d = a_1 + d \cdot t \Rightarrow a_{t+1} = a_1 + d \cdot t$; $a_{t+1} = a_1 + d \cdot [(t+1) - 1]$, which proves that the statement also holds for $n = t+1$. The induction process is complete.

(ii) By part (i) we have established that $a_n = a_1 + (n-1)d$, for every natural number n. So that

$$a_n = a_1 + (n-1)d - md + md;$$

$$a_n = a_1 + [(n-m) - 1]d + md.$$

Again, by part (i) we know that $a_{n-m} = a_1 + [(n-m) - 1]d$. Combining this with the last equation we obtain, $a_n = a_{n-m} + md$, and the proof is complete.

(iii) By part (i) we know that $a_{m+1} = a_1 + [(m+1) - 1]d \Rightarrow a_{m+1} = a_1 + md$; and by part (ii), we have already established that $a_n = a_{n-m} + md$. Hence, $a_{m+1} + a_{n-m} = a_1 + md + a_{n-m} = a_1 + a_n$, and the proof is complete. ■

Remark 4: Note that what Theorem 1(iii) really says is that in an arithmetic progression $a_1, \ldots, a_n$ with a_1 being the first term and a_n being the nth or last term; if we pick two in between terms a_{m+1} and a_{n-m} which are "equidistant" from the first and last term respectively (a_{m+1} is m places or spaces to the right of a_1 while a_{n-m} is m spaces or places to the left of a_n), the sum of a_{m+1} and a_{n-m} remains fixed: it is always equal to $(a_1 + a_n)$, no matter what the value of m is (m can take values from 1 to $(n-1)$). For example, if a_1, a_2, a_3, a_4, a_5 is an arithmetic progression we must have $a_1 + a_5 = a_2 + a_4 = a_3 + a_3 = 2a_3$. Note that $(a_2 + a_4)$ corresponds to $m = 1$, while $(a_3 + a_3)$ corresponds to $m = 2$, but also $a_4 + a_2$ corresponds to $m = 3$ and $a_5 + a_1$ corresponds to $m = 4$.

Likewise, if $b_1, b_2, b_3, b_4, b_5, b_6$ are the successive terms of an arithmetic progression we must have $b_1 + b_6 = b_2 + b_5 = b_3 + b_4$.

The following theorem establishes two equivalent formulas for the sum of the first n terms of an arithmetic progression.

Theorem 2: Let $a_1, a_2, \dots, a_n, \dots,$ be an arithmetic progression with difference d.

(i) The sum of the first (successive) n terms $a_1, \dots, a_n$, is equal to the real number $\left(\dfrac{a_1 + a_n}{2}\right) \cdot n$; we write $a_1 + a_2 + \cdots + a_n = \sum_{i=1}^{n} a_i = \dfrac{n \cdot (a_1 + a_n)}{2}$.

(ii) $\sum_{i=1}^{n} a_i = \left(\dfrac{a_1 + [a_1 + (n-1)d]}{2}\right) \cdot n$.

Proof: (i) We proceed by mathematical induction. For $n = 1$ the statement is obviously true since $a_1 = \frac{1 \cdot (a_1 + a_1)}{2} = \frac{2a_1}{2}$. Assume the statement to hold true for some $n = k \geq 1$. We will show that whenever the statement holds true for some value k of n, $k \geq 1$, it must also hold true for $n = k+1$. Indeed, assume $a_1 + \cdots + a_k = \frac{k \cdot (a_1 + a_k)}{2}$; add a_{k+1} to both sides to obtain

$$\begin{aligned} a_1 + \cdots + a_k + a_{k+1} &= \frac{k \cdot a_1 + a_k}{2} + a_{k+1} \\ &\Rightarrow a_1 + \cdots + a_k + a_{k+1} \\ &= \frac{ka_1 + ka_k + 2a_{k+1}}{2} \end{aligned} \tag{1}$$

But since the given sequence is an arithmetic progression by Theorem 1(i), we must have $a_{k+1} = a_1 + kd$ where d is the difference. Substituting back in equation (1) for a_{k+1} we obtain,

$$\begin{aligned} a_1 + \cdots + a_k + a_{k+1} &= \frac{ka_1 + ka_k + (a_1 + kd) + a_{k+1}}{2} \\ \Rightarrow a_1 + \cdots + a_k + a_{k+1} &= \frac{(k+1)a_1 + k(a_k + d) + a_{k+1}}{2} \end{aligned} \tag{2}$$

We also have $a_{k+1} = a_k + d$, since a_k and a_{k+1} are successive terms. Replacing $a_k + d$ by a_{k+1} in equation (2) we now have $a_1 + \cdots + a_k + a_{k+1} = \dfrac{(k+1)a_1 + ka_{k+1} + a_{k+1}}{2} = \dfrac{(k+1)a_1 + (k+1)a_{k+1}}{2} = (k+1) \cdot \dfrac{(a_1 + a_{k+1})}{2}$, and the proof is complete. The

For $x = 1$, the above derived formula is not valid. However, for $x = 1$; $1 + 2x + 3x^2 + \ldots + nx^{n-1} + (n+1)x^n = 1+2+3+\ldots+n+(n+1) = \frac{(n+1)(n+2)}{2}$ (the sum of the first $(n+1)$ terms of an arithmetic progression with first term $a_1 - 1$ and difference $d = 1$.

The following theorem gives a formula for the sum of the first n terms of a mixed progression.

Theorem 9: Let $b_1, b_2, \ldots, b_n, \ldots$, be an arithmetic progression with first term b_1 and difference d; and $c_1, c_2, \ldots, c_n, \ldots$, be a geometric progression with first term $c_1 = c$ and ratio $r \neq 1$. Let $a_1, a_2, \ldots, a_n, \ldots$, the corresponding mixed progression, that is the sequence whose nth term a_n is given by $a_n = b_n \cdot c_n$, for every natural number n.

(i) $a_n = [b_1 + (n-1) \cdot d] \cdot c \cdot r^{n-1}$, for every natural number n.

(ii) For every natural number n, $a_{n+1} - r \cdot a_n = d \cdot c_{n+1}$.

(iii) If $S_n = a_1 + a_2 + \ldots + a_n$ (sum of the first n terms of the mixed progression), then

$$S_n = \frac{a_n \cdot r - a_1}{r-1} + \frac{d \cdot r \cdot c \cdot (1 - r^{n-1})}{(r-1)^2};$$

$$S_n = \frac{a_n \cdot r - a_1}{r-1} + \frac{d \cdot r \cdot (c - c_n)}{(r-1)^2}$$

(recall $c_n = c \cdot r^{n-1}$).

Proof: (i) This is immediate, since by Theorem 1(i), $b_n = b_1 + (n-1) \cdot d$ and by Theorem 5(i), $c_n = c \cdot r^{n-1}$, and so $a_n = b_n \cdot c_n = [b_1 + (n-1)d] \cdot c \cdot r^{r-1}$.

(ii) We have $a_{n+1} = b_{n+1} \cdot c_{n+1}$, $a_n = b_n c_n$, $b_{n+1} = d + b_n$. Thus, $a_{n+1} - r \cdot a_n = c_{n+1} \cdot (d + b_n) - r \cdot b_n \cdot c_n = d \cdot c_{n+1} + c_{n+1} b_n - r b_n c_n = d \cdot c_{n+1} + b_n \cdot \underbrace{(c_{n+1} - r c_n)}_{0} = d c_{n+1}$, since $c_{n+1} = r c_n$ by virtue of the fact that c_n and c_{n+1} are consecutive terms of a geometric progression with ratio r. End of proof. ■

(iii) We proceed by mathematical induction. The statement is true for $n = 1$ because $S_1 = a_1$ and $\frac{a_1 r - a_1}{r-1} + \frac{d \cdot r \cdot (c - c_1)}{(r-1)^2} = \frac{a_1(r-1)}{r-1} + 0 = a_1 = S_1$. Assume the statement to hold for $n = k$: (for some natural number $k \geq 1$; $S_k = \frac{a_k \cdot r - a_1}{r-1} + \frac{d \cdot r \cdot (c - c_k)}{(r-1)^2}$. We have $S_{k+1} = S_k + a_{k+1} = \frac{a_k \cdot r - a_1}{r-1} + \frac{d \cdot r \cdot (c - c_k)}{(r-1)^2} + a_{k+1} = \frac{a_k \cdot r - a_1 + a_{k+1} \cdot r - a_{k+1}}{r-1} + \frac{d \cdot r \cdot (c - c_k)}{(r-1)^2}(1)$. But by part (ii) we know that $a_{k+1} - r a_k = d \cdot c_{k+1}$. Thus, by (1) we now have,

$$S_{k+1} = \frac{a_{k+1} \cdot r - a_1}{r-1} - \frac{d \cdot c_{k+1}}{r-1} + \frac{d \cdot r \cdot (c - c_k)}{(r-1)^2}$$

$$\Rightarrow S_{k+1} = \frac{a_{k+1} \cdot r - a_1}{r-1} + \frac{-(r-1) \cdot d \cdot c_{k+1} + d \cdot r \cdot (c - c_k)}{(r-1)^2};$$

$$S_{k+1} = \frac{a_{k+1} \cdot r - a_1}{r-1} + \frac{d \cdot r \cdot (c - c_{k+1}) + d \cdot \overbrace{(c_{k+1} - r \cdot c_k)}^{0}}{(r-1)^2}.$$

But $c_{k+1} - r \cdot c_k = 0$ (since $c_{k+1} = r \cdot c_k$) because c_k and c_{k+1} consecutive terms of a statement also holds for $n = k + 1$. ■

(ii) This is an immediate consequence of part (i). Since $\sum_{i=1}^{n} a_i = \frac{n(a_1+a_n)}{2}$ and $a_n = a_1 + (n-1)d$ (by Theorem 1(i)) we have,

$$\sum_{i=1}^{n} a_i = n\left(\frac{a_1 + [a_1 + (n-1)d]}{2}\right),$$

and we are done. ■

Example 1: (i) The sequence of positive integers $1, 2, 3, \ldots, n, \ldots$, is an infinite sequence which is an arithmetic progression with first term $a_1 = 1$, difference $d = 1$, and the nth term $a_n = n$. According to Theorem 2(i) the sum of the first n terms can be easily found: $1 + 2 + \ldots + n = \frac{n\cdot(1+n)}{2}$.

(ii) The sequence of the even positive integers $2, 4, 6, 8, \ldots, 2n, \ldots$ has first term $a_1 = 2$, difference $d = 2$, and the nth term $a_n = 2n$. According to Theorem 2(i), $2 + 4 + \cdots + 2n = \dfrac{n\cdot(2+2n)}{2} = \dfrac{n\cdot 2\cdot(n+1)}{2} = n\cdot(n+1)$.

(iii) The sequence of the odd natural numbers $1, 3, 5, \ldots, (2n-1), \ldots$, is an arithmetic progression with first term $a_1 = 1$, difference $d = 2$, and nth term $a_n = 2n - 1$. According to Theorem 2(i) we have $1 + 3 + \cdots + (2n-1) = n\cdot\left(\dfrac{1+(2n-1)}{2}\right) = \frac{n\cdot(2n)}{2} = n^2$.

(iv) The sequence of all natural numbers which are multiples of 3 : $3, 6, 9, 12, \ldots, 3n, \ldots$ is an arithmetic progression with first term $a_1 = 3$, difference $d = 3$ and nth term $a_n = 3n$. We have $3 + 6 + \cdots + 3n = \frac{n\cdot(3+3n)}{2} = \frac{3n(n+1)}{2}$. Observe that this sum can also be found from (i) by observing that $3 + 6 + \cdots + 3n = 3\cdot(1 + 2 + \cdots + n) = \frac{3\cdot n(n+1)}{2}$. If we had to find the sum of all natural numbers which are multiples of 3, starting with 3 and ending with 33; we know that $a_1 = 3$ and that $a_n = 33$. We must find the value of n. Indeed, $a_n = a_1 + (n-1)\cdot d$; and since $d = 3$, we have $33 = 3 + (n-1)\cdot 3 \Rightarrow 33 = 3\cdot[1 + (n-1)]$; $11 = n$. Thus, $3 + 6 + \cdots + 30 + 33 = \frac{11\cdot(3+33)}{2} = \frac{11\cdot 36}{2} = 11\cdot 18 = 198$.

Example 2: Given an arithmetic progression $a_1, \ldots, a_m, \ldots, a_n, \ldots$, and natural numbers m, n with $2 \leq m < n$, one can always find the sum $a_m + a_{m+1} + \cdots + a_{n-1} + a_n$; that is, the sum of the $[(n-m)+1]$ terms starting with a_m and ending with a_n. If we know the values of a_m and a_n then we do not need to know the value of the difference. Indeed, the finite sequence $a_m, a_{m+1}, \ldots, a_{n-1}, a_n$ is a finite arithmetic progression with first term a_m, last term a_n, (and difference d); and it contains exactly $[(n-m)+1]$ terms. According to Theorem 2(i) we must have $a_m + a_{m+1} + \cdots + a_{n-1} + a_n = \frac{(n-m+1)\cdot[a_m+a_n]}{2}$.

If, on the other hand, we only know the values of the first term a_1 and difference d (and the values of m and n), we can apply Theorem 2(ii) by observing that

$$\begin{aligned}
a_m + a_{m+1} + \cdots + a_{n-1} + a_n &= \Big(\underbrace{a_1 + a_2 + \ldots + a_n}_{\substack{\text{sum of the first}\\ n \text{ terms}}}\Big) \\
&\quad -\Big(\underbrace{a_1 + \ldots + a_{m-1}}_{\substack{\text{sum of the first}\\ (m-1) \text{ terms}}}\Big) \\
\text{by Th. 2(ii)} &= \left(\tfrac{2a_1+(n-1)d}{2}\right)\cdot n \\
&\quad - \left(\tfrac{2a_1+(m-2)d}{2}\right)\cdot(m-1) \\
&= \tfrac{2[n-(m-1)]a_1+[n\cdot(n-1)-\cdot(m-2)\cdot(m-1)]d}{2} \\
&= \tfrac{2(n-m+1)a_1+[n(n-1)-(m-2)(m-1)]d}{2}
\end{aligned}$$

Example 3: (a) Find the sum of all multiples of 7, starting with 49 and ending with 133. Both 49 and 133 are terms of the infinite arithmetic progression with first term $a_1 = 7$, and difference $d = 7$. If $a_m = 49$, then $49 = a_1 + (m-1)d$; $49 = 7 + (m-1)\cdot 7 \Rightarrow \frac{49}{7} = m$; $m = 7$. Likewise, if $a_n =$ then $133 = a_1 + (n-1)d$; $133 = 7 + (n-1)7 \Rightarrow 19 = n$. According to Example 2, the sum we are looking for is given by $a_7 + a_8 + \ldots + a_{18} + a_{19} = \frac{(19-7+1)(a_7+a_{19})}{2} = \frac{13\cdot(49+133)}{2} = \frac{13\cdot 182}{2} = (13)\cdot(91) = 1183.$

(b) For the arithmetic progression with first term $a_1 = 11$ and difference $d = 5$, find the sum of its terms starting with a_5 and ending with a_{13}.

We are looking for the sum $a_5 + a_6 + \ldots + a_{12} + a_{13}$; in the usual notation $m = 5$ and $n = 13$. According to Example 2, since we know the first term $a_1 = 11$ and the difference $d = 5$ we may use the formula we developed there:

$$\begin{aligned}
a_m + a_{m+1} + \ldots + a_{n-1} + a_n &= \frac{2(n-m+1)a_1 + [n(n-1) - (m-2)(m-1)]d}{2}; \\
a_5 + a_6 + \ldots + a_{12} + a_{13} &= \frac{2\cdot(13-5+1)\cdot 11 + [13\cdot(13-1) - (5-2)(5-1)]5}{2} \\
&= \frac{2\cdot 9\cdot 11 + [(13)(12) - (3)(4)]5}{2} = \frac{198 + (156-12)\cdot 5}{2} \\
&= \frac{198+720}{2} = \frac{918}{2} = 459
\end{aligned}$$

The following Theorem is simple in both its statement and proof but it serves as an effective tool to check whether three real numbers are successive terms of an arithmetic progression.

Theorem 3: Let a, b, c be real numbers with $a < b < c$.

(i) The three numbers $a, b,$ and c are successive of an arithmetic progression if, and only if, $2b = a + c$ or equivalently $b = \frac{a+c}{2}$.

(ii) Any arithmetic progression containing a, b, c as successive terms must have the same difference d, namely $d = b - a = c - b$

Proof: (i) Suppose that a, b, and c are successive terms of an arithmetic progression; then by definition we have $b = a + d$ and $c = b + d$, where d is the difference. So that $d = b - a = c - b$; from $b - a = c - b$ we obtain $2b = a + c$ or $b = \frac{a+c}{2}$.

Conversely, if $2b = a + c$, then $b - a = c - b$; so by setting $d = b - a = c - b$, it immediately follows that $b = a + d$ and $c = b + d$ which proves that the real numbers a, b, c are successive terms of an arithmetic progression with difference d.

(ii) This has already been shown in part (i), namely that $d = b - a = c - b$. Thus, any arithmetic progression containing the real numbers a, b, c as successive terms must have difference $d = b - a = c - b$.

Remark 5: According to Theorem 3, the middle term b is the average of a and c. This is generalized in Theorem 4 below. But, first we have the following definition.

Definition 4: Let $a_1, a_2, \ldots, a_n$ be a list (or sequence) of n real numbers(n a positive integer). The arithmetic **mean or average** of the given list, is the real number $\frac{a_1 + a_2 + \ldots + a_n}{n}$.

Theorem 4: Let m and n be natural numbers with $m < n$. Suppose that the real numbers $a_m, a_{m+1}, \ldots, a_{n-1}, a_n$ are the $(n - m + 1)$ successive terms of an arithmetic progression (here, as in the usual notation, a_k stands for the kth term of an arithmetic progression whose first term is a_1 and difference is d).

(i) If the natural number $(n - m + 1)$ is odd, then the arithmetic mean or average of the reals $a_m, a_{m+1}, \ldots, a_{n-1}, a_n$ is the term $a_{(\frac{m+n}{2})}$. In other words, $a_{(\frac{m+n}{2})} = \frac{a_m + a_{m+1} + \ldots + a_{n-1} + a_n}{n-m+1}$. (Note that since $(n - m + 1)$ is odd, it follows that $n - m$ must be even, and thus so must be $n + m$; and hence $\frac{m+n}{2}$ must be a natural number).

(ii) If the natural number is even, then the arithmetic mean of the reals $a_m, a_{m+1}, \ldots, a_{n-1}, a_n$ must be the average of the two middle terms $a_{(\frac{n+m-1}{2})}$ and $a_{(\frac{n+m+1}{2})}$. In other words $\frac{a_m + a_{m+1} + \ldots + a_{n-1} + a_n}{n-m+1} = \frac{a_{(\frac{n+m-1}{2})} + a_{(\frac{n+m+1}{2})}}{2}$.

Remark 6: To clearly see the workings of Theorem 4, let's look at two examples; first suppose $m = 3$ and $n = 7$. Then $n - m + 1 = 7 - 3 + 1 = 5$; so if a_3, a_4, a_5, a_6, a_7 are successive terms of an arithmetic progression, clearly a_5 is the middle term. But since the five terms are equally spaced or equidistant from one another (because each term is equal to the sum of the previous terms plus a fixed number, the difference d), it makes sense that a_5 would also turn out to be the average of the five terms.

Figure 1

Figure 2

If, on the other hand, the natural number $n - m + 1$ is even; as in the case of $m = 3$ and $n = 8$. Then we have two middle numbers: a_5 and a_6

(for $d > 0$)

Figure 3

Proof (of Theorem 4): (i) Since $n - m + 1$ is odd, it follows $n - m$ is even; and thus $n + m$ is also even. Now, if we look at the integers $m, m+1, \ldots, n-1, n$ we will see that since $m + n$ odd, there is a middle number among them, namely the natural number $\frac{m+n}{2}$. Consequently among the terms $a_m, a_{m+1}, \ldots, a_{n-1}, a_n$, the term $a_{(\frac{m+n}{2})}$ is the middle term. Next we perform two calculations. First we compute $a_{(\frac{m+n}{2})}$ in terms of m, n the first term a_1 and the difference d. According to Theorem 1(i), we have,

$$a_{(\frac{m+n}{2})} = a_1 + \left(\frac{m+n}{2} - 1\right) d = a_1 + \left(\frac{m+n-2}{2}\right) d.$$

Now let us compute the sum $\frac{a_m + a_{m+1} + \ldots + a_{n-1} + a_n}{n-m+1}$. First assume $m \geq 2$; so that $2 \leq m < n$. Observe that

$$a_m + a_{m+1} + \ldots + a_{n-1} + a_n$$

$$= \Big(\underbrace{a_1 + a_2 + \ldots + a_m + a_{m+1} + \ldots + a_{n-1} + a_n}_{\text{sum of the first } n \text{ terms}}\Big)$$

$$- \underbrace{(a_1 + \ldots + a_{m-1})}_{\substack{\text{sum of the first } (m-1) \text{ terms} \\ \text{note that } m-1 \geq 1, \text{ since } m \geq 2}}$$

Apply Theorem 2(ii), we have,

$$a_1 + a_2 + \ldots + a_m + a_{m+1} + \ldots + a_{n-1} + a_n = \frac{n[2a_1 + (n-1)d]}{2}$$

and

$$a_1 + \ldots + a_{m-1} = \frac{(m-1)[2a_1 + (m-2)d]}{2}.$$

Putting everything together we have

$$\begin{aligned}
& a_m + a_{m+1} + \ldots + a_{n-1} + a_n \\
= \; & (a_1 + a_2 + \ldots + a_m + a_{m+1} + \ldots + a_{n-1} + a_n) \\
& -(a_1 + \ldots + a_{m-1}) = \frac{n[2a_1 + (n-1)d]}{2} \\
& -\frac{(m-1)[2a_1 + (m-2)d]}{2} \\
= \; & \frac{2(n-m+1)a_1 + [n(n-1) - (m-1)(m-2)]d}{2}.
\end{aligned}$$

Thus,

$$\begin{aligned}
& \frac{a_m + a_{m+1} + \ldots + a_{n-1} + a_n}{n-m+1} \\
= \; & \frac{2(n-m+1)a_1 + [n(n-1) - (m-1)(m-2)]d}{2(n-m+1)} \\
= \; & a_1 + \frac{[n(n-1) - (m-1)(m-2)]d}{2(n-m+1)} \\
= \; & a_1 + \frac{[n^2 - m^2 - n + 3m - 2]d}{2(n-m+1)} \\
= \; & a_1 + \frac{[(n-m)(n+m) + (n+m) - 2(n-m) - 2]d}{2(n-m+1)} \\
= \; & a_1 + \frac{[(n-m)(n+m) + (n+m) - 2(n-m+1)]d}{2(n-m+1)} \\
= \; & a_1 + \frac{[(n+m)(n-m+1) - 2(n-m+1)]d}{2(n-m+1)} \\
= \; & a_1 + \frac{(n-m+1)(n+m-2)d}{2(n-m+1)} = a_1 + \frac{(n+m-2)d}{2},
\end{aligned}$$

which is equal to the term $a_{(\frac{m+n}{2})}$ as we have already shown. What about the case $m = 1$? If $m = 1$, then $n - m + 1 = n$ and $a_m = a_1$. In that case, we have the sum $\frac{a_1+a_2+\ldots+a_{n-1}+a_n}{n}$ = (by Theorem 2(ii)) $\frac{n\cdot[2a_1+(n-1)d]}{2n}$; but the middle term $a_{(\frac{m+n}{2})}$ is now $a_{(\frac{n+1}{2})}$ since $m = 1$; but $a_{(\frac{n+1}{2})} = a_1 + (\frac{1+n-2}{2})d \Rightarrow a_{(\frac{n+1}{2})} = a_1 + (\frac{n-1}{2})d$; compare this answer with what we just found right above, namely

$$\frac{n \cdot [2a_1 + (n-1)d]}{2n} = \frac{2a_1 + (n-1)d}{2} = a_1 + (\frac{n-1}{2})d,$$

they are the same. The proof is complete.

(ii) This is left as an exercise to the student. (See Exercise 23).

Definition 5: A sequence $a_1, a_2, \ldots, a_n, \ldots$ (finite or infinite) is called a **harmonic progression**, if, and only if, the corresponding sequence of the reciprocal terms:

$$b_1 = \frac{1}{a_1}, \quad b_2 = \frac{1}{a_2}, \ldots, b_n = \frac{1}{a_n}, \ldots,$$

is an arithmetic progression.

Example 4: The reader can easily verify that the following three sequences are harmonic progressions:

(a) $\frac{1}{1}, \frac{1}{2}, \frac{1}{3}, \ldots, \frac{1}{n}, \ldots$

(b) $\frac{1}{2}, \frac{1}{4}, \frac{1}{6}, \ldots, \frac{1}{2n}, \ldots$

(c) $\frac{1}{9}, \frac{1}{16}, \frac{1}{23}, \ldots, \frac{1}{7n+2}, \ldots$

7.2 Geometric Progressions

Definition 6: A sequence $a_1, a_2, \ldots, a_n, \ldots$ (finite or infinite) is called a **geometric progression**, if there exists a (fixed) real number r such that $a_{n+1} = r \cdot a_n$, for every natural number n (if the progression is finite with k terms $a_1, \ldots, a_k$; with $k \geq 2$, then $a_{n+1} = r \cdot a_n$, for all $n = 1, 2, \ldots, k-1$). The real number r is called the **ratio** of the geometric progression. The first term of the arithmetic progression is usually denoted by a, we write $a_1 = a$.

Theorem 5: Let $a = a_1, a_2, \ldots, a_n, \ldots$ be a geometric progression with first term a and ratio r.

(i) $a_n = a \cdot r^{n-1}$, for every natural number n.

(ii) $a_1 + \ldots + a_n = \sum_{i=1}^{n} a_i = \frac{a_n \cdot r - a}{r-1} = \frac{a(r^n - 1)}{r-1}$, for every natural number n, if $r \neq 1$; if on the other hand $r = 1$, then the sum of the first n terms of the geometric progression is equal to $n \cdot a$.

Proof: (i) By induction: the statement is true for $n = 1$, since $a_1 = a \cdot r^{\circ} = a$. Assume the statement to hold true for $n = k$; for some natural number k. We will show that this assumption implies the statement to be also true for $n = (k+1)$. Indeed, since the statement is true for $n = k$, we have $a_k = a \cdot r^{k-1} \Rightarrow r \cdot a_k = r \cdot a \cdot r^{k-1} = a \cdot r^k$; but $k = (k + 1 - 1)$ and $r \cdot a_k = a_{k+1}$, by the definition of a geometric progression. Hence, $a_{k+1} = a \cdot r^{(k+1)-1}$, and so the statement also holds true for $n = k$.

(ii) Most students probably have seen in precalculus the identity $r^n - 1 = (r - 1)(r^{n-1} + \ldots + 1)$ to hold true for all natural numbers n and all reals r. For example, when $n = 2$, $r^2 - 1 = (r-1)(r+1)$; when $n = 3$, $r^3 - 1 = (r-1)(r^2 + r + 1)$.

We use induction to actually prove it. Note that the statement $n = 1$ simply takes the form, $r \quad 1 = r \quad 1$ so it holds true; while for $n = 2$ the statement becomes $r^2 - 1 = (r-1)(r+1)$, which is again true. Now assume the statement to hold for some $n = k$, $k \geq 2$ a natural number. So we are assuming that the statement $r^k - 1 = (r-1)(r^{k-1} + \ldots + r + 1)$. Multiply both sides by r:

$$r \cdot (r^k - 1) = r \cdot (r-1) \cdot (r^{k-1} + \ldots + r + 1)$$

$$\Rightarrow \quad r^{k+1} - r = (r-1) \cdot (r^k + r^{k-1} + \ldots + r^2 + r);$$

$$r^{k+1} - r = (r-1) \cdot (r^k + r^{k-1} + \ldots r^2 + r + 1 - 1)$$

$$\Rightarrow \quad r^{k+1} - r = (r-1) \cdot (r^k + r^{k-1} + \ldots + r^2 + r + 1) + (r-1) \cdot (-1)$$

$$\Rightarrow \quad r^{k+1} - r = (r-1) \cdot (r^k + r^{k-1} + \ldots + r^2 + r + 1) - r + 1$$

$$\Rightarrow \quad r^{k+1} - 1 = (r-1) \cdot (r^{(k+1)-1} + r^{(k+1)-2} + \ldots + r^2 + r + 1),$$

which proves that the statement also holds true for $n = k+1$. The induction process is complete. We have shown that $r^n - 1 = (r-1)(r^{n-1} + r^{n-2} + \ldots + r + 1)$ holds true for every real number r and every natural n. If $r \neq 1$, then $r - 1 \neq 0$, and so $\frac{r^n-1}{r-1} = r^{n-1} + r^{n-2} + \ldots + r + 1$. Multiply both sides by the first term a we obtain

$$\begin{aligned} \frac{a \cdot (r^n - 1)}{r - 1} &= ar^{n-1} + ar^{n-2} + \ldots ar + a \\ &= a_n + a_{n-1} + \ldots + a_2 + a_1. \end{aligned}$$

Since by part (i) we know that $a_i = a \cdot r^{i-1}$, for $i = 1, 2, \ldots, n$; if on the other hand $r = 1$, then the geometric progression is obviously the constant sequence, $a, a, \ldots, a, \ldots$; $a_n = a$ for every natural number n. In that case $a_1 + \ldots + a_n = \underbrace{a + \ldots + a}_{n \text{ times}} = na$. The proof is complete. ■

Remark 7: We make some observation about the different types of geometric progressions that might occur according to the different types of values of the ratio r.

(i) If $a = 0$, then regardless of the value of the ratio r, one obtains the zero sequence $0, 0, 0, \ldots, 0, \ldots$.

(ii) If $r = 1$, then for any choice of the first term a, the geometric progression is the constant sequence, $a, a, \ldots, a, \ldots$.

(iii) If the first term a is positive and $r > 1$ one obtains a geometric progression of positive terms, and which is increasing and which eventually exceed any real number (as we will see in Theorem 8, given a positive real number M, there is a term a_n that exceeds M; in the language of calculus, we say that it approaches positive infinity). For example: $a = \frac{1}{2}$, and $r = 2$; we have the geometric progression

$$a_1 = a = \frac{1}{2},\ a_2 = \frac{1}{2} \cdot 2 = 1, a_3 = \frac{1}{2} \cdot 2^2 = 2;$$

The sequence is, $\frac{1}{2}, 1, 2, 2^2, 2^3, 2^4, \ldots, \underbrace{\frac{1}{2} \cdot 2^{n-1}}_{a_n}$.

(iv) When $a > 0$ and $0 < r < 1$, the geometric progression is decreasing and in the language of calculus, it approaches zero (it has limit value zero).
For example: $a = 4,\ r = \frac{1}{3}$.
We have $a_1 = a = 4,\ a_2 = 4 \cdot \frac{1}{3},\ a_3 = a \cdot \left(\frac{1}{3}\right)^2,\ a_4 = 4 \cdot \left(\frac{1}{3}\right)^3$; $4, \frac{4}{3}, \frac{4}{9}, \ldots, \frac{4}{3^{n-1}}$ nth term, $\ldots$.

(v) For $a > 0$ and $-1 < r < 0$, the geometric sequence alternates (which means that if we pick any term, the succeeding term will have opposite sign). Still, in this case, such a sequence approaches zero (has limit value zero).
For example: $a = 9,\ r = -\frac{1}{2}$.
$a_1 = a = 9,\ a_2 = 9 \cdot \left(-\frac{1}{2}\right) = -\frac{9}{2},\ a_3 = 9 \cdot \left(\cdot\frac{1}{2}\right)^2 = \frac{9}{4}, \ldots$

$$9,\ -\frac{9}{2},\ \frac{9}{2^2},\ \frac{-9}{2^3},\dots,\underbrace{9\cdot\left(-\frac{1}{2}\right)^{n-1}=\frac{9\cdot(-1)^{n-1}}{2^{n-1}}}_{n\text{th term}}$$

(vi) For $a > 0$ and $r = -1$, we have a geometric progression that oscillates: $a, -a, a, -a, \dots, a_n = (-1)^{n-1}, \dots$.

(vii) For $a > 0$ and $r < -1$, the geometric progression has negative terms only, it is decreasing, and in the language of calculus we say that approaches negative infinity. For example: $a = 3, r = -2$

$$\begin{aligned} a_1 &= a = 3,\ a_2 = 3\cdot(-2) = -6, \\ a_3 &= 3\cdot(-2)^2 = 12, \dots 3,\ -6,\ 12, \dots, \\ &\underbrace{3\cdot(-2)^{n-1} = 3\cdot 2^{n-1}\cdot(-1)^{n-1}}_{n\text{th term}}, \dots \end{aligned}$$

(viii) What happens when the first term a is negative? A similar analysis holds (see Exercise 24).

Theorem 6: Let $a = a_1, a_2, \dots, a_n, \dots$ be a geometric progression with ratio r.

(i) If m and n are any natural numbers such that $m < n$, $a_n = a_{n-m}\cdot r^m$.

(ii) If m and n are any natural numbers such that $m < n$, then $a_{m+1}\cdot a_{n-m} = a_1\cdot a_n$.

(iii) For any natural number n, $\left(\prod_{i=1}^{n} a_i\right)^2 = (a_1\cdot a_2\dots a_n)^2 = (a_1\cdot a_n)^n$, where $\prod_{i=1}^{n} a_i$ denotes the product of the first n terms $a_1, a_2, \dots, a_n$.

Proof: (i) By Theorem 5(i) we have $a_n = a\cdot r^{n-1}$ and $a_{n-m} = a\cdot r^{n-m-1}$; thus $a_{n-m}\cdot r^m = a\cdot r^{n-m-1}\cdot r^m = a\cdot r^{n-1} = a_n$, and we are done. ■

(ii) Again by Theorem 5(i) we have,

$$a_{m+1} = a\cdot r^m,\quad a_{n-m} = a\cdot r^{n-m-1},\ \text{and } a_n = a\cdot r^{n-1}$$

so that $a_{m+1}\cdot a_{n-m} = a\cdot r^m\cdot a\cdot r^{n-m-1} = a^2\cdot r^{n-1}$ and $a_1\cdot a_n = a\cdot(a\cdot r^{n-1}) = a^2\cdot r^{n-1}$. Therefore, $a_{m+1}\cdot a_{n-m} = a_1\cdot a_n$.

(iii) We could prove this part by using mathematical induction. Instead, an alternative proof can be offered by making use of the fact that the sum of the first n natural integers is equal to $\frac{n\cdot(n+1)}{2}$; $1+2+\dots+n = \frac{n(n+1)}{2}$; we have already seen this in Example 1(i). (Go back and review this example if necessary; $1, 2, \dots, n$ are the consecutive first n terms of the infinite arithmetic progression with first term 1 and difference 1). This fact can be applied neatly here:

$$\begin{aligned} a_1 \cdot a_2 \ldots a_i \ldots a_n &= \text{(by Theorem 1(i))} \\ &= a \cdot (a \cdot r) \ldots (a \cdot r^{i-1}) \ldots (a \cdot r^{n-1}) \\ &= \underbrace{(a \cdot a \ldots a)}_{n \text{ times}} \cdot r^{[1+2+\ldots+(i-1)+\ldots+(n-1)]} \end{aligned}$$

The sum $[1 + 2 + \ldots + (i-1) + \ldots + (n-1)]$ is simply the sum of the first $(n-1)$ natural numbers, if $n \geq 2$. According to Example 1(i) we have,

$$1 + 2 + \ldots + (i-1) + \ldots + (n-1) = \frac{(n-1) \cdot [(n-1)+1]}{2} = \frac{(n-1) \cdot n}{2}.$$

Hence, $a_1 \cdot a_2 \ldots a_i \ldots a_n = \underbrace{(a \cdot a \ldots a)}_{n \text{ times}} \cdot r^{[1+2+\ldots+(i-1)+\ldots+(n-1)]} = a^n \cdot r^{\frac{(n-1)n}{2}} \Rightarrow (a_1 \cdot a_2 \ldots a_i \ldots a_n)^2 = a^{2n} \cdot r^{(n-1) \cdot n}$. On the other hand, $(a_1 \cdot a_n)^n = [a \cdot (a \cdot r^{n-1})]^n = [a^2 \cdot r^{n-1}]^n = a^{2n} \cdot r^{n(n-1)} = (a_1 \cdot a_2 \ldots a_i \ldots a_n)^2$; we are done. ■

Definition 7: Let $a_1, a_2, \ldots, a_n$ be positive real numbers. The positive real number $\sqrt[n]{a_1 a_2 \ldots a_n}$ is called the **geometric mean** of the numbers $a_1, a_2, \ldots a_n$.

We saw in Theorem 3 that if three real numbers a, b, c are consective terms of an arithmetic progression, the middle term b must be equal to the arithmetic mean of a and c. The same is true for the geometric mean if the positive reals a, b, c are consecutive terms in a geometric progression. We have the following theorem.

Theorem 7: If the positive real numbers a, b, c, are consecutive terms of a geometric progression, then the geometric mean of a and c must equal b. Also, any geometric progression containing a, b, c as consecutive terms, must have the same ratio r, namely $r = \frac{b}{a} = \frac{c}{b}$. Moreover, the condition $b^2 = ac$ is the necessary and sufficient condition for the three reals a, b, c to be consecutive terms in a geometric progression.

Proof: If a, b, c are consecutive terms in a geometric progression, then $b = ar$ and $c = b \cdot r$; and since both a and b are positive and thus nonzero, we must have $r = \frac{b}{a} = cb \Rightarrow b^2 = ac \Rightarrow b = \sqrt{ac}$ which proves that b is the geometric mean of a and c. Conversely, if the condition $b^2 = ac$ is satisfied (which is equivalent to $b = \sqrt{ac}$, since b is positive), then since a and b are positive and thus nonzero, infer that $\frac{b}{a} = \frac{c}{b}$; thus if we set $r = \frac{b}{a} = \frac{c}{b}$, it is now clear that a, b, c are consecutive terms of a geometric progression whose ratio is uniquely determined in terms of the given reals a, b, c and any other geometric progression containing a, b, c as consecutive terms must have the same ratio r. ■

For the theorem to follow we will need what is called Bernoulli's Inequality: for every real number $a \geq -1$, and every natural number n,

$$(a+1)^n \geq 1 + na.$$

Let $a \geq -1$; Bernoulli's Inequality can be easily proved by induction: clearly the statement holds true for $n = 1$ since $1 + a \geq 1 + a$ (the equal sign holds). Assume the statement to hold true for some $n = k \geq 1$: $(a+1)^k \geq 1 + ka$; since $a + 1 \geq 0$ we can multiply both sides of this inequality by $a + 1$ without affecting its orientation:

$$\begin{aligned}(a+1)^{k+1} &\geq (a+1)(1+ka) \Rightarrow \\ (a+1)^{k+1} &\geq a + ka^2 + 1 + ka; \\ (a+1)^{k+1} &\geq 1 + (k+1)a + ka^2 \geq 1 + (k+1)a,\end{aligned}$$

since $ka^2 \geq 0$ (because $a^2 \geq 0$ and k is a natural number). The induction process is complete.

Theorem 8: (i) If $r > 1$ and M is a real number, then there exists a natural number N such that $r^n > M$, for every natural number n. For parts (ii), (iii), (iv) and (v), let $a_1 = a, a_2, \ldots, a_n, \ldots$, be an infinite geometric progression with first term a and ratio r.

(ii) Suppose $r > 1$ and $a > 0$. If M is a real number, then there is a natural number N such that $a_n > M$, for every natural number $n \geq N$.

(iii) Suppose $r > 1$ and $a < 0$. If M is a real number, then there is a natural number N such that $a_n < M$, for every natural number $n \geq N$.

(iv) Suppose $|r| < 1$, and $r \neq 0$. If $\epsilon > 0$ is a positive real number, then there is a natural number N such that $|a_n| < \epsilon$, for every natural number $n \geq N$.

(v) Suppose $|r| < 1$ and let $S_n = a_1 + a_2 + \ldots + a_n$. If $\epsilon > 0$ is a positive real number, then there exists a natural number N such that $\left|S_n - \frac{a}{1-r}\right| < \epsilon$, for every natural number $n \geq N$.

Proof: (i) We can write $r = (r-1) + 1$; let $a = r - 1$, since $r > 1$, a must be a positive real. According to the Bernouilli Inequality we have, $r^n = (a+1)^n \geq 1 + na$; thus, in order to ensure that $r^n > M$, it is sufficient to have $1 + na > M \Leftrightarrow na > M - 1 \Leftrightarrow n > \dfrac{M-1}{a}$ (the last step is justified since $a > 0$). Now, if $\left[\!\left[\dfrac{|M-1|}{a}\right]\!\right]$ stands for the integer part of the positive real number $\dfrac{|M-1|}{a}$ we have by definition, $\left[\!\left[\dfrac{|M-1|}{a}\right]\!\right] \leq \dfrac{|M-1|}{a} < \left[\!\left[\dfrac{|M-1|}{a}\right]\!\right] + 1$. Thus, if we choose $N = \left[\!\left[\dfrac{|M-1|}{a}\right]\!\right] + 1$, it is clear that $N > \dfrac{|M-1|}{a} \geq \dfrac{M-1}{a}$ so that for every natural number $n \geq N$, we will have $n > \dfrac{M-1}{a}$, and subsequently we will have (since $a > 0$), $na > M - 1 \Rightarrow na + 1 > M$. But $(1+a)^n \geq 1 + na$ (Bernoulli), so that $r^n = (1+a)^n \geq 1 + na > M$; $r^n > M$, for every $n \geq N$. We are done. ■

(ii) By part (i), there exists a natural number N such that $r^n > \frac{M}{a} \cdot r$, for every natural number $n \geq N$ (apply part (i) with $\frac{M}{a} \cdot r$ replacing M). Since both r and a are positive, so is $\frac{a}{r}$; multiplying both sides of the above inequality by $\frac{a}{r}$ we obtain $\frac{a}{r} \cdot r^n > \frac{a}{r} \cdot \frac{M}{a} \cdot r \Rightarrow a \cdot r^{n-1} > M$. But $a \cdot r^{n-1}$ is the nth term a_n of the geometric progression. Hence $a_n > M$, for every natural number $n \geq N$. ■

(iii) Apply part (ii) to the opposite geometric progression: $-a_1, -a_2, \ldots,$

$-a_n, \ldots$, where a_n is the nth term of the original geometric progression (that has $a_1 = a < 0$ and $r > 1$, it is also easy to see that the opposite sequence is itself a geometric progression with the same ratio $r > 1$ and opposite first term $-a$). According to part (ii) there exists a natural number N such that $-a_n > -M$, for every natural number $n \geq N$. Thus $-(-a_n) < -(-M) \Rightarrow a_n < M$, for every $n \geq N$. ■

(iv) Since $|r| < 1$, assuming $r \neq 0$ it follows that $\frac{1}{|r|} > 1$. Let $M = \frac{|a|}{\epsilon \cdot |r|}$. According to part (i), there exists a natural number N such that $\left(\frac{1}{|r|}\right)^n > M = \frac{|a|}{\epsilon|r|}$ (just apply part (i) with r replaced by $\frac{1}{|r|}$ and M replaced by $\frac{|a|}{\epsilon \cdot |r|}$ for every natural number $n \geq N$. Thus $\frac{1}{|r|^n} > \frac{|a|}{\epsilon \cdot |r|}$; multiply both sides by $|r|^n \cdot \epsilon$ to obtain $\frac{|r|^n \cdot \epsilon}{|r|^n} > \frac{|a| \cdot |r|^n \cdot \epsilon}{\epsilon \cdot |r|} \Rightarrow |a| \cdot |r|^{n-1} < \epsilon$; but $|a| \cdot |r|^{n-1} = |ar^{-n}| = |a_n|$, the absolute value of the nth term of the geometric progression; $|a_n| < \epsilon$, for every natural number $n \geq N$. Finally if $r = 0$, then $a_n = 0$ for $n \geq 2$, and so $|a_n| = 0 < \epsilon$ for all $n \geq 2$. ■

(v) By Theorem 5(ii) we know that,

$$S_n = a_1 + a_2 + \ldots + a_n = a + ar + \ldots + ar^{n-1} = \frac{a(r^n - 1)}{r - 1}$$

We have $S_n - \frac{a}{1-r} = \frac{a(r^n-1)}{r-1} + \frac{a}{r-1} = \frac{ar^n - a + a}{r-1} = \frac{ar^n}{r-1}$. Consequently, $\left|S_n - \frac{a}{1-r}\right| = \left|\frac{ar^n}{r-1}\right| = |r|^n \cdot \left|\frac{a}{r-1}\right|$. Assume $r \neq 0$. Since $|r| < 1$, we can apply the already proven part (iv), using the positive real number $\frac{\epsilon \cdot |r-1|}{|r|}$ in place of ϵ: there exists a natural number N such that $|a_n| < \frac{\epsilon \cdot |r-1|}{|r|}$, for every natural number $n \geq N$. But $a_n = a \cdot r^{n-1}$ so that,

$$|a| \cdot |r|^{n-1} < \frac{\epsilon \cdot |r-1|}{|r|} \Rightarrow$$

$\Rightarrow$ (multiplying both sides by $|r|$) $|a||r|^n < \epsilon \cdot |r - 1| \Rightarrow$
$\Rightarrow$ (dividing both sides by $|r - 1|$) $\frac{|a|\,|r|^n}{|r-1|} < \epsilon$.

And since $\left|S_n - \frac{a}{1-r}\right| = |r|^n \cdot \left|\frac{a}{r-1}\right|$ we conclude that, $\left|S_n - \frac{a}{1-r}\right| < \epsilon$. The proof will be complete by considering the case $r = 0$: if $r = 0$, then $a_n = 0$, for all $n \geq 2$. And thus $S_n = \frac{a(r^n-1)}{r-1} = \frac{-a}{-1} = a$, for all natural numbers n. Hence, $\left|S_n - \frac{a}{1-r}\right| = \left|a - \frac{a}{1}\right| = |a - a| = 0 < \epsilon$, for all natural numbers n. ■

Remark 6: As the student familiar with, will recognize, part (iv) of Theorem 8 establishes the fact that the limit value of the sequence whose nth term is $a_n = a \cdot r^{n-1}$ and under the assumption $|r| < 1$, is equal to zero. In the language of calculus, when $|r| < 1$, the geometric progression approaches zero. Also, part (v), establishes the sequence of (partial) sums whose nth term is S_n, approaches the real number $\frac{a}{1-r}$, under the assumption $|r| < 1$. In the language of calculus we say that the infinite series $a + ar + ar^2 + \ldots + ar^{n-1} + \ldots$ converges to $\frac{a}{1-r}$.

7.3 Mixed Progressions

The reader of this book who has also studied calculus, may have come across the sum,

$$1 + 2x + 3x^2 + \ldots + (n+1)x^n.$$

There are $(n+1)$ terms in this sum; the ith term is equal to $i \cdot x^{i-1}$, where i is a natural number between 1 and $(n+1)$. Note that if $a_i = i \cdot x^{i-1}$, $b_i = i$, and $c_i = x^{i-1}$, we have $a_i = b_i \cdot c_i$; what is more, b_i is the ith term of an arithmetic progression (that has both first term and difference equal to 1); and c_i is the ith term of a geometric progression (with first term $c = 1$ and ratio $r = x$). Thus the term a_i is the product of the ith term of an arithmetic progression with the ith term of a geometric progression; then we say that a_i is the ith term of a mixed progression. We have the following definition.

Definition 8: Let $b_1, b_2, \ldots, b_n, \ldots$ be an arithmetic progression; and $c_1, c_2, \ldots, c_n, \ldots$ be a geometric progression. The sequence $a_1, a_2, \ldots, a_n, \ldots$, where $a_n = b_n \cdot c_n$, for every natural number n, is called **a mixed progression**. (Of course, if both the arithmetic and geometric progressions are finite sequences with the same number of terms, so it will be with the mixed progression.)

Back to our example. With a little bit of ingenuity, we can compute this sum; that is, find a closed form expression for it, in terms of x and n. Indeed, we can write the given sum in the form,

$$\Big(\underbrace{1 + x + x^2 + \ldots + x^{n-1} + x^n}_{(n+1) \text{ terms}}\Big) + \Big(\underbrace{x + x^2 + \ldots + x^{n-1} + x^n}_{n \text{ terms}}\Big)$$

$$+\Big(\underbrace{x^2 + x^3 + \ldots + x^{n-1} + x^n}_{(n-1) \text{ terms}}\Big) + \ldots + \Big(\underbrace{x^{n-1} + x^n}_{2 \text{ terms}}\Big) + \underbrace{x^n}_{\text{one term}}$$

In other words we have written the original sum $1 + 2x + 3x^2 + \ldots + (n+1)x^n$ as a sum of $(n+1)$ sums, each containing one term less than the previous one.

Now according to Theorem 5(ii),

$$1 + x + x^2 + \ldots + x^{n-1} + x^n = \frac{x^{n+1} - 1}{x - 1} \text{ (assuming } x \neq 1 \text{)},$$

since this is the sum of the first $(n+1)$ terms of a geometric progression with first term 1 and ratio x.

Next, consider

$$\begin{aligned} x + x^2 + \ldots + x^{n-1} + x^n &= (1 + x + x^2 + \ldots + n^{n-1} + x^n) - 1 \\ &= \frac{x^{n+1} - 1}{x - 1} - \left(\tfrac{x^i - 1}{x - 1}\right) \end{aligned}$$

Continuing this way we have,

$$\begin{aligned} x^2+\ldots+x^{n-1}+x^n &= (1+x+x^2+\ldots+x^n-1+x^n)-(x+1) \\ &= \frac{x^{n+1}-1}{x-1}-\left(\frac{x^2-1}{x-1}\right). \end{aligned}$$

On the ith level,

$$\begin{aligned} x^i+\ldots+x^{n-1}+x^n &= (1+x+\ldots+x^{i-1}+x^i+\ldots+x^{n-1}+x^n)- \\ -(1+x+\ldots+x^{i-1}) &= \frac{x^{n+1}-1}{x-1}-\left(\frac{x^i-1}{x-1}\right). \end{aligned}$$

Let us list all of these sums:

$$\begin{aligned}
&(1) & 1+x+x^2+\ldots+x^{n-1}+x^n &= \frac{x^{n+1}-1}{x-1} \\
&(2) & x+x^2+\ldots+x^{n-1}+x^n &= \frac{x^{n+1}-1}{x-1}-\left(\frac{x-1}{x-1}\right) \\
&(3) & x^2+\ldots+x^{n-1}+x^n &= \frac{x^{n+1}-1}{x-1}-\left(\frac{x^2-1}{x-1}\right) \\
&\vdots \\
&(i) & x^i+\ldots+x^{n-1}+x^n &= \frac{x^{n+1}-1}{x-1}-\left(\frac{x^i-1}{x-1}\right) \\
&\vdots \\
&(n) & x^{n-1}+x^n &= \frac{x^{n+1}-1}{x-1}-\left(\frac{x^{n-1}-1}{x-1}\right) \\
&(n+1) & x^n &= \frac{x^{n+1}-1}{x-1}-\left(\frac{x^n-1}{x-1}\right),
\end{aligned}$$

with $x \neq 1$.

If we add the $(n+1)$ equations or identities (they hold true for all reals except for $x=1$), the sum of the $(n+1)$ left-hand sides is simply the original sum $1+2x+3x^2+\ldots+nx^{n-1}+(n+1)x$? Thus, if we add up the $(n+1)$ equations memberwise we obtain,

$$1+2x+3x^2+\ldots+nx^{n-1}+(n+1)x^n$$

$$=\quad (n+1)\cdot\left(\tfrac{x^{n+1}-1}{x-1}\right)+\frac{n-(x+x^2+\ldots+x^i+\ldots+x^{n-1}+x^n)}{x-1}$$

$$=\quad (n+1)\cdot\left(\frac{x^{n+1}-1}{x-1}\right)+\frac{(n+1)-(1+x+x^2+\ldots+x^n)}{x-1}$$

$$\Rightarrow\quad 1+2x+3x^2+\ldots+nx^{n-1}+(n+1)x^n$$

$$=\quad (n+1)\cdot\left(\frac{x^{n+1}-1}{x-1}\right)+\frac{(n+1)-\left(\frac{x^{n+1}-1}{x-1}\right)}{x-1};$$

$$1+2x+3x^2+\ldots+nx^{n-1}+(n+1)x^n$$

$$=\quad (n+1)\cdot\left(\frac{x^{n+1}-1}{x-1}\right)+\frac{(n+1)(x-1)-(x^{n+1}-1)}{(x-1)^2};$$

$$1+2x+3x^2+\ldots+nx^{n-1}+(n+1)x^n$$

$$=\quad \frac{(n+1)(x^{n+1}-1)(x-1)+(n+1)(x-1)-(x^{n+1}-1)}{(x-1)^2};$$

$$1+2x+3x^2+\ldots+nx^{n-1}+(n+1)x^n$$

$$=\quad \frac{(n+1)(x-1)\cdot[(x^{n+1}-1)+1]-(x^{n+1}-1)}{(x-1)^2};$$

$$1+2x+3x^2+\ldots+nx^{n-1}+(n+1)x^n$$

$$=\quad \frac{(n+1)(x-1)\cdot x^{n+1}-(x^{n+1}-1)}{(x-1)^2};$$

$$1+2x+3x^2+\ldots+nx^{n-1}+(n+1)x^n$$

$$=\quad \frac{(n+1)x^{n+2}-(n+1)x^{n+1}-x^{n+1}+1}{(x-1)^2};$$

$$1+2x+3x^2+\ldots+nx^{n-1}+(n+1)x^n$$

$$=\quad \boxed{\frac{(n+1)x^{n+2}-(n+2)x^{n+1}+1}{(x-1)^2}}$$

for every natural number n.

geometric progression with ratio r. Hence, we obtain $S_{k+1} = \frac{a_{k+1}\cdot r - a_1}{r-1} + \frac{d\cdot r\cdot(c-c_{k+1})}{(r-1)^2}$; the induction is complete.

The example with which we opened this section is one of a mixed progression. We dealt with the sum $1+2x+3x^2+\ldots+nx^{n-1}+(n+1)x^n$. This is the sum of the first $(n+1)$ terms of a mixed progression whose nth term is $a_n = n\cdot x^{n-1}$; in the notation of Theorem 9, $b_n = n$, $d = 1$, $c_n = x^{n-1}$, and $r = x$ (we assume $x \neq 1$).

According to Theorem 9(iii)

$$\begin{aligned} S_n &= 1+2x+3x^2+\ldots+nx^{n-1} = \frac{(nx^{n-1})\cdot x-1}{x-1} + \frac{x\cdot(1-x^{n-1})}{(x-1)^2} \\ &= \frac{nx^n-1}{x-1} + \frac{x-x^n}{(x-1)^2} - \frac{(nx^n-1)(x-1)}{(x-1)^2} + \frac{x-x^n}{(x-1)^2} \\ &= \frac{nx^{n+1}-nx^n-x+1+x-x^n}{(x-1)^2} = \frac{nx^{n+1}-(n+1)x^n+1}{(x-1)^2}; \end{aligned}$$

Thus, if we replace n by $(n+1)$ we obtain, $S_{n+1} = 1+2x+3x^2+\ldots+nx^{n-1}+(n+1)x^n = \frac{(n+1)x^{n+2}-(n+2)x^{n+1}+1}{(x-1)^2}$, and this is the formula we obtained earlier.

Definition 9: Let $a_1, \ldots, a_n$ be nonzero real numbers. The real number $\frac{n}{\frac{1}{a_1}+\ldots+\frac{1}{a_n}}$, is called the **harmonic mean** of the real numbers $a_1, \ldots, a_n$.

Remark 7: Note that since $\frac{n}{\frac{1}{a_1}+\ldots+\frac{1}{a_n}} = \frac{1}{(\frac{1}{a_1}+\ldots+\frac{1}{a_n})/n}$, the harmonic mean of the reals $a_1, \ldots, a_n$, is really the reciprocal of the mean of the reciprocal real numbers $\frac{1}{a_1}, \ldots, \frac{1}{a_n}$.

We close this section by establishing an interesting, significant and deep inequality, that has many applications in mathematics and has been used to prove a number of other theorems. Given n positive real numbers $a_1, \ldots, a_n$ one can always designate three positive reals to the given set $\{a_1, \ldots, a_n\}$: the arithmetic mean denoted by A.M., the geometric mean denoted by G.M., and the harmonic mean H.M. The arithmetic-geometric-harmonic mean inequality asserts that A.M. $\geq$ G.M. $\geq$ H.M. (To the reader: Do an experiment; pick a set of three positive reals; then a set of four positive reals; for each set compute the A.M., G.M., and H.M. values; you will see that the inequality holds; if you are in disbelief do it again with another sample of positive real numbers.)

The proof we will offer for the arithmetic-geometric-harmonic inequality is indeed short. To do so, we need a preliminary result: we have already proved (in the proof of Theorem 5(i)) the identity $r^n - 1 = (r-1)(r^{n-1}+r^{n-2}+\ldots+r+1)$, which holds true for all real numbers r and all natural numbers n. Moreover, if $r \neq 1$, we have

$$\frac{r^{n-1}}{r-1} = r^{n-1}+r^{n-2}+\ldots+r+1$$

If we set $r = \frac{b}{a}$, with $b \neq a$, in the above equation and we multiply both sides by a^n we obtain,

$$\frac{b^n - a^n}{b-a} = b^{n-1} + b^{n-2}\cdot a + b^{n-3}\cdot a^2 + \ldots + b^2\cdot a^{n=-3} + b\cdot a^{n-2} + a^{n-1}$$

Now, if $b > a > 0$ and in the above equation we replace b by a, the resulting right-hand side will be smaller. In other words, in view of $b > a > 0$ we have,

$$
\begin{array}{c}
(1)\\ (2)\\ (3)\\ \vdots\\ (n-2)\\ (n-1)\\ (n)
\end{array}
\left\{
\begin{array}{l}
b^{n-1} > a^{n-1}\\
b^{n-2}\cdot a > a^{n-2}\cdot a^1 = a^{n-1}\\
b^{n-3}\cdot a^2 > a^{n-3}\cdot a^2 = a^{n-1}\\
\vdots\\
b^2\cdot a^{n-3}\cdot a^2 > a^2\cdot a^{n-3}j = a^{n-1}\\
b\cdot a^{n-2} > a\cdot a^{n-2} = a^{n-1}\\
a^{n-1} = a^{n-1}
\end{array}
\right\}
\begin{array}{ll}
\Rightarrow & \text{add memberwise}\\
 & \\
 & b^{n-1} + b^{n-2}\cdot a + b^{n-3}\cdot a^2 + \dots\\
+ & b^2 a^{n-3} + b\cdot a^{n-2} + a^{n-1}\\
> & n\cdot a^{n-1}
\end{array}
$$

Hence, the identity above, for $b > a > 0$, implies the inequality $\frac{b^n - a^n}{b-a} > na^{n-1}$; multiplying both sides by $b - a > 0$ we arrive at

$$b^n - a^n > (b-a)na^{n-1}$$

$$\Rightarrow \quad b^n > nba^{n-1} - na^n + a^n;$$

$$b^n > nba^{n-1} - (n-1)a^n.$$

Finally, by replacing n by $(n+1)$ in the last inequality we obtain,

$b^{n+1} > (n+1)ba^n - na^{n+1}$,	for every natural number n and any real numbers such that $b > a > 0$

We are now ready to prove the last theorem of this chapter.

Theorem 10: Let n be a natural number and $a_1, \dots, a_n$ positive real numbers. Then,

$$\underbrace{\frac{a_1 \dots + a_n}{n}}_{\text{A.M.}} \geq \underbrace{\sqrt[n]{a_1 \dots a_n}}_{\text{G.M.}} \geq \underbrace{\frac{n}{\frac{1}{a_1} + \frac{1}{a_2} + \dots + \frac{1}{a_n}}}_{\text{H.M.}}$$

bf Proof: Before we proceed with the proof, we mention here that if one equal sign holds the other must also hold, and that can only happen when all n numbers $a_1, \dots, a_n$ are equal. We will not prove this here, but the reader may want to verify this in the cases $n = 2$ and $n = 3$. We will proceed by using mathematical induction to first prove that, $\frac{a_1 + \dots + a_n}{n} \geq \sqrt[n]{a_1 \dots a_n}$, for every natural number n and all positive reals $a_1, \dots, a_n$. Even though this trivially holds true for $n = 1$, we will use as our starting or basc valuc, $n = 2$. So wc first provc that $\frac{a_1 + a_2}{2} \geq \sqrt{a_1 a_2}$ holds truc for any two positive reals. Since a_1 and a_2 are both positive, the square roots $\sqrt{a_1}$ and $\sqrt{a_2}$ are both positive real numbers and $a_1 = (\sqrt{a_1})^2$, $a_2 = (\sqrt{a_2})^2$. Clearly,

$$(\sqrt{a_1} - \sqrt{a_2})^2 \geq 0$$

$$\Rightarrow \quad (\sqrt{a_1})^2 - 2(\sqrt{a_1})(\sqrt{a_2}) + (\sqrt{a_2})^2 \geq 0$$

$$\Rightarrow \quad a_1 - 2\sqrt{a_1 a_2} + a_2 > 0$$

$$\Rightarrow \quad a_1 + a_2 \geq 2 \cdot \sqrt{a_1 a_2}$$

$$\Rightarrow \quad \frac{a_1 + a_2}{2} \geq \sqrt{a_1 a_2},$$

so the statement holds true for $n = 2$.

The Inductive Step: Assume the statement to hold true for some natural number $n = k \geq 2$; and show that this assumption implies that the statement must also hold true for $n = k + 1$. So assume,

$$\frac{a_1 + \ldots + a_k}{k} \geq \sqrt[k]{a_1 \ldots a_k}$$

$$\Rightarrow \quad a_1 + \ldots + a_k \geq k \cdot \sqrt[k]{a_1 \ldots a_k}$$

Now we apply the inequality we proved earlier:

$$b^{k+1} > (k+1) \cdot b \cdot a^k - k \cdot a^{k+1};$$

If we take $b = \sqrt[k+1]{a_{k+1}}$, where a_{k+1} is a positive real and $a = \sqrt[k(k+1)]{a_1 \ldots a_k}$ we now have,

$$\left(\sqrt[k+1]{a_{k+1}}\right)^{k+1} > (k+1) \cdot \sqrt[k+1]{a_{k+1}} \cdot \left(\sqrt[k(k+1)]{a_1 \ldots a_k}\right)^k - k \cdot \left(\sqrt[k(k+1)]{a_1 \ldots a_k}\right)^{k+1}$$

$$\Rightarrow \quad a_{k+1} > (k+1) \cdot \sqrt[k+1]{a_{k+1}} \cdot \sqrt[k+1]{a_1 \ldots a_k} - k \cdot \sqrt[k]{a_1 \ldots a_k}$$

$$\Rightarrow \quad a_{k+1} + k \cdot \sqrt[k]{a_1 \ldots a_k} > (k+1) \cdot \sqrt[k+1]{a_1 \ldots a_k \cdot a_{k+1}}$$

But from the inductive step we know that $a_1 + \ldots + a_k > k \cdot \sqrt[k]{a_1 \ldots a_k}$; hence we have,

$$a_{k+1} + (a_1 + \ldots + a_k) \geq a_{k+1} + k \cdot \sqrt[k]{a_1 \ldots a_k} \geq (k+1) \cdot \sqrt[k+1]{a_1 \ldots a_k \cdot a_{k+1}}$$

$$\Rightarrow \quad a_1 + \ldots + a_k + a_{k+1} \geq (k+1) \sqrt[k+1]{a_1 \ldots a_k \cdot a_{k+1}},$$

and the induction is complete.

Now that we have established the arithmetic-geometric mean inequality, we prove the geometric-harmonic inequality. Indeed, if n is a natural number and $a_1, \ldots, a_n$ are positive reals, then so are the real numbers $\frac{1}{a_1}, \ldots, \frac{1}{a_n}$. By applying the already proven arithmetic-geometric mean inequality we infer that,

$$\frac{\frac{1}{a_1} + \ldots + \frac{1}{a_n}}{n} \geq \sqrt[n]{\frac{1}{a_1} \cdots \frac{1}{a_n}}$$

Multiplying both sides by the product $\left(\frac{n}{\frac{1}{a_1}+\ldots+\frac{1}{a_n}}\right) \cdot \sqrt[n]{a_1 \ldots a_n}$, we arrive at the desired result:

$$\sqrt[n]{a_1 \ldots a_n} \geq \frac{n}{\frac{1}{a_1} + \ldots + \frac{1}{a_n}}.$$

This concludes the proof of the theorem. ∎

7.4 Practice Problems

P1. Determine the difference of each arithmetic progression whose first term is $\frac{1}{5}$; and with subsequent terms (but not necessarily consecutive) the rational numbers $\frac{1}{4}, \frac{1}{3}, \frac{1}{2}$.

Solution: Let k, m, n be natural numbers with $k < m < n$ such that $a_k = \frac{1}{4}$, $a_m = \frac{1}{3}$, and $a_n = \frac{1}{2}$. And, of course, $a_1 = \frac{1}{5}$ is the first term; $a_1 = \frac{1}{5}, \ldots, a_k = \frac{1}{4}, \ldots, a_m = \frac{1}{3}, \ldots, a_n = \frac{1}{2}, \ldots$. By Theorem 1(i) we must have,

$$\left.\begin{array}{l} \frac{1}{4} = a_k = \frac{1}{5} + (k-1)d \\ \\ \frac{1}{3} = a_m = \frac{1}{5} + (m-1)d \\ \\ \frac{1}{2} = a_n + \frac{1}{5} + (n-1)d \end{array}\right\}; \text{ where } d \text{ is the difference of the arithmetic progression.}$$

Obviously, $d \neq 0$; the three equations yield,

$$\left.\begin{array}{ll} (k-1)d = \frac{1}{4} - \frac{1}{5} = \frac{1}{20} & (1) \\ \\ (m-1)d = \frac{1}{3} - \frac{1}{5} = \frac{2}{15} & (2) \\ \\ (n-1)d = \frac{1}{2} - \frac{1}{5} = \frac{3}{10} & (3) \end{array}\right\} \text{ Also, it is clear that } 1 < k; \text{ so that } 1 < k < m < n.$$

Dividing (1) with (2) memberwise gives

$$\frac{k-1}{m-1} = \frac{3}{8}, \Rightarrow 8(k-1) = 3(m-1) \tag{4}$$

Dividing (2) with (3) memberwise implies

$$\frac{m-1}{n-1} = \frac{4}{9} \Rightarrow 9(m-1) = 4(n-1) \tag{5}$$

Dividing (1) with (3) memberwise produces

$$\frac{k-1}{n-1} = \frac{1}{6} \Rightarrow 6(k-1) = n-1 \tag{6}$$

According to Equation (4), 3 must be a divisor of $k-1$ and 8 must be a divisor of $m-1$; if we put $k-1 = 3t$; $k = 3t+1$, where t is a nantural number (since $k > 1$), then (4) implies $8t = m-1 \Rightarrow m = 8t+1$

Going to equation (5) and substituting for $m-1 = 8t$, we obtain,

$$18t = n - 1 \Rightarrow n = 18t + 1.$$

Checking equation (6) we see that $6(3t) = 18t$, which is true for all nonnegative integer values of t. In conclusion we have the following formulas for k, m, and n:

$$k = 3t + 1,\ m = 8t + 1,\ n = 18t + 1;\ t \in \mathbb{N};\ t = 1, 2, \ldots$$

We can now calculate d in terms of t from any of the equations (1), (2), or (3): From (1), $(k-1)d = \frac{1}{20} \Rightarrow 3t \cdot d = \frac{1}{20} \Rightarrow \boxed{d = \frac{1}{60t}}.$ We see that this problem has infinitely many solutions: there are infinitely many (infinite) arithmetic progressions that satisfy the conditions of the problem. For each positive integer of value of t, a new such arithmetic progression is determined. For example, for $t = 1$ we have $d = \frac{1}{60}$, $k = 4$, $m = 9$, $n = 19$. We have the progression,

$$a_1 = \frac{1}{5}, \ldots, a_4 = \frac{1}{4}, \ldots\ldots, a_9 = \frac{1}{3}, \ldots\ldots, a_{19} = \frac{1}{2}, \ldots$$

P2. Determine the arithmetic progressions (by finding the first term a_1 and difference d) whose first term is $a_1 = 5$, whose difference d is an integer, and which contains the numbers 57 and 113 among their terms.

Solution: We have $a_1 = 5$, $a_m = 57$, $a_n = 113$ for some natural numbers m and n with $1 < m < n$. We have $57 = 5 + (m-1)d$ and $113 = 5 + (n-1)d$; $(m-1)d = 52$ and $(n-1)d = 108$; the last two conditions say that d is a common divisor of 52 and 108; thus $\boxed{d = 1, 2, \text{ or } 4}$ are the only possible values. A quick computation shows that for $d = 1$, we have $m = 53$, and $n = 109$; for $d = 2$, we have $m = 27$ and $n = 55$; and for $d = 4$, $m = 14$ and $n = 28$. In conclusion there are exactly three arithmetic progressions satisfying the conditions of this exercise; they have first term $a_1 = 5$ and their differences d are $d = 1, 2$, and 4 respectively.

P3. Find the sum of all three-digit natural numbers k which are such that the remainder of the divisions of k with 18 and of k with 30, is equal to 7.

Solution: Any natural number divisible by both 18 and 30, must be divisible by their least common multiple which is 90. Thus if k is any natural number satisfying the condition of the exercise, then the number $k - 7$ must be divisible by both 18 and 90 and therefore $k - 7$ must be divisible by 90; so that $k - 7 = 90t$, for some nonnegative integer t; thus the three-digit numbers of the form $k = 90t + 7$ are precisely the numbers we seek to find. These numbers are terms in an infinite arithmetic progression whose first term is $a_1 = 7$ and whose difference is $d = 70:\ a_1 = 7,\ a_2 = 7 + 90,\ a_3 = 7 + 2 \cdot (90), \ldots, a_{t+1} = 7 + 90t, \ldots$.

A quick check shows that the first such three-digit number in the above arithmetic progression is $a_3 = 7 + 90(2) = 187$ (obtained by setting $t = 2$) and the last such three-digit number in the above progression is $a_{12} = 7 + 90(11) = 997$ (obtained by putting $t = 11$ in the formula $a_{t+1} = 7 + 90t$). Thus, we seek to find the sum, $a_3 + a_4 + \ldots + a_{11} + a_{12}$. We can use either of the two formulas developed in Example 2 (after example 1 which in turn is located below the proof of Theorem 2).

Since we know the first and last terms of the sum at hand, namely a_3, it is easier to use the first formula in Example 2:

$$a_m + a_{m+1} + \ldots + a_{n-1} + a_n = \frac{(n - m + 1)(a_m + a_n)}{2}$$

In our case $m = 3$, $n = 12$, $a_m = a_3 = 187$, and $a_n = a_{12} = 997$. Thus

$$\begin{aligned} a_3 + a_4 + \ldots + a_{11} + a_{12} &= \frac{(12-3+1)\cdot(187+997)}{2} \\ &= \frac{10}{2}\cdot(1184) = 5\cdot(1184) = 5920. \end{aligned}$$

P4. Let $a_1, a_2, \ldots, a_n, \ldots,$ be an arithmetic progression with first term a_1 and positive difference d; and M a natural number, such that $a_1 \leq M$. Show that the number of terms of the arithmetic progression that do not exceed M, is equal to $[[\frac{M-a_1}{d}]] + 1$, where $[[\frac{M-a_1}{d}]]$ stands for the integer part of the real number $\frac{M-a_1}{d}$.

Solution: If, among the terms of the arithmetic progression, a_n is the largest term which does not exceed M, then $a_n \leq M$ and $a_\ell > M$, for all natural number ℓ greater than n; $\ell = n+1, n+2, \ldots$. But $a_n = a_1 + (n-1)d$; so that $a_1 + (n-1)d \leq M \Rightarrow (n-1)d \leq M - a_1 \Rightarrow n - 1 \leq \frac{M-a_1}{d}$ since $d > 0$. Since, by definition, $[[\frac{M-a_1}{d}]]$ is the greatest integer not exceeding $\frac{M-a_1}{d}$ and since $n-1$ does not exceed $\frac{M-a_1}{d}$, we conclude that $n - 1 \leq [[\frac{M-a_1}{d}]] \Rightarrow n \leq [[\frac{M-a_1}{d}]] + 1$. But n is a natural number, that is, a positive integer, and so must be the integer $N = [[\frac{M-a_1}{d}]] + 1$ Since a_n was assumed to be the largest term such that $a_n \leq M$, it follows that n must equal N; because the term a_N is actually the largest term not exceeding M (note that if $n < N$, then $a_n < a_N$, since the progression is increasing in view of the fact that $d > 0$). Indeed, if $N = [[\frac{M-a_1}{d}]] + 1$, then by the definition of the integer part of a real number we must have $N - 1 \leq \frac{M-a_1}{d} < N$. Multiplying by $d > 0$ yields $d(N-1) \leq M - a_1 \Rightarrow a_1 + d(N-1) \leq M \Rightarrow a_N \leq M$.

In conclusion we see that the terms $a_1, \ldots, a_N$ are precisely the terms not exceeding $[[\frac{M-a_1}{d}]] + 1$; therefore there are exactly $[[\frac{M-a_1}{d}]] + 1$ terms not exceeding M.

P5. Apply the previous problem P4 to find the value of the sum of all natural numbers k not exceeding $1,000$, and which are such that the remainder of the division of k^2 with 17 is equal to 9.

Solution: First, we divide those numbers k into two disjoint classes or groups. If q is the quotient of the division of k^2 with 17, and with remainder 9, we must have,

$$k^2 = 17q + 9 \Leftrightarrow (k-3)(k+3) = 17q,$$

but 17 is a prime number and as such it must divide at least one of the two factors $k-3$ and $k+3$; but it cannot divide both. Why? Because for any value of the natural number k, it is easy to see that the greatest common divisor of $k-3$ and $k+3$ is either equal to $1, 2$, or 6. Thus, we must have either $k - 3 = 17n$ or $k + 3 = 17m$; either $k = 17n + 3$ or

$$\begin{aligned} k = 17m - 3 &= 17(m-1) + 14 \\ &= 17 \cdot \ell + 14 \end{aligned}$$

(here we have set $m - 1 = \ell$). The number n is a nonnegative integer and the number

ℓ is also a nonnegative integer. So the two disjoint classes of the natural numbers k are,

$$\begin{aligned} k &= 3, 20, 37, 54, \ldots \\ \text{and } k &= 14, 31, 48, 65, \ldots \end{aligned}$$

Next, we find how many numbers k in each class do not exceed $M = 10,000$. Here, we are dealing with two arithmetic progressions: the first being $3, 20, 37, 54, \ldots$, having first term $a_1 = 3$ and difference $d = 17$. The second arithmetic progression has first term $b_1 = 14$ and the same difference $d = 17$.

According to the previous practice problem, P4, there are exactly $N_1 = [[\frac{M-a_1}{d}]] + 1 = [[\frac{1000-3}{17}]] + 1 = [[\frac{997}{17}]] + 1 = 58 + 1 = 59$ terms of the first arithmetic progression not exceeding 1000 (also, recall from Chapter 6 that $[[\frac{997}{17}]]$ is really none other than the quotient of the division of 997 with 17).

Again, applying problem P4 to the second arithmetic progression, we see that there are $N_2 = [[\frac{M-b_1}{d}]] + 1 = [[\frac{1000-14}{17}]] + 1 = [[\frac{986}{17}]] + 1 = 58 + 1 = 59$.
Finally, we must find the two sums:

$$\begin{aligned} S_{N_1} &= a_1 + \ldots + a_{N_1} = \frac{N_1 \cdot (a_1 + a_{N_1})}{2} = \frac{N_1 \cdot [2a_1 + (N_1 - 1)d]}{2} \\ &= \frac{59 \cdot [2(3) + (59-1) \cdot 17]}{2} = \frac{59 \cdot [6 + (58)(17)]}{2} \end{aligned}$$

and

$$\begin{aligned} S_{N_2} &= b_1 \ldots + b_{N_2} = \frac{N_2 \cdot [2b_1 + (N_2 - 1)d]}{2} \\ &= \frac{59 \cdot [2(14) + (59-1)17]}{2} = \frac{59 \cdot [28 + (58)(17)]}{2} \end{aligned}$$

Hence,

$$\begin{aligned} S_{N_1} + S_{N_2} &= \frac{59 \cdot [6 + 28 + 2(58)(17)]}{2} \\ &= \frac{59\,[34 + 1972]}{2} = \frac{59 \cdot (2006)}{2} = 59 \cdot (1003) = 59,177. \end{aligned}$$

P6. If S_n, S_{2n}, S_{3n}, are the sums of the first n, $2n$, $3n$ terms of an arithmetic progression, find the relation or equation between the three sums.

Solution: We have $S_n = \frac{n \cdot [a_1 + (n-1)d]}{2}$, $S_{2n} = \frac{2n \cdot [a_1 + (2n-1)d]}{2}$, and $S_{3n} = \frac{3n \cdot [a_1 + (3n-1)d]}{2}$.

We can write

$$S_{2n} = \frac{2n \cdot [2a_1 + 2(n-1)d + (d - a_1)]}{2} \text{ and}$$

$$S_{3n} = \frac{3n \cdot [3a_1 + 3(n-1)d + (2d - 2a_1)]}{2}.$$

So that,

$$S_{2n} = \frac{2n \cdot 2 \cdot [a_1 + (n-1)d]}{2} + \frac{2n \cdot (d - a_1)}{2} \tag{1}$$

and

$$S_{3n} = \frac{3n \cdot 3 \cdot [a_1 + (n-1)d]}{2} + \frac{3n \cdot 2 \cdot (d - a_1)}{2} \tag{2}$$

To eliminate the product $n \cdot (d - a_1)$ in equations (1) and (2) just consider $3S_{2n} - S_{3n}$: equations (1) and (2) imply,

$$3S_{2n} - S_{3n} = \frac{3 \cdot 2n \cdot 2 \cdot [a_1 + (n-1)d]}{2} - \frac{3n \cdot 3 \cdot [a_1 + (n-1)d]}{2} \underbrace{+ \frac{3 \cdot 2n \cdot (d - a_1)}{2} - \frac{3n \cdot 2 \cdot (d - a_1)}{2}}_{0}$$

$$\Rightarrow 3S_{2n} - S_{3n} = \frac{3n \cdot [a_1 + (n-1)d]}{2}$$

but $S_n = \frac{n \cdot [a_1 + (n-1)d]}{2}$; hence the last equation yields

$$3S_{2n} - S_{3n} = 3 \cdot S_n$$

$$\rightarrow \boxed{3S_{2n} - 3S_n + S_{3n}};$$

$$\text{or} \quad 3(S_{2n} - S_n) = S_{3n}$$

P7. If the first term of an arithmetic progression is equal to some real number a, and the sum of the first m terms is equal to zero, show that the sum of the next n terms must equal to $\frac{a \cdot m(m+n)}{1-m}$; here, we assume that m and n are natural numbers with $m > 1$

Solution: We have $a_1 + \ldots + a_m = 0 = \frac{m \cdot [2a_1 + d(m-1)]}{2} \Rightarrow$ (since $m > 1$) $2a_1 + d(m-1) = 0 \Rightarrow d = \frac{-2a_1}{m-1} = \frac{2a_1}{1-m} = \frac{2a}{1-m}$. Consider the sum of the next n terms

$$a_{m+1} + \ldots + a_{m+n} = \frac{n \cdot (a_{m+1} + a_{m+n})}{2};$$

$$a_{m+1} + \ldots + a_{m+n} = \frac{n \cdot [(a_1 + md) + (a_1 + (m+n-1)d)]}{2};$$

$$a_{m+1} + \ldots + a_{m+n} = \frac{n \cdot [2a_1 + (2m+n-1)d]}{2}$$

Now substitute for $d = \frac{2a}{1-m}$: (and of course, $a = a_1$)

$$a_{m+1} + \ldots + a_{m+n} = \frac{n[2a + (2m + n - 1) \cdot \frac{2a}{1-m}]}{2};$$

$$a_{m+1} + \ldots + a_{m+n} = \frac{n \cdot 2a[(1 - m) + (2m + n - 1)]}{2(1 - m)};$$

$$a_{m+1} + \ldots + a_{m+n} = \frac{2an[1 - m + 2m + n - 1]}{2(1 - m)} = \frac{a \cdot n \cdot (m + n)}{1 - m}$$

P8. Suppose that the sum of the m first terms of an arithmetic progression is n; and the sum of the first n terms is equal to m. Furthermore, suppose that the first term is α and the difference is β, where α and β are given real numbers. Also, assume $m \neq n$ and $\beta \neq 0$.

(a) Find the sum of the first $(m + n)$ in terms of the constants α and β only.

(b) Express the integer mn and the difference $(m - n)$ in terms of α and β.

Solution: (a) We have $a_1 + \ldots + a_m = n$ and $a_1 + \ldots + a_n = m$;

$$\frac{m \cdot [2\alpha + (m-1)\beta]}{2} = n \quad \text{and} \quad \frac{n \cdot [2\alpha + (n-1)\beta]}{2} = m,$$

since $a_1 = \alpha$ and $d = \beta$.

Subtracting the second equation from the first one to obtain,

$$\begin{aligned}
2\alpha \cdot (m-n) &+ \beta \cdot [m(m-1) - n(n-1)] = 2n - 2m; \\
2\alpha \cdot (m-n) &+ \beta \cdot [(m^2 - n^2) - (m-n)] + 2(m-n) = 0; \\
2\alpha \cdot (m-n) &+ \beta \cdot [(m-n)(m+n) - (m-n)] + 2(m-n) = 0; \\
2\alpha \cdot (m-n) &+ \beta \cdot (m-n) \cdot [m+n-1] + 2(m-n) = 0;
\end{aligned}$$

$(m-n) \cdot [2\alpha + \beta(m+n-1) + 2] = 0$; but $m - n \neq 0$, since $m \neq n$ by the hypothesis of the problem. Thus,

$$2\alpha + \beta \cdot (m+n-1) + 2 = 0 \Rightarrow \beta(m+n-1) = -2(1+a)$$

$$\Rightarrow \quad m+n-1 = \tfrac{-2(1+a)}{\beta} \Rightarrow m+n = 1 - \frac{2(1+\alpha)}{\beta} = \frac{\beta - 2\alpha - 2}{\beta}.$$

Now, we compute the sum $a_1 + \ldots + a_{m+n} = \frac{(m+n)\cdot[2\alpha+(m+n-1)\beta]}{2}$

$$\Rightarrow \quad a_1 + \ldots + a_{m+n} = \frac{\left(\frac{\beta-2\alpha-2}{\beta}\right) \cdot \left[2\alpha + \left(\frac{\beta-2\alpha-2}{\beta}\right) \cdot \beta\right]}{2};$$

$$a_1 + \ldots + a_{m+n} = \boxed{\frac{(\beta - 2\alpha - 2) \cdot (\beta - 2)}{2\beta}}$$

(b) If we multiply the equations $\frac{m\cdot[2\alpha+(m-1)\beta]}{2} = n$ and $\frac{n\cdot[2\alpha+(n-1)\beta]}{2} = m$ member-wise we obtain, $\frac{m\cdot n\cdot[2\alpha+(n-1)\beta][2\alpha+(m-1)\beta]}{4} = mn$ and since $mn \neq 0$, we arrive at

$$[2\alpha + (n-1)\beta] \cdot [2\alpha + (m-1)\beta] = 4$$

$$\Rightarrow \quad 4\alpha^2 + 2\alpha\beta \cdot (m - 1 + n - 1) + (n-1)(m-1)\beta^2 = 4$$

$$\Rightarrow \quad 4\alpha^2 + 2\alpha\beta \cdot (m+n) - 4\alpha\beta + nm\beta^2 - (n+m)\beta^2 + \beta^2 = 4;$$

$$(2\alpha - \beta)^2 + (m+n) \cdot (2\alpha\beta - \beta^2) + nm\beta^2 = 4.$$

Now let us substitute for $m + n = \frac{\beta-2\alpha-2}{\beta}$ (from part (a)) in the last equation above; we have,

$$(2\alpha-\beta)^2+\left(\frac{\beta-2\alpha-2}{\beta}\right)\cdot\beta\cdot(2\alpha-\beta)+nm\beta^2=4$$

$$\Rightarrow\quad(2\alpha-\beta)^2+(\beta-2\alpha-2)(2\alpha-\beta)+nm\beta^2=4$$

$$\Rightarrow\quad 4\alpha^2-4\alpha\beta+\beta^2+2\alpha\beta-\beta^2-4\alpha^2+4\alpha\beta-4\alpha+2\beta+nm\beta^2=4$$

$$\Rightarrow\quad nm\beta^2+2\alpha\beta-4\alpha+2\beta=4\Rightarrow nm\beta^2=4-2\alpha\beta+4\alpha-2\beta$$

$$\Rightarrow\quad \boxed{nm=\frac{2\cdot(2-\alpha\beta+2\alpha-\beta)}{\beta^2}}$$

Finally, from the identity $(m-n)^2=(m+n)^2-4nm$, it follows that

$$(m-n)^2=\left(\frac{\beta-2\alpha-2}{\beta}\right)^2-\frac{8(2-\alpha\beta+2\alpha-\beta)}{\beta^2}$$

$$\Rightarrow\quad(m-n)^2=\frac{\beta^2+4\alpha^2+4-4\alpha\beta-4\beta+8\alpha-16+8\alpha\beta-16\alpha+8\beta}{\beta^2}$$

$$(m-n)^2=\frac{\beta^2+4\alpha^2-12+4\alpha\beta+4\beta-8\alpha}{\beta^2};$$

$$|m-n|=\frac{\sqrt{\beta^2+4\alpha^2-12+4\alpha\beta+4\beta-8\alpha}}{|\beta|}$$
$$=\frac{\sqrt{(2\alpha+\beta)^2-12+4\beta-8\alpha}}{|\beta|};$$

$$\boxed{m-n=\pm\frac{\sqrt{(2\alpha+\beta)^2-12+4\beta-8\alpha}}{|\beta|}}$$

the choice of the sign depending on whether $m>n$ or $m<n$ respectively. Also note, that a necessary condition that must hold here is

$$(2\alpha+\beta)^2-12+4\beta-8\alpha>0.$$

P9. Prove that if the real numbers $\alpha,\beta,\gamma,\delta$ are successive terms of a harmonic progression, then

$$3(\beta-\alpha)(\delta-\gamma)=(\gamma-\beta)(\delta-\alpha).$$

Solution: Since $\alpha,\beta,\gamma,\delta$ are members of a harmonic progression they must all be nonzero; $\alpha\beta\gamma\delta\neq 0$. Thus

$$3(\beta-\alpha)(\delta-\gamma)=(\gamma-\beta)(\delta-\alpha)$$

is equivalent to

$$\frac{3(\beta-\alpha)(\delta-\gamma)}{\alpha\beta\gamma\delta}=\frac{(\gamma-\beta)(\delta-\alpha)}{\alpha\beta\gamma\delta}$$

$$\Leftrightarrow\ 3\cdot\left(\tfrac{\beta-\alpha}{\beta\alpha}\right)\cdot\left(\frac{\delta-\gamma}{\delta\gamma}\right)=\left(\frac{\gamma-\beta}{\gamma\beta}\right)\cdot\left(\frac{\delta-\alpha}{\alpha\delta}\right)$$

$$\Leftrightarrow\ 3\cdot\left(\frac{1}{\alpha}-\frac{1}{\beta}\right)\cdot\left(\frac{1}{\gamma}-\frac{1}{\delta}\right)=\left(\frac{1}{\beta}-\frac{1}{\gamma}\right)\cdot\left(\frac{1}{\alpha}-\frac{1}{\delta}\right)$$

By definition, since $\alpha,\beta,\gamma,\delta$ are consecutive terms of a harmonic progression; the numbers $\frac{1}{\alpha},\frac{1}{\beta},\frac{1}{\gamma},\frac{1}{\delta}$ must be successive terms of an arithmetic progression with difference d; and $\frac{1}{\alpha}-\frac{1}{\beta}=-d$, $\frac{1}{\gamma}-\frac{1}{\delta}=-d$, $\frac{1}{\beta}-\frac{1}{\gamma}=-d$, and $\frac{1}{\alpha}-\frac{1}{\delta}=-3d$ (since $\frac{1}{\delta}=\frac{1}{\gamma}+d=\frac{1}{\beta}+2d=\frac{1}{\alpha}+3d$). Thus the above statement we want to prove is equivalent to

$$3\cdot(-3)\cdot(-d)=(-d)\cdot(-3d)\Leftrightarrow 3d^2=3d^2$$

which is true.

P10. Suppose that m and n are fixed natural numbers such that the mth term a_m in a harmonic progression is equal to n; and the nth term a_n is equal to m. We assume $m\neq n$.

(a) Find the $(m+n)$th term a_{m+n} in terms of m and n .

(b) Determine the general kth term a_k in terms of $k, m,$ and n.

Solution:

(a) Both $\frac{1}{a_m}$ and $\frac{1}{a_n}$ are the mth and nth terms respectively of an arithmetic progression with first term $\frac{1}{a_1}$ and difference d; so that $\frac{1}{a_m}=\frac{1}{a_1}+(m-1)d$ and $\frac{1}{a_n}=\frac{1}{a_1}+(n-1)d$. Subtracting the second equation from the first and using the fact that $a_m=n$ and $a_n=m$ we obtain, $\frac{1}{n}-\frac{1}{m}=(m-n)d\Rightarrow\frac{m-n}{nm}=(m-n)d$; but $m-n\neq 0$; cancelling the factor $(m-n)$ from both sides, gives $\boxed{\frac{1}{mn}=d}$. Thus from the first equation , $\frac{1}{n}=\frac{1}{a_1}+(m-1)\cdot\frac{1}{mn}\Rightarrow\frac{1}{n}-\frac{(m-1)}{mn}=\frac{1}{a_1}\Rightarrow\frac{m-(m-1)}{mn}=\frac{1}{a_1};\ \frac{1}{mn}=\frac{1}{a_1}\Rightarrow\boxed{a_1=mn}$. Therefore, $\frac{1}{a_{m+n}}=\frac{1}{a_1}+(m+n-1)d\Rightarrow\frac{1}{a_{m+n}}=\frac{1}{mn}+\frac{m+n-1}{mn}\Rightarrow$ $\boxed{a_{m+n}=\frac{mn}{m+n}}$.

(b) We have $\frac{1}{a_k}=\frac{1}{a_1}+(k-1)d\Rightarrow\frac{1}{a_k}=\frac{1}{mn}+\frac{(k-1)}{mn}=\frac{k}{mn}\Rightarrow\boxed{a_k=\frac{mn}{k}}$.

P11. Use mathematical induction to prove that if $a_1,a_2,\ldots,a_n$, with $n\geq 3$, are the first n terms of a harmonic progression, then $(n-1)a_1a_n=a_1a_2+a_2a_3+\ldots+a_{n-1}a_n$.

Solution: For $n=3$ the statement is $2a_1a_3=a_1a_2+a_2a_3\Leftrightarrow a_2\cdot(a_1+a_3)=2a_1a_3$; but a_1,a_2,a_3 are all nonzero since they are the first three terms of a harmonic progression. Thus, the last equation is equivalent to $\frac{2}{a_2} = \frac{a_1+a_d}{a_1a_3} \Leftrightarrow \frac{2}{a_2} = \frac{1}{a_3} + \frac{1}{a_1}$ which is true, because $\frac{1}{a_1}, \frac{1}{a_2}, \frac{1}{a_3}$ are the first three terms of a harmonic expression.

The inductive step: prove that whenever the statement holds true for some natural number $n = k \geq 3$, then it must also hold true for $n = k+1$. So we assume $(k-1)a_1a_k = a_1a_2 + a_2a_3 + \ldots + a_{k-1}a_k$. Add a_ka_{k+1} to both sides to obtain,

$$(k-1)a_1a_k + a_ka_{k+1} = a_1a_2 + a_2a_3 + \ldots + a_{k-1}a_k + a_ka_{k+1} \qquad (1)$$

If we can show that the left-hand side of (1) is equal to ka_1a_{k+1}, the induction process will be complete. So we need to show that

$$(k-1)a_1a_k + a_ka_{k+1} = k \cdot a_1 \cdot a_{k+1} \qquad (2)$$

(dividing both sides of the equation by $a_1 \cdot a_k \cdot a_{k+1} \neq 0$)

$$\Leftrightarrow \frac{(k-1)}{a_{k+1}} + \frac{1}{a_1} = \frac{k}{a_k}. \qquad (3)$$

To prove (3), we can use the fact that $\frac{1}{a_{k+1}}$ and $\frac{1}{a_k}$ are the $(k+1)$th and kth terms of an arithmetic progression with first term $\frac{1}{a_1}$ and ratio d: $\frac{1}{a_{k+1}} = \frac{1}{a_1} + k \cdot d$ and $\frac{1}{a_k} = \frac{1}{a_1} + (k-1)d$; so that, $\frac{k-1}{a_{k+1}} = \frac{k-1}{a_1} + (k-1)kd$ and $\frac{k}{a_k} = \frac{k}{a_1} + k(k-1)d$. Subtracting the second equation from the first yields,

$$\frac{k-1}{a_{k+1}} - \frac{k}{a_k} = \frac{(k-1)-k}{a_1} \Rightarrow \frac{k-1}{a_{k+1}} + \frac{1}{a_1} = \frac{k}{a_k}$$

which establishes (3) and thus equation (2). The induction is complete since we have show (by combining (1) and (3)).

$$k \cdot a_1a_{k+1} = a_1a_2 + a_2a_3 + \ldots + a_{k-1}a_k + a_ka_{k+1},$$

the statement also holds for $n = k+1$.

P12. Find the necessary and sufficient condition that three natural numbers $m, n,$ and k must satisfy, in order that the positive real numbers $\sqrt{m}, \sqrt{n}, \sqrt{k}$ be consecutive terms of a geometric progression.

Solution: According to Theorem 7, the three positive reals will be consecutive terms of an arithmetic progression if, and only if, $(\sqrt{n})^2 = \sqrt{m}\sqrt{k} \Leftrightarrow n = \sqrt{mk} \Leftrightarrow$ (since both n and mk are positive) $n^2 = mk$. Thus, the necessary and sufficient condition is that the product of m and k be equal to the square of n.

P13. Show that if α, β, γ are successive terms of an arithmetic progression, β, γ, δ are consecutive terms of a geometric progression, and γ, δ, ϵ are the successive terms of a harmonic progression, then either the numbers α, γ, ϵ or the numbers ϵ, γ, α must be the consecutive terms of a geometric progression.

Solution: Since $\frac{1}{\gamma}, \frac{1}{\delta}, \frac{1}{\epsilon}$ are by definition successive terms of an arithmetic progression and the same holds true for α, β, γ, Theorem 3 tells us that we must have

$2\beta = \alpha + \gamma$ (1) and $\frac{2}{\delta} = \frac{1}{\gamma} + \frac{1}{\epsilon}$ (2). And by Theorem 7, we must also have $\gamma^2 = \beta\delta$ (3). (Note that γ, δ, and ϵ must be nonzero and thus so must be β.)

Equation (2) implies $\delta = \frac{2\gamma\epsilon}{\gamma+\epsilon}$ and equation (1) implies $\beta = \frac{\alpha+\gamma}{2}$. Substituting for β and δ in equation (3) we now have

$$\gamma^2 = \left(\frac{\alpha+\gamma}{2}\right) \cdot \left(\frac{2\gamma\epsilon}{\gamma+\epsilon}\right)$$

$$\Rightarrow \; \gamma^2 \cdot (\gamma+\epsilon) = (\alpha+\gamma)\cdot\gamma\epsilon \Rightarrow \gamma^3 + \gamma^2\epsilon = \alpha\gamma\epsilon + \gamma^2\epsilon$$

$$\Rightarrow \; \gamma^3 - \alpha\gamma\epsilon = 0 \Rightarrow \gamma(\gamma^2 - \alpha\epsilon) = 0$$

and since $\gamma \neq 0$ we conclude $\gamma^2 - \alpha\epsilon = 0 \Rightarrow \gamma^2 = \alpha\epsilon$, which, in accordance with Theorem 7, proves that either α, γ, ϵ; or ϵ, γ, α are consecutive terms in a geometric progression.

P14. Prove that if α is the arithmetic mean of the numbers β and γ; and β, nonzero, the geometric mean of α and γ, then γ must be the harmonic mean of α and β. (Note: the assumption $\beta \neq 0$, together with the fact that β is the geometric mean of α and γ, does imply that both α and γ must be nonzero as well.)

Solution: From the problems assumptions we must have $2\alpha = \beta+\gamma$ and $\beta^2 = \alpha\gamma$; $\beta^2 = \alpha\gamma \Rightarrow 2\beta^2 = 2\alpha\gamma$; substituting for $2\alpha = \beta + \gamma$ in the last equation produces

$$2\beta^2 = (\beta+\gamma)\gamma \Rightarrow 2\beta^2 = \beta\gamma + \gamma^2$$

$$\Rightarrow \; 2\beta^2 - \gamma^2 - \beta\gamma = 0 \Rightarrow (\beta^2 - \gamma^2) + (\beta^2 - \beta\gamma) = 0$$

$$\Rightarrow \; (\beta-\gamma)(\beta+\gamma) + \beta\cdot(\beta-\gamma) = 0 \Rightarrow (\beta-\gamma)\cdot(2\beta+\gamma) = 0.$$

If $\beta - \gamma \neq 0$, then the last equation implies $2\beta + \gamma = 0 \Rightarrow \gamma = -2\beta$; and thus from $2a = \beta + \gamma$ we obtain $2\alpha = \beta - 2\beta$; $2\alpha = -\beta$; $\alpha = -\beta/2$. Now compute, $\frac{2}{\gamma} = \frac{2}{-2\beta} = -\frac{1}{\beta}$, since $\beta \neq 0$; and $\frac{1}{\alpha} + \frac{1}{\beta} = \frac{1}{-\frac{\beta}{2}} + \frac{1}{\beta} = -\frac{2}{\beta} + \frac{1}{\beta} = -\frac{1}{\beta}$. Therefore $\frac{2}{\gamma} = \frac{1}{\alpha} + \frac{1}{\beta}$, which proves that γ is the harmonic mean of α and β. Finally, by going back to the equation $(\beta-\gamma)(2\beta+\gamma) = 0$ we consider the other possibility, namely $\beta - \gamma = 0$; $\beta = \gamma$ (note that $\beta - \gamma$ and $2\beta + \gamma$ cannot both be zero for this would imply $\beta = 0$, violating the problem's assumption that $\beta \neq 0$). Since $\beta = \gamma$ and $2\alpha = \beta + \gamma$, we conclude $\alpha = \beta = \gamma$. And then trivially, $\frac{2}{\gamma} = \frac{1}{\alpha} + \frac{1}{\beta}$, so we are done.

P15. We partition the set of natural numbers in disjoint classes or groups as follows: $\{1\}, \{2,3\}, \{4,5,6\}, \{7,8,9,10\}, \ldots$; the nth class contains n consecutive positive integers starting with $\frac{n\cdot(n-1)}{2} + 1$. Find the sum of the members of the nth class.

Solution: First let us make clear why the first member of nth class is the number $\frac{n(n-1)}{2} + 1$; observe that the nth class is preceded by $(n-1)$ classes; so since the kth class, $1 \leq k \leq n-1$, contains exactly k consecutive integers, then there precisely $(1+2+\ldots+k+\ldots+(n-1))$ consecutive natural numbers preceding the nth class;

but the sum $1+2+\ldots+(n-2)+(n-1)$ is the sum of the first $(n-1)$ terms of the infinite arithmetic progression that has first term $a_1 = 1$ difference $d = 1$, hence

$$\begin{aligned} 1+2+\ldots+(n-1) &= a_1+a_2+\ldots+a_{n-1} = \frac{(n-1)\cdot(a_1+a_{n-1})}{2} \\ &= \frac{(n-1)(1+(n-1))}{2} = \frac{(n-1)\cdot n}{2}. \end{aligned}$$

This explains why the nth class starts with the natural number $\frac{n(n-1)}{2}+1$; the members of the nth class are the numbers $\frac{n(n-1)}{2}+1$, $\frac{n(n-1)}{2}+2, \ldots, \frac{n(n-1)}{2}+n$. These n numbers form a finite arithmetic progression with first term $\underbrace{\frac{n(n-1)}{2}+1}_{a}$ and difference $d = 1$.

Hence their sum is equal to

$$\begin{aligned} \frac{n\cdot[2a+(n-1)d]}{2} &= \frac{n\cdot\left[2\left(\frac{n(n-1)}{2}+1\right)+(n-1)\right]}{2} \\ &= \frac{n\cdot[n(n-1)+2+n-1]}{2} = \frac{n\cdot[n^2-n+2+n-1]}{2} = \boxed{\frac{n\cdot(n^2+1)}{2}} \end{aligned}$$

P16. We divide 8,000 objects into $(n+1)$ groups of which the first n of them contain $5, 8, 11, 14, \ldots, [5+3\cdot(n-1)]$ objects respectively; and the $(n+1)$th group contains fewer than $(5+3n)$ objects; find the value of the natural number n and the number of objects that the $(n+1)$th group contains.

Solution: The total number of objects that first n groups contain is equal to, $S_n = 5+8+11+14+\ldots+[5+3(n-1)]$; this sum, S_n, is the sum of the first n terms of the infinite arithmetic progression with first term $a_1 - 5$ and difference $d = 3$; so that its nth term is $a_n = 5+3(n-1)$. According to Theorem 2, $S_n = \frac{n\cdot[a_1+a_n]}{2} = \frac{n\cdot[5+5+3(n-1)]}{2} = \frac{n\cdot[5+5+3n-3]}{2} = \frac{n\cdot(7+3n)}{2}$. Thus, the $(n+1)$th group must contain, $8,000 - \frac{n(7+3n)}{2}$ objects. By assumption, the $(n+1)$th group contains fewer than $(5+3n)$ objects. Also $8,000 - \frac{n(7+3n)}{2}$ must be a nonnegative integer, since it represents the number of objects in a set (the $(n+1)$th class; theoretically this number may be zero). So we have two simultaneous inequalities to deal with:

$$0 \le 8,000 - \frac{n(7+3n)}{2} \Leftrightarrow \frac{n(7+3n)}{2} \le 8,000; \quad n(7+3n) \le 16,000.$$

And (the other inequality)

$$\begin{aligned} 8,000 - \tfrac{n(7+3n)}{2} &< 5+3n \Leftrightarrow 16,000 - n(7+3n) < 10+6n \Leftrightarrow 16,000 \\ &< 3n^2+13n+10 \Leftrightarrow 16,000 < (3n+10)(n+1). \end{aligned}$$

So we have the following system of two simultaneous inequalities

$$\left.\begin{aligned} n(7+3n) &\le 16,000 \\ \text{and}\quad 16,000 &< (3n+10)(n+1) \end{aligned}\right\} \begin{matrix}(1)\\(2)\end{matrix}$$

Consider (1): At least one of the factors n and $7+3n$ must be less than or equal to $\sqrt{16,000}$; for if both were greater than $\sqrt{16,000}$ then their product would exceed $\sqrt{16,000}\cdot\sqrt{16,000} = 16,000$, contradicting inequality (1); and since $n < 7+3n$, it is now clear that the natural number n cannot exceed $\sqrt{16,000}$: $n \le \sqrt{16,000} \Leftrightarrow n \le \sqrt{16\cdot 10^3}$; $n \le 4\cdot\sqrt{10^2\cdot 10}$; $n \le 4\cdot 10\cdot\sqrt{10} = 40\sqrt{10}$ so $40\sqrt{10}$ is a necessary upper bound for n. The closest positive integer to $40\sqrt{10}$, but less than $40\sqrt{10}$ is the number 126; but actually, an upper bound for n must be much less than 126 in view of the factor $7+3n$. If we consider (1), we have $3n^2+7n-16,000 \le 0$ (3)

The two roots of the quadratic equation $3x^2+7x-16,000 = 0$ are the real numbers $r_1 = \frac{-7+\sqrt{(7)^2-4(3)(-16,000)}}{6} = \frac{-7+\sqrt{192,049}}{6} =\approx$ approximately 71.872326; and $r_2 = \frac{-7-\sqrt{192,049}}{6} \approx -74.20566$.

Now, it is well known from precalculus that if r_1 and r_2 are the two roots of the quadratic polynomial ax^2+bx+c, then $ax^2+bx+c = a\cdot(x-r_1)(x-r_2)$, for all real numbers x. In our case $3x^2+7x-16,000 = 3\cdot(x-r_1)(x-r_2)$, where r_1 and r_2 are the above calculated real numbers. Thus, in order for the natural number n to satisfy the inequality (3), $3n^2+7n-16,000 \le 0$; it must satisfy $3(n-r_1)(n-r_2) \le 0$; but this will only be true if, and only if, $r_1 \le n \le r_2$; $-74.20566 \le n \le 71.872326$; but n is a natural number; thus $1 < n < 71$; this upper bound for n is much lower than the upper bound of the upper bound 126 that we estimated more crudely earlier. Now consider inequality (2): it must hold true simultaneously with (1); which means we have,

$$\left.\begin{aligned} 16,000 &< (3n+10)\cdot(n+1) \\ \text{and}\quad 1 &\le n \le 71 \end{aligned}\right\}$$

If we take the highest value possible for n; namely $n = 71$, we see that $(3n+10)(n+1) = (3\cdot(71)+10)\cdot(72) = (223)(72) = 16,052$ which exceeds the number 16,000, as desired. But, if we take the next smaller value, $n = 70$, we have $(3n+10)(n+1) = (220)(71) = 15,620$ which falls below 16,000. Thus, this problem has a unique solution, $\boxed{n = 71}$. The total number of objects in the first n groups (or 71 groups) is then equal to,

$$\frac{n\cdot(7+3n)}{2} = \frac{(7)\cdot(7+3(7))}{2} = \frac{(71)\cdot(220)}{2} = (71)\cdot(110) - 7,810.$$

Thus, the $(n+1)$th or 72nd group contains, $8,000 - 7,810 = \boxed{190}$ objects; note that 190 is indeed less that $5n+3 = 5(71)+3 = 358$.

P17. (a) Show that the real numbers $\frac{\sqrt{2}+1}{\sqrt{2}-1}$, $\frac{1}{2-\sqrt{2}}$, $\frac{1}{2}$, can be three consecutive terms of a geometric progression. Find the ratio r of any geometric progression that contains these three numbers as consecutive terms.

(b) Find the value of the infinite sum of the terms of the (infinite) geometric progression whose first three terms are the numbers $\frac{\sqrt{2}+1}{\sqrt{2}-1}$, $\frac{1}{2-\sqrt{2}}$, $\frac{1}{2}$; $\left(\frac{\sqrt{2}+1}{\sqrt{2}-1}\right)+\left(\frac{1}{2-\sqrt{2}}\right)+\frac{1}{2}+\ldots$.

Solution: (a) Apply Theorem 7: the three numbers will be consecutive terms of a geometric progression if, and only if,

$$\left(\frac{1}{2-\sqrt{2}}\right)^2=\frac{(\sqrt{2}+1)}{(\sqrt{2}-1)}\cdot\frac{1}{2} \tag{1}$$

Compute the left-hand side:

$$\begin{aligned}\frac{1}{(2-\sqrt{2})^2} &= \frac{1}{4-4\sqrt{2}+2}=\frac{1}{6-4\sqrt{2}}\\ &= \frac{1}{2(3-2\sqrt{2})}=\frac{3+2\sqrt{2}}{2\cdot(3-2\sqrt{2})(3+2\sqrt{2})}\\ &= \frac{3+2\sqrt{2}}{2\cdot[9-8]}=\frac{3+2\sqrt{2}}{3}.\end{aligned}$$

Now we simplify the right-hand side:

$$\begin{aligned}\left(\frac{\sqrt{2}+1}{\sqrt{2}-1}\right)\cdot\frac{1}{2} &= \frac{1}{2}\cdot\frac{(\sqrt{2}+1)^2}{(\sqrt{2}-1)(\sqrt{2}+1)}\\ &= \frac{1}{2}\cdot\frac{(2+2\sqrt{2}+1)}{(2-1)}=\frac{3+\sqrt{2}}{2}\end{aligned}$$

so the two sides of (1) are indeed equal; (1) is a true statement. Thus, the three numbers can be three consecutive terms in a geometric progression. To find r, consider $\left(\frac{\sqrt{2}+1}{\sqrt{2}-1}\right)\cdot r=\frac{1}{2-\sqrt{2}}$; and also $\left(\frac{1}{2-\sqrt{2}}\right)\cdot r=\frac{1}{2}$; from either of these two equations we can get the value of r; if we use the second equation we have, $\boxed{r=\frac{2-\sqrt{2}}{2}}$.

(b) Since $|r|=\left|\frac{2-\sqrt{2}}{2}\right|=\frac{2-\sqrt{2}}{2}<1$, according to Remark 6, the sum $a+ar+ar^2+\ldots+ar^{n-1}+\ldots$ converges to $\frac{a}{1-r}$; in our case $a=\frac{\sqrt{2}+1}{\sqrt{2}-1}$ and $r=\frac{2-\sqrt{2}}{2}$. Thus the value of the infinite sum is equal to

$$\begin{aligned}
\frac{a}{1-r} &= \frac{\frac{\sqrt{2}+1}{\sqrt{2}-1}}{1-\left(\frac{2-\sqrt{2}}{2}\right)} = \frac{\frac{\sqrt{2}+1}{\sqrt{2}-1}}{\frac{2-(2-\sqrt{2})}{2}} \\
&= \frac{2(\sqrt{2}+1)}{\sqrt{2}(\sqrt{2}-1)} = \frac{2(\sqrt{2}+1)\cdot(\sqrt{2}+1)\cdot\sqrt{2}}{\sqrt{2}\cdot\sqrt{2}\cdot(\sqrt{2}-1)(\sqrt{2}+1)} \\
&= \frac{2\sqrt{2}\cdot(\sqrt{2}+1)^2}{2\cdot(2-1)} = \sqrt{2}\cdot(2+2\sqrt{2}+1) = \sqrt{2}\cdot(3+2\sqrt{2}) \\
&= 3\sqrt{2}+2\cdot\sqrt{2}\cdot\sqrt{2} = 3\sqrt{2}+4 = \boxed{4+3\sqrt{2}}
\end{aligned}$$

P18. (For student who had Calculus.) If $|\rho| < 1$ and $|\beta\rho| < 1$, calculate the infinite sum,

$$S = \underbrace{\alpha\rho}_{\text{1st}} + \underbrace{(\alpha+\alpha\beta)}_{\text{2nd}}\rho^2 + \ldots + \underbrace{(\alpha+\alpha\beta+\ldots+\alpha\beta^{n-1})\rho^n}_{n\text{th term}} + \ldots .$$

Solution: First we calculate the nth term which itself is a sum of n terms:

$$(\alpha+\alpha\beta+\ldots+\alpha\beta^{n-1})\cdot\rho^n = \alpha\cdot\rho^n\cdot(1+\beta+\ldots+\beta^{n-1}) = \alpha\cdot\rho^n\cdot\left(\frac{\beta^n-1}{\beta-1}\right)$$

by Theorem 5(ii). Now we have,

$$\begin{aligned}
S &= \alpha\rho + (\alpha+\alpha\beta)\rho^2 + \ldots + \alpha\cdot\rho^n\cdot\left(\frac{\beta^n-1}{\beta-1}\right) + \ldots \\
S &= \alpha\rho\left(\frac{\beta-1}{\beta-1}\right) + \alpha\rho^2\cdot\left(\frac{\beta^2-1}{\beta-1}\right) \\
&\quad + \ldots + \alpha\cdot\rho^n\cdot\left(\frac{\beta^n-1}{\beta-1}\right) + \ldots
\end{aligned}$$

Note that $S = \lim_{n\to\infty} S_n$, where

$$S_n = \alpha\rho \cdot \left(\frac{\beta-1}{\beta-1}\right) + \alpha\rho^n \cdot \left(\frac{\beta^2-1}{\beta-1}\right) + \ldots + \alpha \cdot \rho^n \cdot \left(\frac{\beta^n-1}{\beta-1}\right);$$

$$S_n = \left(\frac{\alpha\rho}{\beta-1}\right)[(\beta-1) + \rho(\beta^2-1) + \ldots + \rho^{n-1} \cdot (\beta^n - 1)]$$

$$S_n = \left(\frac{\alpha\rho}{\beta-1}\right)[\beta \cdot [1 + (\rho\beta) + \ldots + (\rho\beta)^{n-1}] - (1 + \rho + \ldots + \rho^{n-1})]$$

$$S_n = \left(\frac{\alpha\rho}{\beta-1}\right) \cdot \left[\beta \cdot \frac{[(\rho\beta)^n - 1]}{\rho\beta - 1} - \left(\frac{\rho^n - 1}{\rho - 1}\right)\right]$$

Now, as $n \to \infty$, in virtue of $|\rho\beta| < 1$ and $|\rho| < 1$ we have, $\lim_{n\to\infty} \frac{[(\rho\beta)^n-1]}{\rho\beta-1} = \frac{-1}{\rho\beta-1} = \frac{1}{1-\rho\beta}$ and $\lim_{n\to\infty} \frac{\rho^n-1}{\rho-1} = \frac{1}{1-\rho}$. Hence,

$$S = \lim_{n\to\infty} S_n = \left(\frac{\alpha\rho}{\beta-1}\right) \cdot \left[\beta \cdot \left(\frac{1}{1-\rho\beta}\right) - \left(\frac{1}{1-\rho}\right)\right];$$

$$S = \left(\frac{\alpha\rho}{\beta-1}\right) \cdot \left[\frac{\beta(1-\rho) - (1-\rho\beta)}{(1-\rho\beta)\cdot(1-\rho)}\right] = \frac{\alpha\rho \cdot (\beta-1)}{(\beta-1)\cdot(1-\rho\beta)(1-\rho)};$$

$$\boxed{S = \frac{\alpha\rho}{(1-\rho\beta)\cdot(1-\rho)}}$$

P19. Let m, n and ℓ be distinct natural numbers; and $a_1, \ldots, a_k, \ldots$, an infinite arithmetic progression with first nonzero term a_1 and difference d.

(a) Find the necessary conditions that n, ℓ, and m must satisfy in order that,

$$\underbrace{a_1 + a_2 + \ldots + a_m}_{\text{sum of the first } m \text{ terms}} = \underbrace{a_{m+1} + \ldots + a_{m+n}}_{\text{sum of the next } n \text{ terms}} = \underbrace{a_{m+1} + \ldots + a_{m+\ell}}_{\text{sum of the next } \ell \text{ terms}}$$

(b) If the three sums in part (a) are equal, what must be the relationship between a_1 and d?

(c) Give numerical examples.

Solution: (a) We have two simultaneous equations,

$$\left.\begin{array}{l} a_1 + a_2 + \ldots + a_m = a_{m+1} + \ldots + a_{m+n} \\ \text{and} \\ a_{m+1} + \ldots + a_{m+n} = a_{m+1} + \ldots + a_{m+\ell} \end{array}\right\} \quad (1)$$

According to Theorem 2 we have,

$$\begin{aligned}
a_1 + a_2 + \ldots + a_m &= \frac{m \cdot [2a_1 + (m-1)d]}{2}; \\
a_{m+1} + \ldots + a_{m+n} &= \frac{n \cdot [a_{m+1} + a_{m+n}]}{2} \\
&= \frac{n \cdot [(a_1 + md) + (a_1 + (m+n-1)d)]}{2} \\
&= \frac{n \cdot [2a_1 + (2m+n-1)d]}{2};
\end{aligned}$$

and

$$a_{m+1} + \ldots + a_{m+\ell} = \frac{\ell \cdot [2a_1 + (2m+\ell-1)d]}{2}$$

Now let us use the first equation in (1):

$$\begin{aligned}
\frac{m \cdot [2a_1 + (m-1)d]}{2} &= \frac{n \cdot [2a_1 + (2m+n-1)d]}{2}; \\
2ma_1 + m(m-1)d &= 2na_1 + n \cdot (2m+n-1)d; \\
2a_1 \cdot (m-n) &= [n \cdot (2m+n-1) - m(m-1)]d; \\
2a_1 \cdot (m-n) &= [2nm + n^2 - m^2 + m - n]d;
\end{aligned}$$

According to hypothesis $a_1 \neq 0$ and $m - n \neq 0$; so the right-hand side must also be nonzero and,

$$d = \frac{2a_1 \cdot (m-n)}{2nm + n^2 - m^2 + m - n} \qquad (2)$$

Now use the second equation in (1):

$$\begin{aligned}
\frac{n \cdot [2a_1 + (2m+n-1)d]}{2} &= \frac{\ell \cdot [2a_1 + (2m+\ell-1)d]}{2} \\
\Leftrightarrow \quad 2na_1 + n(2m+n-1)d &= 2\ell a_1 + \ell(2m+\ell-1)d \\
\Leftrightarrow \quad 2a_1 \cdot (n-\ell) &= [\ell(2m+\ell-1) - n(2m+n-1)]d \\
\Leftrightarrow \quad 2a_1 \cdot (n-\ell) &= [2m \cdot (\ell-n) + (\ell^2 - n^2) - (\ell-n)]d \\
\Leftrightarrow \quad 2a_1 \cdot (n-\ell) &= [2m \cdot (\ell-n) + (\ell-n)(\ell+n) - (\ell-n)]d \\
\Leftrightarrow \quad 2a_1 \cdot (n-\ell) &= (\ell-n) \cdot [2m + \ell + n - 1]d;
\end{aligned}$$

and since $n - \ell \neq$, we obtain $-2a_1 = (2m + \ell_n - 1)d$;

$$d = \frac{-2a_1}{2m + \ell + n - 1} \tag{3}$$

(Again, in virtue of $a_1 \neq 0$, the product $(2m + \ell + n - 1)d$ must also be nonzero, so $2m + \ell + n - 1 \neq 0$, which is true anyway since, obviously, $2m + \ell + n$ is a natural number greater than 1).

Combining Equations (2) and (3) and cancelling out the factor $2a_1 \neq 0$ from both sides we obtain,

$$\frac{m - n}{2nm + n^2 - m^2 + m - n} = \frac{-1}{2m + \ell + n - 1}$$

Cross multiplying we now have,

$$\begin{aligned} & (m-n)\cdot(2m + \ell + n - 1) \\ = \ & (-1)\cdot(2nm + n^2 - m^2 + m - n); \\ & 2m^2 + m\ell + mn - m - 2mn - n\ell - n^2 + n \\ = \ & -2mn - n^2 + m^2 - m + n; \\ & m^2 + m\ell - n\ell + mn = 0. \end{aligned}$$

We can solve for n in terms of m and ℓ (or for ℓ in terms of m and n) we have,

$$n \cdot (\ell - m) = m \cdot (m + \ell) \Rightarrow \boxed{n = \tfrac{m\cdot(m+\ell)}{\ell - m}}, \text{ since } \ell - m \neq 0.$$

Also, we must have $\boxed{\ell > m}$, in view of the fact that n is a natural number and hence positive (also note that these two conditions easily imply $n > m$ as well). But, there is more: The natural number $\ell - m$ must be a divisor of the product $m \cdot (m + \ell)$. Thus, the conditions are:

(A) $\ell > m$

(B) $(\ell - m)$ is a divisor of $m \cdot (m + \ell)$ and

(C) $n = \frac{m\cdot(m+\ell)}{\ell - m}$

b) As we have already seen d and a_1 must satisfy both conditions (2) and (3). However, under conditions (A), (B), and (C), the two conditions (2) and (3) are, in fact, equivalent, as we have already seen; so $d = \frac{-2a_1}{2m+\ell+n-1}$ (condition (3)) will suffice.

(c) Note that in condition (C), if we choose m and ℓ such $\ell - m$ is positive and $(\ell - m)$ is a divisor of m, then clearly the number $n = \frac{m\cdot(m+\ell)}{\ell - m}$, will be a natural number. If we set $\ell - m = t$, then $m + \ell = t + 2m$, so that

$$n = \frac{m \cdot (t + 2m)}{t} = m + \frac{2m^2}{t}.$$

So if we take t to be a divisor of m, this will be sufficient for $\frac{2m^2}{t}$ to be a positive integer. Indeed, set $m = M \cdot t$, then $n = M \cdot t + \frac{2M^2t^2}{t} = M \cdot t + 2M^2 \cdot t = t \cdot M \cdot (1+2M)$. Also, in condition (3) , if we set $a_1 = a$, then (since $\ell = m + t = Mt + t$)

$$\begin{aligned} d &= \frac{-2a}{2M \cdot t + (Mt+t) + Mt + 2M^2t - 1}; \\ d &= \frac{-2a}{4Mt + t + 2M^2t - 1}. \end{aligned} \tag{4}$$

Thus, the formulas $\ell = Mt + t$, $n = Mt + 2M^2 \cdot t$ and (4) will generate, for each pair of values of the natural numbers M and t, an arithmetic progression that satisfies the conditions of the problem; for any nonzero value of the first term a.

Numerical Example: If we take $t = 3$ and $M = 4$, we then have $m = M \cdot t = 3 \cdot 4 = 12$; $n = t \cdot M \cdot (1 + 2M) = 12 \cdot (1 + 8) = 108$, and $\ell = m + t = 12 + 3 = 15$. And,

$$d = \frac{-2a}{2m + \ell + n - 1} = \frac{-2a}{24 + 15 + 108 - 1} = \frac{-2a}{146} = \frac{-a}{73}.$$

Now let us compute

$$\begin{aligned} a_1 + \ldots + a_m &= \frac{m \cdot [2a + (m-1)d]}{2} = \frac{12 \cdot \left[2a + 11 \cdot \left(\frac{-a}{73}\right)\right]}{2} \\ &= \frac{12 \cdot [146a - 11a]}{2 \cdot 73} = \frac{6 \cdot (135a)}{73} = \frac{810a}{73}. \end{aligned}$$

Next,

$$\begin{aligned} & a_{m+1} + \ldots + a_{m+n'} \\ &= \frac{n \cdot [2a + (m+n-1)d]}{2} \\ &= \frac{108 \cdot \left[2a + (24 + 108 - 1) \cdot \left(\frac{-a}{73}\right)\right]}{2} \\ &= \frac{108}{2} \cdot \frac{[146a - 131a]}{73} \\ &= \frac{(54)(15a)}{73} = \frac{810a}{73} \end{aligned}$$

and

$$\begin{aligned}
& a_{m+1} + \ldots + a_{m+\ell} \\
= & \frac{\ell \cdot [2a + (2m + \ell - 1)d]}{2} \\
= & \frac{15 \cdot \left[2a + (24 + 15 - 1) \cdot \left(\frac{-a}{73}\right)\right]}{2} \\
= & \frac{15}{2} \cdot \frac{[146a - 38a]}{73} \\
= & \frac{15}{2} \cdot \frac{(108)a}{73} = \frac{(15)(54a)}{73} \\
= & \frac{810a}{73}.
\end{aligned}$$

Thus, all three sums are equal to $\frac{810a}{73}$.

P20. If the real numbers a, b, c are consecutive terms of an arithmetic progression and a^2, b^2, c^2 are consecutive terms of a harmonic progression, what conditions must the numbers a, b, c satisfy? Describe all such numbers a, b, c.

Solution: By hypothesis, we have

$$2b = a + c \text{ and } \frac{2}{b^2} = \frac{1}{a^2} + \frac{1}{c^2}$$

so a, b, c must all be nonzero real numbers. The second equation is equivalent to $b^2 = \frac{2a^2c^2}{a^2+c^2}$ and $abc \neq 0$; so that, $b^2(a^2 + c^2) = 2a^2c^2 \Leftrightarrow b^2 \cdot [(a + c)^2 - 2ac] = 2a^2c^2$. Now substitute for $a + c = 2b$:

$$\begin{aligned}
& b^2 \cdot [(2b)^2 - 2ac] = 2a^2c^2 \\
\Leftrightarrow\ & 4b^4 - 2acb^2 - 2a^2c^2 = 0; \\
& 2b^4 - acb^2 - a^2c^2 = 0
\end{aligned}$$

At this stage we could apply the quadratic formula since b^2 is a root to the equation $2x^2 - acx - a^2c^2 = 0$; but the above equation can actually be factored. Indeed,

$$\begin{aligned}
b^4 - acb^2 + b^4 - a^2c^2 &= 0; \\
b^2(b^2 - ac) + (b^2)^2 - (ac)^2 &= 0; \\
b^2 \cdot (b^2 - ac) + (b^2 - ac)(b^2 + ac) &= 0; \\
(b^2 - ac) \cdot (2b^2 + ac) &= 0 \quad eqno(1)
\end{aligned}$$

According to Equation (1), we must have $b^2 - ac = 0$; or alternatively $2b^2 + ac = 0$. Consider the first possibility, $b^2 - ac = 0$. Then, by going back to equation $\frac{2}{b^2} = \frac{1}{a^2} + \frac{1}{c^2}$

we obtain $\frac{2}{ac} = \frac{1}{a^2} + \frac{1}{c^2} \Leftrightarrow \frac{2a^2c^2}{ac} = a^2 + c^2 \Leftrightarrow 2ac = a^2 + c^2$; $a^2 + c^2 - 2ac = 0 \Leftrightarrow (a-c)^2 = 0$; $a = c$ and thus $2b = a + c$ implies $b = a = c$.

Next, consider the second possibility in Equation (1): $2b^2 + ac = 0 \Leftrightarrow 2b^2 = -ac$; which clearly imply that one of a and c must be positive, the other negative. Once more going back to

$$\frac{2}{b^2} = \frac{1}{a^2} + \frac{1}{c^2}; \; \frac{4}{2b^2} = \frac{1}{a^2} + \frac{1}{c^2}$$

$$\Leftrightarrow \quad \frac{4}{-ac} = \frac{c^2 + a^2}{a^2c^2} \tag{2}$$

$$\Leftrightarrow \quad -4ac = c^2 + a^2; \; a^2 + 4ac + c^2 = 0$$

Let $t = \frac{a}{c}$; $a = c \cdot t$ then Equation (2) yields (since $ac \neq 0$),

$$t^2 + 4t + 1 = 0 \tag{3}$$

Applying the quadratic formula to Equation (3), we now have

$$t = \frac{-4 \pm \sqrt{16-4}}{2} = \frac{-4 \pm 2\sqrt{3}}{2};$$

$$t = -2 \pm \sqrt{3};$$

note that both numbers $-2+\sqrt{3}$ and $-2-\sqrt{3}$ are negative and hence both acceptable as solutions, since we know that a and c have opposite sign, which means that $t = \frac{a}{c}$ must be negative. So we must have either $a = (-2+\sqrt{3})c$; or alternatively $a = -(2+\sqrt{3}) \cdot c$. Now, we find b in terms of c. From $2b^2 = -ac$; $b^2 = -\frac{ac}{2}$; note that the last equation says that either the numbers $-\frac{a}{2}, b, c$ are the successive terms of a geometric progression; or the numbers $-a, b, \frac{c}{2}$ (or any of the other two possible permutations: $a, b, -\frac{c}{2}$, $\frac{a}{2}, b, -c$; and four more that are obtained by switching a with c). So, if $a = (-2+\sqrt{3})c$, then from $2b = a + c$; $b = \frac{a+c}{2} = \frac{(-2+\sqrt{3})c+c}{2} = \frac{(\sqrt{3}-1)c}{2}$. And if $a = -(2+\sqrt{3})c$, $b = \frac{a+c}{2} = \frac{-(2+\sqrt{3})c+c}{2} = \frac{-(1+\sqrt{3})c}{2}$. So, in conclusion we summarize as follows:

Any three real numbers a, b, c such that a, b, c are consecutive terms of an arithmetic progression and a^2, b^2, c^2 the successive terms of a harmonic progression must fall in exactly one of five classes:

(1) $a = b = c$; c can be any nonzero real number

(2) $a = (-2+\sqrt{3}) \cdot c$, $b = \frac{(\sqrt{3}-1)c}{2}$; c can be any positive real;

(3) $a = (2+\sqrt{3})c$, $b = \frac{-(1+\sqrt{3})}{2}c$; c can be any positive real.

(4) $a, b = -\frac{(1+\sqrt{3})}{2}a$, $c = -(2+\sqrt{3})a$; a can be any negative real.

(5) $a, b = -\frac{(1+\sqrt{3})}{2}a$, $c = (-2+\sqrt{3})a$; a can be any negative real.

P21. Prove that if the positive real numbers α, β, γ are consecutive members of a geometric progression, then $\alpha^k + \gamma^k \geq 2\beta^k$, for every natural number k.

Solution: Given any natural number k, we can apply the arithmetic-geometric mean inequality of Theorem 10, with $n = 2$, and $a_1 = \alpha^k$, $a_2 = \gamma^k$, in the notation of that theorem:

$$\frac{\alpha^k + \gamma^k}{2} \geq \sqrt{\alpha^k \cdot \gamma^k} = \sqrt{(\alpha\gamma)^k}.$$

But since α, β, γ are consecutive terms of a geometric progression, we must also have $\beta^2 = \alpha\gamma$. Thus the above inequality implies,

$$\begin{aligned} \tfrac{\alpha^k+\gamma^k}{2} &\geq \sqrt{(\beta^2)^k}; \\ \tfrac{\alpha^k+\gamma^k}{2} &\geq \sqrt{(\beta^k)^2} \\ \Rightarrow \quad \tfrac{\alpha^k+\gamma^k}{2} &\geq \beta^k \\ \Rightarrow \quad \alpha^k + \gamma^k &\geq 2\beta^k, \end{aligned}$$

and the proof is complete.

7.5 Exercise Set

1. Show that if the sequence $a_1, a_2, \ldots, a_n, \ldots$, is an arithmetic progression, so is the sequence $c \cdot a_1, c \cdot a_2, \ldots, c \cdot a_n, \ldots$, where c is a constant.

2. Determine the difference of each arithmetic progression which has first term $a_1 = 6$ and contains the numbers 62 and 104 as its terms.

3. Show that the irrational numbers $\sqrt{2}$, $\sqrt{3}$, $\sqrt{5}$ cannot be terms of an arithmetic progression.

4. If $a_1, a_2, \ldots, a_n, \ldots$ is an arithmetic progression and $a_k = \alpha$, $a_m = \beta$, $a_\ell = \gamma$, show that the natural numbers k, m, ℓ and the real numbers α, β, γ, must satisfy the condition

$$\alpha \cdot (m - \ell) + \beta \cdot (\ell - k) + \gamma \cdot (k - m) = 0.$$

Hint: Use the usual formula $a_n = a_1 + (n-1)d$, for $n = k, m, \ell$, to obtain three equations; subtract the first two and then the last two (or the first and the third) to eliminate a_1; then eliminate the difference d (or solve for d in each of the resulting equations).

5. If the numbers α, β, γ are successive terms of an arithmetic progression, then the same holds true for the numbers $\alpha^2 \cdot (\beta + \gamma)$, $\beta^2 \cdot (\gamma + \alpha)$, $\gamma^2 \cdot (\alpha + \beta)$.

6. If S_k denotes the sum of the first k terms of the arithmetic progression with first term k and difference $d = 2k - 1$, find the sum $S_1 + S_2 + \ldots + S_k$.

7. We divide the odd natural numbers into groups or classes as follows: $\{1\}, \{3, 5\}, \{7, 9, 11\}, \ldots$; the nth group contains n odd numbers starting with $(n \cdot (n-1) + 1)$ (verify this). Find the sum of the members of the nth group.

8. We divide the even natural numbers into groups as follows: $\{2\}, \{4, 6\}$, $\{8, 10, 12\}, \ldots$; the nth group contains n even numbers starting with $(n(n-1) + 2)$. Find the sum of the members of the nth group.

9. Let $n_1, n_2, \ldots, n_k$ be k natural numbers such that $n_1 < n_2 < \ldots < n_k$; if the real numbers, $a_{n_1}, a_{n_2}, \ldots, a_{n_k}$, are members of an arithmetic progression (so that the number a_{n_i} is precisely the n_ith term in the progression; $i - 1, 2, \ldots, k$), show that the real numbers:

$$\frac{a_{n_k} - a_{n_1}}{a_{n_2} - a_{n_1}}, \frac{a_{n_k} - a_{n_2}}{a_{n_2} - a_{n_1}}, \ldots, \frac{a_{n_k} - a_{n_{k-1}}}{a_{n_2} - a_{n_1}},$$

are all rational numbers.

10. Let m and n be natural numbers. If in an arithmetic progression $a_1, a_2, \ldots, a_k, \ldots$; the term a_m is equal to $\frac{1}{n}$; $a_m = \frac{1}{n}$, and the term a_n is equal to $\frac{1}{m}$; $a_n = \frac{1}{m}$, prove the following three statements.

(i) The first term a_1 is equal to the difference d.

(ii) If t is any natural number, then $a_{t\cdot(mn)} = t$; in other words, the terms $a_{mn}, a_{2mn}, a_{3mn}, \dots$, are respectively equal to the natural numbers $1, 2, 3, \dots$.

(iii) If $S_{t\cdot(mn)}$ (t a natural number) denote the sum of the first $(t\cdot m\cdot n)$ terms of the arithmetic progression, then $S_{t\cdot(mn)} = \frac{1}{2}\cdot(mn+1)\cdot t$. In other words, $S_{mn} = \frac{1}{2}(mn+1)$, $S_{2mn} = \frac{1}{2}\cdot(mn+1)\cdot 2$, $S_{3mn} = \frac{1}{2}\cdot(mn+1)\cdot 3, \dots$.

11. If the distinct real numbers a, b, c are consecutive terms of a harmonic progression show that

(i) $\frac{2}{b} = \frac{1}{b-a} + \frac{1}{b-c}$ and

(ii) $\frac{b+a}{b-a} + \frac{b+c}{b-c} = 2$

12. If the distinct reals α, β, γ are consecutive terms of a harmonic progression, then the same is true for the numbers $\alpha, \alpha-\gamma, \alpha-\beta$.

13. Let $a = a_1, a_2, a_3, \dots, a_n, \dots$, be a geometric progression and k, ℓ, m natural numbers. If $a_k = \beta$, $a_\ell = \gamma$, $a_m = \delta$, show that $\beta^{\ell-m}\cdot\gamma^{m-k}\cdot\delta^{k-\ell} = 1$.

14. Suppose that n and k are natural numbers such that $n > k+1$; and $a_1 = a, a_2, \dots, a_t, \dots$ a geometric progression, with positive ratio $r \neq 1$, and positive first term a. If A is the value of the sum of the first k terms of the progression and B is the value of the last k terms among the n first terms, express the ratio r in terms of A and B only; and also the first term a in terms of A and B.

15. Find the sum $\left(a - \frac{1}{a}\right)^2 + \left(a^2 - \frac{1}{a^2}\right)^2 + \dots + \left(a^n - \frac{a}{a^n}\right)^2$.

16. Find the infinite sum $\left(\frac{1}{3} + \frac{1}{3^2} + \frac{1}{3^3} + \dots\right) + \left(\frac{1}{5} + \frac{1}{5^2} + \frac{1}{5^3} + \dots\right)$

$$+\left(\frac{1}{9} + \frac{1}{9^2} + \frac{1}{9^3} + \dots\right) + \dots + \left(\underbrace{\frac{1}{(2k+1)} + \frac{1}{(2k+1)^2} + \frac{1}{(2k+1)^3} + \dots}_{k\text{th sum}}\right) + \dots .$$

17. Find the infinite sum $\frac{2}{7} + \frac{4}{7^2} + \frac{2}{7^3} + \frac{4}{7^4} + \frac{2}{7^5} + \frac{4}{7^6} + \dots$.

18. If the numbers α, β, γ are consecutive terms of an arithmetic progression and the nonzero numbers β, γ, δ are consecutive terms of a harmonic progression, show that $\frac{\alpha}{\beta} = \frac{\gamma}{\delta}$.

19. Suppose that the positive reals α, β, γ are successive terms of an arithmetic progression and let x be the geometric mean of α and β; and let y be the geometric mean of β and γ. Prove that x^2, β^2, y^2 are successive terms of an arithmetic progression. Give two numerical examples.

20. Show that if the nonzero real numbers a, b, c are consecutive terms of a harmonic progression, then the numbers $a - \frac{b}{2}, \frac{b}{2}, c - \frac{b}{2}$, must be consecutive terms of a geometric progression. Give two numerical examples.

21. Compute the following sums:

(i) $\frac{1}{2}+\frac{2}{2^2}+\ldots+\frac{n}{2^n}$

(ii) $1+\frac{3}{2}+\frac{5}{4}+\ldots+\frac{2n-1}{2^{n-1}}$

22. Suppose that the sequence $a_1, a_2, \ldots, a_n, \ldots$ satisfies $a_{n+1}=(a_n+\lambda)\cdot\omega$, where λ and ω are fixed real numbers with $\omega \neq 1$.

(i) Use mathematical induction to prove that for every natural number, $a_n = a_1\cdot\omega^{n-1}+\lambda\cdot\left(\frac{\omega^n-\omega}{\omega-1}\right)$.

(ii) Use your answer in part (i) to show that,

$$\begin{aligned} S_n &= a_1+a_2+\ldots+a_n \\ &= a_1\cdot\left(\frac{\omega^n-1}{\omega-1}\right)+\lambda\cdot\left(\frac{\omega^{n+1}-n\cdot\omega^2+(n-1)\omega}{(\omega-1)^2}\right). \end{aligned}$$

(∗) Such a sequence is called a **semi-mixed** progression.

23. Prove part (ii) of Theorem 4.

24. Work out part (viii) of Remark 5.

25. Prove the analogue of Theorem 4 for geometric progressions: if the $(n-m+1)$ positive real numbers $a_m, a_{m+1}, \ldots, a_{n-1}, a_n$ are successive terms of a geometric progression, then

(i) If the natural number $(n-m+1)$ is odd, then the geometric mean of the $(n-m+1)$ terms is simply the middle number $a_{\left(\frac{m+n}{2}\right)}$.

(ii) If the natural number $(n-m+1)$ is even, then the geometric mean of the $(n-m+1)$ terms must be the geometric mean of the two middle terms $a_{\left(\frac{n+m-1}{2}\right)}$ and $a_{\left(\frac{n+m+1}{2}\right)}$.

Glossary of Symbols

$\mathbb{N}$: The set of natural numbers (or positive integers)
$\mathbb{P}$: The set of positive integers
$\mathbb{Z}$: The set of integers
$\mathbb{Q}$: The set of rational numbers
$\mathbb{R}$: The set of real numbers
$\in$: "belongs to:; or "is a member of"
$\notin$: "does not belong to" or "is not a member of"
$\forall$: "For every"
$\exists$: "There exists"
$\exists|$: "There exists at most one"
$\exists^*$: "There exists exactly one"; or "there exists a unique"
$\nexists$: "There does not exist"
$\ni$: "Such that"
$\cap$: The intersection symbol of sets
$\cup$: The union symbol if sets
$\subset$: "Is a subset of"
$=$: "Is equal to"; or "identical with"
$\equiv$: "Is congruent to"
$|$: "Divides"; the divisibility symbol
$\dagger$: "Does not divide"